FOSSIL FUELS AND THE ENVIRONMENT

Adeniyi A. Afonja

Sineli Books

Mansfield, Texas, U.S.A.; Luton, U.K.

ISBN: 978-0-9985843-3-1

Sineli Books

Mansfield, Texas, U.S.A.; Luton, U.K.
info@sinelibooks.com
Printed in the United States of America
First Impression 2019
Limited edition printed in color

Table of contents

Part 1: Highlights

Part 2: Fossil fuels and the environment

Acronyms and abbreviations

ADB	African Development Bank
AU	African Union
BAT	Best available technologies
BC	Black carbon
BEVs	Battery Electric Vehicles
BrC	Brown carbon
BSER	Best System of Emission Reduction
CBM	Coal-bed methane
CCS	Carbon capture and sequestration/storage
CIS	Commonwealth of Independent States
CF_4	Tetrafluoromethane
CFCs	Chlorofluorocarbons
CFL	Compact fluorescent lighting
CH_4	Methane
CO	Carbon monoxide
CO_2	Carbon dioxide
CO_2e/CO_2eq	Carbon dioxide equivalent
COP	Conference of Parties
COPD	Chronic obstructive pulmonary disease
CPP	Clean Power Plan
ECOWAS	Economic Community for West Africa
EDI	Energy Development Index
EES	Earth-Environment System
EfW	Energy from waste
EGUs	Electric Generating Units
EIA	Energy Information Administration
ENSO	El Nino-Southern Oscillation
EPA	Environmental Protection Agency
ESKOM	Electricity Supply Commission of South Africa
ESPP	East African Power Pool
ESR	Energy star rating
EVs	Electric vehicles
BEVs	Battery electric vehicles
BHEVs	Battery/hybrid electric vehicles
FOLU	Forestry and other land use
GDP	Gross Domestic Product
GHEs	Greenhouse effect
GHGs	Greenhouse gases
GMST	Global mean surface temperature
GNI	Gross National Income
GW	Gigawatt
GWh	Giga-watt-hours
GWP	Global warming potential
HDI	Human Development Index
HEVs	Hydrogen-electric vehicles
HFCs	Hyrofluorocarbons
ICE	Internal Combustion Engine
ICT	Information an Communication Technologies
IAEA	International Atomic Energy Agency
IEA	International Energy Agency

INDCs	Intended Nationally Determined Contributors
IoT	Internet of Things
IPCC	International Panel on Climate Change
IRENA	International Renewable Energy Agency
LCA	Lifecycle analysis
LNG	Liquefied natural gas
LPG	Liquefied Petroleum Gas
MAN	Metropolitan area networks
MSW	Municipal solid waste
MWe	Megawatts energy
MWh	Megawatt-hours
NMVOC	Non-methane volatile organic compounds
NEPAD	New Partnership for Africa's Development
NH	Ammonia
N_2O	Nitrogen dioxide
NO_3	Nitrate
NOx	Nitrogen oxides
O_3	Ozone
OAs	Organic aerosols
OC	Organic carbon
OECD	Organization for Economic Co-operation and Development
OH	Hydroxyl radical
OLR	Outgoing longwave radiation
PBABs	Primary biological aerosol particles
PCBs	Polychlorinated biphenyls
PET	Potential evapotranspiration
PFCs	Perfluorocarbons
PIDA	Program for Infrastructure Development in Africa
PHEVs	Plug-in/conventional hybrids
PM	Particulate matter
PV	Photovoltaic
RF	Radiative forcing
SADC	South African Development Community
SDG	Sustainable Development Goals
SF_6	Sulphur hexafluoride
SOAs	Secondary organic aerosols
SLCF	Short-lived climate forcers
SLCP	Short-lived climate pollutants
SO2	Sulphur dioxide
SPP	Self-Power Providers
SWR	Short wave radiation
Toe	tonnes (metric tons) of oil equivalent
TPEC	Total primary energy consumption
TPES	Total primary energy supply
TSI	The solar irradiance
TWh	Terawatt-hours
UK	United Kingdom
UNDP	United Nations Development Programme
UNEP	United Nations Environmental Programme
UNFCCC	United Nations Framework Convention on Climate Change
USA/US	United States of America

UV	Ultraviolet
VOC	Volatile organic carbon
VRE	Variable Renewable Energy
WAGP	West Africa Gas Pipeline
WAN	Wide area networks
WASAL	World Arid and Semi-Arid Lands
WEC	World Energy Council
WMGHG	Well-mixed greenhouse gases
WHO	World Health Organization
WMO	World Meteorological Organization
WNO	World Nuclear Organization
WRAP	Waste and Resources Action Program
WtE	Waste-to-energy

Preface

Climate change is a major global issue today, and one of the most controversial. The debate is full of well-mixed ideals, myths and realities. The pro-climate change world has an aggressive global campaign machinery against fossil energy use which is believed to be the cause of global warming and climate change. The sceptics are just as strong: either they do not believe that the climate is changing, or they think human activities have nothing to do with it. Infusion of politics inspired by intra-country exigencies has further complicated the issues and it has become difficult to distinguish between realities, facts, and fiction. Ideals and myths are strong weapons for sensitizing and motivating global mitigation action on climate change, but it is also useful to know what is real or possible. For example, 'dump fossil fuels'…….. 'leave fossil fuels in the ground' are prominent environmentalist catch phrases, but the reality is that, in spite of the poor environmental credentials, fossil fuels will remain dominant in the global primary energy mix for decades to come largely because of the prime roles as energy sources and precursors to a very wide range of products which touch lives on virtually all fronts. Furthermore, fossil fuels and derived products take the first and second place respectively in the list of globally traded and export products, dominating the economies of countries across all regions of the world, and the industry is also one of the world's largest employers. Facing realities would help to focus on feasible, sustainable adaptation and mitigation actions that could help the world optimize energy use and deal with the consequences of climate change.

Global warming is one of the major indicators of climate change, and extensive climate database shows that the average global temperature has been rising steadily from pre-industrial 14°C and, although there are many natural phenomena that can cause the Earth to heat up or cool, strong scientific evidence indicates that between 0.8°C and 1.2°C of the rise has been caused by human activities which release heat-trapping greenhouse gases (GHGs) into the atmosphere, mostly in the last seven decades or so. This apparently small temperature rise can have very big impact on the global climate largely because the oceans which occupy around three-quarters of the Earth's surface area absorb over 90% of the anthropogenic heat and much of the carbon dioxide. An increase in the average global temperature by even less than 1°C can have a profound influence on weather and climate in many ways, in particular, wind dynamics and the intensity, frequency and geographical spread of extreme weather. Higher temperatures boost evaporation which dries out soil and increases drought and desertification. Evaporation intensifies as temperatures rise and more moisture in the atmosphere will intensify rainfall. Storms form over warm waters, mainly in the tropical oceans and are energized by rising water temperatures, often rapidly transforming into powerful hurricanes and tornadoes which touch down thousands of kilometers from the origin, often with devastating consequences. Winds take more moisture from the warm oceans and bring more rainfall and flooding when they hit land. As temperatures rise, oceans expand, land and ocean ice melts, ocean waters rise above their normal level and are pushed inland by storms, causing severe flooding. A warmer atmosphere holds more moisture and when the temperatures are below freezing, snowfall can intensify. Extreme precipitation events will become more intense and frequent in many regions of the world, impacting on local weather and causing heavy rainfall in many places. Storms and hurricanes are natural phenomena and catastrophic damages caused by extreme weather have occurred many times over centuries, well before fossil fuel use became established but they occurred at different times often separated by centuries, and were confined to small regions. However, recent extreme weather phenomena have been much more intensive and widespread. The fact that storms need warm waters to form and strengthen means that they will be stronger, more frequent, and more devastating if ocean waters become warmer even by just one or two degrees centigrade because of the size of the oceans. The consequences of anthropogenic pollution go well beyond global warming: oceans acidify, causing significant problems for aquatic life; acid rain degrades structures and plants; ambient pollution by human activities has been identified by the World Health Organization as the world's most serious human hazard, causing over seven million global deaths annually.

Coal has powered global economic and industrial development for over two centuries, joined by oil and natural gas when both became commercially available in the early part of the last century. Together, they have accounted to over 80% of global primary energy supply for decades, and all recent projections indicate that the dominant position of fossil fuels will not change very much for decades to come, simply because there are not many sustainable alternatives to the end-use primary energy derivatives (petrol/gasoline, diesel, jet fuels and fossil electricity) that power the industrial, transportation and building sectors of the global economy. Production processes for most vital building blocks of modern society depend critically on fossil energy and fossil fuel feedstock - primary metals, cement, chemicals, petrochemicals, plastics, pharmaceuticals, fertilizers, cosmetics, polymer fibers (nylon, terylene, carbon fibers etc.), new energy materials, industrial and biomedical materials, auto and aero components, sports equipment, and a very wide range of consumer goods. Over 60% of components of modern planes (Boeing 787, Airbus 380) are manufactured from carbon fibers produced from fossil fuels. Electricity which accounts for over 40% of global primary energy consumption (with around two-thirds filled by fossil fuels) is set to become the most important energy source in all aspects of the global economy and human development. More than 90% of transportation fuels (petrol, diesel oil, aviation fuel, lubricants) are produced from oil. Furthermore, the economies of many nations across all regions of the world depend heavily on production of fossil fuels for use and export, and the industry (upstream and downstream) is one of the largest global employers of industrial labor. The global economy is rising exponentially, projected to double within the next two to three decades; vehicle population will double, and human population will increase by around 30%, all major drivers of energy demand, and all recent projections show that, although renewables will play a more prominent role in power generation, fossil fuels will continue to account for over three-quarters of global primary energy consumption and two-thirds of power generation energy in the next two to three decades.

The main issue with fossil fuels is the carbon content which when combusted forms carbon dioxide, the most abundant (although the least potent) greenhouse gas. While many global mitigation efforts are in place or ongoing (reduction of energy use through improved efficiencies, decarbonization of energy by substitution of renewable/nuclear energy, capture and sequestration of potentially polluting gaseous and solid waste, energy and materials' recycling, etc.), the pace of progress in decarbonizing energy has been slow for many reasons, and low carbon energy accounted for less than 20% of the global primary energy demand in 2018. The world added 33 Gigatonnes of carbon dioxide and other greenhouse gases to the atmosphere in 2018; if pollution continues at this pace, the global average temperature is projected to rise by 2-3°C by the end of the century, with profound consequences for current and future generations. Even if all anthropogenic emissions could be stopped today, the negative consequences of of pollutions so far will continue to resonate for centuries, maybe millennia. Another increasingly problematic source of energy-related pollution is the rapidly growing problem of discarded end-of-life materials, (in particular, plastics and e-waste). Also, around a third of anthropogenic pollutions come from other human activities unrelated to energy, notably agriculture (livestock and rice production, use of fertilizers, etc) and industry, and little attention is currently focused on the significant mitigation possibilities in these sectors.

Technology innovations continuously move the world towards sustainable human development, but there are always unanticipated negative fallouts. Nuclear energy was touted in the 1970s as the energy of the future that would replace fossil fuels because of its sterling low-carbon and energy security credentials. However, the negative environmental impacts of the nuclear waste generated and stored in the last fifty years will remain potent for hundreds, maybe thousands of years. Furthermore, the strong development of fossil energy which accentuated in the nineteen fifties (same period when a steady rise in global average temperature began) did not anticipate the potentially severe negative effects on the environment. This explains why environmentalists are now being challenged by conservationists who have succeeded in stopping or stalling low-carbon hydro, solar, wind, bioenergy projects particularly in the developed world for reasons which include environmental degradation, ecosystem damage, undue competition for vital land resources and infrastructure, noise from windmills, etc. Perhaps the most pragmatic way to move the world towards a sustainable carbon neutral environment without a violent disruption of the global economy and natural ecosystem is to learn to live with and moderate fossil fuel use, and there are many ways. The most promising options are discussed in some depth in this book, notably, improvement in efficiencies across the whole spectrum of energy production and use which could reduce energy consumption and associated emissions by around 40%, decarbonization of electricity and agriculture, reduction of people's lifestyle carbon footprint, and enhancement of the natural carbon cycle through reforestation. Various projections and models have shown that the cumulative effects of these measures can stop and reverse the growth rate of anthropogenic pollutions within a decade or two, but this would require much stronger and resilient country policies and actions that succumb less to internal political and economic exigencies, and better coordination of international actions compared with the set goals in the Paris-2015 Protocol.

Literature on atmospheric pollution and climate change is prolific and global action by governments and non-governmental organizations is extensive, while the voice of sceptics is also becoming increasingly loud. The climate change issue has become highly politicized and polarized, with people rejecting the ample scientific evidence that does not align with their political motives and preferences. It is interesting to note that most of the arguments used by the pro-climate change activists and the sceptics are often supported with extracts from the same literature: all that is needed is to quote a statement out of context. Furthermore the subject of climatology is very complex and comprehension is often beyond the capability of non-scientists. The primary objective of this book is to sift through extensive recent publications and existing global scientific and technical database on environmental pollution and climate change, and present the important facts and conclusions as objectively as possible and in a public-friendly manner. Potential pathways to a decarbonized energy world are discussed in some depth, in particular, the roles of improved efficiencies in energy production and use and across all sectors of the global economy, a major rise in the share of renewable energy particularly in power generation, much stronger penetration of electric vehicles in the global auto market, moderation of people's choices and lifestyles that enhance energy demand, and strong, resilient policies that survive political cycles. The major issues are summarized as highlights in the first part, largely devoid of technical jargon, while the scientific basis is presented in some depth in the second part for the benefit of those who are interested. Hopefully, the presentation will help clarify many of the often confusing and controversial statements in public debates about environmental pollution, weather and climate change.

Adeniyi A. Afonja
Professor Emeritus
(Materials & Energy engineering)

Dedication

This book is dedicated to the global body of environmental scientists, climatologists, and public health specialists whose works provided the very rich and comprehensive database from which the author has drawn extensively. The works of numerous international organizations, notably the United Nations (UN), the International Energy Agency (IEA), the International Panel on Climate Change (IPCC) and the World Health organization (WHO), have provided the world with objective, scientific, technical, medical and political views on the potentially serious and enduring impacts of human-induced climate change on humankind and the ecosystem. They have also provided response options, potential pathways to a sustainable global environment, and continuous evaluation of progress. Numerous governments are doing invaluable work in evolving intervention policies which need to be stronger, more resilient, more enduring and proliferated across all regions of the world because the effects of climate change are global and largely independent of the source of pollution. Non-governmental organizations are also doing very valuable work by sensitizing the world to the implications of uncontrolled pollutions and holding governments accountable for their actions or inactions. However, the current focus of intervention and mitigation actions is on energy-related emissions but other human activities which generate a third of the total global anthropogenic emissions need to come under the radar as well, in particular, agriculture. Also, non-governmental organizations need to extend their actions beyond governments to include the ordinary people whose ultimate lifestyle choices and excesses fuel the rapid growth in the demand for energy and products of agriculture which together account for around 80% of human-related emissions.

About the author

Professor Adeniyi A. Afonja is a professor emeritus, materials science and engineering, with over 50 years of experience in university education, research and consulting in materials and energy engineering. He received his B.Sc. in mechanical engineering from Kings College, University of Durham, United Kingdom, M.Sc. in industrial metallurgy and management techniques and Ph.D in mechanical engineering from Aston University, Birmingham, also in the United Kingdom. He also holds a diploma in iron and steel technology from UNIDO-USSR in-plant training program. Professor Afonja is a Chartered Engineer and Fellow of the Institute of Metals/Materials (UK & Nigeria). He is a recipient of many Fellowships, including Commonwealth Fellowship (UK), Senior Commonwealth Fellowship (UK), Fulbright Fellowship (USA), Senior Fulbright Fellowship (USA), United Nations Development Program (UNIDO) Fellowship, Research Council of Canada Fellowship. He has been a visiting Fellow, Senior Fellow, visiting Professor in many universities and research establishments, including Massachusetts Institute of Technology, (USA), University of Wisconsin-Whitewater (USA), National Energy Research Laboratories, CRE/CANMET (Canada), Coal Research Establishment, Stoke Orchard (UK), Northern Carbon Research Laboratories, University of Newcastle (UK). He has authored over a hundred articles and several books on engineering materials and energy. He retired in 2005 but is still actively involved in research and consulting.

Part 1
Highlights

Energy and Human Development

Human development has always been tied to the availability of appropriate energy. Industrial and economic development which began with the Industrial Revolution of the late 1700s stimulated the transition from hand tool production and animal power to new mass manufacturing processes, a development which would not have been possible without the emergence of the coal industry. The availability of coal energy led to a revolution in agriculture, textile and metal production, and transportation. Advances in agricultural techniques and industrial production led to major improvements in the quality of life, which in turn stimulated growth in energy demand and the development of mass production technologies. Economic and social development tend to go hand-in-hand with energy sector transformation, and urbanization, a major consequence of economic growth, rapidly transforms the traditional agrarian economy into industrialization and a knowledge-based economy. This shift requires modern energy to thrive, while also providing the incentive for investment in modern energy infrastructure. Not many nations have succeeded in growing their economies and human development without the dual strategy of modernizing the economy and developing a sustainable modern energy infrastructure.

The global economy has been transforming from rural/agrarian to urbanized modern production, earning power and quality of life have been rising, stimulating expansion of energy infrastructure, and giving prominence to the interdependence of energy and human development. The gross domestic product (GDP) could double within the next two decades, with increased rural-urban migration and high levels of economic growth which means rising living standards and upward class mobility. The global fleet of vehicles will likely double to around 2 billion by 2040; global population will grow from 7 billion today to around 9 billion; and 1.2 billion mostly in the developing world will still lack access to adequate modern energy. A dynamic growth of secure and sustainable energy is critical to fueling these projected changes in the global dynamics and promoting universal access to modern energy.

Access to modern energy is a prerequisite for economic, technological and human development, and is essential for humanity to thrive because it stimulates the development of modern infrastructure that in turn grows large-scale economic enterprises, promotes urbanization and raises the earning power and overall development of the people. This development transforms 'access' to 'consumption' and further stimulates demand for energy and development of additional modern energy infrastructure. Economic and social development thus tends to go hand-in-hand with energy sector transformation. Economic activities promote urbanization which in turn helps to move labor from rural agrarian subsistence economies to high-income wage economies that promote social mobility and allow people to afford modern energy and devices that use energy. Without this development, human development remains low. Many countries are energy poor, with around 80% of the world's population living in countries where average modern energy consumption is less than the United Nations' minimum of 1.9 metric tons of oil equivalent (or 100 Gigajoules of energy) per person per year. Most of the countries below this minimum are also in the lowest quarter of the United Nations' human development ranking.

The World Energy Council (WEC) defines three core dimensions that determine energy sustainability, often referred to as *the energy trilemma.* It identifies a triple challenge of ensuring that the world has access to affordable, secure and reliable energy supplies while reducing energy-related anthropogenic emissions to moderate the risk of environmental pollution: *energy security, energy equity, and environmental sustainability,* three major challenges which most countries struggle to reconcile. This critical goal promotes transition to a more sustainable, environmentally-sensitive energy future. Access to modern energy, in particular, electricity and modern domestic energy, is a prime requirement for the achievement of economic growth, human development, and environmental sustainability. Modern energy is an important determinant of human health, as it plays a critical role in healthcare delivery capabilities and in the development of clean and safe household environment. In recognition of this, the United Nations set a target of 'sustainable modern energy for all by 2030' as part of the sustainable development goals adopted by 193 nations in 2015. Even before then, most nations had already developed strategies for improving their modern energy infrastructures, focusing on access to electricity. The number of people without access to electricity fell to 1.1 billion (about 15% of the global population) in 2016 from 1.7 billion in 2000. The electricity access deficit is overwhelmingly concentrated in sub-Saharan Africa where 62.5% (609 million people) are still without electricity.

Although access to electricity is the most prominent as an explicit objective of most national development strategies, access to modern household energy for cooking and heating is even more urgent because of its direct impact on household health. However, compared with access to electricity, little has changed about access to modern cooking energy over the last two decades: around 2.8 billion people (38% of the world's population) still lack access to clean household energy, same as in 2000. Developing Asia and sub-Saharan Africa dominate the global totals, and, although the numbers for developing Asia are larger, their share of the population is significantly lower than in sub-Saharan Africa (43% compared with around 80%). Nearly 3 billion people worldwide still rely on the use of traditional solid biomass while another 300 million cook with kerosene and coal, with serious environmental consequences. Around 68% live in developing Asia while another 30% live in sub-Saharan Africa.

All recent outlooks and projections for the global economic/GDP growth and energy demand over the next two to three decades indicate that both will be driven by the developing world due to rapid population growth and increasing prosperity in the regions. Increasing urbanization and emergence of a large and growing middle class in the regions will largely shape the global economic and energy demand trends. Over 80% of the expansion in world economic output will occur in the developing regions, with China and India accounting for around half of the growth. The prospects for Africa are less bright (in spite of enormous wealth of potential energy resources) due to weak productivity and a highly-stressed modernization effort caused by the rapidly expanding population. Most of the countries in the region survive on export of raw minerals and then spend all the returns and more in importing the products of processed minerals. While the region will account for almost half of the increase in global population, the contribution to the GDP growth will be less than 10%. The United Nations projects that the population of Africa will double to 2.4 billion by 2050 compared with its population in 2016 and about 50% of all newborns worldwide will be born in Africa by the end of the century. Economic development that raises the level of education and economic prosperity in the region will be crucial because it promotes rural-urban migration and smaller families. Even with the projected rapid increase in access to modern energy over the text two or three decades, around two-thirds of the global population mostly in the developing world, led by sub-Saharan Africa will still remain below the minimum per capita modern energy consumption, a critical propellant for human development.

Global energy resources, demand and supply

Energy is the prime mover of economic development and availability in sustainable and acceptable forms will to a large extent control the pace of world economic and human development in the foreseeable future. The world is richly endowed with many natural primary energy resources - renewables (solar, wind, hydro, biomass, geothermal, tidal, gravitational), and non-renewables (fossil coal, oil, natural gas, and nuclear). The problem is how to manage these resources to provide a balanced mix to satisfy the strongly growing world energy requirements in a sustainable manner that does not continue to degrade the environment. Only four of the global primary energy resources are available everywhere, abundant, free and inexhaustible: solar, wind, gravitational and biomass energy. Hydro and geothermal energy sources are also free, but not every country has access to these resources. Fossil fuel and nuclear energy resources are non-renewable, and global distribution is inequitable, with relatively few countries holding most of the resources. All natural energy sources are classified as *primary energy* and, while some can be used directly - solar, wind, coal, natural gas, biomass, - most end-use energy sources are products of technological processes that convert them into useable forms required for powering machinery, automobiles, etc. These are known as secondary, converted or end-use energy sources, and include petrol/gasoline, diesel, jet fuel, synthetic liquid and gaseous fuels, lubricants, heat and electricity. Without these energy sources, the rate of global economic and human development over the last 300 years or so would not have been achievable.

Fossil fuels

Fossil fuels (coal, oil and gas) were formed from pre-historic plants and animals that lived millions of years ago. When they died, they got buried under layers of swamp mud, rock and sand. They decomposed slowly into organic materials and sank to deeper depths, often hundreds of meters and sometimes over a thousand meters deep, subjected to high temperatures and pressures. The formation of fossil fuels (coal, oil and gas) is a continuous though exceptionally slow process and only mature deposits which were initiated around 300 million years ago are being exploited currently. However, with the exponentially increasing rate of exploitation, mature deposits may become increasingly rare. Furthermore, the potential effects of human activities such as deforestation and ocean pollution on the natural processes of formation, especially on the initiation of potentially suitable sites (mires) and the microorganisms that break down the carbonaceous deposits are not clear. However, in view of the fact that deposits have been formed continuously for hundreds of millions of years and the potentially disruptive of human activities intensified only in the last 200-300 years or so, it is possible that many currently immature deposits will become available to future generations.

In reality, all primary energy sources are renewable, the difference being in the time scale: while it would take hundreds of millions of years to replace fossil fuels being exploited currently, solar, wind, hydro, geothermal are continuously available and are therefore classified as renewables. Fossil fuels constitute around 97% of the proved global non-renewable primary energy reserves. On average, oil and gas reserves are adequate to meet global demand for around fifty years (up to 120 years in some countries). Coal will last longer, over a hundred and fifty years on the average (up to 350 years in some countries). However, fossil fuel resources that have been discovered and proved are believed to be only a fraction of the total global resources, due to limitations in currently available exploration and exploitation technologies. Most of the world's oil and gas reserves are off-shore and current exploration and drilling technologies are capable of no more than about 3 kilometers depth while half of the oceans which represent about three quarters of the planet are much deeper. Emerging technologies are locating new resources and providing access to reserves that were considered unrecoverable previously. For example, the recoverable coal reserves using currently available advanced mining technologies are believed to be less than 5% of the total global resources. Deposits that were previously regarded as inaccessible or uneconomical but which meet other required criteria are now being gasified *in-situ* to generate heat and electricity. Fossil fuels have supplied virtually all the global primary energy requirements since the Industrial Revolution of the late 1600s, accounting for over 80% in the last fifty years or so. However, the global distribution of resources is inequitable: just five countries hold over 60% of global oil reserves (Venezuela, Saudi Arabia, Canada, Iran, Iraq); five countries (Russian Federation, Iran, Qatar, Turkmenistan, Saudi Arabia) also hold over 60% of the world's natural gas reserves; while five countries (United States, Russian Federation, China, Australia, India) hold nearly 80% of the total global coal reserves. It should be noted however that global proven reserves of these primary energy resources are believed to be only a small proportion of resources which are yet to be located, proven and moved to reserves.

Nuclear energy

The inequitable distribution of non-renewable energy, the need for energy equity and security, and pressures of environmental sustainability (which implies more energy for less carbon emissions) constitute formidable challenges for most countries. Most of the growth in global primary and end-use energy demand will be in emerging countries which will also account for most of the growth in GDP expected to double over the next two to three decades and population which is projected to increase by around 25%. For many of these countries, *energy equity* is the prime challenge and local resources, (in particular coal, the most polluting fossil fuel) are the first choice. Developed nations face a different challenge: *energy security* which has become prominent since the global oil market crisis of the 1970s and energized the development of nuclear energy for peaceful applications. Nuclear power generation became well-established in the 1960s and grew at a fast pace over the next two decades, fast-tracked by the global energy crisis. For three decades, the OECD dominated nuclear power generation, accounting for nearly 93% of the total global nuclear electricity generation capacity in 2015. However, two serious accidents (Three Mile Island in 1979, Chernobyl in 1986) marked the beginning of a steady decline which reached its lowest after another accident in Fukushima, Japan in 2011. Nuclear share of global power generation declined from peak 18% in 1996 to under 11% in 2011 and, in spite of the fact that Japan has reactivated 5 of its 42 nuclear power reactors and some other countries have commissioned new plants, nuclear share of global power generation was still below 11% in 2018.

Sixteen countries depend on nuclear energy for at least one quarter of their electricity, and the United States leads the world in terms of nuclear power output, accounting for nearly a third, but the proportion in the domestic power generation fuel mix was less than 20% in 2018. In contrast, nearly 80% of the total power output in France in the same year was from nuclear plants. Also, many states that do not have nuclear power plants depend in part on nuclear-generated power through regional transmission grids. However, the future of nuclear power generation is unclear. Japan is one of the leading countries in the deployment of nuclear generation capacity, the third largest in the world. Prior to the Fukushima accident in 2011, the country filled around a quarter of its electric power demand with nuclear electricity, but this dropped to less than 2% in 2013 and, although some plants are already back in operation and many more are being evaluated, it is unlikely that share of nuclear power in the country's energy mix will ever rise to the pre-accident level again, especially with the country's increased interest in renewable energy and investment in liquefied natural gas projects in other countries.

The United States, the world's largest nuclear power generator and producer of a third of the world's nuclear electricity from its nearly a hundred reactors, has shut down around a third in the last two years and initiated construction of only two new plants; United Kingdom shut down 30, Germany 29, Japan 18, although most of the plants shut down permanently still had many years of useful life. Projections for the United States indicate that the share of nuclear energy in power generation will decline from the current 19% to 12% by 2050. Also, many other countries that have historically accounted for the majority of nuclear power development are de-commissioning ageing nuclear power plants and slowing down on new construction, Russia being the only exception, with 24 commissioned or planned new nuclear power plants as of 2019. However the urgent need to meet growing demand for power and global pressure to decarbonize electricity have been fueling interest in nuclear power generation in the developing world, in particular Asia over the last two decades, and many plants are either under construction or planned. Forty three plants were either commissioned, under construction or planned in China in 2019, 14 in India, 4 in Egypt.

The fact that a nuclear plant can operate continuously for years anywhere in the world, fueled by relatively small quantities of imported uranium is a major attraction as a means of rapidly increasing power supply to meet the growing local demand especially in the developing world. Also, although uranium (the primary fuel for nuclear energy) constitutes less than 3% of the total world energy reserves (in metric tons of coal equivalent), only small quantities are required to generate enormous energy: one metric ton of uranium energy is equivalent to about 20,000 metric tons of coal energy. Uranium resources are also distributed inequitably, with over half of the proven global reserves located in Australasia. However, there is growing concern on the ability of emerging nations to manage extremely dangerous spent fuel rods, nuclear waste, and accidents, considering the experience of the developed nations most of which are simply stockpiling waste dating back to the nuclear projects of the 1940s: there is still no environmentally sustainable method of nuclear waste disposal and serious environmental issues of the nuclear accidents dating back forty years are still being dealt with today. Substantial nuclear waste has been buried or dumped in the oceans and, although ocean dumping was prohibited in 1994, the damage is already done because dumped waste which may remain potent for hundreds or thousands of years was encased in concrete or copper canisters which could disintegrate in just years and release toxic waste into the oceans. Furthermore buried waste could be disturbed and exposed by earth movements or contaminate aquifers. However, China and India which will host most of the proposed new plants are nuclear powers with significant experience in nuclear technology.

Renewable energy

Renewable energy resources are natural resources that can be converted to useful energy - solar, wind, biomass, tidal, geothermal, hydropower. They are abundant, widely available, free and, if fully exploited could supply all the global primary energy requirements. However, formidable challenges, notably lack of appropriate technologies, intermittency, weak/shifting policy instruments, stiff competition with well-established fossil fuels, and change inertia have so far restricted the total contribution to global annual primary energy requirements to less than 20% mostly in heat and power generation. Renewables are the fastest growing primary energy sources, used mainly for steam raising and power generation, meeting a quarter of global energy demand growth in 2018, with wind energy accounting for 36% of the growth in renewables-based power output. Modern bioenergy (solid, liquid or gaseous fuels, excluding the traditional use of biomass) was responsible for half of the renewable energy growth, and provided four times the contribution of solar photovoltaic (PV) and wind combined, used mostly in final energy consumption to deliver heat in buildings and industry. Although hydropower remains the largest renewable resource, meeting around 16% of global electricity generation in 2018 and accounting for over 70% of total renewables' contribution, the share of other renewables, in particular solar and wind will grow very rapidly over the next two decades. Renewables will account for more than 70% of electricity generation growth over the next few years, led by solar PV, followed by wind, hydropower and bioenergy but the individual contribution to global primary energy demand will still be low: solar and wind accounted for only 7% in 2018, bioenergy 3%. Although the growth rate is relatively low compared with the power sector, the heat sector (which includes heating for buildings and industry) accounts for the largest overall share of renewables' contribution to global primary energy, meeting about 9% of the heating sector energy demand, projected to grow to about 12% over the next few years. Bioenergy (solid, liquid, gaseous) will lead the growth in renewable consumption over the next few years, accounting for around 30% (IEA, 2019c).

China and the United States led the highest growth ever of renewable share of primary energy consumption in 2018, contributing around 50% of the increase in renewables-based electricity generation, followed by the European Union, India and Japan. However, in spite of the unprecedented growth rate in the last few years, renewables accounted for less than 15% of the global primary energy supply and 17% of the total final energy end-use consumption in 2017, shared equally between modern renewables and traditional biomass. Most recent projections indicate that renewables will still contribute no more than 20-25% to global electric primary energy consumption by 2040, but some optimistic projections predict a possible rise to as high as 35% by 2030. Furthermore, there are very wide variations by country: many countries have zero renewable power deployment, while some, like Denmark source up to 60% of their electricity from renewable energy. Countries, notably China and India are investing aggressively in renewable power generation, with China targeting 40% share by 2040. Hydroelectricity has dominated renewable energy power generation for decades but its contribution is projected to reduce significantly due to increased use of solar and bioenergy. Renewable energy increased by 4% in 2018, accounting for almost one-quarter of global energy demand growth. The power sector led the gains, with renewables-based electricity generation increasing at its fastest pace this decade. Solar PV, hydropower, and wind each accounted for about a third of the growth, with bioenergy accounting for most of the rest. Renewables covered almost 45% of the world's electricity generation growth, second only to coal, and accounted for about 25% of global power output in 2018. Recent projections indicate that renewable share in power generation could rise to 30-35% by 2040. However, the International Energy Agency Sustainable Development Scenario projects that renewable share needs to rise to 50-55% with solar and wind accounting for around 60% in order to set the world on the path to a sustainable environment. China accounted for over 40% of the growth in renewable-based electricity generation, followed by Europe, which accounted for 25%. The United States and India together contributed another 13% (IEA, 2019c).

The development of renewable energy, with the exception of hydroenergy has been slow despite the extensive availability of renewable resources, due to the intermittent nature of these resources and dependence on weather by most of them. Draught affects water levels for hydropower generation and cultivation of crops for biomass energy; solar and wind are weather-dependent; biofuels are competing with agriculture for scarce land and infrastructure;

and resistance by conservationists is growing. The bulk of renewable energy, (in particular, biomass, solar and wind) is used for electric power generation but the intermittent nature and lack of cost-effective power storage make it difficult to integrate solar and wind power (variable renewable energy, VRE) with national electricity grids. Also, the very low utilization capacities (20% and 30% for solar and wind respectively) are major constraints to their combined share in power generation, which was only about 7-8% in 2018, projected to grow to about 12% in 2040, while the total contribution of all renewables including hydropower will rise to only about a third. In the last two decades or so, efforts have been made to develop bi-fuel hybrid power plants which combine wind, solar, biomass or tidal power with fossil fuel. Some hybrid solar-gas plants are already in commercial operation in some developed economies, in particular, the United States and Europe-28. Renewable energy is deployed whenever available and the intermittent gaps are filled by natural gas or coal. Also, utility renewable installations store surplus energy in deep-cycle lead-acid or lithium-ion battery banks for use when solar/wind energy is not available. This enables stand-alone solar and wind power systems to serve buildings, small communities, and critical installations. An ambitious hybrid power project is under development in Los Angeles, California, USA, expected to be the world's largest and cheapest 530-megawatt solar-battery storage power plant. About 400 megawatts will be consumed in real time over about 12 hours of daytime while the balance will be stored in an 800 megawatt-hour battery storage system (equivalent of two hours of daytime consumption) to partially supply the overnight requirements while natural gas and hydroenergy fill the balance.

Technology advances, particularly those that unlock efficiencies, and falling costs in recent years have been driving the adoption of renewable energy, particularly in the developing world, with the power sector leading the way. The global power sector share of renewable energy grew by an average of 8% per year over the last ten years, setting a record of around 25% share of power generation in 2018. Weighted average levelized cost of PV solar and off-shore wind power generation now fall within the fossil fuel range, and projections show that both will become even more competitive over the next few years. Global investment in renewable energy has also been increasing, much of it in the developing world. In recent years, developing countries have been attracting the majority of renewable energy investments, with China alone accounting for about one-third of the global total. Most of the investments have been in PV solar and wind, which together accounted for about 90% of the global total. Solar PV is particularly suitable for utility power generation, serving small off-grid communities, commercial units, or household power generation, enabling consumers to produce power for their own needs and feed surplus energy into the grid. Utility scale installations are becoming increasingly competitive with new fossil-fuel power generation, even without subsidies. Solar power offers a unique opportunity to promote net zero *carbon footprint* (the amount of anthropogenic carbon release associated with a unit of activity), or *carbon neutrality* because the total amount of energy used by the building is roughly offset by the amount of renewable energy generated on the site.

About 60% of the total energy emitted by the Sun reaches the Earth's surface. If just 0.1% of this energy can be converted into heat at an efficiency as low as 10% it could supply four times the global electric power demand. However, solar radiation is available in the appropriate intensity and for adequate duration in relatively few global locations, mostly in regions hosting emerging countries that do not have the wherewithal to harness and deploy solar-powered power generation on any massive scale. One major exception is Morocco where the world's largest solar concentrating power plant was commissioned recently. China, Japan, United States, and Germany also have strong support policies promoting extensive deployment of solar power, from large-scale to small stand-alone units supplying homes, critical installations and small communities. The State of California in the United States recently passed a law requiring all new buildings to feature solar roof panels.

Wind is available everywhere in the world and if just 1% could be effectively utilized, it could provide electricity equivalent to the capacity of all global electric power generating plants in the world today. However, the intensity in many locations varies widely, and there are relatively few locations where the wind is sufficiently strong and available for enough time to power electric generation plants. Even then, the load factor is low, 15-40% compared with 75-90% for thermal plants. Wind power is useful only when the intensity and consistency are sufficient to drive a wind turbine, and the best locations are offshore mostly in remote locations where obstructions are minimal. Wind power contributed 5-6% to the total global electric power generation in 2018, although there were significant variations between countries. For many countries, wind power has become a pillar in their strategies to phase out fossil and nuclear energy. For example, wind power produced 43% of Denmark's and 13% of Germany's electricity in 2018, and an increasing number of countries - Ireland, Portugal, Spain, Sweden, Uruguay - have reached double digit wind power share.

The Earth's crust stores an enormous amount of energy, believed to have originated largely from radioactive decay of minerals but also during the Earth's formation. The core mantle boundary which is about ten kilometers deep is believed to be about 4,000°C, hot enough to melt rocks and set up a geothermal gradient between the core and the Earth's surface. High pressures force molten rocks towards the Earth's surface, and eruptions of volcanoes and hot lava occur frequently in some regions of the world. Underground water that comes in contact with the heat becomes heated and penetrates to the Earth's surface as warm springs, and when hot enough, they heat homes, swimming pools, etc. In some locations the temperature could be as high as 380°C, enough to produce superheated steam which can be captured for electric power generation, and appropriate technologies are already available. It is estimated that the upper ten kilometers of the Earth's crust contains 50,000 times as much energy as found in all the World's fossil fuels combined. Geothermal energy is abundant in areas located around the boundaries of tectonic plates which make up the Earth's outermost shell also known as the lithosphere. Countries which are rich in geothermal energy include North and Latin American countries, African countries, Japan, Australia and Russia. Geothermal energy has been used since the ancient times for domestic hot water production and space heating, and, currently is being used extensively for space and industrial process heating, heating of pools and greenhouses, and water treatment. However, focus has shifted in the last decade or two to electric power generation and commercial power plants are operating in around 25 countries of the world including countries located along the Great Rift Valley of Africa (Kenya, Ethiopia, Uganda), and others located around the East Mediterranean. The world's largest geothermal plant (850 megawatts) is located in California, USA, and hundreds of new plants are being developed around the world. El Salvador and Iceland currently derive nearly 30% of their electricity from geothermal energy and Kenya plans to source over half of its electric power requirements from the resource within the next decade. New technologies are being developed which involve drilling to the hot rock layer, fracturing the rock and pumping water into the crack to generate superheated steam for driving steam turbines in power generating plants above ground. It is believed that when relevant technologies become fully developed, geothermal energy could supply the entire electricity requirements of many countries, including the United States.

Global primary and end-use energy mix: status and outlook

Three key, interdependent drivers determine how the world uses energy: technology, policy and consumer preferences, although the interplay varies depending on local circumstances which may change from time to time. Deployment of new technologies facilitates the achievement of more output for less energy, leading to cheaper consumer goods; strong and enduring policies (both motivating and punitive) tend to incentivite faster deployment of technologies and shape consumer preferences; demand for energy depends significantly on choices which consumers make in their daily lives, which in turn are shaped by technology and policy. Global energy demand is growing rapidly, driven largely by the growing economy and population. The global economy could more than double over the next twenty years or so, driven largely by the increasing prosperity of fast-growing economies of emerging nations, while global population will rise by around 25-30% and global population of automobiles will double by 2040. However, energy demand has been decoupling from these drivers over the last decade or so, and is projected to increase by only a third, mitigated largely by the rapidly decreasing energy intensity and increasing energy efficiencies across the whole spectrum of the global economy. Much of the growth in primary energy demand will be accounted for by the emerging nations, with a rise in share from around 60% in 2016 to 70% in 2040. Fossil fuels provided 80% of the global primary energy in 2018, with little change in the last five decades, and, by all recently published projections, will still provide around 78-80% in 2040. Coal accounted for 26% of the total global primary energy supply in 2018, oil 31%, natural gas 23%, nuclear 5%, hydro 3%, biomass and waste 10% and other renewables (biofuels, solar, wind, geothermal) 2% (IEA, 2019i). Increasing efficiency and reduction of energy intensity will likely slow down growth in demand in the developed countries but demand in developing countries will continue to grow due to the expected rapid growth in industrialization, especially the heavy industries such as primary metals (iron and steel, copper, aluminium etc.), chemicals, petrochemicals and cement production mostly located in emerging nations, all of which have high energy intensity and relatively small scope for efficiency improvement.

The role of coal in the global primary energy mix has been aptly described by the International Energy Agency as *"a tale of two worlds with climate action policies and economic forces leading to closing of coal power plants in some countries while coal continues to play a part in securing affordable energy in others ....Ultimately, despite significant media attention being given to divestments and moves away from coal, market trends are proving resistant to change."* Power generation accounted for 40-42% of the global primary energy demand in 2018, filled by about 67% of combustibles 98% of which were fossil fuels. In spite of its poor environmental credentials, coal is currently the dominant fuel in global electricity generation and in countries across all regions of the world, supplying up to 98% of electricity in some countries. While coal's contribution to the power generation mix is expected to decline from 39% in 2018 to around 28-31% on the average by 2040, many nations across all regions will still depend largely on coal for power generation. Global coal power generation increased by around 3%, and accounted for about 40% of the additional power generation worldwide in 2017. Although coal share in power generation had been falling for several years prior to 2017, there has been a significant rise in the last two years, with coal accounting for 39% in 2018, roughly the same share over the last two decades. By comparison, share of natural gas was 23%, hydro 16%, nuclear 10%, wind 5%, oil 3%, biomass and waste 3%, solar photovoltaics 2%. South East Asia produced nearly 70% of the total global output of coal in 2018 and it remains the primary fuel in a region that will account for most of the future growth of primary energy and electricity demand. Furthermore, around 15% of global coal demand fills other demands apart from power generation: iron and steel, non-metallic minerals (notably cement, petrochemicals), residential, commercial and industrial heating and steam-raising, and many industrial and agricultural applications. Copper is a critical metal in the electrical and automotive industries and China is the world's largest producer of refined copper, accounting for nearly 40% of the total global production in 2017. Chile, the second largest produced only 10.4%, Japan 6.3%, USA 4.6%. Also, Asia (led by China) was the source of two-thirds of the world's supply of aluminium in 2017. Refining of both primary metals requires enormous electrical energy which is supplied largely by coal-fired power plants. Oil is a critical energy resource for transportation across all regions of the world, and use of natural gas is projected to rise, particularly in power generation over the next two decades because of its low pollution emissions compared with other fossil fuels, and also due to increasing globalization of supply driven by a steep rise in liquefied natural gas production. Just two regions: Asia Pacific and North America accounted for 60% of the global consumption of oil and coal in 2018.

Primary energy is used in a variety of ways: oil is transformed into a variety of fuels (petrol/gasoline, diesel oil, jet fuel) by refining; coal, natural gas, biomass, nuclear power are transformed into electricity by power plants; coal, natural gas and biomass are used directly in heating; coal, natural gas and biomass are transformed into synthetic fuels; coal and natural gas provide energy for many industrial processes such as cement and primary metals' production; and coal, natural gas and oil are feedstocks for the production of a wide variety of chemicals, and petrochemicals which are precursors to a very wide range of consumer products including pharmaceuticals, cosmetics, plastics, polymer fibers (nylon, terylene, carbon fibers etc.), syngas, fertilizers, etc. Presently, power generation accounts for 40-42% of total global energy end-use and demand is expected to rise as new technologies evolve, many more countries industrialize, and more people in emerging nations gain access to electricity. The final consumers of end-use energy are industry, transportation, buildings, and a few others. Industry (including power generation and fuel feedstock) accounted for about half of the total global consumption in 2018, buildings for 29% and transport for 21%.

Electricity is the world's fastest growing end-use energy, growing more than three times as fast as any other energy, and twice as fast in emerging countries compared with developed countries, fueled by rapid global GDP and population growth, rising urbanization and middle class population growth in emerging countries, and also by rising global digital economy, electrification in transportation, building and industry, and other technological changes. Renewables are fast becoming the technology of choice for power generation due largely to the very positive environmental credentials, falling costs and supportive government policies. The share of renewables in power generation could increase by 50-60% over the next two decades. However, the combined contribution of solar and wind will remain relatively low, rising from the current 6-8% to less than 20% unless the formidable challenges of variability and the need for massive investments in flexible grid and demand-response technologies such as smart meters and energy storage are resolved. Demand for electricity is expected to rise by up to 60%, accounting for nearly all of the global total energy demand growth over the next two decades, and nearly all the growth will be accounted for by emerging nations, with Asia Pacific accounting for around 60%. In 2018, fossil fuels provided 66-70% of the primary energy used for power generation and, although a significant shift in mix is projected, with the contribution of natural gas rising at the expense of coal and oil, all recent projections show that fossil fuels will still account for around 65% of the total global power generation and around 80% of global primary energy demand in 2040, about the same contribution as in 2005.

Primary energy demand is projected to grow by about 35% by 2040, across all end-use sectors, with the industrial sector accounting for around half of the overall increase. Nearly all of the growth in energy demand will come from fast-growing developing economies, with China and India and other emerging Asia accounting for around two-thirds of the growth. Renewables will grow faster than any other primary fuel source, accounting for about 40% of the expected total growth in primary energy demand over the next two decades. Around two-thirds of the increase in primary energy demand will be used to generate power to meet the exponentially increasing demand for access to power in the developing world. Oil is the world's leading fuel, accounting for around 34% of total global primary energy consumption in 2018. Oil is refined into gasoline, diesel oil, jet fuel, etc., and around 90% of the products fill about 95% of fuel demand by the transportation sector. Non-OECD countries accounted for around two-thirds of the demand, with Asia accounting for over one third, and China alone consuming 22%. Although some growth in the share of transportation energy by biofuels, natural gas and electricity is expected, the prospects of any significant impact on the dominance of oil over the next few decades are not very bright.

Energy-related emissions and the environment

The environment sustains life on earth in many ways, including regulation of global temperature, weather, climate, and there are many natural control mechanisms in place. Although many of the events that are experienced on earth - weather and climate change, severe weather, draught - are in fact natural phenomena, there is ample scientific evidence that human activities particularly in the last century are disrupting many of the natural control systems and making the negative impacts on the Earth more widespread, intense and severe.

The Earth's natural environment

The Earth is central to a global system commonly defined in terms of four major spheres: *lithosphere*, *hydrosphere*, *atmosphere*, and *biosphere*. The hydrosphere may be sub-divided into fresh and frozen water, hence the common fifth member: *cryosphere*. The lithosphere comprises the Earth's crust (from the surface to a depth of about 100 km), and the upper part of the solid mantle that extends to a depth of about 3000 km. The hydrosphere comprises water (liquid or frozen) that exists under or over the surface of planet Earth - oceans, lakes, streams, glaciers, ground waters. The lower atmosphere (*troposphere*) comprises layers of gases that extend from the Earth's surface to about 15-20 km into space, and hosts most of the natural activities that determine the local weather - wind, clouds, precipitation, etc. The next atmospheric layer which extends to about 50 km is the *stratosphere* which plays a critical role in the regulation of temperature on earth and screening off much of the Sun's potentially dangerous rays. Although these are the two most important layers that make life on earth possible, there are several other layers above the stratosphere which have different characteristics and different influences on the Earth's natural environment. The biosphere refers to parts of the land, sea, and atmosphere that host all living organisms, microorganisms and plants. The atmosphere (in particular, the first two layers) sustains life on Earth through the supply of oxygen needed by most organisms for respiration, carbon dioxide required by vegetation (plants, algae etc.) for photosynthesis, regulation of the Earth's temperature, weather, climate, and protection of the Earth from potentially damaging effects of the Sun's ultraviolet radiation. The four major spheres that make up the ecological system are characterized by intricate intra and interactions, all interdependent and interconnected. The interaction between these spheres to a large extent determines the weather and climate, the functions are controlled by natural regulatory processes and the boundaries between them may be clear or ill defined. For example, while the boundaries between organisms and vegetation are fairly well defined there are no clear boundaries between air, water, climate, energy, radiation, etc. The environment that is controlled by natural forces which have the capacity to dilute, absorb or dissipate pollutants is known as *natural environment*. The natural environment is imperfect in many ways, for example, extreme weather, flooding, volcanoes, earthquakes are all natural occurrences which have always been part of the ecosystem, caused by known or unknown natural phenomena.

The Earth-Environment systems have co-existed for millions of years and the way they interact determines the sustainability of life on Earth. The Sun generates enormous energy through sustained natural chemical processes; physico-chemical activities in the atmosphere moderate the Sun's energy that reaches the Earth's surface, without which life would not be sustainable; the Earth's rotation around the Sun determines climate; the winds move global surface energy around, picking up moisture from the oceans and dumping rain on land, largely controlling the weather. Many natural processes like the Earth's rotation, gravity, internal structure, and other unknown phenomena can alter both the weather and climate on earth. Extensive archeo-geological evidence shows that some desert areas of the world including the Sahara desert were at one time fertile lands, many current land areas were once occupied by oceans, and some of the oceans today were dry lands. The Earth-Environment systems have natural processes of regulating events on Earth that sustain life - temperature, rainfall, human and plant respiration, protection from damaging radiation from the Sun, etc. About two-thirds of the solar short wave radiation reaching the Earth daily is absorbed and the balance is reflected into space as infrared long-wave radiation. Part of the reflected energy passes through the atmosphere back into space while the balance is trapped in the stratosphere by a mixture of minor gases known as *greenhouse gases (GHGs)*, and reflected back to Earth as appropriate, to control global surface temperatures. This natural process ensures that a significant part of the infrared energy reflected from the Earth's surface in day time is absorbed and retained in the stratosphere. When the Earth's surface temperature drops at night, heat is radiated from the gas layer back to Earth. These processes help regulate the Earth's temperature, keeping it at around 13-15°C. Without this regulation, temperatures on the Earth would be around minus 18°C, much too cold to sustain life. The lower part of the stratosphere also contains ozone gas which, though present in minor quantities, filters the dangerous components of the Sun's ultraviolet rays that could cause serious health problems on earth.

The Earth's anthropogenic environment

Environmental pollution is the introduction into the atmosphere of contaminants which can alter significantly the natural balance of the environment systems, with potentially serious consequences. Pollution can be in many forms, including gases, particulates, chemical compounds, or in energy form - heat, noise, light, radioactivity, etc. There are many naturally occurring contaminants - emissions coming from volcanoes, sand storms, vegetation decays, etc., but the environment has in place a system of processing, which is crucial to the sustenance of life on Earth. For example, carbon dioxide exhaled by humans and animals is utilized by plants that produce oxygen, carbohydrates and proteins which humans and animals need to survive. However, human activities have accentuated emissions in the last two hundred years or so, coming from energy use, agriculture, livestock production, etc, thereby disrupting the natural balances. The environment contaminated by human-related pollution is known as the *anthropogenic environment*. Around 68% of anthropogenic emissions come from energy production and use, 12% from agriculture and livestock production, 6% from industrial processes, and 14% from large-scale biomass burning, post-burn decay, landfills, and indirect emissions coming from reactions between anthropogenic chemical compounds. Advances in atmospheric science have identified many of the variables which control the Earth's natural phenomena, and concluded that many human activities, particularly in the last two centuries or so can interfere with the natural processes by which equilibrium is maintained, with undesirable consequences, and environmental pollution is the main focus.

Natural greenhouse gases comprise mainly carbon dioxide, methane and nitrous oxide, all of which are also produced by human activities, thereby raising their concentrations and heat-trapping capability, and the ability to reflect more heat back to earth, causing global warming. Moisture is also a major greenhouse gas but the natural processes of regulation make it non-anthropogenic. Carbon dioxide accounts for around 90% of energy-related emissions, and has become the benchmark against which the environmental impacts of all other anthropogenic greenhouse gases are rated, although the gas is the least potent in terms of environmental damage per unit concentration. This system has established a combined measure known as *carbon dioxide equivalent (CO₂eq or CO₂e)* that represents all anthropogenic emissions. In effect, any reference to carbon dioxide in environmental discourse implies a basket of all the major anthropogenic gases. The carbon dioxide equivalent allows the potency of different greenhouse gases to be compared on a common basis, rated with reference to carbon dioxide which is assigned unity.

Human activities pollute the environment in many ways, but the most potentially hazardous are: enhancement of greenhouse gases, causing global warming; depletion of the protective ozone layer, causing harmful rays of the Sun to reach the Earth; ambient pollution which releases toxic compounds into the immediate human environment, causing serious health issues and other ecosystem problems; and dumping of solid wastes which are harmful to both the environment and life on earth. Records of global anthropogenic emissions from 1850 and numerous scientific reports show that the atmospheric concentrations of carbon dioxide (CO_2eq) have been increasing gradually over the past century compared with pre-industrial levels but there has been a significant rise in the rate of increase from the 1950s when industrial activities intensified. Significant increases have also occurred in the levels of the other main greenhouse gases - methane (CH_4) and nitrous oxide (N_2O). Pre-industrial levels of CO_2eq which ranged between 180 and 280 ppm had increased to 407.4 ppm in 2018. Human activities release greenhouse and tropospheric pollution gases and particulates (mainly carbon dioxide, nitrogen oxides, sulphur oxides, methane, and black carbon) into the atmosphere, largely from energy production and use, but also from net land use (deforestation means lower carbon dioxide intake for photosynthesis and higher concentration in the atmosphere) and agriculture (mainly CH_4 and N_2O from livestock, rice cultivation and use of fertilizers). Some industrial processes not related to energy also produce nitrous oxide, fluorinated gases and ozone. Energy-related pollution accounts for around two-thirds of global anthropogenic pollutions: release of greenhouse gases which enhance the natural concentration and cause global warming in the stratosphere; emissions of F-gases (mainly from refrigeration and air conditioning) which deplete the ozone layer in the lower part of the stratosphere; tropospheric (ambient) emissions of particulate and hazardous chemical gases and compounds including ground level or 'dirty' ozone which have instant negative impact on local weather, visibility, public health, animal, aquatic and plant life, and vegetation.

Not all of greenhouse gases make an equal contribution to the greenhouse effect, or equal impact on the environment and residence times in the atmosphere vary significantly. Anthropogenic greenhouse gases may remain in the atmosphere for very short periods or for as long as 150 years. *Global Warming Potential (GWP)* is a quantified measure of the globally averaged relative radiative forcing impacts of a particular greenhouse gas. It is a cumulative radiative forcing, (both direct and indirect effects) integrated over a period of time from the emission of a unit mass of gas relative to some reference gas which is carbon dioxide. Carbon dioxide being by far the largest anthropogenic gas emission in terms of quantity (around 90%), is assigned a unit global warming potential (GWP) over all time, and serves as the baseline. In effect, GWP of any other greenhouse gas is a measure of how well the gas absorbs reflected solar energy from the Earth, preventing it from immediately escaping into the atmosphere and eventually heating up the Earth, compared with carbon dioxide. The higher the GWP, the more positive the radiative forcing (RF), and the more warming the gas causes (some pollutions can also cause negative forcing or cooling depending on their location in the atmosphere). As discussed above, carbon dioxide equivalent (CO_2eq) is a method of placing emissions of various radiative forcing agents on a common footing in accounting for their relative effects on the environment. It describes, for a given mixture and amount of greenhouse gases, the amount of CO_2eq that would have the same global warming ability, when measured over a specified time period, usually 100 years. Most quoted CO_2eq values represent a basket of greenhouse gases listed in the Annex A to the Kyoto Protocol which was the first major global mitigation action on anthropogenic greenhouse gas emissions. For example, the GWP of one molecule of methane (CH_4) has 20 times the impact of a molecule of carbon dioxide for a 100-year time scale; nitrous oxide (N_2O) 300 times; ground-level ozone 2,000 times; and a chlorofluorocarbon molecule (a gas used in refrigeration and air conditioning) has from 13,000 to 20,000 times the impact of a molecule of carbon dioxide.

Ozone depletion

Natural ozone is concentrated in a thin layer mostly in the lower portion of the stratosphere (the atmospheric layer that is immediately above the troposphere that is in direct contact with the Earth), where it plays a key role in screening out potentially dangerous ultra-violet (UV-A and UV-B) radiation from the Sun by using the energy to dissociate into unstable atoms and then recombining continuously thereby converting dangerous energy into less harmful heat energy. Without this natural protection, the negative impact on life on earth would be severe since these rays have been shown to cause many ailments including skin cancer in humans. There is ample scientific evidence confirming that human activities are producing chemical gases known collectively as F-gases (mainly halocarbons used in refrigeration, air conditioning and some other industrial processes). These gases end up in the lower part of the stratosphere and are capable of reacting with ozone, causing depletion and exposing the Earth to dangerous radiation. This is evident from satellite images showing 'holes' in the natural ozone layer. Furthermore, some of them are also very potent greenhouse gases. The Kyoto Agreement has largely eliminated the use of F-gases in the developed world, but use is still prevalent in emerging economies. Since there are no physical country or regional boundaries in the stratosphere, as long as the use of these highly potent gases continues anywhere in the world, the negative impacts on the environment will be global.

Carbon intensity

Carbon Intensity (also called carbon intensity index or emission intensity, or specific emissions index) is the amount of metric tons of carbon dioxide or equivalent produced per unit of output or an activity, for example per megajoule of energy produced, per capita or per GDP, or even per human activity like driving a car over a hundred kilometers. Population and GDP are the major determinants of primary energy consumption, and therefore a country's energy-related emissions, but two factors may mitigate carbon intensity, namely *energy intensity* and *fuel mix*. Energy intensity/GDP gives some indication of a country's level of economic development and the extent of adoption of energy-efficient technologies, but simply moving away from high-energy intensity production, like is happening in many developed countries can also lower energy intensity. On the other hand, emerging countries which are increasingly hosting primary production of goods that fill most of the world's demand, such as primary metals, polymers, cement, chemicals, petrochemicals will have high energy intensities largely because of the inherent energy-intensive nature of these production processes, but also because these countries often cannot afford to adopt the most efficient technologies. Emissions intensity also depends on fuel mix (carbon content of energy consumed). Coal has the highest carbon content of fossil fuels, followed by oil. Natural gas has the lowest carbon content, while most renewables and nuclear energy have relatively low carbon contents. The developed countries are decarbonizing energy by moving away from coal and substituting less carbon-intensive natural gas and renewables, mainly for power generation. On the contrary, coal is the only readily available and affordable fuel (for power generation, steam raising, home heating, etc.) in many emerging countries, hence carbon intensity will be relatively high. Carbon intensity is useful in comparing the effect of greenhouse gases from various sources on the environment. Values of carbon intensity can be calculated for electric power plants per unit of energy produced using different fuels, for heat generating units per joule of heat, and for various production processes. For example, the emissions intensity of wood is 0.39 $kgCO_2$/kWh, corresponding values for other primary fuels are coal 0.36-0.34 depending on the maturity, crude oil 0.26, natural gas 0.22, diesel oil 0.27, petrol/gasoline 0.25. Carbon intensity calculations also make it possible to compare the environmental impact of different power generation systems: the carbon intensity of a coal-fueled power plant is 888-1054 metric tons of CO_2eq/GWh (on a lifetime basis) depending on coal grade and comparative average values for other fuels and energy sources are oil 733, natural gas 499, solar PV 85, biomass 45, nuclear 29, wind 26, hydro 26.

Carbon footprint

Carbon footprint literally means an estimate of the amount of direct or indirect contribution of greenhouse gas emissions from any source over a time frame (usually one year), for example, events, products, organizations or persons. A comprehensive, quantitative assessment based on this definition would not be possible because of inadequate knowledge about the complex interactions between contributing natural and anthropogenic processes. For example, every person inhales oxygen and exhales carbon dioxide, plants use carbon dioxide from the atmosphere for photosynthesis but they are the primary sources of carbon in fossil fuels, plant/animal carbohydrates and proteins, and there are many other natural sources of greenhouse gases. However, virtually every daily activity contributes to a person's anthropogenic carbon footprint: use of all household goods and materials which require energy for manufacture, energy use in home heating and cooling, transportation, etc. A carbon footprint is measured in metric tons of carbon dioxide equivalent (tCO_2.eq)/year, and carbon footprint calculations consider all six of the Kyoto Protocol greenhouse gases: carbon dioxide (CO_2), methane (CH_4), nitrous oxide (N_2O), hydrofluorocarbons (HFCs), perfluorocarbons (PFCs) and sulphur hexafluoride (SF_6). Carbon footprint makes it possible to compare the relative negative impact of fuels, other energy sources, manufacturing processes, indeed all human anthropogenic activities on the environment. Carbon footprint values can be calculated for virtually any human activity: manufacturing, industrial, commercial, home energy use;

use of consumer goods, commercial and personal transportation; recycled, incinerated or dumped waste; etc. Per capita carbon footprints for countries in 2018 range from 0.1 metric ton/yr for Mali to 16.1 for United States, 5.6 for United Kingdom, and 38.2 for Qatar.

Carbon neutrality

Carbon is a critical element that sustains life on earth. It is the primary element in the human-animal-plant lifecycle. Plants process carbon dioxide from the atmosphere into carbohydrates (around 50% carbon), animals feed on plants to produce protein (75-85% carbon), and humans feed on both, exhaling carbon dioxide into the atmosphere; humans, animals and plants die, returning carbon to the Earth. As discussed earlier, carbon dioxide is a major greenhouse gas which helps regulate temperature on earth, but there is an optimal concentration beyond which it becomes anthropogenic. Carbon neutrality is an environmental mitigation effort which seeks to balance atmospheric input and output of carbon dioxide in the natural carbon cycle in order to preserve the natural concentration, by actions that seek to offset anthropogenic carbon dioxide. Examples include massive planting of trees which can utilize much of the emissions from energy generation, or installation of solar power units that can meet fully the power needs of homes and possibly feed excess power to public grids. An industrial enterprise can become carbon neutral by capturing and sequestrating all the carbon dioxide produced and other more complex possibilities include offsetting carbon dioxide produced in one area of activity by actions such as efficiency improvement and lower carbon energy production (such as those that involve material and heat recycling) that reduce emissions from other areas, or purchasing of carbon credits from other enterprises which have lower or negative carbon footprint. People can also strive to be carbon neutral by optimizing energy and use in homes and transportation, and actively supporting *refuse, reduce, reuse, recycle* policies on consumer products.

Lifecycle carbon footprint

Total Life-Cycle Analysis (LCA) is an accounting process that considers the environmental impact of all stages of a particular process or product chain. For example, LCA for an energy resource involves consideration of emissions at every stage from production to utilization, from material and fuel mining, manufacturing of components, construction, installation, de-commissioning, to waste management. A typical cradle-to-grave LCA of fossil fuels begins with extraction of oil, natural gas or coal and ends with electricity or fuel delivered to the consumer. Lifecycle analysis of carbon intensity is a very useful method of comparing anthropogenic emissions from generating plants using different fuels. Calculations of the total life cycle emissions of a fuel take into account emissions at the mining stage, transportation, processing, ultimate utilization and reused or discarded solid or heat by-products. Emissions from the production of materials of construction such as steel and cement used in dam construction, materials used in the transportation and distribution of natural gas and fuel oil are also taken into account. For example, solar energy is considered a zero emission energy resource in the traditional evaluation process, however, life-cycle-analysis presents a different picture: the overall carbon footprint is significant because enormous energy (usually supplied by coal-fired plants) is required for the production of silicon used in the manufacture of photovoltaic (PV) solar cells, collectors, associated equipment, building and infrastructure, the copper used extensively in solar power generation etc. In effect, there is nothing like near-total generation of electricity from zero carbon sources (solar, hydroelectric, nuclear, wind) as claimed by some developed countries and environmentalists - all energy sources have carbon footprint. It is also a common assumption that gas-fired power generating plants emit around 50% lower environmental pollutants compared with coal. However, total life-cycle analysis shows that, although the carbon footprint of natural gas at point-of-use is significantly lower than that of coal, the ultimate carbon footprint depends on losses in production and transportation. Excessive release of gas (mainly highly potent methane) into the atmosphere during production, flaring and transportation can narrow or even eliminate the relative advantage of natural gas compared with coal especially when the coal plant is of comparable efficiency. In fact, it is estimated that, unless emissions of methane (the main component of natural gas) at the gas production stage can be kept below 2-3%, there is no relative advantage, since methane has twenty times more heat-trapping capability than carbon dioxide which is the predominant pollutant in coal-fired plants. Life cycle analysis also involves the quantification of all emissions in the chain, including emissions saved (displacement emission) where for example the by-products of one process are used to replace products of another process, thereby avoiding the emissions associated with those replaced products. Steel, aluminium, copper, lead, tin and indeed most metals are 100% recyclable. For example, the energy required in processing recycled steel is about 30% less than required for processing virgin steel raw materials. Recycled aluminium only consumes about 10% of the energy required to produce the material from bauxite ore. In effect recycling of material and waste heat can significantly lower the ultimate carbon footprint of many products. In summary, the true assessment of the environmental impact of a product should not only take account of direct emissions from the manufacturing and transportation to retailers but must also consider a host of indirect emissions such as those caused by the production of the raw materials used in the production of the good.

Energy-related emissions

Energy-related emissions come from the three major sectors of the global economy: Industry, building and transportation. Power generation, the most polluting energy enterprise is classified with industry but related emissions are distributed between the three sectors on the basis of electricity use. The energy supply sector, from extraction to end-use is the largest consumer of primary energy and, because many of the processes are inefficient, the sector is also the largest contributor to anthropogenic greenhouse gas emissions which account for over two-thirds of the global total. Virtually all energy production (drilling, mining), conversion, transportation, distribution and utilization processes produce substantial anthropogenic gaseous and particulate emissions. Production processes degrade landscapes, contaminate groundwater, oceans, rivers and aquifers. Coal mining, preparation and transportation (in particular, opencast mining) damage large areas of land and emit particulate dust and methane; around 150-170 million cubic meters of natural gas (mainly methane) are flared worldwide every year in oil and gas production (equivalent to one-quarter of the United States' or 30% of the European annual gas consumption); and there are significant losses during transportation of all fossil fuels. Gas flaring emits black carbon and about 400 million metric tons of CO_2eq per year; significant amounts of oil and gas escape into the environment during production and transportation (fugitive losses). Fossil and nuclear power plants contaminate land masses, rivers and oceans with heat and radioactive solid waste; acid rain formed by reaction between anthropogenic acidic gases and moisture causes environmental problems which damage infrastructure, plant and vegetation, and threaten ecosystems of rivers, lakes and oceans; solid waste (in particular, polymer and electronic waste) and traditional use of biomass are causing increasingly serious environmental problems worldwide. Most of these pollutants are hazardous and have been associated with human illnesses.

Emissions from fossil fuel use

Fossil fuels power global economic growth and human development, and currently account for about 80% of the total world's primary energy use (oil 31%, coal 26%, gas 23% in 2018). In spite of global efforts to develop less carbon-intensive energy resources, the share of fossil fuels within the world energy supply has remained relatively unchanged over the past four decades and all recent projections show that fossil fuels will continue to dominate the primary energy mix in the foreseeable future, still accounting for around 78-80% in 2040 and beyond. Fossil fuels power global electricity generation (around two-thirds), and, together with electricity, supply nearly all the energy required by the three exponentially growing sectors of the global economy: industry, transportation and building. Efficiency gains across all sectors of the economy and all regions have been moderating growth of energy demand that would have been commensurate with the fast growth of the global economy and population, and projections indicate a modest growth of around 35% over the next two decades, compared with over 100% and 25-30% growth in the economy and population respectively.

Over 40% of the total global primary energy consumption is used to produce electricity, all forms of generation have significant environmental impact on air, water and land, and nearly half of the total global energy-related anthropogenic emissions come from electricity and heat generation. Electricity is central to all aspects of economic and human development: manufacturing, food production, healthcare, conducive household environment, etc., and is an important component of every person's environmental footprint. Electricity is the world's fastest-growing form of end-use energy consumption and demand is expected to grow by about 70% over the next two decades, accounting for around two-thirds of the global primary energy increase over the period, with developing and emerging nations accounting for most of the growth. These countries depend on local deposits of low-grade coals, the most polluting of all fossil fuels (lignite coal has around twice the carbon footprint of natural gas), and all recent projections predict that they will also account for most of the future growth in GHG emissions. Energy-related emissions rose to a historic high of 33.1 Gigatonnes CO_2eq in 2018, a nearly 2% (569 million metric tons) rise from 2017, equivalent to the total emissions from international aviation, and China, India and the United States accounted for 85% of the net increase. Just two countries : the United States and China accounted for around 45% of the total global energy-related emissions and fossil fuels filled around 80% and 85% respectively of the primary energy requirements in both countries. However, coal accounted for 60% of the primary energy consumption in China compared with around 13% in the United States. Oil was the greatest pollutant in the 1970s but has been overtaken by coal. Also, the regional distribution of emissions has changed significantly: in the early part of the 20th century, virtually all emissions originated from the United States and Europe, today together they account for less than 30%.

Emissions from power generation

While emissions from all sectors of the economy and from all fossil fuels increased in 2018, the power sector accounted for nearly two-thirds of emissions growth and coal used for power generation accounted for nearly a third of the emissions, mostly in Asia. Emissions from electricity and heat generation in emerging and developing countries have doubled in the last two decades, compared with a significant fall in industrialized countries. China alone accounted for around two-thirds of the emissions' increase in 2018, coming mainly from large increases in coal-fired power generation and energy-intensive primary materials, chemicals, chemicals, petrochemicals and cement production.

Fossil fuels (mostly coal, 39% and natural gas, 22%) supplied 64% of global power generation in 2018, and the most dramatic growth has been in China. Expected shifts in energy mix to less carbon-intensive sources (wind, solar, nuclear, natural gas) will help reduce primary energy use per unit of power output and the CO_2eq intensity of delivered electricity by more than 30% over the next two decades or so, but most of the reduction will be accounted for by the industrial nations who will consume the lowest electricity. Although emissions from power generating plants account for over 40% of the total energy-related emissions, emissions from specific plants depend on the technology adopted and fuel, and are unique to the individual facility because efficiency counts, but fossil fuels (coal, oil and natural gas) are by far the highest pollutants. Coal-fired power plants fueled by young coals (lignite and sub-bituminous) are the most polluting, producing around fifty times the emissions from renewable and nuclear power plants for the same power output, and these are the cheapest and most widely available grades in the developing world. Coal-fired power plants are the largest contributors to global energy-related emissions, accounting for about a third (over 10 Gt) in 2018. In fact there has been a rise in the demand for coal in the last two years after a significant fall in the previous two years and coal-fired power plants were the single largest contributors to the growth in emissions observed in 2018. The International Energy Agency (IEA, 2019i) assessed the impact of fossil fuel use on global temperature increases, and found that emissions from coal combustion were responsible for over 0.3°C of the 1°C increase in global average annual surface temperatures above per-industrial levels, making coal the single largest source of global temperature increase. Nevertheless the contribution of coal to global energy-related emissions has reduced to 30% from around 44% in 2016, due partly to closure of inefficient plants but also to commissioning of more technologically advanced coal power plants .

Natural gas is less polluting at the generating plant but more polluting than coal at the production and transportation stages. As discussed earlier, when assessed on a lifetime cycle basis, natural gas could be as highly polluting as mature coal depending on efficiency along the production and transportation chain. The relative advantage is the potential for capture of natural gas pollution at the production stage. Natural gas is rapidly replacing coal as the fuel of choice for power generation, particularly in Europe and North America driven by strong environmental policies and, in the case of United States, abundant supplies of cheap natural gas. However, because coal is relatively abundant and affordable in the emerging regions, in particular Asia, the fuel outcompetes all other power generation energy sources and the impact of natural gas use on energy decarbonization efforts has not been as significant as desirable. Furthermore, although many aging coal-fired plants in the developed world are being retired and very few are being commissioned, most of the plants in the emerging world are relatively new, 12 years old on the average, meaning they have decades to go before reaching their planned end of production in about 30 to 50 years. Also, India plans to increase coal-fired power generation by around 25% over the next two to three years and China accounts for about 55% of the total global coal power capacity. While China is shutting down small highly-polluting coal power plants, new, more efficient coal power plants are being built or planned, with a total capacity of around 300 GW, more than the combined capacity of the United States' 350 coal power plants. China is also building many more coal-to-fuel and coal-to-chemical plants and around two thousand coal power plants are currently under construction or planned in different parts of the world, mostly in the emerging economies but also in Poland, Australia. Most are expected to come on stream in the next few years and will increase the total global coal power capacity by around 30%. Nevertheless, it is estimated that switching from coal to gas reduced coal consumption by about 60 million metric tons in 2018, helping to avert 95 million metric tons of carbon emissions, decreasing total energy-related emissions by about 15%. The switch was most significant in China and the United States, each reducing emissions by 40-45 million metric tons. Furthermore, most new coal-fired power projects all over the world are adopting high-efficiency clean-coal technologies which include high-technology combustion systems, coal gasification, and emissions capture and processing.

Emissions from industry

Energy and industry have been closely interdependent since the Industrial Revolution, and their future remains intertwined. Energy powers all kinds of industries - manufacturing, non-manufacturing/services, construction, agriculture that produce food and goods for mankind. On the other hand, consumer demand for the wide range of varied products of industries drives energy demand, provides the impetus for massive investments in energy infrastructure, and propels continuous development of new energy technologies. Energy cost is always a major component of virtually any industrial production process, and this has provided a strong incentive for a persistent pursuit of higher efficiencies across the energy production and end-use spectra. Industrial energy demand growth would be at par with economic growth but for the extensive gains in manufacturing and other energy end-use efficiencies over the last two decades or so. Energy-intensive manufacturing (iron and steel, copper, cement, paper, food, glass, chemicals, primary metals) accounted for nearly 70% of the energy use by the sector in 2018, and the chemicals industry was the fastest growing user of fossil fuels in the industrial sector, both for energy and as feedstock for the production of chemicals and petrochemicals. Much of the increase in energy-related anthropogenic pollution in the last two decades or so has been driven by large increases in the production of these energy-intensive materials, in particular, steel, cement, chemicals and petrochemicals which depend primarily on fossil fuels. Most of these heavy industries are now in emerging and developing countries, many of which have also expanded very rapidly their coal production and coal-fueled power generation capacities. All recent projections indicate that these countries will account for nearly all of the increase in industrial primary energy consumption and carbon dioxide emissions over the next two decades. Production is often inefficient and energy intensity is high, lower-carbon energy options are few, more energy is required for a unit output compared with developed countries, and, inevitably, energy-related emissions remain high. However, there are significant regional differences within the emerging regions: while Asia accounted for 53% energy-related emissions in 2018, Africa contributed only 4%.

Over half of the primary energy consumption by industry is not used as fuel but as feedstocks for the production of polymers, chemicals, petrochemicals, fertilizers, lubricants, etc. In spite of projections of over 100% growth in the global economy over the next two decades or so, mostly in emerging regions, efficiency gains and growing use of less carbon intensive energy and manufacturing technologies will help reduce emissions from the sector relative to GDP by 35-50%, and contribute to a nearly 45% decline in the carbon intensity of global GDP. The industrial sector energy demand is higher than any other end-use sector, and accounts for 50-55% of total global final energy use. However, in spite of the leadership in end-use energy consumption, the sector accounts for only 19% of emissions from fuel combustion (25% when fossil fuel use as feedstock is taken into account and 36% when emissions from heat and power generation are allocated on the basis of end use). The developed world is transiting to low energy-intensive economy (light manufacturing and services), leaving the emerging world to host energy-intensive industries such as primary metals' extraction and processing, cement, chemicals, petrochemicals, fertilizer production from mainly fossil energy sources. Furthermore, manufacturing is 2 to 3 times more carbon and energy intensive than services, and, within manufacturing, energy intensity of production of primary materials could be two orders of magnitude higher than light manufacturing. This explains why most of the future energy-related emissions will come from the emerging regions. It should be noted also that the developed world is becoming increasingly dependent on the emerging countries for the supply of primary materials, thereby driving rapid expansion of production facilities and infrastructure.

Emissions from transport

Transportation currently accounts for about 25% of the global CO_2eq emissions. The world's fleet of vehicles will at least double by 2040, to around 2 billion, and most of the growth will be in emerging nations, with fossil fuels (mainly oil and other liquid fuels) currently supplying about 95% of the sector's fuel demand. However, growth in demand for transportation fuels will be modest, 30% or so, responding to increasing efficiency of conventional vehicles across all regions of the world, and the share of fossil fuels is projected to decline to about **88%** as the use of alternative fuels slowly increases. Electric vehicles have a strong potential for decarbonizing the transportation sector, depending on the sources of energy used in the manufacture and operation. Currently fossil fuels account for most of the manufacturing energy of electric vehicles and routine battery charging and unless there is a fast and significant transition to lower carbon power sources, the lifecycle carbon footprint could be twice as high as for comparable conventional internal combustion engine vehicles. Penetration of electric vehicles into the global auto market is projected to rise from 5 million in 2018 to 150-350 million (currently 5%, increasing to 6-18% of global vehicle population) by 2040 (different levels of optimism and assumptions), mostly in the light-duty vehicle sector assuming current problems of high cost, support infrastructure, low and unstable global oil market prices, strong competition from increasingly fuel-efficient and environment-friendly conventional cars, and consumer preferences can be surmounted.

Emissions from Buildings

The buildings (residential and commercial) sector is the least polluting of all energy end-use sectors, accounting for only about 6-9% of the total global energy-related emissions, although the contribution is much higher (27%) when emissions from generation of heat and electricity used in the building sector are factored in. The housing sector currently accounts for around 20% of the total global delivered energy, projected to rise by a modest 20% by 2040, despite the expected rapid growth in global population, urbanization, economic status of the population, and an increasingly digital world. Non-OECD nations will account for around 40% of the total growth, with China and Africa alone accounting for about 30% of the growth in demand, but energy efficiency rise - energy efficient buildings, fittings, appliances, consumer products, and increasing awareness among consumers of the multiple benefits of optimized energy use - will play a critical role in moderating the demand growth in the sector, limiting rise to about 30% across all regions over the next twenty years or so. Residential electricity use will account for around 90% of the growth and is projected to rise by about

75% by 2040, driven largely by around 150% increase in non-OECD countries (about 250% rise in India and Africa). However, the overall contribution of the building sector to global energy-related emissions will likely remain relatively low.

Non-energy Emissions

Around a third of the global anthropogenic emissions come from non-energy related sources: forestry and land use, agriculture and livestock production, and non-energy related industrial processes. Agriculture, forestry and other land use (AFOLU) activities account for 12-13% of carbon dioxide, 44% of methane and 82% of nitrous oxide emissions from human activities globally. Agricultural activities (biogenic sources) such as wet rice cultivation, soil fertilization, and vegetarian livestock production release significant quantities of methane into the atmosphere. For example, the digestive process of a cow emits about a quarter of a kilogram of methane into the atmosphere daily. Significant methane emissions also emanate from decaying vegetation, especially in landfills, hydro dams, artificial lakes and reservoirs. Agriculture accounts over a third of non-energy emissions, non-energy industrial processes about 20%, with the balance coming from other sources. Non-energy anthropogenic emissions have received relatively little attention even though they account for nearly a third of the global total emissions, yet there is substantial scope for reduction, notably in agriculture, use of landfills for waste disposal, uncontrolled incineration of waste, bush burning, etc.

Consequences of Environmental Pollution

Evidence abounds showing that the Earth system has always responded and will continue to respond to natural forcings - solar, volcanic, orbital, natural variations in atmospheric composition. For these reasons, many phenomena associated with climate change today have always been part of natural occurrences which have caused drastic changes in various parts of global environment over time. For example, there is ample evidence that some of the current deserts were once occupied by water, many drylands and coastlands today were once under the oceans, and massive land movements, volcanoes and tsunamis are all natural processes which impact on the environment. However, these impacts were restricted to relatively small sections of the world. There is strong scientific evidence that human activities particularly since the Industrial Revolution, are making the impact on mankind of some of these natural phenomena increasingly more widespread, more common, more intense and more severe. Furthermore, recent scientific evidence shows that the rise in global temperatures over the past 150 years has been far more rapid and widespread than any warming period in the past 2,000 years and none of the events that are often quoted by global warming sceptics is remotely equivalent either in degree or extent to the warming over the last few decades. The pre-industrial warming events affected only limited regions of the planet at a time unlike the warming of the last five decades or so which has been planetwide. The unusual heat waves and cold spells also affected limited regions at different times often separated by centuries. As discussed earlier, events in three layers of the atmosphere have critical but different impacts on life on earth: the middle atmosphere (stratosphere), the lower-middle atmosphere (lower stratosphere), and the lower atmosphere (troposphere or ambient atmosphere). Some of these risks may be existential to civilization, posing permanent, large, negative, and irreversible consequences to humanity. While the negative impacts of stratospheric pollution are gradual and global, the effects of tropospheric pollution on human health and well-being of the ecosystem are largely local and instant.

Middle stratospheric pollution

The Earth's climate system is powered by solar radiation, around 56% of which reaches the Earth's surface in the visible part of the electromagnetic spectrum. Global temperature is determined by the balance between incoming solar energy and the radiation reflected back into space by the Earth's surface, and has been relatively constant over many centuries. Natural fluctuations in solar energy output due to the Earth's rotation, and consequent fluctuations in the amount of incoming short wave radiation (SWR) can cause changes in the energy balance, so can human activities that release emissions into the atmosphere: gases which modify natural concentrations of greenhouse and ozone gases, and aerosols which cause severe human health problems. As discussed earlier, activities of greenhouse gases in the stratosphere regulate the global average temperatures which have been around 14°C for over a century up to 1980 or so. Without this natural control, the Earth would be too cold to sustain life. However, there are wide temperature variations across the globe, from around zero in the Antarctica to around 50-60°C in the Libyan desert but there is ample scientific evidence that the average global decadal temperature has increased by nearly around 0.6°C in the last four decades. This may seem insignificant but small changes in the Earth's average temperature can have very big impacts on the global climate. Perhaps the most important indicator of climate change, is global warming, caused largely by anthropogenic GHG emissions which end up in the stratosphere (middle atmosphere), trap and reflect more heat than normal back to earth. A consistent rise of the global average temperature by even less than 1°C can have potentially severe and diverse consequences on earth - extreme weather; warming, rising and acidifying oceans; desertification, poor crop yields, etc.

Weather and climate change

Weather describes the conditions of the atmosphere at a certain place and time with reference to temperature, pressure, humidity, wind, and other key parameters, collectively known as meteorological elements. Weather is also described by the presence and movement of clouds, precipitation, and the occurrence of spatial phenomena, such as thunderstorms, dust storms, tornados, etc. Climate takes into account all the above variables over a period of time ranging from months to years, projected in either direction to thousands and millions of years in order to arrive at a global climate pattern. Weather is largely local and can change from hour to hour, day to day, month to month, or year to year. Climate on the other hand, generally refers to an aggregate of weather statistics over much wider areas for a decade or more. Natural fluctuations in solar energy output due to the Earth's rotation, and consequent fluctuations in the amount of incoming shortwave radiation (SWR) can cause changes in the Sun's energy that hits the Earth's surface and the amount that is reflected, thereby altering the balance either way, so can human activities that release emissions of gases and aerosols (anthropogenic emissions) into the atmosphere which enhance the natural concentrations of greenhouse or deplete ozone concentration. However, while the natural greenhouse effect either heats up or cools the Earth as necessary, anthropogenic greenhouse gases almost always heat up the Earth, while depletion of ozone caused by pollution enables hazardous rays to reach the Earth. The most familiar features of weather are the day-to-day and day-to-night variations in temperature, heat waves, cold spells, humidity, precipitation, rainfall, windiness, atmospheric pressure, and cloud structure, while increasing global average temperature is the most important cause and indicator of climate change. While weather can change from day-to-day, even within the day, climate change happens slowly over decades, hundreds, maybe even thousands of years. Global climate is taken as the average across the world, and could be in terms of average temperature or precipitation patterns.

Any or combinations of the Earth-Environment natural phenomena can cause weather or climate change, resulting in extreme weather, desertification, negative health impacts, etc., and there are many natural processes which can cause global warming or cooling, with significant impact on global climate. For example, the movement of tectonic plates, volcanic activities, and changes in the Earth's orbital axis can cause major climatic change, resulting in the appearance and disappearance of long periods of warm or cold global weather, and this explains why the Earth has experienced many periods of ice ages over the last one million years, some lasting several hundred years. El Niño (warm phase) and La Niña (cold phase) are opposite phases of a natural climate cycle known as the El Niño-Southern Oscillation (ENSO) cycle. ENSO swings between warmer and cooler seawater in the tropical pacific and frequently causes climate pattern across the region to swing back and forth every three to five years on the average, each swing lasting around one year or more, leading to significant differences in the average ocean temperatures, winds and surface pressure, and rainfall across parts of the tropical Pacific. During El Nino, rainfall is below average in the western tropical Pacific around Indonesia and above average in the central and eastern areas where ocean temperatures are higher than average. The effects of La Nina on temperatures and rainfall across the region are reversed. The cause of ENSO is unclear and the frequency can be quite irregular. Although the main effects are localized, the side effects can impact on weather in far away countries that border the Pacific Ocean, causing severe flooding in North America and draughts in nations that border the western Pacific Ocean. In fact, El Nino is regarded as the Earth's most influential climate pattern. There is also ample geological and paleo-climatological evidence that oceans and land mass have interchanged over time as a result of earthquakes and other massive land movements. For example, the Sahara desert is believed to have been covered by ocean thousands of years ago during the warm, wet age and many areas were fertile land. However, while many changes in weather and climatic patterns observed today may have been caused by natural phenomena, there is ample scientific evidence that human interference can exacerbate the spread, frequency, intensity and consequences. For example, recent research has shown that the physical processes in the ocean and atmosphere that produce strong El Ninos are being supercharged by warmer ocean waters caused by human-induced climate change and will become more frequent, stronger, causing more extreme weather events (severe storms, heatwaves, floods) particularly in the United States and other North American countries (Wang *et al.*, 2019).

As discussed earlier, the Earth's climate is largely determined by the radiant energy (sunlight) received from the Sun. The Earth may absorb all the energy received, or some of it in which case the balance is reflected back into space. The balance between the absorbed and reflected solar energy determines the average global temperature which in turn largely determines the climate. Radiative forcing (RF) is a quantitative measure of the difference between energy received by the Earth from the Sun and the amount radiated back into space (net change in energy balance in response to some external perturbation). Positive RF means that the Earth receives more energy than it radiates and negative RF implies that the Earth loses more energy into space than it receives. Net energy gain (positive RF) will cause global warming while negative RF means that the Earth cools. The RF concept is valuable for comparing the influence on global mean surface temperature (GMST) of most agents affecting the Earth's radiation balance. There is ample scientific evidence to support the belief that the RF effect of energy-related emissions is positive, and the average global temperature is rising, exemplified by warming oceans and rise in water levels, increased flooding, increasingly intensive extreme weather, melting arctic ice, etc. Furthermore, human activities (land clearing, construction, agriculture) have changed and continue to change the Earth's surface topography and atmospheric composition. Some of these changes have direct or indirect effects on the energy balance of the Earth and wind movement, and are thus potential drivers of weather and climate change.

Extreme weather (which may be largely localized) is one that is rare at a particular place and/or time of the year. Climate change, whether driven by natural or human forcings, can lead to changes in the frequency and intensity of extreme weather events, such as extreme precipitation events or warm spells, draught, heavy rainfall, flooding, heat waves, etc. The effects of climate change are much more diversified and widespread than weather changes. An increase in the average global temperature can influence weather and climate in many ways, in particular, spread, intensity and frequency of extreme weather. Higher temperatures boost evaporation which dries out soil and increases *drought* and *desertification*. Evaporation intensifies as temperatures rise and more moisture in the atmosphere will intensify rainfall and flooding. At present, single extreme events cannot generally be directly attributed to anthropogenic influence, but it has been established that human activities are contributing to global warming which in turn can exacerbate such extreme events as thunderstorms and flooding. Hurricanes start as strong winds over warm tropical oceans near the equator, with air picking up moisture from the warm waters of the oceans and rising due to fall in density. The depression that is formed between the rising air and the ocean surface is filled by fresh air which picks up moisture and rises. The rising wet, warm air cools off and the water in the air forms clouds. This process is repeated until thick clouds are formed and the complex system of forces within the cloud could start off a spin. Storms formed north of the equator spin counterclockwise while those formed in the south spin clockwise, determined by the Earth's magnetic field. The spinning storm begins to move slowly under the influence of winds, which are continuously energized by increasingly warm waters. Storms can travel over thousands of kilometers across oceans, picking up strength and speed from warm oceans and, by the time they make a landfall, they may have transformed into powerful hurricanes and tornadoes with speeds ranging from around 100-400 kilometers an hour. Such powerful storms can cause catastrophic damage to structures, causing significant fatalities, dump heavy rains and cause heavy flooding, especially in the coastline areas. It is interesting to note that the most severe impacts of hurricanes and tornadoes occur far away from the origin, just like the El Nino. Storms and hurricanes are natural phenomena and catastrophic damages caused by extreme weather have occurred many times, in many places over centuries, well before fossil fuel use became widespread. However, the fact that storms need warm waters to form and strengthen means that they will be stronger, more frequent, and more devastating if ocean waters become warmer as a result of human activities. Also, the combined effect of expansion of warmer ocean waters and melting ice caps and glaciers could raise ocean levels by up to two meters by the end of the century, to the point that many coastal towns and cities in parts of the world become submerged, as is currently being predicted for some coastal areas of Europe and the United States.

Draught and desertification

Draught is an extended period of deficient rainfall relative to the statistical multi-year average (long-term mean) for a region - a season, a year or several years. Desertification on the other hand is defined as a process of land degradation in arid, semi-arid and dry sub-humid regions, resulting from various factors, including climatic variations and human activities. The underlying cause of most draughts can be related to changing weather patterns manifested through excessive build up of heat on the Earth's surface, meteorological changes which result in reduction of rainfall and reduced cloud cover, all of which result in greater evaporation rates. Extended draught could lead to desertification and there are many natural environmental phenomena that can cause draught. However, the spread, persistence and resultant effects of draught are aggravated by human activities such as deforestation which alters the soil's natural albedo (difference between the Sun's energy that is absorbed or reflected); overgrazing and poor cropping methods, all of which reduce water retention of the soil; and improper soil use strategies and conservation techniques which lead to soil degradation.

Consequences of climate change

The natural greenhouse gases (GHGs) reside in the middle atmosphere and the concentrations regulate the Earth's temperature. Human activities have been shown to increase the concentrations of most of the gases, notably carbon dioxide, methane and nitrous oxide, thereby enhancing their heat-trapping ability and the amount of heat reflected back to earth, and increasing the global average temperature. The global climate system has been monitored over the last hundred years or so, especially since the 1950s when sophisticated instrumentation and satellite technology became available, and the changes observed have led to the conclusion that global average temperature, the main indicator of climate change, is rising, with potentially serious consequences for life on earth. If the trend continues, the average global temperature could increase by up to 2°C by the end of the century, with dare consequences for current and future generations. Apart from global warming, there are many other negative effects of anthropogenic pollution, notably warming and acidification of the oceans, acid rains, tropospheric smog, etc., all of which could have potentially devastating effects on life on earth. The recognition of the fact that natural phenomena can cause significant positive or negative climate forcing (heating or cooling of the Earth) has facilitated current conclusions about climate change which are determined based on comparison of current phenomena with paleo-climatic database to determine whether or not they are unusual. Perhaps the most convincing evidence of climate change over the last 200 years or so is the result of in-depth analysis of observational records of the atmosphere, land, ocean and cryosphere systems over the period. There is incontrovertible quantitative evidence that the atmospheric concentrations of GHG gases, in particular, carbon dioxide (CO_2), methane (CH_4) and nitrous oxide (N_2O) have increased substantially over the last hundred years or so; land and sea surface temperatures have increased; oceans have acidified; observations from satellites and *in situ* measurements indicate significant reductions in glaciers, Arctic sea ice and ice sheets; and data on radiative budget and ocean heat content suggest a small imbalance; satellite data on atmospheric temperatures show increases in tropospheric temperatures and decreases in stratospheric temperatures. The fact that the stratosphere and troposphere exhibit opposing temperature changes is a strong indication that the Sun is not the main driver (otherwise the response would both be either positive or negative). The opposing response is consistent with known anthropogenic GHG effects.

There is extensive data dating back some two hundred years on multiple independent climate indicators, from high up in the atmosphere to the depths of the oceans. They include changes in surface, atmospheric and oceanic temperatures, glaciers, snow cover, sea ice, sea level and atmospheric water vapor. Many independent expert atmospheric science research groups notably the International Panel on Climate Change (IPCC, 2013, 2018) have analyzed these extensive databases and all have come to the same conclusion that the world has warmed significantly since the 19th century when industrial and human development activities intensified. Although each year and even decade is not always warmer than the last, strong data-based evidence shows that global surface temperatures have warmed substantially since 1900, and more than 90% of the excess energy absorbed by the climate system since at least the 1970s has been stored in the oceans. As the oceans warm, winds, storms are energized and transformed into powerful hurricanes, ocean waters expand, accounting for much of the observed rise in sea levels over the past century. Melting of glaciers and ice sheets due to increasing global temperature also contributes to the rising levels of the oceans. A warmer world is also a moister one, because warmer air can hold more water vapor, and various global analyses show that specific humidity, which measures the amount of water vapor in the atmosphere, has increased over both land and the oceans, which means that rains will be more intense and flooding more severe. Regional climates - monsoon systems, tropical phenomena, cyclones - are the complex result of processes that vary strongly with location and so respond differently to changes in global-scale influences. Monsoons are the most important modes of seasonal climate variation in the tropics, and are responsible for a large fraction of the annual rainfall in many regions. Their strength and timing are related to atmospheric moisture content, land-sea temperature contrast, land cover and use, atmospheric aerosol loadings, and other factors. Increasing global temperature will intensify monsoons in the future, affecting larger areas, because atmospheric moisture content increases with temperature. In summary, anthropogenic pollution is not the only cause of climate change or extreme weather but is a significant contributor notably by making the Earth warmer. The effects of global warming such as warmer oceans and rising sea levels can make bad storms, rainfalls, flooding, heat waves and cold periods worse and much more extensive and destructive. Droughts and wildfires will become more intensive and widespread as a result of changes in evapo-transpiration and stronger winds, crop yields will be lower and the impact on human health would be profound.

The global environment

The global environment is a single, unified, continuous entity and there are no physical country or regional boundaries. While the major effects of ambient pollution are localized and often short-lived, the consequences of stratospheric pollution are global and enduring. Gases in the stratosphere are relatively well mixed, hence the group is often referred to as well-mixed greenhouse gases (WMGHG). Their life span is in decades or centuries and they are relatively evenly mixed and evenly spread, with changes in concentrations occurring slowly over decades. One important point about stratospheric pollution which is often missed but needs to be emphasized is the fact that there are no hard atmospheric boundaries between nations and regions, therefore global forcings per unit emission and emission metrics for greenhouse gases do not depend on the geographic location of the source of emission, and the impact of events in one part can resonate in the farthest corner of the world. In effect, any part of the world could suffer the consequences of stratospheric pollution arising from the farthest regions. For example, even though Europe and North America have made significant advances in decarbonizing energy which may have resulted in cleaner local environment, the negative impacts of stratospheric pollution from exponentially rising coal power generation and energy-intensive manufacturing in Asia will be global: both Europe and North America have had unusually severe weather for most of the last decade. Furthermore, even if all anthropogenic pollution were to stop today, most of the benefits will not become evident for decades, maybe centuries, and future generations will be the greatest beneficiaries.

Health implications of climate change and environmental pollution

Climate change has been at the center of global discussion on environmental pollution and, although there are many natural phenomena which can cause climate change or generate pollution, it is a well established fact that human activities are interfering with the natural control processes and aggravating the consequences. Global warming which is the most prominent consequence of climate change and emanates from the stratosphere has many health implications, notably illnesses, injuries and fatalities caused by extreme weather (heat waves, tornadoes, flooding, etc). The negative impacts of climate change are global, gradual and enduring, maybe for hundreds of years, but the impacts of ambient pollutions which are largely local can be instant and more devastating because they occur mostly in the part of the atmosphere (troposphere) that is in direct contact with life on earth.

Lower-stratospheric pollution and ozone layer destruction

As discussed briefly earlier, natural ozone is concentrated in a thin layer mostly in the lower portion of the stratosphere, from approximately 15 to 30 km above Earth's surface. The concentration is highest (2-8 parts per million) in the 20-40 km altitude range, although the thickness of the layer varies seasonally and geographically. Ozone in the Earth's stratosphere is created by the Sun's ultraviolet rays (with wavelengths shorter than 240 nm) which supply energy for the dissociation of some of the natural atmospheric oxygen molecules constantly being introduced into the atmosphere by photosynthesis. The molecules split into two highly unstable oxygen atoms which react with oxygen molecules to form ozone, (an exothermic reaction) with the release of heat energy. The net effect of the above two reactions is the formation of two molecules of ozone from three molecules of oxygen, accompanied by the conversion of light energy to heat energy. The ozone formed is unstable and absorbs more of the Sun's ultraviolet rays which causes it to decompose into oxygen molecules and atoms, with the release of heat energy. This continuous set of reversible reactions involving the breakdown and formation of ozone with the accompanying absorption and dissipation of energy is known as the *ozone-oxygen cycle or* Chapman cycle. There is no net ozone depletion because the process produces atomic oxygen that reacts with molecular oxygen to form another ozone molecule. A natural ozone layer is thus created in the lower part of the stratosphere. These reactions play a critical role in utilizing potentially harmful ultraviolet (UV-A & B) radiations from the Sun and converting to less harmful heat energy, thereby preventing all of UV-A and most of UV-B radiation from reaching the Earth's surface and causing human ailments. Although exposure to UV radiation can be beneficial, notably the production of vitamin D which is essential to human health, over-exposure can lead to serious health issues, including cancer, blinding eye diseases, and premature aging. Human activities in the last few decades, notably in refrigeration and air conditioning have been releasing some chemical gases often grouped as halocarbons or F-gases into the atmosphere. Apart from the fact that some of these gases are potent greenhouse gases and heat trappers, they also react with the natural ozone, thereby lowering its concentration and creating pathways (ozone holes) for dangerous UV rays to reach the Earth's surface. One CFC molecule can destroy about a hundred thousand ozone molecules and reduce the concentration in the atmosphere, causing ozone holes. Concerted global efforts in the last two decades coordinated by the United Nations have led to a significant reduction in the use of fluorocarbons mostly in the developed world and there are signs that the ozone layer is healing gradually but use of CFC gases is still prevalent in the developing world and effective action needs to spread across all regions of the world.

Tropospheric/ambient/aerosol pollution and human health

The lowest level of the atmosphere (troposphere) is in direct contact with humankind and there are many natural processes which cause tropospheric pollution (aerosols), notably volcanoes, sandstorms, earthquakes, lightening-induced bush fires. However, many outdoor/indoor human activities also contribute significantly to ambient pollution, releasing black carbon, carbon dioxide, nitrous oxide, ozone, and many other potentially harmful chemical compounds into the atmosphere. Aerosols originating from large volcanic eruptions and human activities enter the troposphere but, depending on prevailing wind dynamics, can also reach the stratosphere. Those that reach this level sediment out of the stratosphere within a year or two but, in that short period, can cause stratospheric cooling. However, most aerosol emissions remain in the lower atmosphere for a short time ranging from a few days to a few weeks, causing smog and health issues. Because aerosols, notably black carbon are distributed unevenly in the atmosphere, they can heat and cool the climate system in patterns that can drive changes in weather, in particular, precipitation. Much of the pollution in the troposphere comes from local activities including power generation, transport and urban traffic logjams, local mining and industrial activities, agriculture, bush burning, domestic energy use, and use of traditional biomass.

Ambient emissions - black carbon; organic carbon; carbon, nitrogen and sulphur oxides; ozone, and many other toxic compounds - impact more directly and instantly on life on earth, causing smog over cities, acid rain, toxic household environment, multiple health issues, and having negative impacts on animal, aquatic and plant life. Exposure to ambient pollution can trigger new cases of diseases or exacerbate pre-existing human health conditions and the effects on wildlife, aquatic life, and vegetation can also be devastating. The World Health Organization (WHO, 2018) has identified aerosol pollution as the world's largest single environmental risk and estimates that 4.2 million global premature deaths in 2016 were linked to ambient (outdoor) air pollution, mainly from non-communicable cardiovascular and respiratory diseases: stroke, chronic obstructive pulmonary disease (COPD), lung cancer, asthma, bronchitis, and other acute respiratory infections, with low- and middle-income countries accounting for a very large proportion. Indoor pollution arising from inefficient use of traditional and fossil fuels (wood, coal, biomass, kerosene) is just as deadly, accounting for around 3 million deaths, and women and children are the most vulnerable (a total of over seven million deaths annually from ambient outdoor and indoor pollution). Close to half of deaths due to pneumonia among children under 5 years of age (about one million deaths a year) are caused by particulate matter (soot) from household pollution. Nearly 90% of air-pollution-related deaths occur in the poorer countries, with nearly 2 out of 3 occurring in WHO's South-East Asia and Western Pacific regions. The most vulnerable are women, children and older adults, and those with pre-conditions all of who tend to spend more time indoors where the concentration of the most harmful particulate matter is highest. Particulate matter is considered the most dangerous component of ambient pollution and the fine particles below 2.5 microns in size ($PM_{2.5}$) pose the greatest problems because they can bypass the body's natural defenses, penetrating deeply into the lungs and bloodstream. Children are particularly sensitive to poor air quality because their lungs are proportionally larger than those of adults in relation to their body weight, hence they breathe more. Also, their immune system is still developing and they are less able to fight health issues that arise from indoor polluted air. Furthermore, babies and other young ones tend to stay around their mothers during cooking, thereby getting exposed to the highest concentrations of aerosols.

Outlook for energy-related emissions

The global economy is rising, population growth is dynamic, and the quest for human development is strong, all powered by energy. The world is faced with a major challenge of providing adequate energy to meet exponentially increasing global demand in an environmentally sustainable manner (more energy with lower carbon emissions). The International Energy Agency (IEA) has stated that energy must be fully decoupled from global economy and the rise in energy-related emissions must be reversed within the next two decades if the world is to embark on the track towards the achievement of the set goal of limiting the rise of global temperature by the end of the century to 2°C, preferably 1.5°C. As discussed earlier, even this small temperature rise could have a catastrophic impact on the global climate. Also, the International Panel on Climate Change (IPCC, 2018) recently assessed all available scientific, technical and socio-economic literature relevant to global warming and concluded (with a high degree of confidence) that, if pollution continues at the current rate, global warming will likely reach 1.5°C above pre-industrial levels between 2030 and 2052. The maximum temperature reached will depend on the point at which net zero global anthropogenic emissions is achieved and sustained. Even then, warming from the cumulative net global emissions up to the time of net zero emissions will persist for centuries to millennia and will continue to cause further long-term changes in the climate system such as sea level rise and associated impacts. Also, the International Energy Agency (IEA, 2019f) projects that achievement of the 1.5°C target will require the deployment of much stronger and integrated and coordinated global action and policy instruments to further reduce the projected level of emissions for 2040 (based on current and expected actions) by around 40%.

Outlook for global primary energy demand

Developing an outlook for energy even over the period of a few years can be very complex and inaccurate because of the many unpredictable variables and necessary assumptions. Nevertheless, tools including extensive database and software packages have greatly improved accuracy over the last two decades or so. Most projections consider several scenarios: Emerging transition (ETS) which takes into account current, expected and anticipated national and global policy instruments and technology developments over the next two decades or so, and faster, even faster transition scenarios (FTS, EFTS) which identify potential options for repositioning the world on a feasible pathway towards achieving sustainable environmental targets. Primary energy consumption grew at a rate of nearly 3% in 2018, the fastest rate over the last decade, almost double its 10-year average, and fossil fuels filled nearly 85% of the demand. Just three countries: China, United States and India accounted for more than two-thirds of the global increase in energy demand. Inevitably, carbon emissions grew by 2%, the fastest growth rate in nearly a decade. While the growing economy was a significant trigger, much of the energy demand increase may be weather-related: the unusual severity of cold weather and heat waves (both potential consequences of climate change) and a significant rise in demand for heating and air conditioning across the world's major centers of energy demand. Most of the rise in energy demand was for natural gas and electricity consumption but there was also a significant rise in the demand for coal for electricity generation and heating even in the developed world.

The global economy (GDP) is expected to more than double over the next two decades, leading to a significant rise in energy demand, although the continuing decline in energy intensity (energy used per unit of GDP) driven by increasing efficiency in energy production and utilization will likely moderate energy demand and limit rise to no more than about 30-35%. All recent published outlooks and projections by reputable independent international organizations show that fossil fuels will still account for around 80% of global primary fuel consumption for the next two to three decades. This demonstrates clearly the complexity of mitigating climate change which requires provision of more energy for less carbon, and shows that the world is on an unsustainable path towards the declared goals for a sustainable environment. Despite the continuing rapid growth in renewable energy in 2018, it provided only a third of the increase in power generation and accounted for only a quarter of the total primary energy use for power generation in 2018. Decarbonizing energy is the surest path towards a sustainable environment but there are formidable challenges: while the industrial world is set to continue and elevate its achievements over the last decade in lowering energy carbon footprint, most of the growth in energy demand over the next two decades or so will be in the emerging economies and fossil fuels, in particular coal, will account for most of the consumption. These regions will host most of the world's energy-intensive production capacity for the primary building blocks of the global economy: primary metals, cement, chemicals, petrochemicals, fertilizers which depend primarily on coal as energy sources (heat and electricity) as well as feedstocks. Furthermore, most of these countries cannot afford to adopt currently available technologies for decarbonizing energy. In effect, expected significant drop in energy-related emissions in the developed world will be largely neutralized by major rises in the emerging nations. China is already adopting cleaner coal technologies which compare favorably with natural gas in terms of carbon intensity but other countries in the region will need international assistance.

Effect of global economic, social and exigential forces on fossil energy use

Another major environmental mitigation challenge is the prime role of fossil energy in the global economy. The United States has recently emerged as the world's largest producer of oil and gas and more than doubled its oil output over the last decade to about 12 million barrels a day, driven by the wide deployment of fracking technology. The country is also poised to becoming one of the world's largest exporters of fossil fuels. Crude oil is the world's most valuable export product, followed by refined petroleum oils (from crude oil), then automobiles. The global oil and gas export market is dominated by the Middle Eastern countries, Russia and the United States, while coal exports are dominated by Asian countries, Russia Australia and the United States. The United States holds the world's largest reserves of coal and even though the country is making effort to reduce internal coal use, the export drive has picked up in recent years. The economies of Russia and many other European and emerging countries depend primarily on the production and export of oil and gas, while many countries (Indonesia, China, India, Australia, Russia, Poland, South Africa) rely heavily on the production and export of coal. Extensive inter-country and inter-regional natural gas pipeline networks are in service already or under construction and the global market for liquefied natural gas is growing very rapidly, dominated by Australia, Qatar, the United States, Russia, Iran, Canada. The economies of many emerging countries depend critically on exportation of fossil fuels, accounting for nearly all the foreign exchange earnings. Also, every major oil and gas company in the world and many Asian coal producers are still committing to projects that do not align with the Paris Climate Change Agreement which targets limiting the increase in global average temperatures to 1.5 degrees Celsius by the end of the century. This is because fossil energy production and export are major driving forces of the global economy and across all regions of the world in terms of revenue (taxes and foreign exchange earnings) and employment generation across the whole spectrum of production, servicing, conversion and distribution. It is difficult to imagine that, for environmental reasons, any country (developed or emerging) can contemplate a reduction in fossil energy production, local use and export because of the obvious and potentially severe negative impact on its economy. Furthermore, the fluctuating dynamics of the global oil market which is closely tied to the global economy impact negatively on global efforts to decarbonize energy: every time oil prices decline, investments in renewable energy use, energy efficiency improvements and energy decarbonization also decline. Fossil energy use has also become a major political pawn in some developed countries, causing significant instabilities in the deployment and sustenance of effective energy-related environmental mitigation policies that survive political cycles. Policies are developed and reversed depending on the local exigencies and political climate. It is interesting to note that the United States, the largest polluter ever, currently the world's current second largest source of energy-related emissions and one of the leading architects of the Paris-2015 Protocol is in the process of pulling out of the Agreement and reversing most of the country's existing environmental mitigation policies because of the effects on its economy.

Another potentially significant challenge to global efforts to decarbonize energy is the effect of projected slower growth of power demand in the developed world due largely to improving efficiencies. This is expected to slow the speed with which renewables can penetrate power generation since it is hard for a new renewable power plant to compete commercially against existing fossil fuel facilities. Therefore penetration of renewables in the power fuel mix will depend on the pace at which existing power stations are retired. Furthermore, the intermittency of most renewable energy resources and therefore the need for a fossil fuel-powered base are problematic: although solar and wind units can power small communities and installations, the main contribution to the global energy demand and consumption is still through integration with fossil-fueled power grids. While solar and wind power plants are already competing with gas-fired peaking plants to manage short-run fluctuations in supply and demand even across country borders, conventional power plants remain the main sources of power grid stability and system flexibility. Another issue with renewable energy, in particular, bioenergy is the enormous amount of resources required, notably land and irrigation resources which are also in critical demand for the production of food considered more urgent in view of the fast global population growth rate. However, in spite of all these potential obstacles, there will be significant decarbonization of energy in the coming decades but this will be due mainly to shifts in energy mix and improved efficiencies across all sectors of the global economy.

There is growing doubt on the commitment of the over a hundred countries that signed the Paris-2015 Agreement and submitted country plans, considering the recent monumental and consequential policy reversal by the United States which will inevitably impact negatively on the commitment of most other countries. While Europe-28 is very much on course because of the central policy institution and enforcement strategies, there is little to show that other countries are making much progress in meeting their commitments which in any case are insufficient to place the world on the path to a sustainable environment. China, the world's largest polluter which accounted for 28% of the global energy-related emissions in 2018 has released laudable and comprehensive energy decarbonization plans but the country appears to be moving in the opposite direction, driven by local exigencies of providing power to the teaming and increasingly middle-class population, while also producing more power to the increasingly global Chinese manufacturing industry. Various projections show that China needs to reduce its coal power capacity by over 40% from current levels in order to meet the reductions required to hold global warming at the target limit. On the contrary, the country has been expanding its coal use for energy and as chemicals/petrochemicals feedstock very rapidly. China is pursuing a policy that is clearly out of alignment with the Paris-2015 Agreement. The country has been responding to strong pressures from organized industries to increase the coal power plant fleet by 40%. In the past one year, developed countries cut their coal power capacity by around 8 Gigawatts; in the same period, China added 43 Gigawatts. Also, plants under construction and those under suspension and likely to be revived will eventually add another 148 Gigawatts to the total coal power capacity, almost the same as the entire 150-GW coal generating capacity of the European Union. Furthermore, China is promoting and financing around a quarter of all coal power plants proposed or under construction in other countries, mostly in the developing world.

Outlook for global energy-related emissions

Events in two regions which contributed three-quarters of the total energy-related emissions in 2016: Asia (53%) and Americas (22%) and accounted for 85% of the net increase in emissions in 2018, will largely determine the global outlook on anthropogenic emissions over the next two to three decades. China alone accounted for 28% while the United States contributed about 15%. China's outlook to 2050 shows that a combination of efficiency improvements, technology implementation, fuel substitution, and emission control policies will hold emissions level flat in the next thirty years, overall, and in all sectors of the economy, with the exception of industry where an annual growth rate of around 0.6% is projected. However, the country is under severe pressure to provide more primary energy and electricity for its extensive energy-intensive industrial sector as well as the increasingly middle-class population, and coal is the only major primary energy available locally (natural gas resources are also being developed rapidly). The country currently operates more than half of the world's coal power plants and many more are under construction or planned. All recent projections show that China will remain the world's largest emitter of energy-related pollution in 2040. The United States has been shutting down coal-fired and nuclear power plants in favor of natural gas-fired plants largely because the unit cost of coal-powered electricity can no longer compete in view of the increasingly stringent environmental regulations and declining costs of natural gas. Not even the recent policy reversals in favor of coal production and use have succeeded in stemming the rapid downturn in the industry. Primary energy consumption is expected to remain flat over the next twenty years, and so will energy-related emissions, although there will be significant sector variations. While emissions in the transportation and building sectors will continue to fall, contribution by the power sector will remain flat, but there will be significant rise in the industrial sector emissions.

Outlook for energy use and emissions in the emerging world

The global dynamics of primary energy consumption have changed significantly in recent years and an even more dramatic change is expected. Just two decades ago, Europe and North America accounted for more than 40% of global primary energy demand compared with 20% for developing economies in Asia. This situation is projected to completely reverse by 2040: Asia will account for half of the growth of natural gas, 60% of the rise in wind and solar PV, more than 80% of the increase in oil and nearly all of the growth in wind, coal and nuclear energy. Asia will also dominate power generation: already, six of the world's top ten power companies in terms of installed capacities are Chinese utilities. Asia's share of oil and gas trade will rise from around half of the current world total to more than two-thirds by 2040. The share size of China in terms of economy and population means that any changes in energy use make a significant impact on the global scenario. China's economy is slowing down and also transiting from high-intensity manufacturing to less energy-intensive production. Furthermore, growth of the renewable and nuclear energy sectors has been rising, hence a gradual decrease in energy demand rate is expected, and energy intensity will also start to decrease. Overall, there should be a significant fall on CO_2eq emissions of the country by 2040. Significant shifts from coal to natural gas, renewable energy and nuclear power are projected for the country, and the consequent effects on CO_2eq emissions are expected to be substantial because of the size of the country's economy. China's choices will play a major role in determining global energy-related emission trends, in particular, the country's transition towards a more services-based economy and deployment of low-carbon energy, accounting for around 40% of new global installed solar PV and wind capacity over the next twenty years or so. This would be a significant positive development in the global emissions mitigation effort since the country is likely to account for around 40% of the expected growth in electricity demand, the highest source of energy-related pollution over the period. The country is also projected to overtake the United States by 2030 as the world's largest producer of low-carbon nuclear electricity.

China currently leads the world in the global EV market, accounting for about 25% of the total global sales in 2017, and will account for about 40% of the projected global investment in electric vehicles by 2040. It is unlikely that the country can do much to reduce coal use and emissions from industry considering the extensive use of coal for power generation and fossil fuels both as energy sources and as feedstock for primary metals, non-metals and chemicals/petrochemicals production, and the major global demand for its products. However, decarbonizing electricity could have a significant impact in its transportation sector, particularly the electric vehicles sub-sector which currently depends heavily on coal electricity for manufacture and operation. On the contrary, energy demand by India continues to grow at more than 2.5 times that of China, representing more than a third of the global increase, and most of the growth will be filled by coal and nuclear power. However, India's contribution to the global total emissions in 2016 was only a quarter of China's emissions, and recent projections show that China will still account for two times India's emissions in 2050.

Around 80% of the global population consuming less than 100 Gigajoules of energy per capita per year (the minimum that stimulates healthy human development growth) live in the emerging world which will also account for most of the world's population growth over the next two decades or so. Rapid growth of the economies and urbanization, expected massive growth of the middle class population and the desire for higher living standards will drive energy demand across the regions, although the demand will be partly offset by efficiency gains and declining energy intensity. The declared UN policy goal of modern energy for all by 2030, which may be overoptimistic, will require around 25% more energy than projected in the emerging transition scenario of the International Energy Agency in order to reduce energy poverty to about a third by 2040. A rapid rise in the deployment of power generation plants in the developing world is expected over the next two decades and most will be powered by fossil fuels. Asia Pacific accounted for 75% of global use of coal in 2018, and most of it was produced within the region. The coal was used for the production of electricity, heat, steel, copper, aluminium, cement, chemicals and petrochemicals, fertilizers, much of which ended up in the developed world.

Various recent projections expect an exponential growth of electric vehicle from the current 5 million to 150-250 million by 2030-40. Lithium-ion battery and electric motors are the main components of EVs both of which require input of some special metals, notably copper, lithium, nickel, cobalt, platinum and palladium. Copper is already an important metal for the electrical sector of the global economy and electrical systems of internal combustion vehicles, and an additional 3 million metric tons a year will be required to fill the future needs of the EV industry. Nickel is an important steel alloying element and is also used in many other industries but an additional 1.3-1.5 million metric tons a year will be required by the EV industry by 2040. The other metals are critical stockfeeds of many industries but current requirements are relatively small. However demand is expected to rise 3-5 fold by 2040 because of the needs of the EV industry. The expected steep rise in demand for these metals raises several environmental issues: nearly all the metals are currently being sourced mainly from the emerging world, mining is relatively crude, inefficient and highly polluting; extraction and refining are energy-intensive, powered mainly by coal-fueled electricity. The battery manufacturing process accounts for around 40% of the energy required for producing an EV, again filled mostly by coal electricity since Asia is currently the primary source. There are plans for some European manufacturing facilities but most of the critical metals will still come from the developing world where there are relatively few environmental regulations. Mining of these metals which is often open-cast is already creating major environmental issues including toxic mineral dust, contamination of rivers and water supplies and destruction of vegetation and landscapes. Mineral experts doubt that current and projected production of these metals can meet the expected demand in 2040 but, more importantly, production of EV batteries and required materials will account for a significant proportion of the expected rise in global energy demand-related anthropogenic emissions over the next two decades, most of which will come from the emerging regions.

Another potential environmental issue with electric vehicles is the non-availability of appropriate technologies for the disposal of discarded lithium-ion batteries. When an EV battery loses a significant part of its electricity storage capacity, typically 20%, it can no longer power a vehicle effectively and has to be changed. Estimates vary but current batteries which cost as much as a medium-sized ICE vehicle ($15,000-20,000 to replace) will need to be changed after around 150,000 kilometers or ten years, which means that there could be up to 30 million discarded used batteries by 2040, assuming around 10% of the expected 250-350 million EV vehicles on the roads would have needed battery replacement. However, the battery still has high power storage capacity which makes second-life use possible, for example in stationary electricity storage applications, notably for surplus power in utility-scale intermittent renewable energy installations such as in buildings, thereby increasing the lifetime use of the battery by around 70% before reaching end-of-life. While lead-acid batteries used in ICE vehicles are a hundred percent recyclable, there is currently no viable technology for lithium-ion batteries. Lithium-ion chemistry is very complex, dangerous, and very sensitive to heat, often leading to fires and explosions, and there have been many fire incidents in airplanes, electric vehicles, laptops, cell phones, etc. Furthermore, there is currently no standardization of either the chemistry or the technology, which makes it difficult to develop appropriate recycling technologies. Apart from the fact that li-ion batteries contain very expensive and carbon-intensive materials such as lithium, cobalt, nickel, copper, graphite which could be recovered by recycling batteries after second-life use, unrecycled, discarded batteries will create new environmental issues in the next two decades unless effective recycling technologies emerge. Furthermore, recycling discarded batteries will reduce the battery greenhouse emissions attributable to the vehicle on a per-kilometer basis by about

42% (ICCT, 2018) while also reducing the demand for virgin raw materials.

Effect of climate change on global energy use

Extensive scientific data indicate (with a high degree of confidence) that human-related emissions have been the dominant cause of the observed global warming since the mid-20th century. The Earth's surface temperature is projected to rise over the 21st century under all possible emission scenarios, and there is substantial scientific evidence in support of the theory that a warming global environment will make extreme weather more frequent and intense, occurring in many more places than usual. It is very likely that heat waves will occur more often and last longer, flooding will be more frequent and severe, oceans will continue to warm and acidify, ocean levels will continue to rise, extreme weather storms and tornadoes will become more frequent and devastating because they are powered by heat derived from warming oceans. Extreme precipitation events will become more intense and frequent in many regions of the world, impacting on local weather. All these events will continue to fuel demand for more energy for adaptation including heating and cooling and, without major progress in decarbonization, higher anthropogenic emissions.

The significant rise in energy use and associated emissions in the last two years is believed to be closely linked with the unusual frequency of extreme heat waves and cold weather, particularly in Europe and North America. More energy was needed for heating and air conditioning, filled mainly by fossil fuels. The International Energy Agency projects a steep rise in global population of air conditioners from its current 1.6 billion to around 4 billion by 2050, with most of the growth coming from emerging economies. Over half will be accounted for by just two countries – China and the United States. Of the 2.8 billion people currently living in the hottest parts of the world, only 8% have air conditioners. As populations grow and prosperity rises particularly in the hotter regions, the demand for air conditioners and adequate energy to drive them will rise, putting enormous strain on electricity grids around the world and driving up local and global emissions. Already, air conditioners and fans account for about 20% of the total electricity use in buildings around the world, equivalent to 10% of all global electricity consumption (IEA/OECD 2018). Increased demand of power for air conditioners pushes up not only overall power needs, but also the need for generation and distribution capacity to meet demand at peak times, placing further stress on the power system. Space cooling can represent as much as 70% of peak residential demand and providing electricity capacity to meet the peak demand is very expensive because it is used for only limited periods, and annual associated emissions are estimated at around 1130 million metric tons. Also, the quality and efficiency of air conditioners currently in use worldwide vary widely by a factor of 2 to 3 and there are no standards. While improvements in building design can reduce the need for air conditioning to some extent, a global drive for efficiency standards for cooling systems could reduce energy demand for air conditioning and associated emissions by around 45% by 2050.

Mitigation of environmental pollution: between hope and reality

There is little doubt that human activities are doing enormous damage to the global environment, with consequences which go beyond climate change and impact directly on human health and the ecosystem. Global interest has been pushed on many fronts by many bodies, notably the United Nations, The World Health Organization and the International

Energy Agency. However, there has been a lot of disinformation that has tended to blur the boundary between hope and reality. For example, fossil fuel use has been identified as the major source of anthropogenic pollution and it is the wish of everyone that the world should move completely away from use and adopt less carbon intensive energy. However, the in-depth analysis presented above shows that this is unlikely to happen any time soon, almost certainly not in two to three decades to come because of issues with potential alternatives and the prime position of fossil fuels in the global economy. Also, the rising global demand for many vital building blocks of modern society - primary metals, cement, chemicals, petrochemicals, plastics, pharmaceuticals, fertilizers, new energy materials, industrial and biomedical materials, consumer goods - all of which depend critically on fossil fuels will continue to sustain their dominance in the world's primary energy mix. Furthermore, as discussed earlier, the fossil energy industry is not only one of the major employers in the world, the economies of many countries across all regions of the world depend critically on production and export of oil, coal, natural gas. All recent projections from numerous reputable independent bodies, notably the International Energy Agency, (an autonomous organization) indicate that the world will still depend on fossil fuels for nearly 80% of its primary energy requirements over the next two to three decades. Although renewable energy share of global primary energy consumption will rise significantly, the main impact will be in power generation. While it is not intended to detract from the very laudable efforts on numerous political and non-governmental activities sensitizing the public and targeting total elimination of fossil fuels by 2030-40, awareness of the reality helps to focus on more realistic goals, in particular, *adaptation* and *mitigation*. While adaptation focuses on the development of coping strategies for the consequences of energy-related environmental pollution, mitigation targets reduction in global energy demand through improved efficiencies and fuel substitution. Also, current global focus is on public policy-driven mitigation but it is vital to build public awareness on what people can do to reduce individual carbon footprint. After all, consumer choices are the main drivers of industrial production and energy demand. This is discussed in some detail in a later section.

Environmental Pollution: Adaptation and Mitigation options

There is little doubt that the Earth's average temperature has been increasing steadily in the last hundred years or so, with potentially severe negative consequences. What is not so clear is the extent to which all the observed changes can be attributed to human activities. However, there is strong scientific evidence that nature's control systems in terms of regulating the energy balance of the Earth and its environment, are being compromised significantly by anthropogenic emissions from various human activities. There is little doubt that climate change will make tornadoes and heat waves much more likely and much more severe, energized by rising global temperature. It is noteworthy that Europe-28 which is the most successful region in energy decarbonization and environmental pollution reduction has also been one of the most hard-hit regions by severe weather in recent years, reinforcing the fact that climate change knows no national or regional boundaries. The debate is ongoing on who is liable for degrading the environment and what should be the consequences, a development which is discussed briefly in this section. However, all available projections show that fossil fuels will remain dominant in the global primary energy mix for decades to come and the most urgent reality is that mankind must seek to manage use in an environmentally sustainable manner by fast-tracking movement towards a carbon-neutral world (highly decarbonized energy and agriculture, and reforestation) and developing coping mechanisms for dealing with the undesirable consequences (that are impacting in real time already) on survival and health of people and the ecosystem. Two options that have gained prominence in recent times are *adaptation* and *mitigation*. Adaptation and mitigation are complementary strategies for managing environmental pollution: while adaptation focuses on minimizing the negative consequences of climate change on people, mitigation seeks to develop and promote ways of moderating human actions that cause climate change.

Liability for climate change

Global debate is strengthening on who should be held responsible for degrading the environment and what should be the price to be paid, and the focus is on the global energy companies, with lawsuits beginning to emerge. The first issue is that no science has proved conclusively that human activities are entirely responsible for climate change. As discussed earlier, there are many natural phenomena and processes that have been causing climate change and extreme weather well before human civilization. Therefore it would be difficult to apportion blames between mankind and nature. There is however no doubt that human activities especially in the last century or so have been making the consequences of climate change more intensive and more widespread. While it is true that energy production and use account for two-thirds of anthropogenic pollutions most of the balance is accounted for by other human activities, especially agriculture. There is little doubt that energy companies have prioritized profit at the expense of sustainability of the environment over time and need to be held accountable through more punitive tax policies, but the entire world is in fact liable. The agricultural sector which accounts for around 15% of pollutions is virtually off the radar; consumers who are the ultimate beneficiaries of all the products of energy and agricultural technologies are either unaware that their activities, choices and preferences account for most of the pollutions from energy use, or are unprepared to review and modify their choices in an environmentally sustainable matter. Not many are willing to modify their lifestyles to optimize energy use (moderation of home and transportation energy use, consumer goods' use, support for recycling, etc); livestock production is the main source of pollution from agriculture, mainly methane which is one of the most potent of the greenhouse gases but not many are prepared to give up meat or reduce consumption for the sake of the environment. Soil fertilization is the other main source of pollution from agriculture but more than half of the world population would starve without it. Perhaps the best way to make the world pay for environmental degradation and moderate their lifestyle choices is to upgrade the current carbon taxes into *attribution taxes* which should cover energy and agricultural production and spread to every consumer good and agricultural product.

Adaptation to climate change

Adaptation options are largely localized and within the jurisdiction of local and state governments, hence the impact is more immediate compared with most mitigation options which depend on both intra- and international actions, and take much longer time before benefits become evident. It is difficult to quantify the impact of climate change on human and natural systems for many reasons: the nature of hazards, exposure and vulnerability can vary widely across regions and socio-economic stratifications; furthermore, there are non-climatic stressors that can influence vulnerability, exposure and ability to adapt to changing situations. Adaptation measures are designed to minimize potential consequences of phenomena such as extreme weather and ambient pollution on life. Potential adaptation strategies include improving the design of structures to withstand extreme weather, multi-level response strategies to disasters, provision of early warning systems, emergency evacuation plans, provision of storm shelters, mobilization of first responders, etc. For example, the intense heat waves of recent years particularly in Europe and North America, believed to have shattered records and caused many deaths have stimulated action on coping strategies. The level of preparedness was low even in developed countries considering that less 5% of homes in Europe which has historically had a temperate climate are air-conditioned, compared with over 90% in the United States, and relatively few public transportation have air conditioning. Many potentially vulnerable utilities were severely disrupted: expanded rail lines, disruption of control systems, system collapse due to unusual spikes in electricity demand, traffic gridlocks due to signal failures, power outage, etc. Current measures include critical review of emergency response measures for public transportation and utilities, creation of public cool rooms, extension of public swimming pool hours, strengthening of local and national emergency response strategies, closure of schools, activation of help lines, and public awareness campaign.

Adaptation policies need to be targeted to be effective. For example, urban areas hold more than half of the world's population and most of its built assets and economic activities. A high proportion of greenhouse gas and aerosol emissions is also generated by urban-based activities and residents. Potential adaptation strategies include improving the design of buildings and structures to withstand extreme weather, development of shelters, effective plans for emergency medical and food supplies, multi-level response strategies to disasters, and building of awareness on what people can do to reduce personal carbon footprints, such as embracing and supporting recycling, energy conservation in transportation and across the home front, and a drastic change in individual lifestyle. Design of urban coping strategies needs to take account of the social stratification: while the wealthy often can take care of themselves, the majority of urban dwellers are usually low to middle income people many of who depend a lot on public welfare and utility systems. Rural areas have different vulnerability to disasters, for example, obstructions that could slow down tornados are relatively few and the impact of touchdowns can be very severe. Also, while many of the more prosperous urban dwellers can afford personal transportation which facilitates quick evacuation from impending weather disasters, and air conditioners in the event of heat waves, there are few escape options for poor rural people. Furthermore, most disaster response systems are based in urban areas and help can be slow in arriving in case of disasters. Rapid urbanization and migration have peaked in many regions, and poverty/extreme poverty rates in rural areas are falling, with the exception of sub-Saharan Africa where rates are rising. Many in the region live on subsistence agriculture, and access to fertile land and adequate rainfall are crucial. Most of these people are already subjected to many other non-climatic stressors and any disruption of their livelihood, for example, as a result of draught can be devastating. However, many rural communities devise their local adaptation strategies since there is often very limited institutional support.

Mitigation of climate change

The international Energy Organization (IEA) and many other independent organizations including the International Panel on Climate Change (IPCC) have projected that the rise in carbon emissions needs to stop by 2030, followed by sustained decline in order to place the world on a path towards the goal of limiting global temperature rise to 2°C or below by the end of the century, which requires a major decoupling of energy use from carbon emissions. This goal was adopted at the Paris-2015 conference on the environment coordinated by the United Nations in which countries submitted plans for action. However, all indications suggest that current and proposed mitigation efforts are inadequate and much greater action than currently in place or planned would be required to avoid the potentially disastrous consequences of climate change. Furthermore, the commitment of many nations to even the inadequate planned country actions is very much in doubt. The International Energy Agency (IEA, 2019f; 2019i) has proposed a viable path (Faster Transition Scenario, FTS) towards the Paris-2015 target which will require an additional 40% reduction in emission projections compared with Emerging Transition Scenario (ETS) by 2040. The ETS projections (also known as New Policies Scenario, NPS or 'business as usual') already take into account all existing, expected and anticipated technologies and policy instruments. The IEA proposal is often referred to as 'Sustainable Development Scenario (SDS) which outlines an integrated approach to achieving internationally agreed objectives on climate change, air quality and universal access to modern energy. The projected emissions in the New Policies Scenario will be about 36 Gt in 2040 and need to drop to around 18 Gt in the Sustainable Development Scenario. The Agency also proposes an Even Faster Scenario (EFS) that could limit the global temperature rise to below 2°C, possibly as low as 1.5°C. The International Renewable Energy Agency (IRENA, 2018) has shown that a profound and radical transformation of the power sector beyond current plans and policies would be crucial, one that enhances efficiency and fast-tracks decarbonization of energy. Cumulative emissions must be reduced by around 500 Gigatons (equivalent to around 15 years of annual global emissions at the current rate) by 2050 to meet the sustainable environment target. In order to achieve this goal, electricity which is expected to increase its share in total energy end-use from 20% currently to 50% by 2050 must be heavily decarbonized preferably by fast-tracking the share of renewable energy from around 25% currently to around 86% of power generation (the most optimistic of recent projections predict 30-35%) and its share of primary energy consumption from the current 25% to around 65%. This could reduce emissions in the energy sector by as much as 60%. Potential pathways to these ambitious targets are discussed in the following section.

Mitigation involves development of aggressive and effective policy and technology strategies for reducing the growth rate of environmental pollution and, while there are many options, a detailed analysis of the potential benefits of many mitigation options shows that four options will be the key drivers of future energy-related anthropogenic emissions' reduction across all regions of the world: *efficiency improvement in energy production and end-use; decarbonizing energy; promotion of waste-to-energy (recycling); and strong, resilient policy instruments* (in order of potential). Building public awareness on the potential contribution of people to climate mitigation efforts, in particular, reduction of personal carbon footprint through rationalization of lifestyle choices is also vital. Potential areas include helping to reduce household and transportation energy use and associated emissions, and embracing and supporting municipal recycling and energy conservation. The International Energy Organization (IEA, 2019j) estimates that, under the Sustainable Development Scenario (SDS) which is needed to meet energy and climate goals, improved energy efficiency will account for about 42% of emissions savings while energy decarbonization (fuel switching to renewables, natural gas, nuclear; heat and materials' recycling; and carbon capture) will account for about 51%.

Improving energy efficiency and intensity

Energy-related emissions mirror the trends in energy consumption. Increasing energy efficiency results in lower energy intensity (amount of energy consumed per GDP or indeed any activity including fuel consumed by a vehicle over a unit distance). In spite of the projected virile economic and population growth and associated rise in energy demand, improved energy efficiency will continue to slow down the rate of growth of energy demand

and emissions, and all recent projections show that emissions could fully decouple from energy demand by 2040 if current efforts on energy efficiency improvement are sustained and upgraded. In a recent study presented in two reports, the International Energy Agency (IEA, 2017g; 2019h) developed the Efficient World Scenario which shows that, in spite of the projected growth of GDP by 100%, human population by 20-25% and vehicle population by around 100% over the next two decades, improved efficiency could moderate the growth in energy demand significantly, perhaps to as low as 30-35%. Since most of the technologies needed are already available today, strong and enduring policy instruments will be critical drivers. Achievement of this goal could result in a peak in energy-related greenhouse gas emissions within the next few years, followed by a fall by around 12% in 2040, equal to over 40% of the abatement required to be in line with Paris-2015 Agreement targets. The IEA report (2019h) identifies the significant potential of broadening and deepening global efforts on energy efficiency, thereby unlocking multiple benefits across all end-use sectors. Virtually all the technologies required to achieve this goal are already available and energy efficiency alone could cause greenhouse gases to peak within the next few years, a key target of the Paris-2015 Agreement on climate change.

Energy efficiency has been evolving as a key resource, with multiple benefits for economic and social development across all economies. Apart from the obvious benefits of enhanced energy efficiency: reduced energy demand and lower greenhouse gas emissions, investment in energy efficiency can provide many different benefits to many different stakeholders, such as freeing investment saved from lower energy use for other urgent economic and human development issues. One of the main challenges is the promotion of widespread deployment of a wide range of technologies which are already available, and many new technologies that are currently at different stages of development, deployment of which could reduce significantly energy use and related emissions in the future. Transformational change in the energy sector can be a very long process due largely to '*transformational inertia*' or '*status quo mentality.*' This explains why many power plants all over the world are still operating at 30-35% efficiency when technologies have been available for decades that could be deployed through upgrades and retrofits to raise efficiencies to 45% and above for power plants and as high as 80% for energyplexes, with substantial economic benefits.

Improved efficiency of any energy production process or end-use translates to lower energy intensity and a significant reduction in related CO_2eq emissions. Flaring is a glaring example of inefficiency in oil and gas production: apart from wasting valuable energy that could be utilized in power generation, flaring is a significant source of GHG emissions (mostly methane) and carbon black. Improved efficiencies in energy production and the end-use sectors (industry, transportation, building) or fuel substitution drive down energy intensity, carbon intensity, and carbon footprint. Deployment of new technologies not only improves efficiency which translates to cost savings, lower energy consumption and reduced pollution for a given output, it also facilitates compliance with increasingly stringent environmental regulations. The International Energy Agency (IEA, 2017k) estimates that much stronger deployment of technologies that are familiar and available at commercial scale today can reduce significantly future emissions from the energy sector. However, stronger policy instruments would be required to motivate change. Without the strong environmental pollution regulations introduced in many developed countries, the current downward trend in emissions in OECD countries would have been difficult to achieve.

All sectors of the primary energy value chain have efficiency improvement potentials: coal, oil and gas exploration, production and transportation, power generation, energy transmission and distribution, industrial operations, transportation, commercial, and domestic use, with significant reductions in emissions. The energy production and conversion sector is the largest energy user, and also the least efficient, losing around 40% of the feed primary energy in the production, conversion, transmission and distribution stages. Estimates vary widely but it is believed that between 30% and 50% of primary energy produced worldwide is wasted through inefficient production processes, transmission and utilization. Projections indicate that up to 40% of future global primary energy requirements could come from energy savings through improved efficiencies in production and utilization. It is estimated also that energy efficiency represents about 40% of the reduction potential of anthropogenic greenhouse gases. In fact, energy efficiency is now often referred to as *'first fuel'* because of its immense potential for reducing end-use energy and carbon intensities, and reducing exploitation rates of global primary energy reserves. There is a growing awareness that improving efficiency in energy production and end-use is the most promising mitigation option that helps decouple GDP/population growth and energy-related emissions. Global total energy use per GDP peaked in 2007 and has been declining since then, due largely to improving efficiencies, a development which has also impacted significantly on the third dimension of the global energy trilemma - *energy security* (the others are *energy access* and *environmental sustainability*). As a result of declining demand for primary energy largely propelled by increased efficiency, many countries were able to cut coal, oil and gas imports by as high as 30% in 2016, thereby improving internal energy security and saving substantial financial resources.

Every sector of the economy has significant potentials for exploiting the benefits of enhanced energy efficiency: industry, transportation, building. Industry accounts for over half of the total end-use energy and there is extensive scope for efficiency improvement. For example, about 40% of the global electric power and 70% of industrial power are used to run electric motors in a very wide variety of applications ranging from driving industrial equipment to air conditioning, domestic fans, microwaves and washing machines, with efficiencies as low as 30%. When other losses along the energy chain are taken into account, overall efficiency could be as low as single digits. Another example is the enormous waste of energy in power generation, transmission and use: about 320 units of thermal energy are required at the power generation stage to produce one unit of useful electric light to the traditional incandescent bulb. In effect, energy equivalent to 319 units is lost in generation, transmission, distribution, wiring and fittings. Investment in more efficient power generation, transmission and distribution technologies could reduce fuel input by around 30%, while conversion to more efficient compact fluorescent and LED lighting could save up to 80% of fuel input required to generate the required electricity.

Improving efficiency and the resultant decrease in CO_2eq intensity of energy use will help stem emissions growth in spite of growth in population and GDP. The International Renewable Energy Agency (IRENA) estimates that efficiency gains need to move much more rapidly, resulting in a fall in energy intensity to about a third of current levels by 2050, which would translate to a drastic fall of total global primary energy demand to below current levels by that year. However, this is only feasible if the efficiency gains cut across all regions and all sectors of the global economy. Even though China's GDP rose by about 1,000% from 1990 to 2015, energy efficiency gains kept the rise in CO_2eq emissions to about 300%. Emissions in OECD countries have remained relatively flat for the last two decades due to efficiency gains, and are projected to decline by about 20% by 2040, but emissions from the non-OECD countries will rise by around 50% over the same period despite a 40% gain in efficiency across the emerging economies. However, the growth rate of less than 1% will be much lower than the 3%/year from 1990 to 2015. The prospects of lowering energy-related emissions depend largely on ability to reduce energy demand through improved efficiency, and opportunities vary across sectors of the global economy. Mitigation actions that target efficiency improvement in the power and end-use sectors across all regions of the world, in particular the emerging economies are the most promising in reducing global energy-related emissions.

Efficiency improvement in power generation

Efficiency improvement in power generation translates to lower fuel consumption per kilowatt-hour of generated power and lower emission. Around 65% of the total global electric power generation is powered by fossil fuels but power plants operating around the world are only 30% to 40% efficient, in particular, coal-fired power plants, the most polluting. In effect, around 60-70% of the carbon energy content of the feed fuel is wasted, even though the full load of emissions is released into the environment. Added to this are significant transmission and distribution losses which could be 10-40% of the generated power. Coal-fired power plants are the least efficient because many were built decades ago. Also, most coal-fired power plants utilize sub-critical technology, whereas much more efficient ultra- and super-critical technologies that could raise efficiencies to around 50% have been available and in commercial use for decades. It is estimated that one percent increase in efficiency of conventional pulverized coal-fired plants results in 2 to 3% reduction in CO_2eq emissions, and highly efficient modern coal plants emit up to 40% less CO_2eq than the older plants. Newer power plants and many retrofitted installations now operate at around 40% efficiency; adoption of pre-gasification of coal could raise efficiency to around 60% and lower emissions significantly while making carbon capture much easier because carbon dioxide is released at high pressure.

Much of the efficiency improvement in power generation can be achieved even in existing fossil-powered plants by adopting simple technologies that are already available (best available technologies, BAT), such as retrofitting a coal-fired power plant with more efficient boilers or burners, enriching combustion air with oxygen, gasifying coal, or conversion to less carbon-intensive natural gas. However, even at the higher levels of efficiency there is still a tremendous amount of energy lost in the generation process which could be reduced through the adoption of other readily available technologies such as more advanced combustion optimization technologies and control systems, and waste heat recycling. The economic and ecological benefits of recovering and utilizing more of the energy in the input coal or natural gas are also enormous, and many technologies are already available, for example, co-generation plants, energyplexes and other advanced power generation technologies could return efficiencies as high as 80% (Afonja, 2017).

Efficiency improvement in the industrial sector

Industrial energy intensity (amount of energy required per unit product) has been declining in recent years (about 20% since 2000) due to wider adoption of newer, more energy efficient technologies across all sectors of the global economy and in all regions of the world. There has also been a notable shift in the regional pattern of industrial energy demand in favor of emerging regions which now use 65-70% of global industrial energy. Five energy intensive sectors - chemical and petrochemical, iron and steel, aluminium, cement and pulp and paper - account for 67% of the world's total industrial energy consumption, and are located mostly in emerging countries. Very few innovative energy-saving technologies have emerged in these sectors in recent years but the fact that many of the facilities are new or have been recently expanded/upgraded, provided some opportunity to adopt more efficient technologies.

However, opportunities for efficiency gains in the industrial sector are still very high, and it is estimated that application of current best available technologies (BATs) could reduce energy use in the energy-intensive industries by 10% to 25%. Emissions are closely tied to energy use especially when fossil fuels are the main sources, and the industry sector which consumes over 50% of end-use energy currently also accounts for around 20% of energy-related emissions, and several recent projections expect a rise of around 20% in emissions from the sector by 2040, due largely to increasing use of fossil fuels both as energy sources and feedstocks for chemicals and petrochemicals production, manly in the emerging countries. Options for mitigation include process and plant modification, waste and energy recycling, more efficient process control, and adoption of proven but under-utilized technologies. Recycling particularly of primary metals, paper and plastics can also result in significant energy savings compared with production from virgin raw materials.

Efficiency improvement in the transport sector

Transportation accounts for about 27% of final energy consumption and 24% of energy-related pollution, and growth will be mitigated by rising transportation efficiencies. Nearly all the global transportation fuel requirements were filled by fossil oil in 2018, with renewables accounting for only 4%, and 96% of the renewable share was filled by biofuels, while electric vehicles accounted for only 2%. Projections show that growth of renewable share in transportation will remain very modest over the next decade or two. Improvement in fuel economy of transportation vehicles is considered one of the most cost-effective ways to reduce GHG emissions from the sector, for example, the average kilometers covered per liter of fuel is expected to double by 2040. This should reduce demand for oil and transport-related pollution in 2040 substantially. However, the population of vehicles is growing exponentially, expected to double by 2035-40, and the bulk of the growth will be in the emerging world where vehicles are much less fuel-efficient and pollution control is lax. Furthermore, the increasing demand for personal transportation is fueling the growth in highly-polluting single-stroke, 2-3 wheeled motorcycles, and growth in diesel-fueled, highly polluting commercial trucking will also be dynamic.

The impact of energy efficiency in the transportation is well below the potential, due largely to the instability in the global oil market. Significant improvements in fuel efficiency of conventional vehicles combined with low global oil prices have led to faster increase in sales, in particular, of relatively less efficient large passenger vehicles such as sport utility vehicles (SUVs) and family vans in many countries, thereby dampening the global rate of improvement in passenger vehicle fuel efficiency. Another problem is the relative lack of improvement in the truck sub-sector which accounts for around 43% of total oil consumption for road transport, and uses the bulk of diesel, the most polluting of fossil end-use fuels. While many countries have mandatory efficiency standards for passenger vehicles, only a few have fuel economy standards for commercial trucks in place. Economic growth stimulates demand for personal transport, commercial trucking, aviation, marine and rail transportation, and the majority of the growth across all the sub-sectors will occur in non-OECD countries, consistent with the expected growth in GDP and population. While energy consumption by the global transportation sector is expected to grow by only 9% or so by 2040, (mitigated substantially by projected significant drop in demand by OECD countries) the demand in non-OECD countries will grow by around 41%. The sector is the main user of the liquid products of oil refineries - gasoline, diesel, jet fuel, - accounting for around 95% of the total global consumption in 2018. This share is expected to reduce by 6-10% over the next two decades due to a projected increase in the use of alternative fuels (biofuels, natural gas, electricity) which will each provide about 5% of transport demand in 2040. The global fleet of vehicles will nearly double over the next 20-25 years, so will commercial trucking, air and marine transportation, and most of the increase will be in non-OECD countries which are projected to account for over 60% of the total global transportation fuel consumption by 2040. Energy demand by the transportation sector has the least growth rate of the three end-use sectors due to improvements in vehicle fuel efficiency which has doubled in the last two decades or so, notably in OECD countries, and is projected to rise by another 50-55% over the next two decades. As a result, transportation energy use for the region is projected to decline in spite of substantial growth in vehicle population, but in non-OECD countries, increased demand across all modes of passenger transportation and light-commercial and heavy trucking will outpace improvements in fuel efficiency. Furthermore, most of the vehicles that can no longer operate in the developed world due to unacceptable emissions are exported to the emerging nations.

Efficiency improvement in the building sector

Although the building sector is the least polluting of all the major sectors of the global economy, electricity is the main energy source and the sector will account for much of the growth in global electricity demand over the next two or three decades, mostly in the developing world. Global growth in building energy use has been dynamic, increasing by a factor of two over the last two decades or so, largely driven by economic growth mainly in the major emerging economies: in 2017, buildings were responsible for about 30% of global final energy use. However efficiency has been improving steadily over the last two decades and has fully decoupled from the growth of floor space, population and economic output. It is estimated that efficiency improvements helped to reduce growth in energy demand in the sector by around 12% over the period and much of the savings have come from space heating and lighting. The fastest-growing end-uses in buildings are space cooling and appliances, increasing energy intensity per floor area by 80% and energy consumption by appliances by nearly 60% over the last two decades, compared with around 20% for space heating, water heating, cooking, lighting (IEA, 2019h). There has been an exponential growth in household connected devices over the last decade or so: smart LED lamps, sensors, actuators, cameras, appliances, gateways, estimated at around 4 billion in 2016 and projected to rise by about 1 billion over the next couple of years. While these devices may enhance the quality of life, they are increasingly becoming significant building energy consumers due to the large size (around ten per household projected to grow to 50 over the next few years), but also because of large variations in efficiencies of devices many of which are low. Furthermore, most of these devices remain idle but connected for around 90% of the time (Networked standby) and could account for around 10% of the household electricity use.

Policy-driven building energy efficiency improvements have continued to rise, in particular on building envelopes (energy-efficient design of layout, structure, materials, utilities, lighting etc) as well as equipment: heating, ventilation and air conditioning (HVAC) and appliances. However, various recent studies have shown that the building sector is not on track to achieve global climate commitments and stronger policies would be required on building and household equipment efficiency codes. It is estimated that energy for lighting could be reduced by around 75% by switching from incandescent and halogen lighting to LED and CFL lighting, and substantial efficiency improvement opportunities abound in home and water heating and air conditioning which currently account for the bulk of household energy consumption but are regulated by few policy efficiency codes. Integration of renewable energy such as solar systems with existing and new buildings where feasible could also help reduce grid energy consumption and decarbonize the building sector. Energy efficiency is now one of the primary selling points of new homes and appliances, many older homes are being upgraded and inefficient appliances are being replaced. Electrical fittings and appliances are increasingly becoming energy efficient and smart control systems allow for remote control and optimization of energy use in homes. These developments are driving builders and manufacturers to continuously improve on efficiencies. However, this needs to extend to networked devices many of which are currently of low quality and efficiency. Furthermore, many home appliances (new and used), in particular, refrigerators, air conditioners, microwave ovens that can no longer meet emission standards in the developed world end up in emerging countries. As discussed earlier, the negative impact of pollutions from these regions on climate will resonate across all regions of the world.

Decarbonizing energy

Carbon is the primary energy content of fossil fuels and bioenergy, derived through dead plants and animals in the natural carbon cycle but the fuels also contain hydrogen which is combustible. In general, the fuels are burned to raise steam or heat which can be harnessed to energize many operations including power generation, home and industrial heating, and transportation internal combustion engines. The main product of combustion is carbon dioxide which also accounts for by far the largest share of energy-related emissions. Coal has the highest carbon content of all fossil fuels and, because of highly inefficient combustion technologies much more coal needs to be burned than necessary per unit energy output. If for example oxygen-enriched air or pure oxygen is used in combustion instead of ordinary air, efficiency could increase by double digits, thus enhancing the enormous potential of efficiency gains as a key factor in mitigating energy-related emissions as well as lowering the carbon footprint of coal significantly. Retrofitting coal and natural gas power plants with more efficient burners can also lower GHG emissions significantly. Decarbonization of energy is a very promising emissions' abatement option and, with efficiency improvement, they can account for around 90% of the emissions reduction required to set the world on the right path towards achieving the Paris-2015 target. Energy can be decarbonized in several ways, notably through efficiency improvement from production through generation to end-use which has been discussed above; decarbonization of power generation; carbon trading, capture, use and storage; and recycling.

Decarbonizing power generation

Power generation is a major target for decarbonizing energy since the sector accounts for the largest share (around 42%) of energy-related emissions. It is estimated that a significant improvement through the adoption of more efficient technologies and substitution of coal with renewables (in particular, solar and wind), nuclear energy or natural gas could decarbonize energy substantially and cut emissions from the sub-sector by up to 50%. Power plants have a life span of 30-40 years (many power plants worldwide are close) and, while many undergo retrofits over their life span to improve operating efficiencies, most of the investments in recent times have been in new power plants, featuring new advanced technologies that not only improve efficiencies, but also facilitate compliance with increasingly stringent worldwide environmental control regulations. Coal is the backbone of power generation in many countries across all regions and has been responsible for more than 40% of global energy-related emissions growth since 2000. Energy-related emissions reached a record high of 33.1 Gigatons CO_2eq in 2018 and inefficient, low-technology coal-fired power plants accounted for over 30%. The majority of them were installed in the emerging economies in the last decade or so, decades younger than their average economic lifetime of around 40 years. Upgrading from sub-critical operation to super-critical and ultra-super-critical boiler technologies which have been available commercially for more than a decade could decarbonize fossil-fueled power generation substantially. Also, investments in renewable energy power generation technologies (in particular, wind and solar) which are readily available commercially and are becoming increasingly competitive could further mitigate pollution from the power sector. Increased substitution of coal with natural gas could also decarbonize energy significantly depending on the lifecycle carbon footprint of the specific gas supply. Renewable energy is the world's fastest growing primary energy resource, rising by nearly 15% in 2018. However renewables are still a small share of the global power generation mix relative to coal and, considering that the developing regions which depend heavily on coal will account for most of the robust future growth in power demand and generation, the contribution of renewables to energy decarbonization is likely to be low over the next few decades. Most recent projections predict a rise in share of renewables in power generation to 25-35% over the next two decades, effectively doubling the current contribution. While this will lower emissions from power generation significantly, the sector will remain the largest contributor to global energy-related emissions over the next two to three decades because the total contribution of renewables to power generation in 2018 was only 25% (hydropower, 16%; wind, 5%; solar PV, 2%; biomass/waste etc., 4%; coal, 38%; gas, 23%; oil, 3%; nuclear, 10%) (IEA, 2019i). Projections show that renewables' share will grow rapidly over the next decade or two to 30-35% or even higher, driven largely by solar and wind, and coal share could drop to below 30%.

The power sector is growing at a phenomenal rate, nearly 4% in 2018 which is one of the fastest growth rates in the last two decades, accounting for around half of the growth in global primary energy consumption. Although the developing world drove most of the growth accounting for around 80%, the growth rate in the developed world was also significant, believed to have been weather-driven, particularly in the United States where power demand grew by 3.7%. The fuel mix in the global power system has remained flat for the last two decades, with the contribution of fossil fuels virtually unchanged, and projections indicate insignificant change over the next two decades. Decarbonzing the power sector is crucial if electrification is to play a leading role as a pathway towards a global lower carbon energy system. Considering that the future of nuclear energy is tenuous and the power sector accounts for the largest share of energy-related emissions, a dramatic shift to renewable energy power generation could be the prime mover of the transition to this scenario but it would require a dramatic rise from the current level of 25% or so to around 85% by 2050, driven mainly by solar and wind energy.

Decarbonizing the industry sector

Although decarbonizing of the power sector currently offers the greatest opportunities for environmental mitigation by efficiency improvement and fuel substitution, all other areas of the economy also have significant possibilities. Natural gas is playing increasingly significant role as full or partial replacement for coal in power generation, primary metals and cement production, and steam-raising. Also, since industry is a major consumer of electricity, decarbonization of the power sector would help lower emissions from industry significantly, so would substantial increase in efficiency in as many areas of the sector as possible. Efficiency gains translate to lower energy consumption and carbon intensities of industrial processes per unit of output. Only about a third of industrial energy use is currently covered by mandatory efficiency policies, with coverage highest in China, India and Japan, and only about a quarter of global electric motors (the largest consumer of industrial electricity) is currently regulated in many major economies. Doubling the current coverage would greatly reduce carbon intensity since electric motors are primary equipment in most industrial processes. Most OECD countries and China are moving away from energy-intensive production to low-intensity production and services, while this would help decarbonize industry in some regions, it may amount to merely moving intensive emissions to other regions since the global demand for the products of intensive manufacturing will continue to rise, in particular, iron and steel, copper, aluminium, chemicals, petrochemicals, cement, fertilizers. As discussed earlier, the negative impacts of anthropogenic emissions on climate anywhere in the world will spread to everywhere else since around two-thirds of the world is covered by oceans which connect all regions and will harbor most of the global rise in temperature. Energy savings potentials are highest in the light-energy sectors such as food and beverage, textile, pharmaceutical, and assembly manufacturing, and could account for around 70% of total energy savings in the global industrial sector. Efficiency improvement in the energy-intensive sector has been very slow, around 5% over the last twenty years or so. However, adoption of several already available technologies (BATS) could raise efficiency gains to around 25% and reduce energy intensity by up to 50% by 2040, notably recycling of primary metals which has the potential to cut energy requirements by 60-90% compared with production from virgin ores, fuel substitution, heat recovery and reuse, and recycling of other materials such as plastics and paper. For example, fly ash, usually a waste product from coal combustion is now used increasingly to replace cement in concrete mixtures, thereby reducing the cement required and saving the energy that would have been used and the emissions that would have been discharged into the atmosphere during its production. However, strong regulatory instruments would be required, such as subsidies, tax policies, direct provision of materials recycling services and public awareness campaign.

Decarbonizing the transport sector

The transportation sector accounts for about a quarter of energy-related emissions, mostly from oil-fueled internal combustion engines (ICs) which power most vehicles currently. It is widely accepted that faster growth of the electric vehicle market is a key pathway to decarbonizing transportation, provided the electricity for manufacture and operation is from low-carbon sources, and global annual sales which have been strong, (concentrated within passenger cars, light-duty trucks and public buses) increase much faster, by a factor of 5 or so. Projections based on Emerging Transition Scenario predict a rise in the share of electric vehicles in global vehicle population, up from 5 million in 2018 to around 350 million by 2040, of which around 300 million are passenger cars. This is equivalent to about 15% of the projected population of all cars and 12% of light-duty trucks. However, in spite of the high relative advantages of electric vehicles (EVs) compared with combustion powered vehicles (better fuel economy, zero anthropogenic tailpipe emissions, and potentially lower lifecycle carbon footprint depending on the source of electricity), the penetration of electric vehicles into the global auto market relative to well established IC vehicles has been very slow and will remain so over the next two decades or so. Achieving the projected growth rate (which would still be grossly inadequate for decarbonizing transportation) will depend on aggressive solution to numerous potential problems, notably the cost of batteries, inadequacy of charging facilities especially for users who cannot have home-based charging facility, and, crucially, consumer interest which may be dampened significantly by uncertainty and increasing fuel economy in conventional vehicles. Cars that can deliver around 25 km/liter (60 miles/US gallon) are already on the market (some latest small models claim 80 miles/gallon), and many recent models have features that are the main selling points of EVs, including stop-start capability. Projections predict a further rise of 50-55% in ICE fuel efficiency by 2040. Also, the latest diesel vehicles are fitted with advanced emissions control systems, and actually have lower CO_2eq emission levels per kilometer compared with petrol engines since they cover around 30% more distance for the same volume of fuel, although emissions of nitrogen oxides are still marginally higher.

Most of the growth in EV market will be accounted for by China, Europe, North America and Japan, while nearly all the growth in vehicle population will be in emerging countries. Furthermore, the electric vehicle market is very sensitive to global oil prices which have been low for some years, thereby fueling the growth of internal combustion engine vehicles at the expense of EVs. For example, fuel prices in the United States are at the lowest for years due largely to increasing production of oil and gas by fracking and significant drops in operating costs, hence the sales of SUVs and other larger ICE vehicles have risen sharply, compared with Europe where fuel is very expensive and the smaller ICE vehicles are favored. It should be noted also that internal combustion engine technologies have been around for over a hundred years and the very strong research, development and technology innovations base will no doubt pose a formidable challenge to the proliferation of electric vehicles unless much more comprehensive and varied support policies than currently anticipated emerge. For example, some countries, particularly in Europe have introduced punitive taxes on ICEs and are offering strong incentives to promote the EV market. Even then, the impact on global sales of EVs has been minimal: China which has minimal policy instruments accounted for nearly half of the global sales of EVs in 2018, United States 24%, and Europe which has the most robust policy support, 22% which translates to only 0.4% of all the 320 million cars and light commercial vehicles (LCV) on European roads in that year. It is noteworthy however that European sales of EVs rose by 33% compared with 2017. If this double-digit rise could be sustained over the next decade, EV share of car and LCV sales could rise to around 50%, raising the population to around 10% of the European vehicle fleet.

The current and projected rise in global prosperity will lead to a shift from high-occupancy public transportation to personal vehicles, thereby reducing the global transport load factor (average number of passengers per vehicle), increasing demand for transportation fuel, and increasing road congestion which in turn leads to higher emissions. Furthermore, China will remain the largest market for EVs over the next decade and the most important source of lithium-ion batteries, the most expensive component of EVs. Since the country relies heavily on coal-fueled power generation and its electricity has one of the highest carbon footprints in the world, the potential role of EVs in decarbonizing global transport sector may be significantly compromised. Natural gas and biofuels are also featuring increasingly in transportation, driven largely by strong public policy instruments: petrol/gasoline in some countries contains up to 10% of biofuel-sourced ethanol and buses are operating on natural gas. Brazil pioneered biofuels technology and has been producing ethanol fuel from sugarcane for around a hundred years but the global fuel crisis of the 1970s and the need for energy security stimulated the launching of a national policy which led to the establishment of the National Alcohol Program (PROALCOOL), with strong support policies that included the introduction of vehicles that could run on pure ethanol (normal ICEs can only run on petrol containing 10-15% ethanol). The country remained the leading ethanol producer until overtaken by the United States in 2006. However, sugarcane ethanol is superior to products from starch and vegetable oil which are the precursors in the United States and Europe respectively. Sugarcane ethanol can provide lifecycle savings in greenhouse gas emissions of the order of 70% -100% compared with conventional petrol, but savings in emissions of corn ethanol are lower, between 20%-50%. Furthermore, converting sugarcane to ethanol is highly efficient because the electricity required is generated by burning the crop's crushed residue (bagasse). In spite of the expected rapid market penetration of electric vehicles over the next two to three decades, the net decline in demand for fossil oil in transportation will be minimal: oil will still account for around 85% of transportation fuels in 2040, down from 94% currently, with natural gas, electricity and biofuels each providing around 5%. The EV30@30 Scenario, (the assumed trajectory for transportation decarbonization consistent with the International Energy Agency Sustainable Development Scenario) is perhaps the most optimistic projection of the global population of EVs over the next two decades. It targets a market penetration rate that raises the population of EVs from 5 million in 2018 to more than 250 million in 2030, an average of additional 20 million EVs on the road per year. It is estimated that this phenomenal growth rate would save around 250 million metric tons CO_2eq that would have been emitted by a fleet of ICEs of equivalent size. However, even this size of market penetration will have little impact on global demand for oil: the reduced oil demand would be only about 4 million barrels a day compared with the global consumption of about 100 million barrels (including biofuels) in 2018 (IEA, 2019b; Statistica, 2019). In effect, the greatest impact of EV penetration will be local, reducing smog over the world's cities and mitigating the negative impact of ambient pollution on human and ecosystem health.

Decarbonizing the building sector

The building sector (commercial and residential) is the least polluting of all the major sectors of the global economy, accounting for less than 10% of the total global energy-related emissions. However, the sector is second only to industry in electricity consumption and when emissions from the power sector are shared by sector use, emission contribution from the building sector rises to around 25%. The building sector can contribute to global effort to decarbonize energy in two key ways: reduction in electrical energy consumption through improved efficiency of use, and phasing out traditional biomass use. As discussed earlier, substantial efficiency gains have been achieved over the last decade or two through significant innovations in the design of building envelopes, fittings and appliances, but more needs to be done especially in heating and cooling systems which account for the largest share of energy consumption in the building sector. Phasing out of traditional biomass use will be slow due largely to the fact that the emerging world accounts for most of the use and there are no viable or affordable alternatives. However, a lot can still be done to improve efficiencies of biomass production and burning equipment. For example, most processes for producing charcoal from hard wood have single digit efficiencies, and many wood and charcoal burning appliances are crude and inefficient. The United Nations and World Health Organization are supporting the development and proliferation of more efficient charcoal kilns and cooking stoves in the developing world.

Carbon trading, capture, use and storage

Carbon trading (emissions trading, cap-and-trade) is a market-based tool designed to limit GHG emissions mainly from the industrial and commercial sectors sectors of the economy. The system is designed to penalize high emitters and compensate low polluters. A limit (cap) is set for allowable emissions and any excess incurs a tax while emissions that are lower than the limit attract credit. Those that are over the limit can buy credit from low-polluting enterprises to offset the excess emissions. Trading gives companies a strong incentive to save money by cutting emissions in the most cost-effective ways. Caps are divided into units, typically metric tons and distributed to companies freely or by auction. Companies that cut their pollution faster can sell allowances to companies that pollute more, or save them for future use. Caps are determined by the scheme's governing body which may be mandatory or voluntary and revised upwards or downwards regularly to moderate supply and demand forces which primarily control pricing. Member companies that do not have enough allowances to cover their emissions must either make reductions or buy another company's spare credits. The system has been very effective in controlling GHG emissions in the European Union and is spreading to other regions. In the EU system, a cap is set on the total amount of certain greenhouse gases that can be emitted by installations covered by the system, and is reduced over time so that total emissions fall. Within the cap, companies receive or buy emission allowances which they can trade with one another as needed. They can also buy limited amounts of international credits from emission-saving projects around the world. After each year a company must surrender enough allowances to cover all its emissions, otherwise heavy fines are imposed. If a company reduces its emissions, it can keep the spare allowances to cover its future needs or else sell them to another company that is short of allowances.

The International Energy Agency (IEA) has set a cap on emissions and a timetable for achievement if the 2°C or below emissions target is to be achieved. Working with the World Bank and the International Monetary Fund, targets have been set for specific industries. The iron and steel industry which accounts for 7-9% of all direct fossil fuel emissions and also the world's largest industrial source of energy-related emissions has been set a target of 65% reduction standard by 2050. However a recent evaluation report shows that most companies in the industry are far behind target. Many cleaner iron and steel technologies are already available (direct reduction, carbon capture, natural gas injection, heat and materials recycling) or under development but most new iron production technologies and innovations are too limited in output compared with the traditional blast furnace process or too expensive to deploy because they would raise production costs by 20% to 30%. While a few companies are embracing innovations, have set targets that achieve carbon neutrality by 2045 and appear to be on course, most others either have no targets of are well behind in developing and adopting low-carbon technologies. However increasing carbon prices particularly in Europe are forcing companies to comply or slash production and suffer major financial losses. Unfortunately, relatively few countries mostly in Europe and East Asia are enforcing carbon pricing, while the major pollution emitters - China, United States, Russia - lag behind because of lax policy instruments.

Carbon capture, use and storage (CCUS) or carbon capture and sequestration (CCS) is an emerging technology for capturing and storing as much of energy-related emissions as possible at source by stripping flue emissions from fossil fuel powered generation and other industrial processes. Technologies are already available and many more are under development for the capture and safe storage of carbon dioxide and particulates produced in major industrial operations, in particular, petroleum production, power generation, iron and steel, cement and chemicals/petrochemicals production. Carbon capture technologies either separate CO_2 from waste gases before or after combustion (pre- and post-combustion capture). In pre-combustion capture, coal is gasified to produce synthetic gas (syngas), a mixture of hydrogen and carbon dioxide, which is further reformed with steam to produce almost pure hydrogen which is then used for driving a turbine to produce electricity. The process not only reduces CO_2 production but also offers an opportunity to capture the little amount in the reformed syngas at high efficiency because the gas is at high pressure, thereby virtually eliminating pollution from coal-fired power plants. Gasified coal has comparable or even better carbon footprint credentials than natural gas because it contains neither carbon dioxide nor methane which are the predominant compounds in natural gas. Large-scale coal gasification installations are operating or under construction in many parts of the world primarily for electricity generation, or for production of chemical feedstocks. The hydrogen obtained from coal gasification can be used for various purposes such as making ammonia, powering a hydrogen economy, or upgrading fossil fuels. Coal-derived syngas can also be converted into transportation fuels such as gasoline, diesel or methanol fuel/additive through additional treatment. Also, syngas can be liquified for use as a fuel in the transport sector. Post-capture technology processes flue gases from coal or natural gas combustion to recover CO_2, and NO_x gases, but the process is less efficient compared with pre-capture technology. A third option (oxy-fuel combustion technology) is designed to minimize production of carbon dioxide in fossil fuel combustion by using oxygen-enriched air or recycled flue gases instead of the conventional air for combustion.

Carbon capture, use and sequestration/storage (CCUS) is a very promising technology for significantly reducing energy-related GHG emissions. However, although capture technologies are well developed and efficient, deployment in both existing power plants and new projects is expensive and operation costs are higher. Furthermore, options for storage and re-use are still very limited. Most carbon capture demonstration projects are in North America and the world's first large-scale adoption of carbon capture technology in the power sector was commissioned in Canada in 2014. Several demonstration plants are operating in other countries and some commercial projects are under construction, particularly in the United States. Other countries - Japan, Australia, South Korea, the Middle East - are also actively considering CCUS projects. Many CCUS projects are integrated with oil and gas production because the captured gas is either used to pressurize oil wells to enhance production, or stored permanently in exhausted wells. About 30 million metric tons of GHG gases are captured annually, with ExxonMobil accounting for around 40%. However, the International Energy Agency (IEA, 2019j) estimates that this needs to rise to around 24 billion metric tons per year (800% scale-up), accounting for 7% of the cumulative emissions reductions needed globally by 2040 for positioning the world on the path to a sustainable environment.

The viability of CCUS technology is not in doubt, especially as a very promising method of significantly reducing energy-related GHG emissions, and a key technology option to decarbonize the power sector and industry. The potential is particularly high in countries with a high share of fossil fuels in electricity production. However, there are formidable obstacles to widespread deployment: retrofitting existing plants or including carbon capture in new plant design could drive up capital and operating costs by up to 30%, and many emerging nations that will host most of the expected new fossil-fueled power plants cannot absorb the added costs. Another major constraint is what to do with the captured gas, since current options are very limited. Also, finding suitable storage sites, preferably near the source is a major problem, while transportation over long distances for safe disposal (in disused mines and other suitable underground sites) is too expensive and often impractical for wide adoption. Furthermore, technologies for capturing carbon dioxide from combustion effluents and from the air and conversion into useful products or benign state are at early stages of development. Progress in resolving these issues will largely determine the reductions in anthropogenic emissions that can be achieved over the next 2-3 decades through carbon capture. Currently over 40 carbon capture projects are operating around the world, under construction or consideration but they are mostly tied to petroleum production which uses the gas to pressurize wells and enhance oil production.

Many research and development projects are also ongoing, targeting the enhancement of the natural carbon cycle by promoting the use of the excess anthropogenic carbon dioxide in the atmosphere by plants to produce carbohydrates and proteins, using the Sun's energy, thereby moving the world towards net zero and negative emissions). The Intergovernmental Panel on Climate Change (IPCC) has determined that achieving the ambitions of the Paris Agreement to limit future temperature increases to 1.5 degrees will require more than just an acceleration of efforts to reduce emissions; it may also require the deployment of technologies to actually remove carbon from the atmosphere. There are many options under development but the most mature carbon dioxide removal technology is bio-energy with CCS, or BECCS. The technology involves enhancement of the natural carbon cycle by growing biomass which extracts carbon dioxide from the atmosphere as it grows, harvesting and processing to chemical products or other forms of energy, and storing the captured carbon dioxide underground or converting it to useful products. Possible applications of BECCS include: dedicated or co-firing of biomass in power plant; combined heat and power; cement plants, pulp and paper mills; lime kilns; ethanol plants; biogas refineries; and biomass gasification plants. Certain biomass conversion processes, including fermentation and gasification technologies, generate high-purity carbon dioxide streams as an intrinsic part of the process, thus providing lower-cost capture opportunities. The IPCC found that in pathways with limited or no temperature overshoot, up to 400 Gigatonnes of BECCS could be required this century. Currently, there is one large-scale BECCS facility in operation, at the Illinois Industrial Carbon Capture and Storage facility in the United States, capturing and storing one million metric tons of carbon dioxide per year.

Aforestation and reforestation also have significant potential for removing carbon dioxide from the atmosphere. Land is both a source and a sink of natural and anthropogenic greenhouse gases (GHGs) and plays a key role in the exchange of energy, water and aerosols between the land surface and the atmosphere. Changes in land cover (deforestation, aforestation, reforestation) directly affect the Earth's surface temperature by altering the natural balances of moisture and heat exchange with the atmosphere. Land exposure through human activity will alter the natural albedo and cause the global land mass to absorb more of the Sun's energy than it radiates, resulting in global warming, draught and desertification. These will increase the vulnerability of land ecosystems, biodiversity and society to climate change and weather and climate extremes. Although it is difficult to separate natural and anthropogenic fluxes, it is estimated that net carbon dioxide emissions of 5-7 Gigatonnes per year come from land use and land use change (IPCC, 2019). In effect, degraded and desertified land areas warm the world and affect the climate, while climate change exacerbates land degradation, particularly in low-lying coastal areas, river deltas, drylands and in permafrost areas, with severe consequences for life in the areas. Sustainable land management can contribute significantly to reducing the negative effects of these multiple stressors. It is estimated that human activities have destroyed about half of the world's population of trees that existed before the rise of human civilization. Planting billions of trees across the world in land areas that were previously degraded (reforestation), sparsely vegetated and not used for agriculture (aforestation) is perhaps the most effective and cheapest way of reducing the concentration of carbon dioxide in the atmosphere. Trees are very effective at taking carbon dioxide out of the atmosphere through photosynthesis, in some places offsetting human emissions of carbon dioxide by 30 percent. Carbon dioxide is the precursor to plant carbohydrates and proteins which feed both humans and animals.

About 15 billion trees are cut down yearly across the world to make room for agriculture and for use by industry. Reforestation not only provides a natural sink for carbon dioxide, it results in cooling due to enhanced evapotranspiration which in turn can result in cooler days particularly in the tropical regions, and a reduction in frequency, intensity and duration of heat-related events such as heat waves. It is estimated that a worldwide planting program could remove two-thirds of all the emissions from human activities without encroaching on crop land or urban areas (land and oceans already absorb about 55% of CO_2 produced by human activity). A recent study (Bastin *et al.*, 2019) estimates that about 11% of the global land area could host around 1.2 trillion native trees, which means increasing the estimated global population of trees by about 40%, and the tropical areas are particularly suitable. This proposal may sound ambitious but presents an essential and achievable vision and a potentially important blueprint for country and global public policies as well as private sector initiatives. Although forest restorations could take up to a hundred years to bring about the full effect of removing around 200 billion metric tons of carbon dioxide from the atmosphere (bioenergy crops and shrubs take only months), future generations can look forward to a much less hostile environment and the devastating effects of pollution-induced climate change. The right trees planted in the right locations could capture around 200 Gigatonnes of carbon dioxide, equivalent to about two-thirds, of all anthropogenic carbon dioxide burden since the Industrial Revolution. Tree planting initiatives already exist, notably the Bonn Challenge, a German initiative endorsed and extended by the New York Declaration on Forests at the 2014 UN Climate Summit. It has become a global effort to bring 150 million hectares of the world's deforested and degraded land into restoration by 2020, and 350 million hectares by 2030. It should be noted however that as trees mature, the capacity to sequester and use carbon dioxide from the atmosphere declines, hence sustainable forest management strategies which include harvesting and replanting economic trees will be vital.

Natural peat bogs (ponds and marshlands filled with rotted plant matter) are regarded as natural high-carbon ecosystems estimated to hold over 40% of all stored carbon, more than the carbon stored in all other vegetation types, including the world's forests. While mature vegetation eventually reach saturation and cease to absorb carbon dioxide, peat bogs can continue to sequester for centuries. However, peat bogs are currently being drained and cleared for agriculture and peat has many uses as fuel, in oil spill control, etc. Environmentalists have advocated that global carbon-curbing plans should include conservation of natural pit bogs. It should be noted however that peat bogs are the precursors to coal formation. There are many other carbon sequestration technologies under development, notably the use of carbon dioxide to fertilize ocean algae which are valuable bioenergy feedstock, machines that pull out carbon dioxide from the atmosphere for sequestration or use, conversion of carbon dioxide to benign natural carbonate ores, photo-electrochemical production of synthetic natural gas (syngas) from water and carbon dioxide using solar energy, or reacting carbon dioxide with hydrogen under very high pressures to produce methyl alcohol and many other fuels.

Recycling

The world generates about 3 million metric tons of waste daily, and the rate is projected to double by 2025. Nearly half is organic waste and about a third is made up of metals, plastics and paper. Around 80% is used for landfills and the balance is recycled, incinerated to generate electricity and heat, or used for composting. The United States, Japan, China, and many countries in Europe have extensive waste-to-energy (WtE) programs and most claim substantial reductions in anthropogenic emissions. In 2014, the United States recycled and composted about a third of the 258 million metric tons of municipal solid waste (MSW) generated, resulting in a reduction of over 181 million metric tons of carbon dioxide equivalent emissions, comparable to the annual emissions of nearly 40 million passenger cars. Over 33 million metric tons were combusted with energy recovery (which reduced fossil fuel use), while 133 million metric tons were landfilled. The United Kingdom Waste and Resources Action Program (WRAP) claims a more modest but still substantial reduction of 10-15 million metric tons a year of emissions, amounting to nearly 50 million metric tons between 2010 and 2015.

Recycling is the process of converting discarded waste to useful products, sometimes different from the original products, with significantly less energy than would have been required for making the same products from virgin raw materials. There is also the additional advantage of conserving natural raw materials and the associated production and processing emissions. Furthermore, recycling slows down the need for exploiting low-grade ores, a more energy-intensive process that releases higher emissions. Recycling requires significantly less energy per kilogram of material produced than primary production, and also decreases the negative impact of mining, processing and transportation of ores. The process offers higher environmental benefits and lower environmental impacts compared with other methods of waste disposal. Different products offer different opportunities and gains: metals are inherently recyclable and the same product can be manufactured from recycled material, with significant gains. For example, producing aluminium soft drink cans from recycled material saves around 95% of the energy required to make the same product from virgin bauxite ore and saves four metric tons of bauxite ore and the associated mining and processing emissions per metric ton of recycled aluminium. Every metric ton of recycled aluminium reduces CO_2eq emissions by about 14 metric tons compared with production from virgin ore (EPA, 2011). Recycling one metric ton of paper saves about 17 trees and also saves one metric ton of emissions. Many other metals and materials - iron, steel, copper, lead, silver, paper, glass - are recyclable, with less dramatic but very significant energy savings and GHG emissions reduction when recycled feedstock is used. One major constraint to recycling is the development of policy-driven appropriate strategies for collection, sorting, pretreatment, etc.

Waste-to-energy (WtE) or energy-from-waste (EfW) is the process of generating energy in the form of electricity or heat from the primary treatment of waste either by direct combustion, or by conversion to biofuels such as methane, methanol, hydrogen or other synthetic fuels. Modern incinerators can convert heat to electricity at 14-28% efficiencies, but if configured as part of co-generation which combines electricity generation with space heating, efficiencies higher than 80% are possible. Also, strict emission regulations have forced manufacturers of incinerators to produce plants with low emissions of nitrogen and sulphur oxides, as well as particulates. Apart from the fact that conversion of waste to useful energy makes economic sense, the more important advantage is the avoidance of uncontrolled incineration or dumping of waste, particularly plastic waste which releases hazardous gaseous/particulate products into the atmosphere when incinerated, or fails to degrade for many years when used for landfills. Landfill is the most popular disposal system worldwide, but decomposition of the organic matter releases significant amounts of carbon dioxide and methane into the environment and plastics may not degrade for decades.

Waste-to-energy technologies can contribute to global climate mitigation by significantly reducing emissions from other methods of waste disposal. Many developed countries are adopting incineration to produce power and heat, a system which cuts CO_2eq emissions by more than 50% compared with landfill systems. New, more efficient waste gasification technologies are emerging, and adoption could cut emissions by two-thirds. Also, using the energy generated by WtE plants will reduce the demand for energy from fossil fuels and eliminate the associated emissions. A modern WtE plant can produce carbon emission savings in the range 100 to 350 kg CO_2eq per metric ton of waste processed, depending on waste composition and amount of heat and electricity produced. Even greater savings up to 800 kg CO_2eq per metric ton of waste can be achieved if WtE completely replaces landfilling (WEC,

2016a). Emissions (gas-phase and particulate) from WtE plants which feature incineration are comparable to coal-fired power plants, but those which adopt gasification produce significantly lower emissions. In any case, strict environmental regulations in many developed countries have made it mandatory for both coal and incineration plants to adopt a series of process units for cleaning flue gas and disposal of post-combustion residue in an efficient and environmentally friendly manner.

Polymers are products of fossil fuel-sourced petrochemicals and, since invention in the 1950s, have taken over virtually every aspect of human life, from packaging to clothing. One of the most attractive properties is that they are unreactive, which makes them suitable for food and chemicals storage. Unfortunately, this property also makes it difficult to dispose of polymer waste. Most polymers are not biodegradable and many can remain intact for several hundred years in landfills. When incinerated, polymers release a lot of energy, carbon dioxide, black carbon and chemical aerosols into the atmosphere. Furthermore, the bulk of raw plastic waste from the developed countries is being exported to emerging nations where they are sorted manually and washed by small enterprises under unhealthy conditions and recycled or incinerated, usually in unregulated environment, thus releasing potentially dangerous aerosols into the atmosphere and contributing to ambient pollution. This is happening because policy instruments in the developed countries which promote local processing are weak and could amount to merely transferring an environmental problem from one region to another. Plastics offer less but still significant opportunities for recycling compared with metals and paper, primarily because of the diversity of types with very divergent thermo-chemical characteristics. Most plastics cannot be recycled to produce the original product, but can be reprocessed into other useful products, with significant savings in energy and reduction in anthropogenic emissions. Plastics are produced from fossil fuels and considering the exponential increase of plastic use in the last five decades, their share of global primary energy use has been increasing and disposal has become a major global problem.

Societal mitigation

Carbon footprint calculation has become a powerful tool for assessing the impact of human (including personal) behavior on global warming. Every consumer good or fuel use has a carbon footprint: the foods and goods that humans buy everyday, bottled drinks, polymer shopping bags, travels, household energy use, transport energy all have carbon footprints which must be accounted for in calculating personal contribution. For example, ten liters of petrol or diesel burnt in a car, or of oil used in home heating contribute 23-27 kg of carbon dioxide to the atmosphere; the manufacture of three empty one-liter plastic bottles of water/soft drink or twenty plastic shopping bags releases about 1 kg of CO_2-eq into the atmosphere. In effect, reducing personal carbon footprint can reduce people's contribution to global anthropogenic emissions significantly. The world average of personal carbon footprint was about 5 metric tons/person/year of CO_2-eq in 2018, but there were wide variations between countries. For example, Qatar (38.2), Canada (16.1), Australia (16.8) USA (16.1), China (8.0), UK (5.6), Uganda (0.1) (Crippa *et al.*, 2019; Wikipedia, 2019). With over 7 billion people in the world today a reduction of annual personal contribution to emissions by just 20% could have reduced the world total by around 7 Gigatons of CO_2-eq, equivalent to over one-fifth of global energy-related emissions in 2018.

Potential areas of personal carbon footprint reduction are many, and both the environment and personal finances benefit: rationalizing energy use across the entire spectrum of personal activities such as opting for energy-efficient homes and appliances when buying or leasing homes; upgrading the energy efficiency of existing homes through improvement of insulation; replacement of inefficient appliances and lighting fixtures with energy-smart systems; switching off idle appliances, opting for energy-efficient internal combustion or electric vehicles; planning journeys such as shopping to eliminate unnecessary trips; pooling vehicles; using public transport when feasible; etc. Strong support for municipal recycling programs can also make a significant impact on personal carbon footprint. On line shopping which is becoming increasingly popular has the potential to cut personal carbon footprint significantly since just one delivery van can eliminate the need for several hundred trips to shops. Also, voluntary adoption of reusable bag shopping will eliminate the energy required to manufacture several hundred disposable bags (from fossil fuels) per person per year and the associated carbon emissions, thereby further reducing personal carbon footprint.

Environmental issues have become very topical and political in many countries, with environmental activists promoting awareness of society and pushing all sectors of the economy to offer energy-efficient products. However, there is an urgent need for strong policy instruments and non-governmental organization activities that target the improvement of public awareness on potential personal contributions to global environmental mitigation efforts. It is noteworthy that the youth are becoming increasingly interested in climate change, after all, they are the ones likely to take the greater share of the catastrophic consequences. Youth activism is becoming increasingly common in many countries and they are succeeding in raising the level of public consciousness, forcing political parties around the world to develop innovative policies. The youth have successfully organized global strikes and protests, joined by millions of workers around the world. The latest global climate strike: "you will die of old age, we will die of climate change" led by a high-school teenager inspired action in around 140 countries across all regions of the world, with around 5 million school children, workers of major companies joining in strikes and protests. It is interesting to note that employers actually encouraged their staff to join in the strike, and school boards all over the world declared a day-off to enable students join in the strike. Environmental issues are dominating politics in many countries, Green parties dominated by the youth are emerging as strong political forces, particularly in Europe and they are succeeding in making climate change a very prominent election issue.

Strong and sustained policy instruments that survive political cycles

Society tends to ignore or resist change that targets the status quo and some compelling force is always necessary to stimulate compliance. This is the primary driver of the advertisement industry which largely employs motivational strategies. However, statutory regulations and strong policy instruments that employ both motivational and enforcement strategies are required at three different levels of political governance to achieve any significant results in the global quest to mitigate environment pollution: municipal, national and international. The effectiveness of environmental mitigation actions depends critically on resilient policy instruments, (both supportive and punitive) which transcend political cycles.

Municipal level mitigation

Municipal governments can effectively institute and enforce regulations that help pollution abatement, in particular, on activities that have direct impact on their areas of governance. Some of the regulations may be national but many are designed to address specific local issues, notably transportation pollution control and refuse management. Imposition of punitive taxes on or outright prohibition of polluting vehicles is already having a significant impact on efforts to clean up city environments and regulations on waste disposal and recycling of reusable items such as plastics, metal products, paper and electronic waste, shopping bags, compostable household waste, are becoming increasingly common. It is estimated that a trillion single-use plastic bags are used worldwide annually (around 2 million every minute). More than 480 billion of plastic bottles (water, soft drinks, etc.) were used in 2016 across the world, projected to double by 2040. Fewer than half of the bottles were collected for recycling and only 7% of those collected were turned into new bottles, the balance being discarded or used in landfill. Most of the discarded bottles fragment and end up in rivers and oceans, causing serious environmental and ecological problems. It is estimated that between 5 million and 13 million metric tons of discarded plastic have ended up in the world's oceans, ingested by fish and other aquatic life, and some of it is already showing up in the human food chain. Dead whales with up to ten kilograms of ingested plastic have been found on shores in different parts of the world. Apart from being potentially toxic, ingested plastics cannot be usefully assimilated into the food chain. Another problem with plastics is the fact that they do not degrade easily and, although exposure to strong ultraviolet rays from the Sun can embrittle some types, causing them to break up and possibly end up in the atmosphere as aerosol, or in the oceans, other types can remain intact for several hundred years. Many municipalities especially in Europe and the USA have achieved phenomenal success in restricting the use of one-cycle shopping bags simply by introducing a small tax. A recent policy by the United Kingdom to impose just 5 pence on plastic shopping bags provided to customers by a few major retailers has had a dramatic effect, cutting demand by 85% in just six months, and the few large supermarkets that were involved in the pilot project reported very substantial cost savings. Most shoppers now carry reuseable shopping bags. The pilot project is being expanded to other outsources of shopping bags, drinking straw, and sale of bottled drinks. Many other countries have reported similar dramatic gains and others particularly in Europe and North America also have in place or are introducing policies that help stimulate the R[4] culture: *refuse, reduce, reuse, recycle.*

National mitigation

Mitigation of energy-related emissions requires strong national policies in addition to local efforts, particularly in relation to enforcing emissions standards in power generation and transport. Introduction of stringent power generation and vehicle emission standards and support policies for electric vehicles are all having significant effects in cutting down emissions, forcing inefficient power plants to upgrade or close down, and polluting vehicles off the roads. Other policy interventions promote renewable power generation, in particular, solar and wind, and the introduction of carbon tax whereby the government sets a price that emitters must pay for each ton of greenhouse gas emissions they emit beyond a set limit. This drives businesses and consumers to take steps, such as switching fuels, adopting new technologies to reduce their emissions to avoid paying the tax, or trading carbon. Carbon tax includes an emissions tax which is based on the quantity an entity produces and a tax on goods or services that are generally greenhouse gas-intensive such as a carbon tax on fossil fuel-sourced energy. Carbon trading discussed earlier is another effective policy instrument and is becoming the instrument of choice for controlling greenhouse gas emissions. The scheme is one of the prime GHG emissions' controls used by The European Union which leads the world in the development of strong energy-friendly policy instruments. All 28 member states are obliged to adopt emission laws and regulations irrespective of the internal political dynamics. The United States also has strong national policies on the environment but enforcement powers are limited due to the largely independent status of states, most of who have their own policies based on local realities. For example, it is difficult to enforce the stringent national emission codes in states that depend heavily on coal for power generation and employment, although economic forces are at play, forcing coal power plants to convert to cheaper natural gas or close down and driving coal mines to bankruptcy. China, one of the largest sources of energy-related pollution has strong mitigation policies but the pressures of power demand from the very large primary industry and growing middle class population have resulted in many policy reversals. Many coal plants have been closed, but many more are under construction, although they are adopting more efficient technologies. India, another significant polluter has strong renewable energy policy but coal remains the primary energy source for its growing primary industry and teaming population who want access to modern energy.

As discussed earlier, forests are important parts of the natural carbon cycle and act as effective carbon sinks (biosequestration). It is estimated that 4 billion hectares of forest ecosystems (about 30% of the global land area store large reservoirs of carbon, together holding more than double the amount of carbon in the atmosphere (FAO, 2005). Forests also minimize disruption of the natural albedo, and act as barriers that slow down potentially damaging winds. Although the climate protection role of forests is in no doubt, it is complex to determine how much of the forest carbon sink and reservoir can be managed to mitigate atmospheric CO_2 buildup, and in what way. It is estimated however that each intact hectare of the world's tropical forests across Africa, Amazonia and Asia traps around 0.6 metric tons of carbon a year, which adds up to about 5 billion metric tons, equivalent to about a sixth of the total global energy-related emissions in 2018. Many countries municipalities across all regions of the world have strong reforestation programs, not only for environmental reasons but also to restore economically valuable trees and regain valuable land devastated by excessive deforestation. Reduction of current global deforestation rates by 50% by 2050 would be a major contribution to a sustainable environment.

International mitigation

Many international organizations are playing key roles in promoting global discourse on climate change and energy-related emissions. The United Nations (UN), International Energy Agency (IEA), World Energy Council (WEC), Intergovernmental Panel on Climate Change (IPCC), Energy Information Administration (EIA), ExxonMobil, BP, World Health Organization (WHO), United Nations Environmental Protection Agency (UNEP), European Union, and many more have been providing extensive information and data which have become vital resources for country and international actions. The World Health Organization has brought into full global focus the consequences of ambient outdoor and household pollution, which have a more immediate and severe impact on the health of life on earth than climate change. The 1992 United Nations' conference in Rio de Janeiro was perhaps the first major global initiative to address the problem of environmental pollution. The meeting developed an International Environmental Treaty (United Framework Convention on Climate Change (UNFCCC) which has become a major stimulant and guideline for country action. The thrust of the Treaty was the stabilization of greenhouse gas concentrations in the atmosphere at a level that would prevent dangerous anthropogenic interference with the climate system. The parties to UNFCCC treaty have met annually since 1995 as Conference of the Parties (COP) to assess progress in dealing with climate change. The meeting in Kyoto in 1997 developed a Kyoto Protocol which established legally binding obligations for developed countries to reduce their greenhouse gas emissions. Six greenhouse gases were identified which, if reduced could drive down global warming significantly. From 2005 the conferences have also served as the Meetings of Parties of the Kyoto Protocol (CMP). The latest meeting was held in Paris in November 2015, attended by 197 parties and the Paris Agreement, a global treaty on the reduction anthropogenic emissions was adopted. Signatories were obliged to establish National Greenhouse Gas Inventories and develop strategies for control and removal by 2020. The Paris-2015 Protocol set a goal of limiting the increase of the world's average temperature to no more than 1.5-2 degrees centigrade above pre-industrial levels.

One major achievement of international action on pollution mitigation coordinated by the United Nations is the mobilization of country and regional actions. Many countries, particularly in the industrialized world have put in place voluntary measures to reduce environmental pollution. These measures are often referred to as Intended Nationally Determined Contributors (INDCs). Europe-28 leads the world in terms of regional action to balance the energy trilemma: energy equity, energy security and energy sustainability. In 2008 the European Union (EU) committed to climate and energy goals to be reached by 2020. These targets known as '20-20-20' targets aim to achieve a 20% cut in emissions of GHG compared with 1990 levels; a 20% share of the final energy consumption coming from renewables; and a 20% increase in energy efficiency. Already, the Union has achieved an average of 25% renewable power generation and some EU countries have set even higher goals. For example, Denmark leads the world in renewable power generation: in 2016, wind and solar accounted for 44% (42% wind and 2% solar), and the country expects to achieve around 70% by 2022. Germany set a national emissions' reduction target of 40% and a 55% cut in emissions by 2030. Also, the share of renewable energy (mainly solar and wind) in the country's electricity consumption should rise from around 30% in 2017 to 65% by 2030, and coal-fired power generation will be phased out completely. On the other hand, Poland is one of the top ten producers of coal in the world, and the second largest user in Europe, after Germany. Coal-fired power plants currently generate more than 80% of electricity in Poland and the coal industry is a major employer. Furthermore, coal is used extensively in home heating, not only in Poland, but in many other European countries. According to a recent report by the WHO, 33 cities in the country featured in the list of the 50 most polluted cities on the continent. For Poland and many other countries which still rely heavily on coal, (China, India, Australia, Germany, South Africa, and some states in the US), energy security remains the prime goal and balancing the energy trilemma is a major struggle. The Polish government acknowledges that coal will continue to play a prime role in the country's primary energy mix in the foreseeable future, but the country has developed extensive plans to cut environmental pollution by increasing energy efficiency and decarbonizing the transport system. Many of the country's power plants are old and are being replaced by much more efficient coal-fired power plants and nuclear power could also play a significant role in the country's future energy supply mix.

France derives nearly 80% of its electricity from nuclear power from 58 nuclear reactors, the second largest nuclear power fleet in the world after the United States. The country is also the world's largest exporter of electricity to neighboring countries due to the very low cost of generation. Much of the country's reactor fleet is reaching the end of its lifetime (average, 30 years) and there has been no clear policy on de-commissioning or replacement, nor a feasible trajectory to renewable energy in such a short time, considering that renewables account for only 15% of the country's power generation fuel mix currently. In spite of the country's significantly low-carbon electricity mix arising from nuclear power generation, France released the Energy Transition for Green Growth Act in 2015, which set goals to cut the share of nuclear power from 78% presently to 50% by 2025, while also reducing greenhouse emissions by 40% in 2030. Environmental issues have become very prominent in French politics and this explains the inherent policy instability. For some years now, the country's energy policy has not survived political cycles, for example, the current government has reversed the previous government's decision to reduce nuclear power generation capacity and determined that nuclear is "the most carbon-free way to produce electricity, along with renewables." Also, a progressively increasing tax regime on CO_2eq emissions was introduced to fast-track decommissioning of coal-fired power plants. This decision appears the most pragmatic considering that the French nuclear industry employs over 200,000 people and the pace of renewable energy deployment is slow. However, considering the antecedents, it is unclear whether this policy will survive for long.

The United States and China lead the world as potential INDCs - they are also the two leading sources of CO2eq pollution, together, they accounted for about 45% of global emissions in 2018. However, both countries have committed to INDCs and are introducing strong policies to meet set targets. In 2015, the United States Environmental Protection Agency (EPA) introduced a Clean Power Plan (CPP) to reduce emissions by 26-28% below 2005 levels by 2025 (EIA, 2016). Around 80% of emissions in the US is energy-related, the balance coming from other sources such as agriculture, land use and forestry. Two of the largest sources of energy-related emissions are the transportation and electric power sectors. The CPP targets power plants which are the largest sources of carbon pollution, accounting for around one-third of all greenhouse gas emissions. Emission performance rates (Best System of Emission Reduction, BSER) were established for existing fossil fuel-fired electric generating units (EGUs) - electric utility steam generating units and stationary combustion turbines. The CPP reflects the different needs of different states and each state is given the flexibility to choose how to meet the set goals. The CPP, if implemented, was projected to reduce U.S. emissions by 0.5 billion metric tons by 2040 (EIA, 2018). One major flaw in the United States' CPP action plan which was signed into law in 2015 was its obvious focus in eliminating coal-fired power plants. It would be near impossible for existing plants to meet the stringent emission control standards, and, for the many states which depend heavily of coal for power generation and the coal industry, the CPP was unacceptable. In any case, most states already have local emission mitigation policies, designed for their specific local conditions. Many states have instituted legal actions against the federal government, and succeeded in stalling implementation, even before the recent action by the current administration to repeal the law and reverse or temper several other environmental mitigation policies, notably automobile efficiency regulations and policies targeting inefficient lighting, regulations on gasoline-ethanol blending, emissions from power generation plants. However, even without the CPP, various state policies, rising use of renewables, increasingly competitive natural gas pricing, and negative economic forces in the coal industry have been driving down the use of coal in power generation, with a decline of about 16% between 2010 and 2017. A further decrease of about 35% is expected by 2030, after which it levels off. Nearly 300 coal power plants (40% of the country's coal power generation capacity) have closed down in the last ten years and, in spite of the current administration's efforts to save the coal industry, fifty of them closed in the last three years due to economic forces and many of the leading coal companies have collapsed. In summary, the CPP has the potential to reduce the US power generation emissions by about 30% through 2050, but its future is in doubt, considering the fierce resistance by many states and efforts by the current administration to cancel the initiative and reverse many other national policies targeting environmental pollution. It should be noted also that the United States which accounts for the second highest level of global energy-related pollution in 2018 and championed the Paris-2015 Accord has withdrawn from the Agreement, reversed policies designed to discourage the use of coal for power generation, and is on course to reverse many efficiency improvement standards that are in place. This demonstrates clearly the urgent need for resilient policies that survive political cycles. Fortunately most states in the country have continued to pursue vigorously energy decarbonization goals in line with the Paris-2015 Protocol. Nevertheless, the withdrawal of the country that has contributed more to global energy-related pollution than any other in the last century, and played a pivotal role in evolving the Agreement will likely have a cooling effect on the commitment of many other countries who are already facing the challenge or reconciling energy equity with environmental sustainability.

China which surpassed the United States as the world's largest CO_2eq emitter in 2008 has also set a goal of 20% emission reduction and 20% of non-fossil energy use by 2030, and the main drivers will be solar and nuclear energy. Although the country's economy grew by around 7% in 2017, emissions increased by just 1.7%, clearly mitigated by increasing renewables deployment, faster coal-to-gas/nuclear power switching and replacement of many existing coal power plants with more efficient plants. Many other European OECD countries have also set ambitious INDC targets. In fact, 146 national climate change panels presented draft INDCs at the United Nations Climate Change Conference in Paris in 2015. However, a recent progress assessment by the International Panel on Climate Change (IPCC, 2018) indicates that most INDCs are unlikely to meet or exceed the planned targets, considering the intra-country economic and political dynamics. Asia accounted for two-thirds of the growth in global emissions in 2018, due largely to the growth in the economies of the region, and all recent projections show an even faster growth over the next two decades or so. In spite of declared efforts to decarbonize energy in the region, it is difficult to see a clear path towards achieving this goal, considering that coal is a major primary energy and economic resource in the region. All recent projections indicate that the region will account for most of the growth of energy consumption and energy-related emissions over the next two decades or so.

Events in the emerging economies, in particular Asia Pacific will largely determine the extent of success in global energy-related emissions' abatement over the next two decades or so. The region has accounted for most of the emissions' growth in the last decade and this situation is expected to be sustained over the next two decades. Most countries in the region including Japan and Australia depend on coal for 70 - 85% on of power generation and accounted for over half of the global energy-related emissions in 2018. The region also accounted for three-quarters of total global emissions from industry from metals, minerals and chemicals/petrochemicals production. Furthermore, Asia has caught up with the Americas which had been historically the largest producer of transport emissions. With each contributing about 2.5 Gt CO_2eq to the global total of 8 Gt CO_2eq in 2016. Since the region will account for most of the future growth in vehicle population, electricity demand and energy-intensive industrial production over the next two decades, projections show that nearly all the growth in energy-related emissions will also come from there. While several countries in the region have policies targeting decarbonization of energy, the strong drive for energy equity and security will make any significant achievement unlikely. Furthermore, the region has become the host for energy-intensive primary industries and depository of toxic solid waste from the developed world. As discussed earlier, the environment has no hard country or regional boundaries, most energy-related GHG emissions end up in the stratosphere where they enhance the heat trapping capabilities of greenhouse gases and warm up the Earth. The gases in the stratosphere are fairly well-mixed and the net effect across the globe is fairly uniform, irrespective of the origin. While local efforts to decarbonize energy will help to clean up the local environment, no country or region will be immune to the negative impact of emissions from any other region on the global climate. It is vital therefore that international efforts focus more on assistance to this region and all other emerging economies.

In spite of the commendable global effort to mitigate energy-related pollution, it does not appear the world is on course for achieving the Paris-2015 goal. All recent projections show that emissions will grow by around 30-35% over the next two decades reaching around 36-37 Gt CO_2eq in 2040, in spite of the numerous country and regional mitigation policies in place and anticipated. While the growth rate will be significantly lower compared with the last two decades, the impact will not be sufficient to reach the Paris-2015 goal. Projections indicate that, taking into account current and planned mitigation policies, global average temperature will likely rise by about 3°C in 2100, with severe consequences for future generations. Much stronger policies that can reduce emissions by a further 50% over the next two decades will be required to fully decouple energy use and emissions, and set the world on track for limiting global warming to no more than 2°C (preferably lower) above the pre-industrial level by the end of the century.

A recent publication by the International Energy Agency (IEA, 2019g) examines in-depth the complexity of reconciling the divergent dimensions of the tumultuous global oil markets, geopolitical tensions, carbon emissions and climate targets, and the goal of providing electricity for the 850 million people around the world who currently lack access. The publication presents a set of scenarios that explore different possible futures, the actions or inactions that bring them about and the interconnections between different parts of the system. The report projects what would happen if the world continues along its present path (Current Policies Scenario), determines that even the New Policies Scenario that takes account of proposed and other likely new policies will be inadequate in stopping and reversing energy-related environmental pollution: the scope and momentum of proposed new policies are not enough to offset the effects of an expanding global economy and growing population. The rise in emissions will slow down but, with no peak before 2040, the world falls far short of shared sustainability goals. The report also presents a feasible pathway to a sustainable environment (Sustainable Development Scenario) which requires much more rapid and widespread changes across all parts of the energy system. Considering the complexity of the world's current and future energy needs, there are no simple or single solutions: multi-dimensional actions will be required, notably rapid adoption and scale-up of multiple low-carbon energy technologies and improved efficiencies across the whole spectrum of energy production and use, which lead to much more efficient and cost-effective energy technologies that must spread across all regions of the world to make any significant, positive impact on climate change.

Part 2
Fossil Fuels
and the Environment

Chapter 1

Energy and human development

1.1 INTRODUCTION

The search for sources of energy to complement human physical energy is as old as mankind. The Sun's energy was perhaps the first available source and was used mainly for drying; wood was converted to heat energy for cooking; bows and arrows, slings, and traps which convert human physical energy to dynamic energy were in common use for hunting and wars from the early times; wind energy was used to propel boats; and geothermal energy provided warm water. Coal was collected from outcrops and converted to heat energy for many applications including smelting and shaping metals since the early times. Archaeological evidence has been uncovered which proved that early Mankind made fires from gas leaks and oil seeps from underground. The Chinese were believed to have been smelting copper using coal sourced from the northeastern part of the country some 3000 years ago, and lumps of asphalt formed from underwater seeps from the Dead Sea were in common use in the Middle East thousands of years ago. The Native Americans collected oil from the surface of streams and lakes for use as medicine, for waterproofing their canoes, and as adhesives and mortar binders.

Oil and gas also have a very long historical background comparable to coal. The ancient Sumerians, Assyrians and Babylonians used crude oil and asphalt collected from large seeps from below the ground around the Euphrates river as mortar and for waterproofing some five to six thousand years ago, and the ancient Egyptians used liquid oil as fuel for lamps, medicine for treating wounds, and in embalming thousands of years ago. French explorers who arrived in America in early seventeenth century discovered that natives were igniting gases that were seeping into and around some lakes. Natural gas seeps were first discovered in Iran and several other parts of the Middle East several thousand years ago and provided the fuel for the 'gods of fire' worshiped in ancient times in these areas. Coal, oil and gas are commonly referred to as 'fossil fuels' because they were all formed from preserved animal and plant remains buried under the Earth or the seas hundreds of millions of years ago. The first move to mass-produce and trade fossil fuels began with coal in England around the beginning of the 12th century AD and the first major underground coal mine was opened in Scotland in 1575. More coal mines developed in other parts of England, spreading to continental Europe and innovative mining techniques evolved rapidly. Coal was discovered in Eastern North America in the 18th century and mines quickly spread to other parts of the country.

Prior to this development, coal that was in use came from small scale opencast or bell pit mines and provided heat for homes and small local industries. The commercialization of coal production fueled the growth of mass production of machinery and goods, which marked the beginning of the Industrial Revolution. Availability of mass-produced coal had a profound effect on the iron and steel industry which had previously operated on expensive and increasingly scarce charcoal. It became possible to scale up production and achieve the temperatures needed to produce higher grade steel, which in turn stimulated the growth of a wide range of industries, in particular, production of farm and textile machinery, a development which elevated farming and cottage industries to large-scale industrial businesses, and promoted the growth of communities, the humble beginning of urbanization.

Starting from around the first to the end of the 15th century AD, the potential of coal-fueled steam generation as a power source had been explored in theory and practice and many inventors developed crude steam jacks and small steam-driven machines. The first major milestone was the invention of a steam-powered pump, patented by Thomas Savery in 1698. This invention solved an intractable problem of mine flooding faced by British miners. The steam pump was used extensively to pump out water from mines but it was extremely inefficient and was only effective to a few meters depth. Improved designs were developed by Newcomen (1712) and James Watt (1765), Oliver Evans (1805), and several other inventors. Steam engines were driving railways and boats by the beginning of the 19th century and, around fifty years later, the first steam-powered automobiles were invented. Another milestone in the history of coal-fired steam power was the invention of the steam turbine by the British engineer, Charles Parsons in 1880. Steam generated in a boiler was injected at ordinary pressure into a turbine chamber which housed a rotor fitted with slanting blades and coupled to an alternator. The impact of steam on the rotor blades drove the rotor which in turn drove the alternator at high speed. This invention is still the core of thermal electric power generation today and involves firing a boiler with fossil, bio-, nuclear, solar, wind, or geothermal energy to generate steam, the only major innovation being the pressurization of the steam which enables modern systems to work at much higher efficiencies. While reciprocating steam engines have largely gone into oblivion, the steam turbine is still the bedrock of electric power generation worldwide, irrespective of the fuel source.

The humble beginning of the oil and gas industry dates back to around 350 AD when the Chinese were using drilling bits attached to bamboo poles to drill and recover oil from depths of several hundred meters. In 1846, Canadian geologist Abraham Gesner distilled coal to obtain a clean-burning lamp fuel which he named kerosene. The fuel was cheaper and cleaner than the traditional whale oil fuel which was then in common use. A few years later, Lukasiewicz of Poland developed a process for obtaining kerosene from the more readily available crude oil, also known as rock oil. The first rock oil mine was built in Poland in the early 1850s, followed by a refinery in Baku, Russia a few years later which supplied most of the world's fuel requirements. The first commercial oil and gas well in North America was drilled in Oil Springs, Ontario, Canada in 1858, followed by a major discovery in Pennsylvania in 1859 when a considerable amount of oil seeped from a 20-meter deep dug water well (CEC, 2012). A method was devised to pump up the oil to the surface and this method is still the basis for modern crude oil drilling technologies. The developments in the mid-nineteenth century are widely regarded as the humble beginnings of the modern history of crude oil drilling and petroleum (petro-leum means rock oil) refining industries. Gas is usually found in association with oil and most modern oil fields also yield gas as a by-product but there are many independent, mostly untapped dry gas fields in many parts of the world. Estimates indicate that the world's gas resources may be considerably larger than the oil reserves.

The discovery and recovery of oil in substantial quantities marked the turning point in the world energy scene and revolutionized transportation technology at the turn of the last century. Coal began to be replaced by oil in many applications because of the positive features of oil: it is abundant and, relative to coal, easy to produce, handle and transport. Also, compared with coal, the negative effects of recovery and utilization on the environment are low. Gas is also becoming very prominent as an alternative energy resource in many applications, in particular, electric power generation. Several other developments have continued to change the world energy mix over the years, in particular, the development of hydro energy and nuclear energy. Global oil and gas reserves are concentrated in only a few geographical zones and most needs worldwide are met through imports, unlike coal which is much more widely distributed and has been found in commercial quantities in over a hundred countries in every region of the world. This explains why, in spite of the well known negative impact on the environment, coal remains very prominent, and will probably continue to play a prime role in the global energy mix for decades to come, particularly for power, cement and iron and steel production.

Before the Industrial Revolution, the world relied solely on human and animal labor, and traditional bioenergy, all of which severely limited economic growth. However, availability of fossil fuels in commercial quantities provided the wherewithal for fast-tracking growth at an exponential rate which has been sustained to date. All technology developments and innovations depend primarily on availability of the right energy in appropriate abundance, for example, auto and air travel, telecommunication, materials production, industrial machinery, home appliances, healthcare, etc. For decades, the world has relied on fossil fuels for around 80% of primary energy and electricity requirements, and the situation is not expected to change very much in the foreseeable future.

1.2   ENERGY AND HUMAN DEVELOPMENT

Human development has always been tied to the availability of appropriate energy. Industrial and economic development which began with the Industrial Revolution of the late 1700s and stimulated the transition from hand tool production and animal power to new mass manufacturing processes, would not have been possible without the development of the coal industry. The availability of coal energy led to a revolution in agriculture, textile and metal production, and transportation. Advances in agricultural techniques and mass industrial production led to major improvements in the quality of life, which in turn stimulated growth in energy demand and the development of industrial and agricultural mass production technologies. The global economy has been transforming from rural/agrarian to urbanized modern production, earning power and quality of life have been rising, stimulating expansion of energy infrastructure, and giving prominence to the interdependence of energy and human development.

The United Nations Development Program (UNDP, 2016) defined human development as *"all about human freedom: freedom to realize the full potential of every human life, not just of a few, nor of most, but of all lives in every corner of the world - now and in the future."* Up to the late 1980s, National Income accounting was the primary tool used by development economists to compare human development across the world. This changed in 1990 when the United Nations Development Program (UNDP) introduced a people-centered indicator: the Human Development Index (HDI) - a composite measure not only of economic advances but also improvements in human well-being. Variables include life expectancy, education and standard of living. The concept takes into account the current social, economic and political well-being as well as the basic survival needs and future development potential of communities around the world. The basic variables considered initially were life expectancy, education and standard of living as indicated by the gross national income (GNI) per capita and gross domestic product (GDP). The HDI has become a very powerful measure of human development and a basis for comparing countries on a zero to one scale. In realization that the index did not fully indicate the status of human development in emerging countries, the Human Poverty Index (HPI-1) has been introduced to take into account more variables, in particular, probability of death at birth or not surviving to age 40; adult literacy; the percentage of the population without access to clean water; and the percentage of children who are underweight for their age.

The relationship between energy and human development is bi-directional. Primary energy has become a prerequisite for economic and industrial development, both of which are major propellants for human development. Access to energy is vital for the basic human needs: food and shelter, healthcare, transportation, education, safe water, communication services, social development, etc. Modern energy infrastructure empowers large-scale economic and industrial enterprise development. On the other hand, progress in these areas induces rural-urban migration, and leads to higher labor productivity, rising incomes, off-farm employment, all of which allow people to afford modern homes, appliances and equipment and services such as electricity, natural gas, modern cooking fuel, automobiles, etc. These developments raise the demand for modern energy, and increased energy demand stimulates investment in new technologies for exploration and exploitation of modern energy resources. Adequate access to and affordability of modern energy are particularly crucial at the early stages of development and largely determine the quantum and pace of human development.

1.3 GLOBAL ENERGY TRILEMMA

The global economy is growing and GDP could double within the next two decades, with increased rural-urban migration; high levels of economic growth mean rising living standards and upward class mobility, and global fleet of vehicles will likely increase by around 80% to 1.8 billion by 2040; global population will grow from 7 billion today to around 9 billion; and 1,2 billion will still need access to modern energy. A dynamic growth of secure and sustainable energy is critical to fueling these projected changes in the global dynamics, and promoting universal access to modern energy. Energy demand should grow at roughly the same rate as economic growth but projected efficiency gains in energy production and end-use across the economy, and adoption of energy-efficient technologies will moderate growth in energy demand to around 35% over the period.

The World Energy Council (WEC, 2016d) defines three core dimensions that determine energy sustainability, a triple challenge of ensuring that the world has access to affordable and reliable energy supplies while reducing energy-related anthropogenic emissions to moderate the risk of environmental pollution:

- Energy security
- Energy equity
- Environmental sustainability

The extent to which an individual country can reconcile these often conflicting key dimensions known as 'Energy Trilemma' determines its prosperity, economic growth and human development. The World Energy Council (WEC) identifies three key interconnected policy areas needed to balance the Energy Trilemma: coherent, predictable energy policies that can be sustained across the political cycles, stable regulatory and legal frameworks for long-term investment, and public and private partnerships that stimulate research, development and innovation. WEC ranks 125 countries annually, overall and per dimension in terms of their ability to provide a secure, affordable and environmentally sustainable energy system.

1.3.1 Energy security

Access to adequate energy is a crucial determinant of a country's economic growth and prosperity, and every country has access to several of the primary sources – oil, natural gas, coal, biomass, hydro-energy, wind energy, solar energy, nuclear energy. However, global oil and gas resources which constitute around 60% of the total world's primary energy supply are located in a few countries, but coal is much more widely spread. Around 70% of global reserves of oil and 50% of gas resources are located in the Middle East and North-Central America (Figure 1.1). Most of the oil and gas reserves are located away from countries with the highest demand, and in politically unstable regions. This explains why oil and gas trade have become prominent pawns in geo-politics.

Global fuel reserves by region

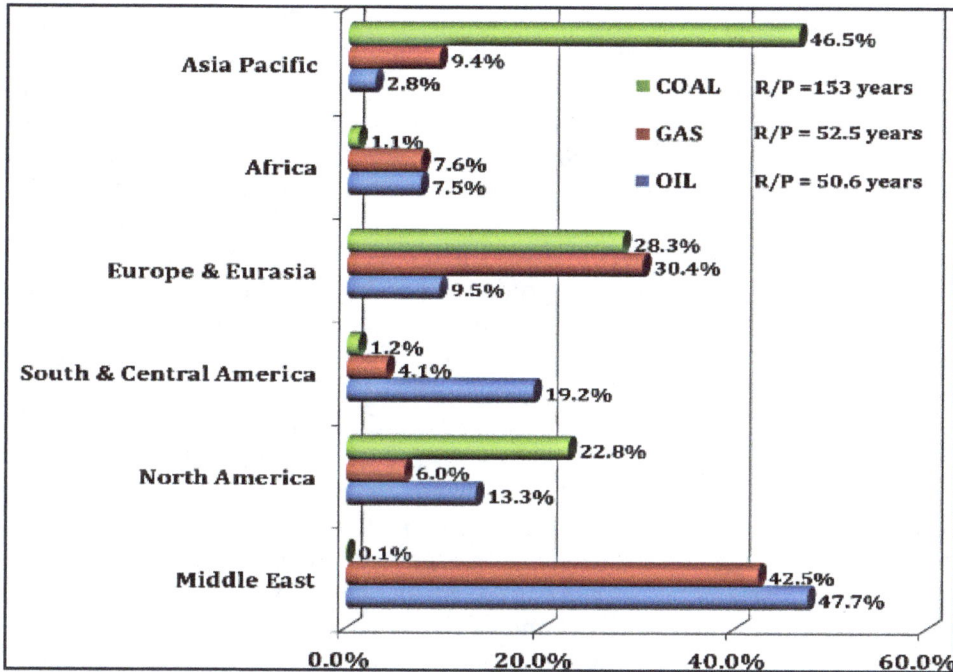

Six countries the hold the largest reserves of fossil

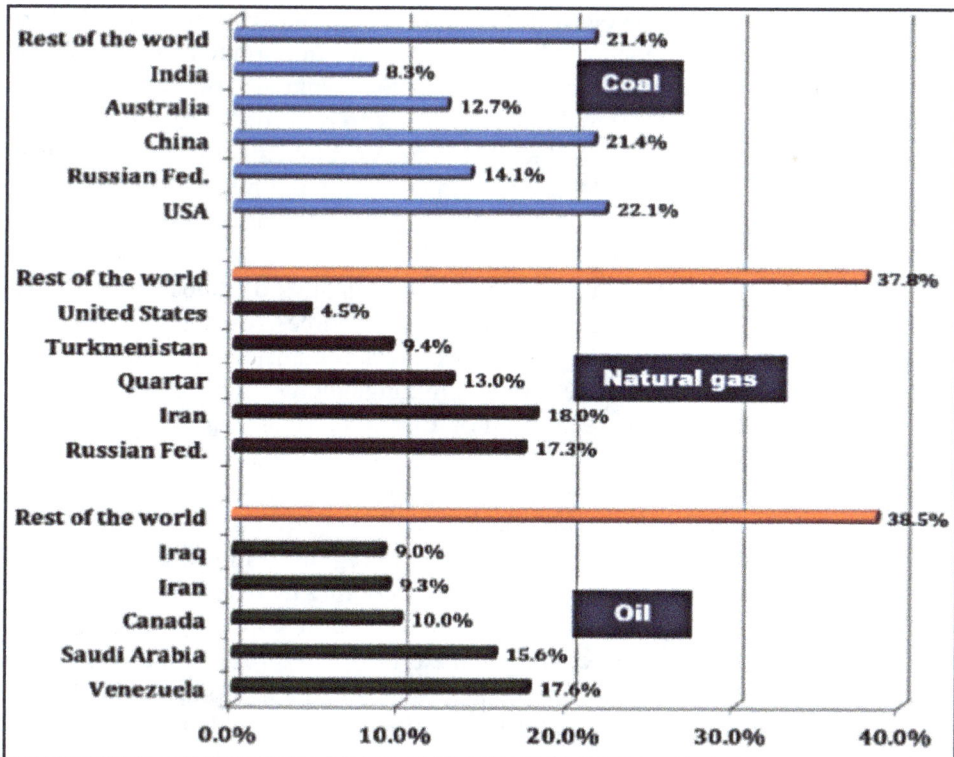

Figure 1.1 Global reserves of fossil fuels by region and main countries, 2016. (R/P is the ratio of reserves to annual production in 2016) *(BP, 2017).*

The global oil crisis of the early 1970s accentuated the need for domestic energy security and most countries have strategic plans for the management of primary energy supply from domestic and external sources, notably the development of domestic energy infrastructure to meet current and future local energy demand, and stockpiling, thereby ensuring substantial insulation from external supply source disruptions. The need for energy security stimulated the development and proliferation of nuclear power which is largely independent of developments in the global oil market. Nuclear plants only need small quantities of fuel to run for years and can be externally sourced while most other inputs are locally available.

The world economy has been growing rapidly over the last decade or so, driven largely by growth in productivity (GDP per person), and the projected rapid growth in the economies of many emerging countries. Energy production and use are becoming more efficient, thereby driving down energy demand per unit of GDP (energy intensity) across all sectors of the economy and all regions of the world. Global population, an important driver of energy demand, has doubled from 3.7 billion in 1970 to 7.4 billion in 2015, with the increase concentrated in Africa, India, Southeast Asia and the Middle East where most of the projected economic and energy demand growth will also occur (Figure 1.2). The latest projections indicate that global population will reach nearly ten billion by 2050 (UN, 2019) and developing countries will account for most of the growth. Continued rapid population growth and the implied rapid growth in energy demand present formidable challenges for sustainable development, particularly in the developing world, and holds important implications for economic and social development and environmental sustainability. Growth of energy-intensive industries, rising urbanization and growth in social status in these regions will likely fuel a rapid growth in energy demand by around a third over the same period.

The inequity in the global distribution of fossil primary energy reserves, coupled with the strategic role of energy in development have made the security of primary energy sources a leading priority for all nations of the world, and an important factor in global oil market dynamics. Emerging countries that hold most of the resources seek to maximize income from sale of crude oil in the international market, but spend most of the earnings importing finished products, including refined oil products from the developed countries. The developed countries also use strategic reserves to control global oil pricing by holding large stocks when oil is cheap and releasing to the international market as appropriate to force down prices. Furthermore, rapid deployments of advanced oil prospecting and production technologies particularly in North America are changing the dynamics of global oil and gas production. While only a few countries have oil and gas resources, over a hundred countries have substantial coal reserves and many more resources remain unexplored. For these countries, coal has become the main resource for modernizing and proliferating internal modern energy supply, and ensuring energy security. For example, Asia has been leading the world's economic growth for two decades and the rate is expected to be faster over the next two decades, yet the region has very limited oil and gas resources, but holds a substantial proportion of the global coal resources. It is inevitable therefore that, for strategic and security reasons, coal will remain the primary fuel of the region in the foreseeable future, although solar and nuclear power deployments are also becoming increasingly significant. It should be noted also that the economies of many countries across all regions of the world depend critically on the production, processing and export of fossil fuels: the United States, Russia, Australia, China, Indonesia, Norway, Poland, the United Kingdom, and many countries in the Middle East and Africa. It is difficult to imagine that any of these countries will ever shut down production because of the environment. In fact, most of the countries are expanding their fossil fuel infrastructures internally and across country borders.

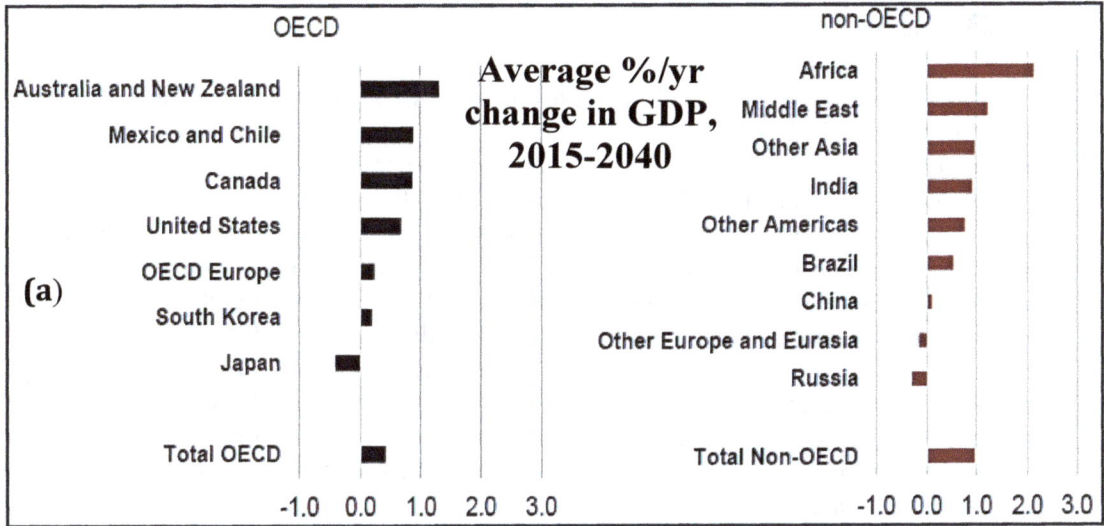

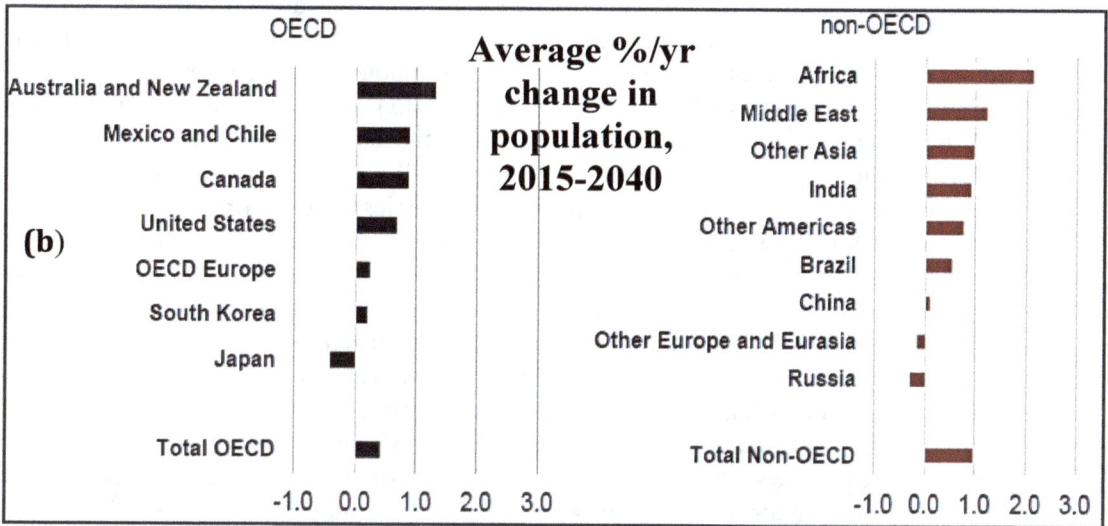

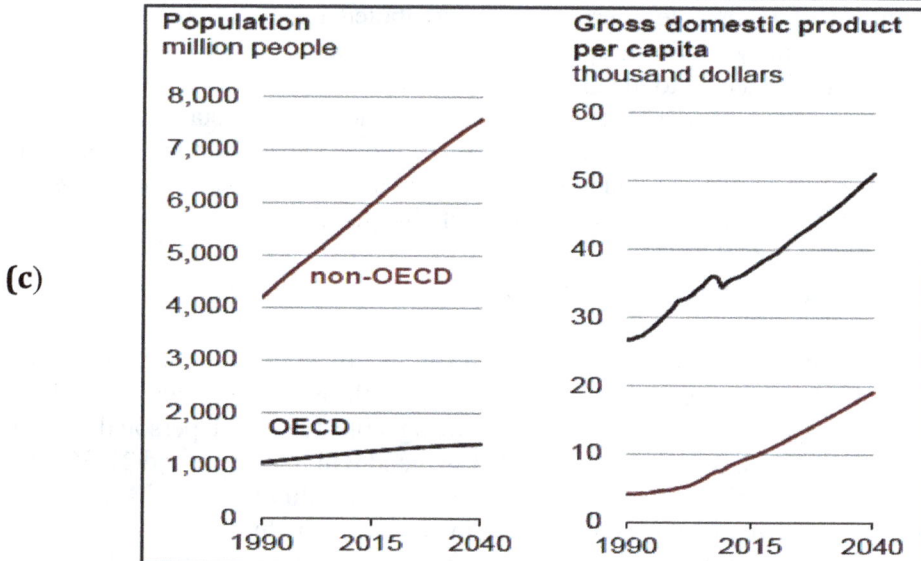

Figure 1.2   Projected global economic variables by region, 2015-2040 (a) economic growth (b) population growth (c) population and GDP per capita *(EIA 2017).*

1.3.2 Energy equity

Over 80 % of the global population resides in non-OECD countries and the growth rate over the next twenty years is expected to be more than twice the rate of the total OECD population (EIA, 2017). Strong economic growth, increased access to energy, and rapidly growing populations and urbanization lead to rising demand for energy. Technology-enabled urban lifestyles demand more electricity and transportation. The growth of the middle class, rising incomes, and more electricity-enabled appliances and machines contribute to global electricity demand, expected to double by 2060 (WEC, 2016b). One in five people globally lack access to electricity currently, but there are wide regional disparities, with the largest concentration in the non-OECD world. Many countries in the emerging world, especially in sub-Saharan Africa, face enormous pressure to increase energy production at a pace that exceeds the local population's growth to meet demand and provide modern energy for the estimated 1.1 billion people who currently do not have access to modern energy services, but this has been a formidable challenge.

Energy equity (access to modern energy by all) becomes sustainable, only if it is affordable, powered by rapid economic growth, urbanization, and funding capacity. Energy consumption in non-OECD countries is expected to increase by about 41% between 2015 and 2040 in contrast to a 9% increase in OECD countries (EIA, 2017). Around 1 billion vehicles including 229 million commercial vehicles will be added to the global vehicle population, mostly in emerging nations, and growth in industrialization will be driven largely by energy-intensive primary industries in these regions. For many countries in the group, providing adequate modern energy for these projected developments are very formidable goals. Many countries with low gross domestic product (GDP) and low rankings on the energy equity dimension are in a dilemma on how to promote productive access, affordability and at the same time create a conducive environment for investment in energy infrastructure. Some countries try to do this by introducing energy subsidies but this has not been very successful. While they may be beneficial in the short-term, subsidies can be costly to deploy, difficult to manage, contentious to remove, and most often do not benefit the poor for whom it is intended. Furthermore, they do not stimulate private sector investment in the long-term, and infrastructure provided mainly by the public sector tends to be unprofitable and poorly maintained. Possible solutions include the development of renewable energy-fueled distributed generation, stand-alone units for households and businesses, mini-grid/off-grid systems to serve rural communities that cannot be cost-effectively connected to the grid, and introduction of innovative policies and business models that can attract private sector investment. Deployment of stand-alone solar energy systems has great potential since many of the countries are located in areas of high and sustained solar energy intensity, and upgrades to more advanced systems that serve communities, institutions, commerce and industry are reachable goals.

1.3.3 Environmental sustainability

The projected dynamic growth of the economies and population largely in the emerging nations over the next two decades will stimulate growth in urbanization, rising living standards across all regions, with more people being able to afford personal transport and appliances which require energy. Global energy demand will rise by around 25-35% and fossil fuel use will remain dominant, accounting for nearly 80% in 2040. (ExxonMobil, 2018). Most of the industrial growth in the emerging nations will be in the energy-intensive primary industries - iron and steel, cement, non-ferrous metals, chemicals. In effect, delivering on the projected increase in energy demand in ways that mitigate the risk of environmental pollution will be a major challenge for many countries over the next two decades or so, especially when choice of potential primary energy sources is limited as is the case in many developing countries. Feasible options include continuing shift to less carbon intensive power generation whenever possible, and increased energy efficiency in every sector of the economy. Effective proliferation of these options across regions could stop the rise in CO_2 emissions by the 2030s despite increasing energy use. The major challenge is the fact that most of the countries in the emerging economies that will account for most of the future growth of the global population, economy, and energy demand, do not have much choice in terms of energy mix, neither can they afford to pay for high-efficiency technologies.

1.3.4 Global energy trilemma ranking

The World Energy Council (WEC) rates countries around the world annually on the basis of performance, based on the three core dimensions of the energy trilemma. Ratings (A – C) are awarded on performance in each of the three cores, as well as overall scores (between 1 and 10). About 125 countries are then ranked to reflect the extent to which each country has succeeded in reconciling the triple challenge of energy security, equity and sustainability, a critical goal that promotes transition to a more sustainable, environmentally-sensitive energy future (Figure 1.3). The European Union led the world in 2019 as it had done for the previous five years, claiming nine of the top ten positions in the overall rating and five of the top ten with triple-A scores. On the other hand, only five countries in the bottom twenty are not in Africa. The key challenge faced by sub-Saharan Africa is modern energy access, despite significant energy resources and renewables' potential. Around 65% of the total population still lack access to electricity and most still depend on traditional bio-energy for domestic needs. Most of the countries are unable create a conducive business environment that could attract investment, or build the institutional capacity needed to unlock their resource potentials. Furthermore, the fast rate of population increase severely limits the impact of investments in development infrastructure

1.4 ENERGY AND SUSTAINABLE DEVELOPMENT GOALS

Energy has long been recognized as the prime mover of sustainable human development, but its inclusion in the United Nations Sustainable Development Goals (SDGs) adopted by the 193-member countries in 2015 has elevated this vital resource to the forefront of development goals. It was a clear recognition of the critical role of modern energy as the prime mover of all the other human development goals - economic growth, health, education, etc. The document sets a target to ensure access to affordable, reliable, sustainable and modern energy for all by 2030 (IEA 2017b). The provision of secure, affordable and modern energy for all citizens is central to the set development goals which focus on poverty reduction, improvements in health and education, and gender equality. Economic and social development tend to go hand-in-hand with energy sector transformation. Urbanization, a major consequence of economic growth, rapidly transforms the traditional agrarian economy into industrialization and a knowledge-based economy. The shift from traditional to industrial economy requires modern energy to thrive, while also providing the incentive for investment in modern energy infrastructure. Not many nations have succeeded in growing their economies and human development without the dual strategy of modernizing the economy and developing a sustainable modern energy infrastructure.

TOP TEN COUNTRIES

1. Switzerland
2. Sweden
3. Denmark
4. United Kingdom
5. Finland
6 France
7. Austria
8. Luxemburg
9. Germany
10. New Zealand

ENERGY
SECURITY

ENVIRONMENTAL
SUSTAINABILITY

ENERGY
EQUITY

NORTH AMERICA

EUROPE

LATIN AMERICA
AND CARIBBEAN

ASIA

MIDDLE EAST AND
GULF STATES

AFRICA

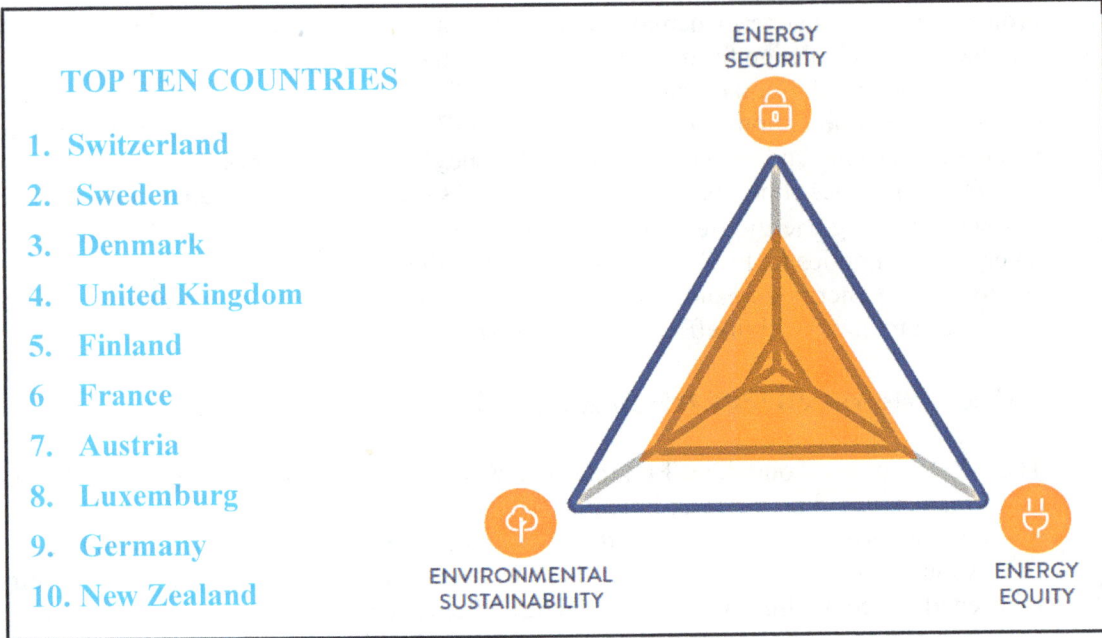

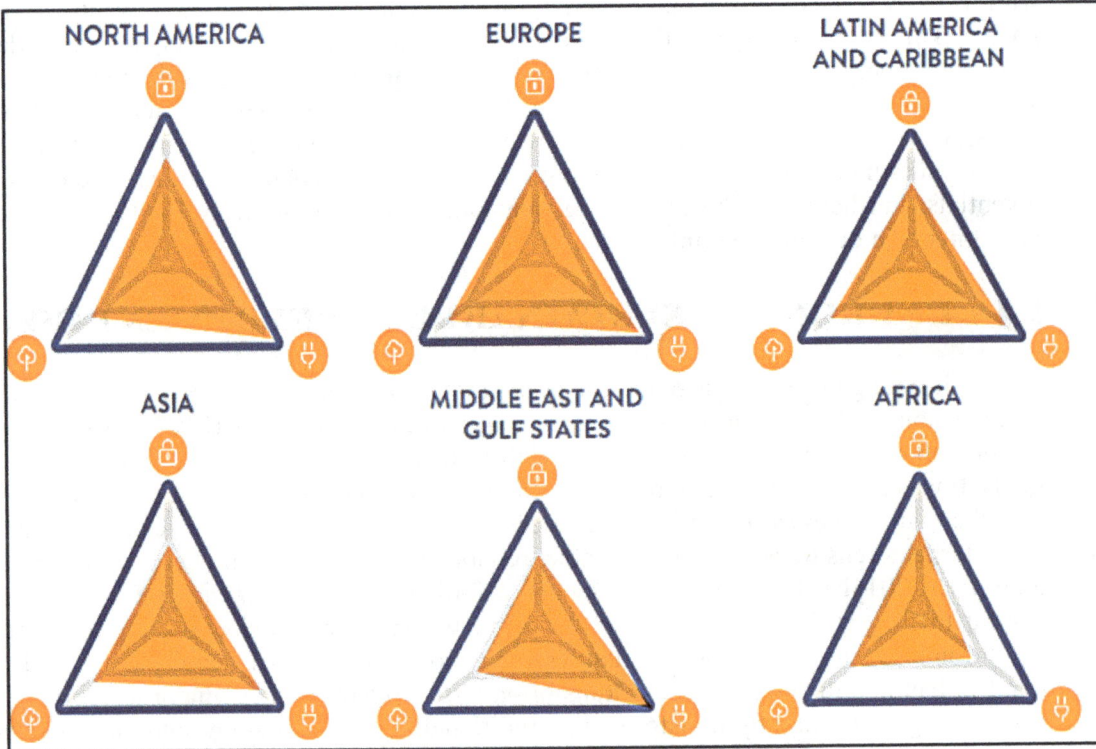

Figure 1.3   Global Energy Trilemma ratings, 2019 *(WEC, 2019).*

1.4.1   Energy inequity

There are wide variations in energy consumption between developed and developing countries, and between the rich and poor within countries, consistent with variations in human development. In fact, per capita modern energy consumption has become a major variable for assessing the level of human development in countries of the world (Figure 1.4). The USA per capita primary energy consumption is the highest in the world, ten times the value for Africa and non-OECD Asia, and nearly four times the world average. The disparity is even greater for electricity - the average American consumes twenty times more electricity than the average African and over four times the world average. While the country's share of the world's population is only 4.6%, it accounts for 24% of the world's energy consumption and over 30% of GDP. On the contrary, the least developed countries (LDCs) with 10% of the world's population account for about 1% of energy consumption and a mere 1% of the world's GDP. This demonstrates to a large extent the symbiotic relationship between energy use and human development. USA is also the world's largest economy and ranks among the top ten countries with the highest human development, while twenty of the lowest thirty in the 2016 United Nations Development Program (UNDP) 2017 HDI table are in sub-Saharan Africa. It is also interesting to note that the first (Norway) and six of the ten in the top group of countries with 'very high human development' are members of EU-28.

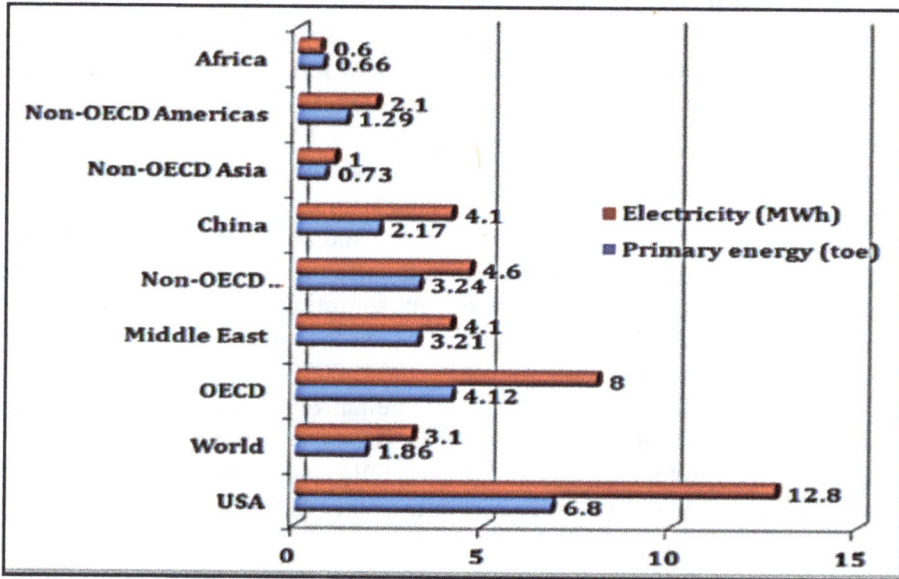

Figure 1.4 Per capita primary and electrical energy consumption in 2015 by region (Data from) (toe = metric tons of oil equivalent; MWh = megawatt.hour) *IEA, 2017b).*

Around 3 billion people worldwide still live on basic bioenergy - wood, farm waste, charcoal etc. Without transition to modern fuels and energy systems, prospects for human and economic development are severely limited and occupation remains largely agrarian. Even if there is access to modern energy, most people who remain in rural areas cannot afford it. In effect, access to modern energy must stimulate the development of modern infrastructure that in turn grows large-scale economic enterprises, promotes urbanization and raises the earning power of the people. This development transforms 'access' to 'consumption' and further stimulates demand for energy and development of additional modern energy infrastructure.

The United Nations Human Development Index (HDI) establishes the relationship among energy use, economic growth and social development but the synergy between energy use and human development is not fully taken into account, in particular, the absolute amount of energy used per capita and the share of modern energy services (especially electricity), both of which are crucial in early stages of development. Clearly, human development and energy access are closely linked and the substantial gains in living standards, industrial and technology development of the last two centuries were made possible by the rapid expansion of access to affordable energy in sufficient quantities that helped to stimulate productivity. Since fossil fuels currently supply over 80% of global energy and all projections indicate no significant change over the next two or three decades, the major challenge will be the development and proliferation of strategies and competence to mitigate the negative environmental impact of fossil energy use.

1.4.2 Energy poverty

Access to basic energy is a prerequisite for human development and is fundamental to fulfilling economic growth and basic social needs - food production, health, safe water, transportation communication, education. Per capita energy consumption has been used widely in evaluating the stage and pace of human development, and is an important variable in the determination of the UNDP HDI Index. Results showing trends for around 150 countries over many years are published annually by the UNDP. There is a very strong correlation between per capita energy consumption and the Human Development Index for all countries (Figure 1.5). The world's average energy consumption is about 1.9 metric tons of oil equivalent per person per year and lower values are regarded as energy poverty. In effect, all Africa and non-OECD Asia are energy poor (Figure 1.4). Very few countries with per capita energy use of less than 2 metric tons of oil equivalent have a HDI score of more than 0.7. Per capita consumption of energy has been used widely for identifying rich and poor countries but has been found to be inadequate for developing countries because the use of traditional bioenergy, especially wood and charcoal which often dominates energy consumption, particularly in countries of Asia and Africa. This has prompted the development of another measure: Energy Development Index (EDI) by the World Energy Council (WEC, 2004; 2010). The EDI is a composite measure of energy use in developing countries and of their progression in the use of modern energy services. The WEC analysis established a strong relationship between modern energy consumption (EDI) and human development (HDI), and access to primary modern energy has become a major variable in classifying countries into rich/poor, developed/developing groups (Figure 1.6).

Three key indicators of energy use were identified for placing developing countries on the human development curve:

- Per capita energy consumption

- The share of modern energy services in total energy use

- The share of the population with access to electricity in their homes

The United Nations classify people living on less than $2 a day as poor and application of this standard to developing countries establishes a strong relationship between per capita energy use and poverty level (Figure 1.5). Per capita energy demand in the richest countries within the group where less than 5% of the population is classified as poor is ten times higher than in the poorest countries where more than 75% are poor.

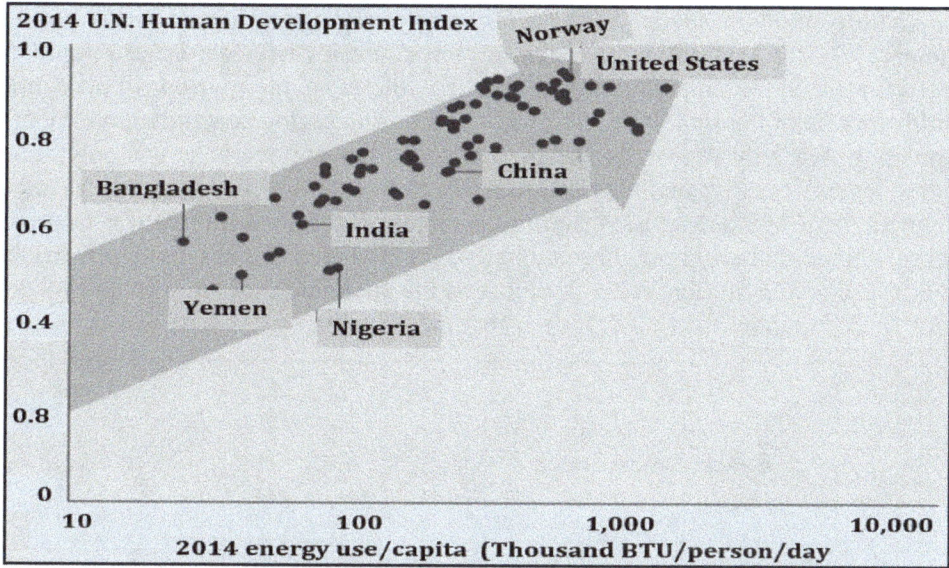

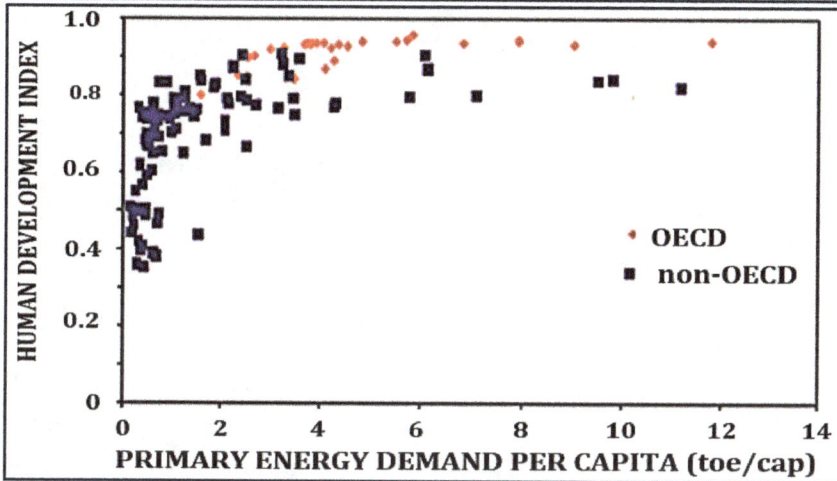

Figure 1.5 Human Development Index (HDI) and energy demand per capita *(IEA/UNDP (2004; ExxonMobil 2017)*.

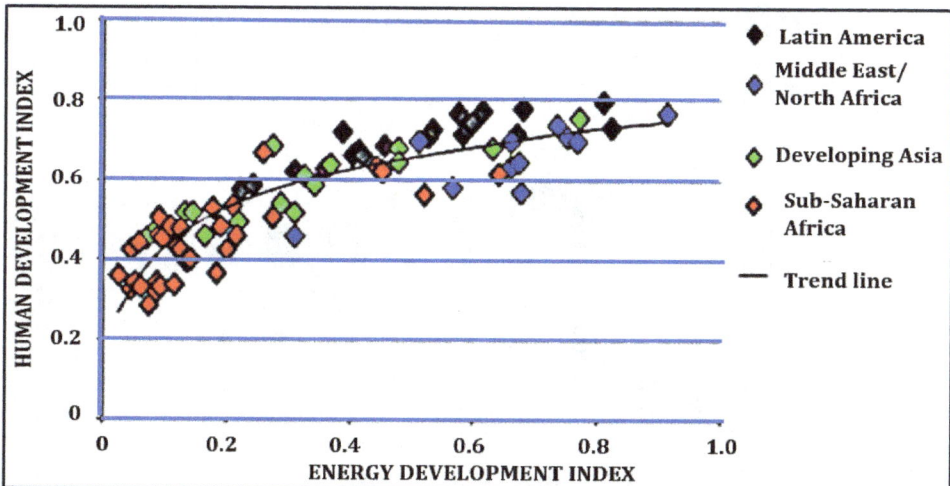

Figure 1.6 Comparison between the new Energy Development Index and the Human Development Index in 2010 *(OECD/IEA , 2012)*.

The share of modern energy in the total energy use is also lowest in the poorest countries (Figures 1.7 & 1.8). Furthermore, the top ten countries with the largest population without electricity are in the emerging world. Africa is the most energy-poor of all continents in the world, yet one of the richest in energy resources. The region accounts for 13% of the world's population, but only 4% of its energy demand. Energy poverty is not only about access to modern energy but also includes affordability and reliability, and is not restricted to emerging countries alone but also an issue across developed economies where there is universal energy access. Although there is no internationally accepted measure of energy poverty in developed economies, one definition in common use is the spending measure (the proportion of households spending more than 10% of their income on energy).

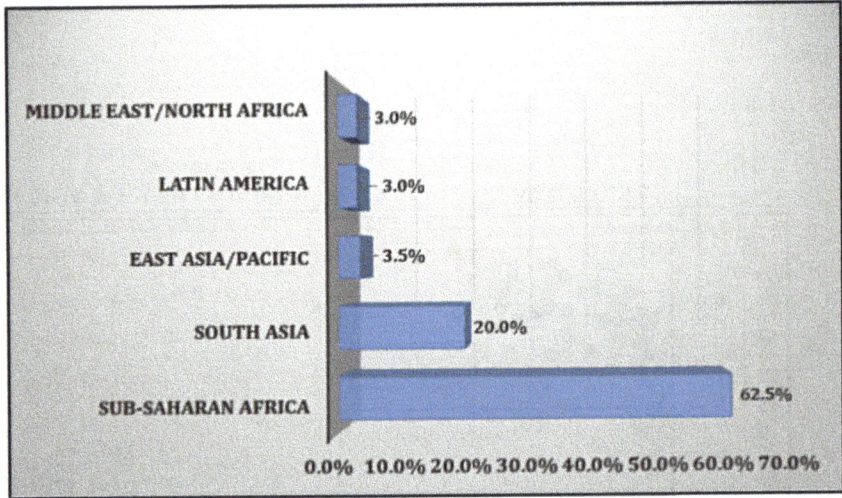

Figure 1.7   Regions of the world with least access to electricity (World Bank, 2017; *(OECD/IEA, 2012)*.

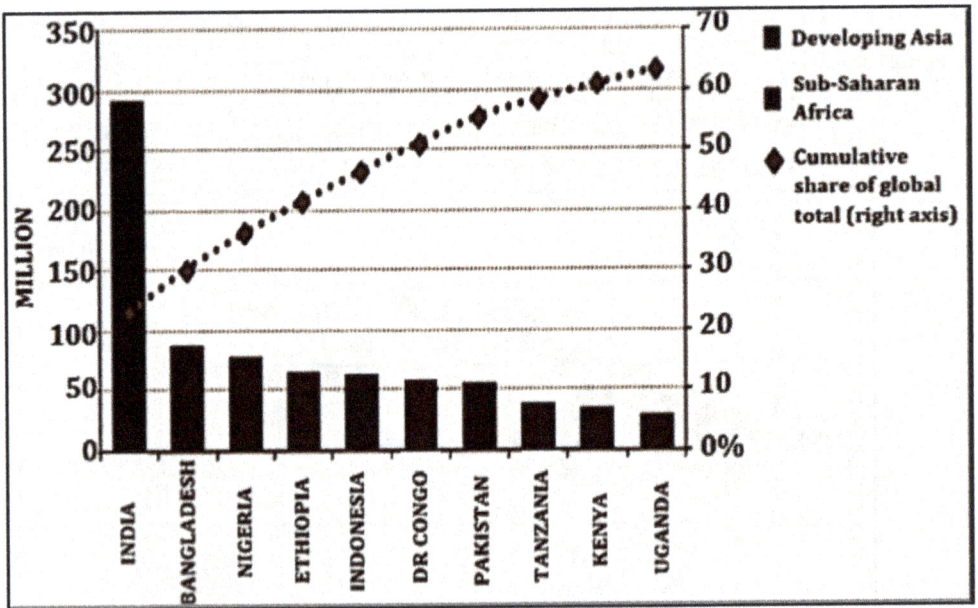

Figure 1.8   Top ten countries with the largest population without access to electricity in 2010 (World Bank, 2017; *(OECD/IEA, 2012)*.

A recent study by the International Energy Agency across developed countries estimated that about 200 million people, over 15% of the total population in developed economies suffer from energy poverty (Figure 1.9), (IEA 2017b). The study covered OECD countries for which fairly reliable data was available, and focused on electricity and energy used in homes (for heating, cooling, cooking, etc., but excluding mobility). It is interesting to note that most countries in the developed world have no data on energy poverty or any coherent policy for addressing the issue, probably because energy utilities are largely private enterprises and pricing is controlled by competition.

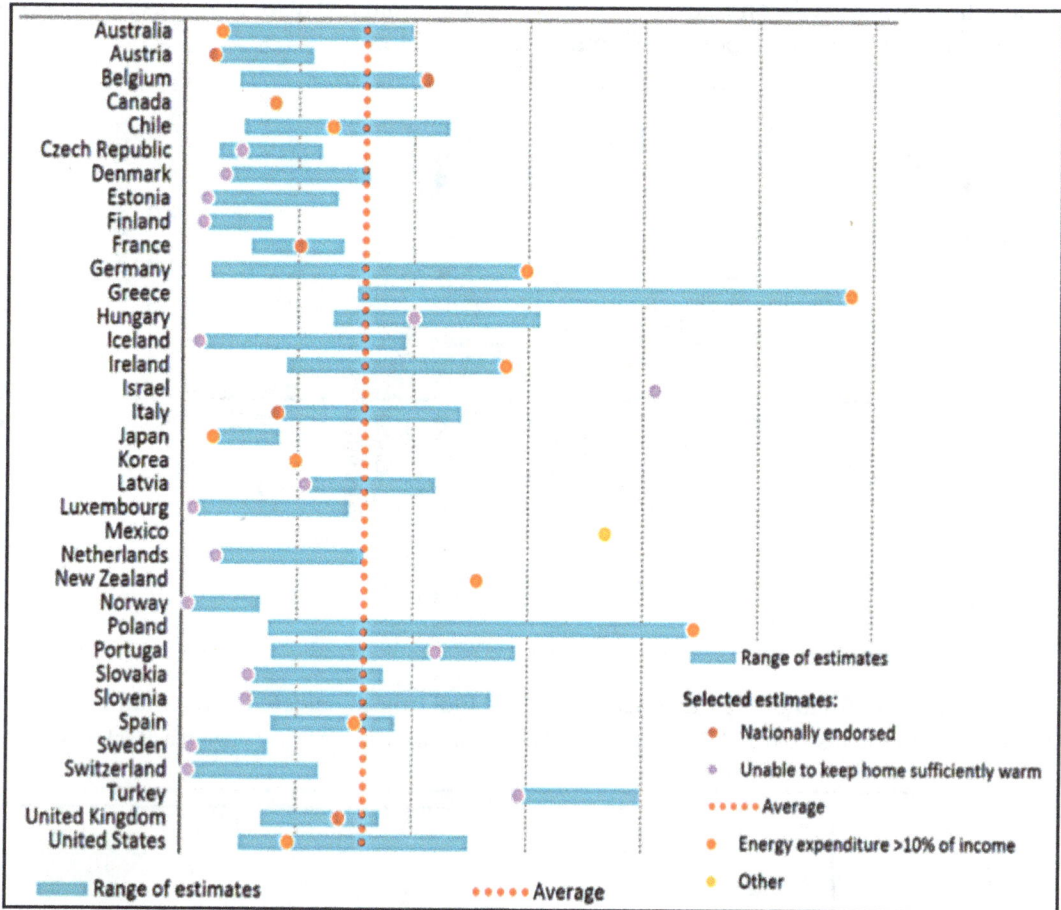

Figure 1.9 Share of population living in energy poverty in developed economies in 2015. *(IEA 2017b).*

1.4.3 Between access and consumption

The International Energy Agency defines energy access as *"a household having reliable and affordable access to both clean cooking facilities and to electricity, which is enough to supply a basic bundle of energy services initially, and then an increasing level of electricity over time to reach the regional average"* (IEA 2017b). Clean energy means modern fuels and technologies, including electricity, natural gas, liquefied petroleum gas (LPG), improved cookstoves, etc. This definition is now being universally adopted as the benchmark to measure progress of any country towards the sustainable energy goals.

Most studies on energy and development assume that access to modern energy stimulates human development. However, there is a growing body of evidence that it does not, unless it first propels economic and industrial development and provides employment opportunities which empower affordability of the modern equipment (refrigerator, air conditioner, car, etc.) that consume energy. In effect, urbanization and industrialization are preconditions for moving labor from rural agrarian subsistence economies to high-income wage economy that allows people to move to higher income levels and acquire appliances that use modern energy. It is important also to distinguish between traditional and modern energy. Most poor and developing countries rely heavily on traditional energy (wood, biomass) also known as 'wood economy' which is consistent with poverty (Figures 1.10 & 1.11).

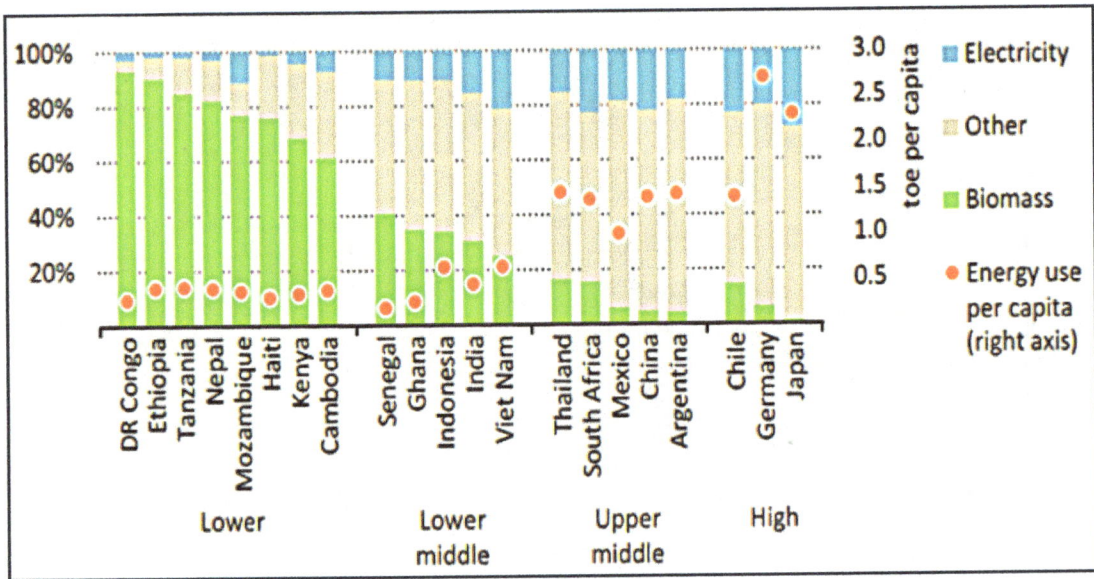

Figure 1.10   Final energy use per capita and fuel mix in selected low, middle and high-income countries, 2015. Notes: 'Other' includes coal, oil, natural gas and other renewables; toe = metric ton of oil equivalent. *(EIA 2017).*

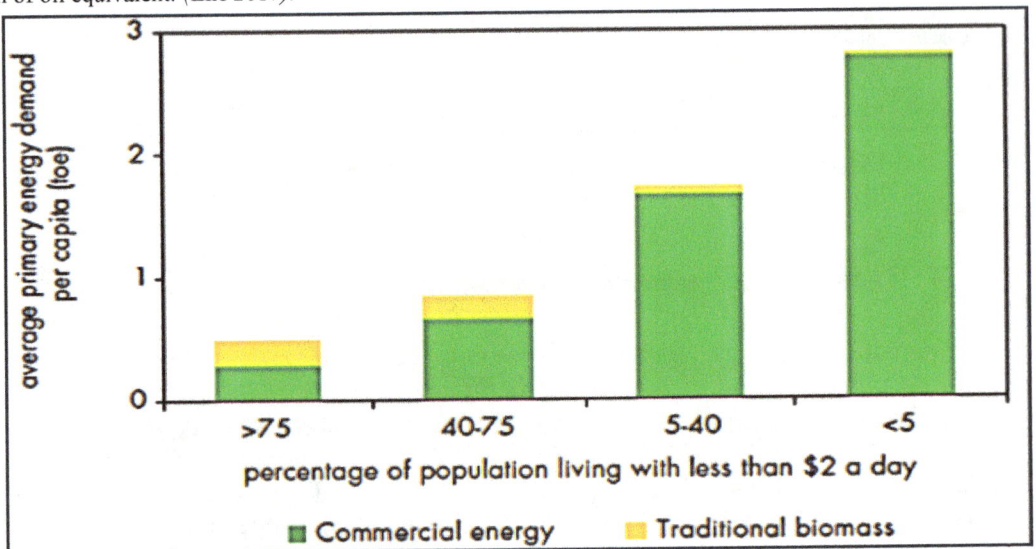

Figure 1.11   Average primary energy demand per capita and population living on less than $2 a day, *(IEA; World Bank, 2004).*

For countries that still rely heavily on wood economy, the chances of sustainable economic and industrial growth beyond subsistence level are remote. In fact, no country has ever moved from poverty to affluence without transiting from wood to modern energy (Breakthough Institute, 2016). In effect, access to modern energy is a prerequisite for sustainable economic and industrial growth which in turn increases the earning power of people and promotes modern energy consumption. Without this development, human development remains low. About three billion people (around 40% of the global population) mostly in Asia and Sub-Saharan Africa still live in the wood economy and the negative effects on all aspects of human development are enormous.

The relationship between rising incomes and rising energy modern consumption is bi-directional (Bhata and Angelou, 2014). Modern energy infrastructure enables large-scale economic enterprise that creates opportunities for high incomes while rising incomes allow people to afford modern energy, resulting in higher demand and consumption of energy. Increase in modern energy consumption stimulates further growth in energy infrastructure that fuels economic growth. It is on record that, in many countries currently classified as energy poor, a significant proportion of the population do not consume energy not because they have no access but because they cannot afford it. In summary, energy access only contributes to human development in so far as it supports economic and industrial development that could create high-income employment opportunities that enable more people to afford modern energy. In effect, development policies should focus on providing modern energy for productive enterprise development which in turn stimulates investment in large-scale, cost-effective power generation and distribution facilities. (Pachauri et al, 2012).

1.4.4 Between access, availability and productivity

Universal access to modern energy is a prime mover of economic and human development, but only if it is put to productive uses that create goods, services and wealth. If available energy is only sufficient for lighting and charging cell phones, it is unlikely to stimulate any economic growth, or change the social status of any household, nor will unreliable supply or energy that is too expensive to be affordable by a significant proportion of the population. It is important therefore that there must be sufficient and reliable modern energy that will be available when needed and can be used in income generation activities such as agriculture, industry, mining, and commercial activity, as well as for educational and health services. In general, as the economy shifts from subsistence farming and trade to more productive modern enterprises, demand for energy rises and a larger share is devoted to productive activities (Figure 1.12).

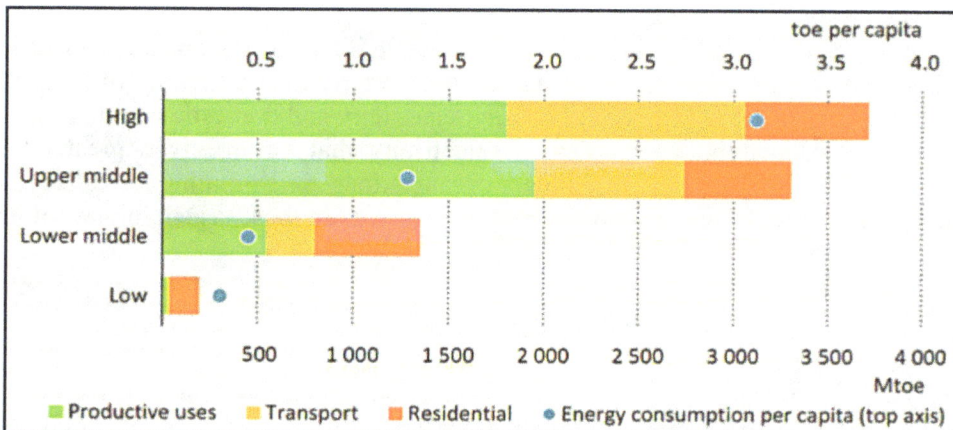

Figure 1.12 Total final energy consumption by income group, 2016 *(IEA, 2017b).*

Clearly, the proportion of total final energy that is used for productive activities is a strong indication of the extent of the progress of a country towards modernizing and growing its economy. Many countries in the developing world which have substantial modern energy installed capacity, cannot deliver power much of the time because of system synchronization and maintenance issues. This is particularly common in sub-Saharan Africa where most major utilities are poorly-maintained public enterprises. It should be noted therefore that sustainable modern energy availability in itself is not sufficient to stimulate economic growth and industrial development, it must be accompanied by the development of local capability to operate and maintain complex modern equipment, which in turn requires development of educational institutions, technological and financial knowledge capacity, and modern infrastructure. Strong and sustainable policy instruments that survive political cycles are also necessary in order facilitate access to finance and stimulate enterprise development.

1.4.5 Access to productive modern energy

Throughout modern history, every country that succeeded in growing its economy to an advanced state, has required secure access to modern energy. Growing economy and prosperity requires modern energy to provide all-round improvements in the quality of life, and sustain services such as lighting, heating, transport, quality healthcare and education. Small solar power projects may meet many of the basic needs such as lighting and charging cell phones, but there must be an avenue for growth over time to a level which can support productive enterprises, otherwise the beneficiaries will remain energy poor. While there is no internationally accepted definition of *'modern energy access'*, the International Energy Agency has identified four main indicators, all of which are crucial to economic and social development (IEA, 2014):

- Household access to a minimum level of electricity.

- Household access to safer and more sustainable (i.e. minimum harmful effects on health and the environment if possible) cooking and heating fuels and stoves.

- Access to modern energy that enables productive economic activity, e.g. mechanical and power for agriculture, textile and other industries.

- Access to modern energy for public services, e.g. electricity for health facilities, schools and street lighting.

The above elements of 'access' are only sustainable if they are technically available, adequate, reliable, convenient, safe, and affordable (often referred to as *'quality of access'*). The IEA definition emphasizes two types of energy: electricity and domestic cooking/heating/lighting energy. Unfortunately, it is access to electricity that has received greater national and international attention, but development organizations are beginning to realize that access to modern domestic energy for cooking and heating is even more urgent in view of its enormous negative impact on household health.

1.4.5.1 *Access to electricity*

Of all modern energy types, electricity access is the most prominent as an explicit objective of national development strategies, because it is a critical enabler of development, especially at the household and community levels. Furthermore every industry and commercial enterprise depends on availability of adequate and reliable power supply to function. Virtually all aspects of the United Nations human development goals are only achievable if adequate electricity is available, accessible and affordable: economic prosperity, health, education. A large set of activities that can improve human welfare and development depend on access to adequate and sustainable electric power supply. It is important also to distinguish between the needs of urban and rural areas. While many people living in urban areas have the economic means to pay for electricity, most in rural areas cannot afford it, even if it is available, hence electrification models for rural areas need strong policy support in order to be able to provide affordable electricity in sufficient quantities that can support the growth of small-scale enterprises such as mechanized agro-processing, which in turn will raise the economic status of rural areas. The International Energy Agency estimates that 1.1 billion people globally, were without access to electricity in 2016 and many more suffered from poor and irregular supply (IEA 2017b). More than 95% of those living without electricity were in developing countries in Asia and sub-Saharan Africa (Figure 1.13a). Asia has made significant progress in the last two decades, moving access from 69% in 2000 to 89% in 2016 and accounting for three-quarters of the growth in access since 2011. North Africa has also made major progress, with all countries in the sub-region having nearly 100% access both in urban and rural areas. However, access growth in sub-Saharan Africa has been slow and uneven. Population growth has continued to outpace power supply efforts and electrification rate in the region in 2016 was only 43%. However, there were significant variations between countries: while access in South Africa was 86%, only 12% of Liberians had access to electricity (Figure 1.13b).

Around 80% of those lacking access to electricity across sub-Saharan Africa were in rural areas but, again, there were wide variations: about 83% of rural South Africa had access compared with less than 1% in several other countries in the region. The total installed capacity in sub-Saharan Africa is around 90 GW, half of which is in South Africa. Off-grid and mini-grid systems (like in Ghana and Rwanda) provide most of the rural electricity while centralized grid electricity caters more for the urban areas. Power supply to the minority that have a grid connection is often inadequate and unreliable. Energy transmission and distribution infrastructures are old, inefficient and poorly maintained, leading to frequent power outages, lack of synchrony between generation, transmission and distribution, and power transmission/supply losses of 10-45%. This has fueled the widespread adoption of highly polluting private power generation, running on gasoline or diesel. Despite the fact that grid power is much cheaper, most commercial operations in many countries of the region rely on private power, with grid power as back-up. Although the contribution of the private power sector (Selfpower providers SPP) is not included in international energy supply calculations, the capacity could be comparable to the national grid supply. In fact, in a country like Nigeria, the SPP output is estimated to be at least twice as large as the national grid supply.

Economic and development growth worldwide is increasingly being powered by electricity, which currently accounts for around 20% of end-use energy and has become the fastest growing end-use power. Global electricity demand is projected to rise by 60% by 2040, accounting for 55% of the world's total energy demand growth, and demand by non-OECD countries is expected to double (ExxonMobil, 2018, IEA, 2018). The fastest growth will be in the residential and commercial end-use sector, rising by 70%, while growth in the industrial sector will be lower, around 50%. Although demand by the transport sector will more than double during the period, due largely to the expected growth in the electric vehicle market, it will account for only 2% of the total end-use electricity demand. The number of people who have access to electricity is not representative of those who have it available when they need it. Installed generating capacity is often not indicative of available electric power because poor management, maintenance and high energy losses can make a significant difference between generating capacity and useful power available in real time to consumers. In many developing countries, up to 50% of installed capacity is often idle due to inefficient operating strategies and poor maintenance culture.

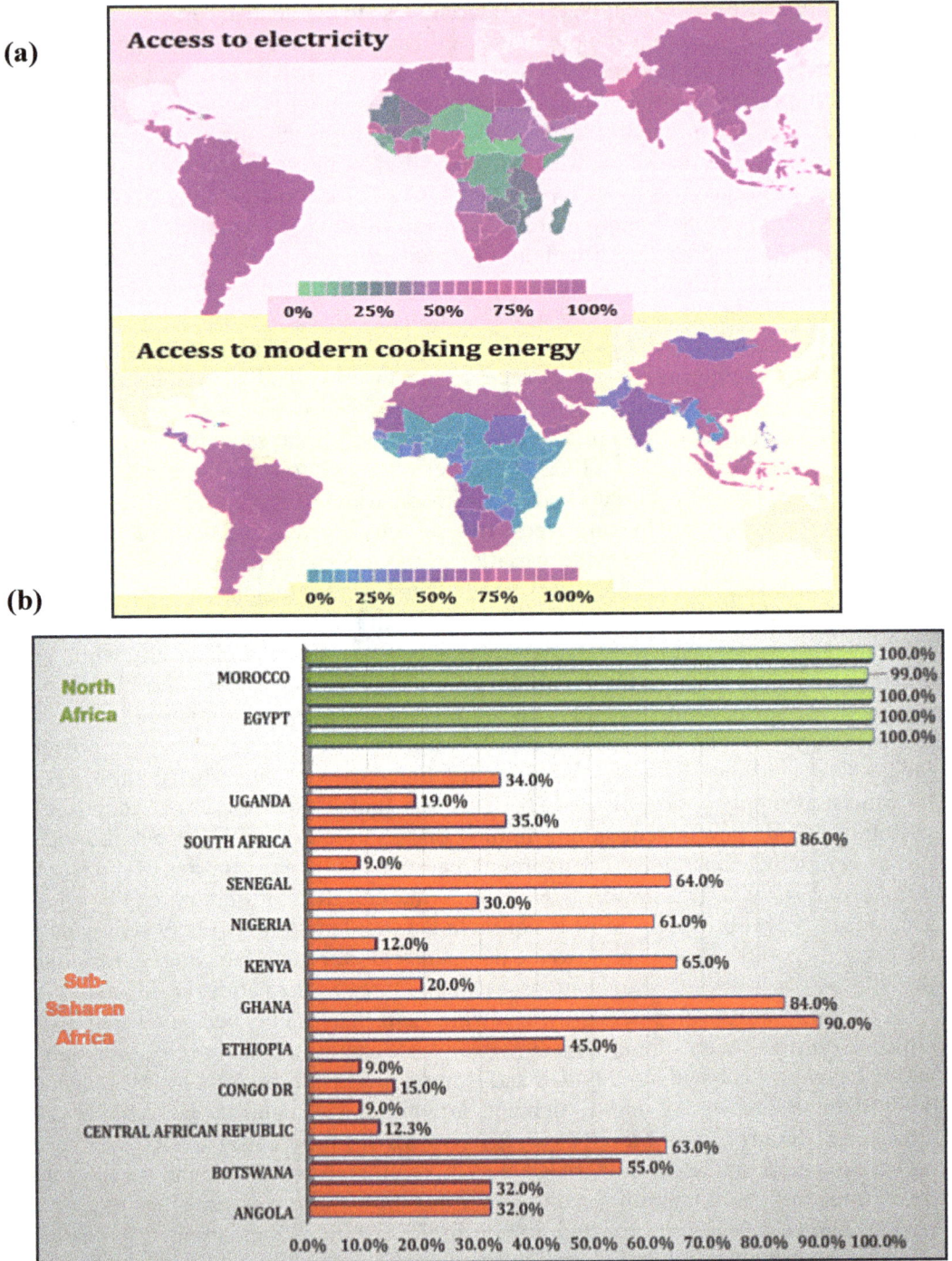

Figure 1.13 (a) Countries with poor access to electricity and clean cooking energy in 2016 (b) Access to electricity by some African countries in 2016 *(IEA 2017b)*

1.4.5.2 *Access to modern cooking energy*

Nearly 3 billion (38% of the global population), and 50% of the population in developing countries, lack access to clean cooking energy (Figure 1.13a). They rely on wood, charcoal and agro-waste burned in traditional stoves in poorly ventilated space. In many sub-Saharan countries, more than 90% of households rely on traditional energy and just five countries - Nigeria, Ethiopia, DR Congo, Tanzania and Kenya - account for around half of the population in the region using solid biomass for cooking (Figure 1.14). However, this does tend to obscure the fact that, in most of the other countries in the region, more than half, and in many, above 90% of the population, rely on solid biomass for cooking. Also, according to the World Health Organization (WHO), around three-quarters of those dependent on solid biomass for cooking live in rural areas. The toxic environment created in households by this cooking method has been linked to around 2.8 million premature deaths annually. Although significant progress has been made in Asia in the last two decades promoting the use of modern cooking energy, in particular, liquefied petroleum gas (LPG), the average global number without access has not changed significantly over the period, due largely to rapid population growth outstripping progress in many countries in the emerging world.

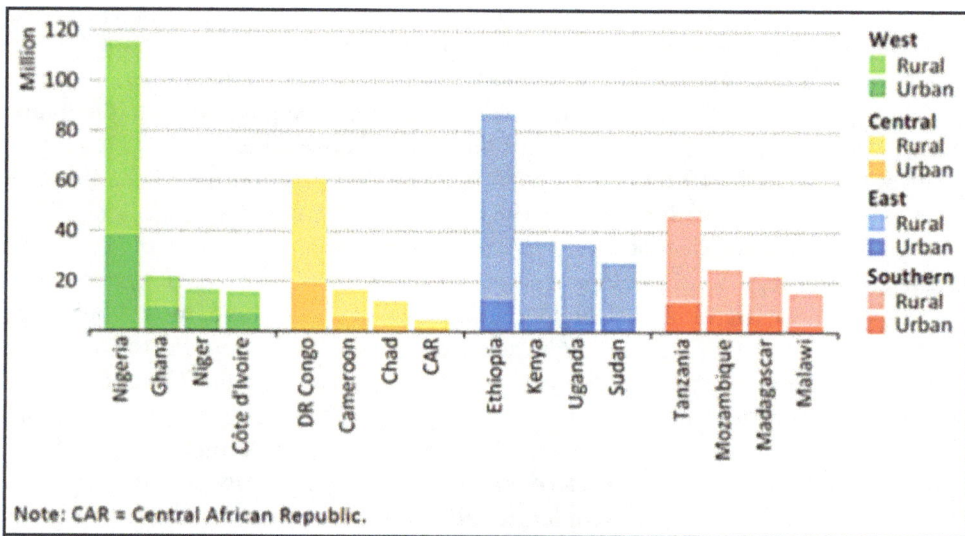

Figure 1.14   Largest populations relying on the traditional use of biomass for cooking in sub-Saharan Africa by sub-region, 2012 *(IEA, 2014).*

1.5   MITIGATING ENERGY POVERTY AND LOW ACCESS TO MODERN ENERGY

Modern energy is one of mankind's most important assets and access is fundamental to good standards of living and human development. Many fundamental forces shape energy needs and access in nations around the world but the most important are population growth, urbanization, economic growth, and demographic shifts. Non-OECD countries host seven-eighths of the world's population and will account for much of the future growth, yet average per capita access to energy is only 2.8 toe compared with 8 toe for OECD countries. Comparative values for access to electricity are 1.9 and 4.4 MWh respectively (Figure 1.4).

Economic, industrial and human development are all energy-driven, and most developing countries appreciate the fact that provision of adequate, stable modern energy is a prerequisite for any sustainable development plan, but many lack the resources needed to stimulate growth in energy infrastructure. While some have options on the potential sources, others have relatively few. Many countries of the developing world have remained largely agrarian and most of the population live in rural areas. In most countries that have developed over the last two hundred years or so, access to modern energy, electricity and transportation was the main propellant that moved the vast majority of their populations off rural farming into the urban areas where access to stable, high wages and consequent affordability of modern energy-consuming technologies fueled the demand for even more energy. In effect, urbanization and industrialization have always been preconditions for mitigating energy poverty.

Even within the energy-poor world, there is a significant stratification. While the prospects of energy-fueled development are very bright for China and India over the next few decades, projections for Africa indicate that there will be little change. Africa has enormous primary energy resources - coal, oil, gas, hydro, solar, geothermal - but lacks the political stability and effective leadership, both of which are crucial for holistic and sustainable development. For example, Nigeria has extensive reserves of oil, gas, hydro, solar energy resources, and is second only to Russia on the world's list of countries that flare associated gas (a prime energy resource) during oil production, yet per capita access to energy is one of the lowest in the world. It is significant to note however that the country has made good progress in the last decade or so, in gathering some of associated gas for power generation, but, even then, major policy flaws, lack of policy continuity, and security issues have made the gas supply to the many dependent power plants located in different parts of the country unreliable.

Prior to the early 1980s, China was in a similar position as Africa but the change of leadership and goals which started in 1981 propelled the country to the second largest global economy in just three decades, and a formidable fuel that drives the global economic and industrial machinery. The developing countries of Asia and North Africa have made significant progress on electrification for all in the last decade or so and most could reach the full electrification goal within the next decade. On the contrary, progress towards both universal electrification and modernization of household energy in sub-Saharan Africa (with the exception of South Africa) has been slow. Only about 40% currently have access to electricity on the average, but some countries in the region have single digit access. Transition from traditional domestic energy to modern energy has also been slow, and the region accounts for around 80% of the 2.8 billion people globally who still rely on traditional household energy (solid biomass, kerosene and coal). Clearly, the United Nations' goal of 'modern energy for all by 2030' would be difficult to achieve without major national, regional and international policy instruments which articulate strategies for targeting the rural areas where most of the poor live, identifying appropriate technologies, and attracting international and private sector investments.

1.6 ENERGY AND DEVELOPMENT OUTLOOK

Energy poverty is not only an impediment to human development, it has also been associated with human health. The World Health Organization estimates that nearly 3 million premature deaths per year are due to reliance on traditional cooking methods (biomass, charcoal, coal), and the use of candles, kerosene and other polluting fuels for lighting, often in enclosed space with poor ventilation. Household pollution also exacerbates the health issues of people with pre-conditions such as asthma and chronic obstructive pulmonary disease (COPD). Sustainable economic development is impossible without availability of appropriate modern energy which goes beyond providing basic needs, to productive uses in economic enterprises such as agriculture and industry. Historically, economic development has been propelled by a shift away from agrarian based economy towards industrialization and knowledge-based economy, a development which changes the quality, quantum and mix of energy required for sustenance. These structural changes in the economy motivate urban migration and raise earning power which requires more modern energy and provides the incentive for energy sector expansion and modernization. Economic growth promotes a shift from the use of traditional biomass and inefficient stoves to modern energy, in particular, electricity (Figure 1.12).

1.6.1　Global energy demand outlook

There are many recent projections of global demographics, economic growth and fuel demand by sector and region, (ExxonMobil, 2018; WEC, 2016; EIA, 2017; BP, 2018; IEA, 2017c), and there is broad agreement on the expected rapid growth in the global economy and population, both of which are major drivers of energy demand, and are expected to double over the twenty to thirty years. There is also general agreement that increasing efficiency across all sectors of the economy will mitigate the rise in primary energy demand, limiting it to no more than about a third of the current level. Most of the projections also highlight the special case of sub-Saharan Africa which has the lowest level of access to modern energy of all regions of the world, and consider various scenarios on the potential obstacles to the achievement of global targets on human development and access to modern energy. Some of the projections are presented in Figures 1.15 and 1.16, and the broad conclusions are summarized in Box 1.1.

1.6.2　Modern energy for all by 2030 outlook

Access to modern energy, in particular, electricity and modern domestic energy, is mandatory for the achievement of economic growth, human development, and environmental sustainability. In recognition of this, the United Nations set a target of 'energy for all by 2030' as part of the sustainable development goals adopted by 193 nations in 2015. Even before then, most nations had already developed strategies for improving their modern energy infrastructures, focusing on access to electricity. The number of people without access to electricity fell to 1.1 billion (about 15% of the global population) in 2016 from 1.7 billion in 2000. The electricity access deficit is overwhelmingly concentrated in sub-Saharan Africa where 62.5% (609 million people) lack access (Figure 1.7).

About 100 million people globally have been gaining access to electricity per year over the last few years, and the progress has been most rapid in developing Asia, led by China and India. Access in the region increased to 89% in 2016, up from 67% in 2000. China reached full electrification in 2015 and several other countries in the region are well on track to achieve full electrification by 2030. Many developing countries in Latin America and the Middle East are also on course. However, the situation in sub-Saharan Africa is less optimistic: only 40-43% currently have access to electricity and, while several countries in the region (South Africa, Ethiopia, Gabon, Ghana and Kenya) will reach or are on track to reach universal access to electricity by the target date, most of the countries in the region will lag behind. Around 90% of the current 674 million in the region who currently lack access to electricity (mostly in rural areas), will still be without access in 2030 (Figures 1.17 to 1.19).

In contrast with the progress being made in universal electrification, little has changed about access to modern cooking energy over the last two decades: around 2.8 billion still lack access to clean cooking, same as in 2000 (IEA 2017a). One third of the global population (around 2.5 billion) still rely on the use of traditional solid biomass while another 300 million cook with kerosene and coal. Around 68% live in developing Asia while another 30% live in sub-Saharan Africa. Although significant progress is being made in Asia (many are switching to liquefied petroleum gas in urban areas), around 2.3 billion worldwide, over 900 million in sub-Saharan Africa, will still remain without access to clean cooking in 2030.

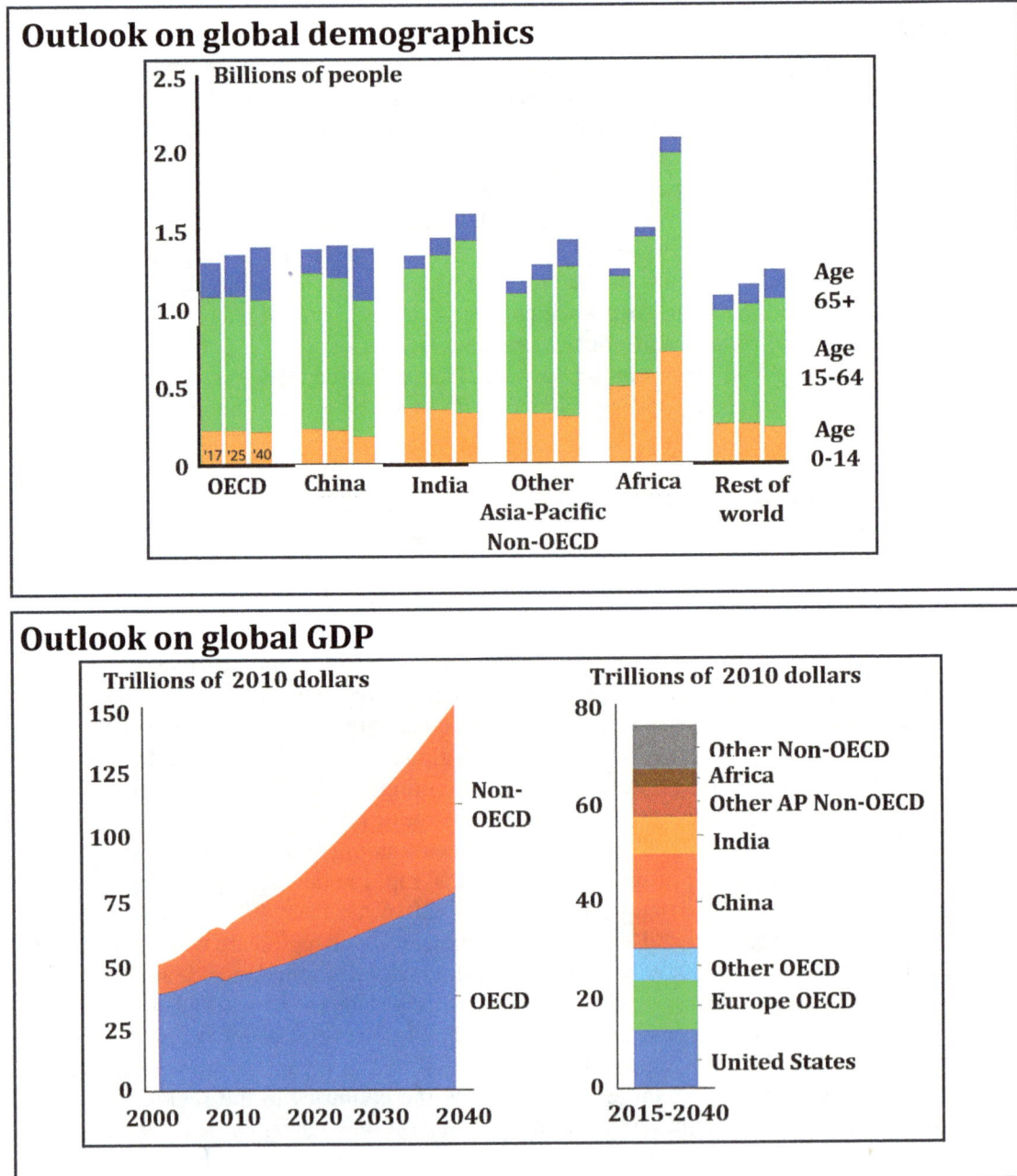

Outlook on global demographics

Outlook on global GDP

Figure 1.15a Outlook on world's demographics and global GDP *(ExxonMobil, 2017).*

(a) Outlook on primary energy by fuel, sector and region

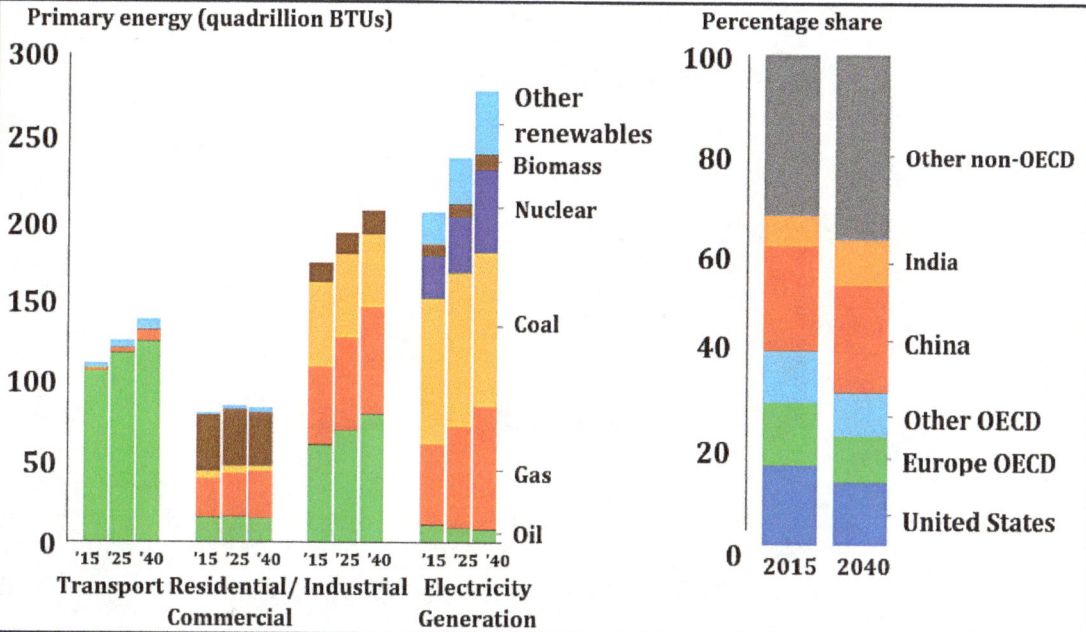

(b) Outlook on primary energy and power generation fuel mix

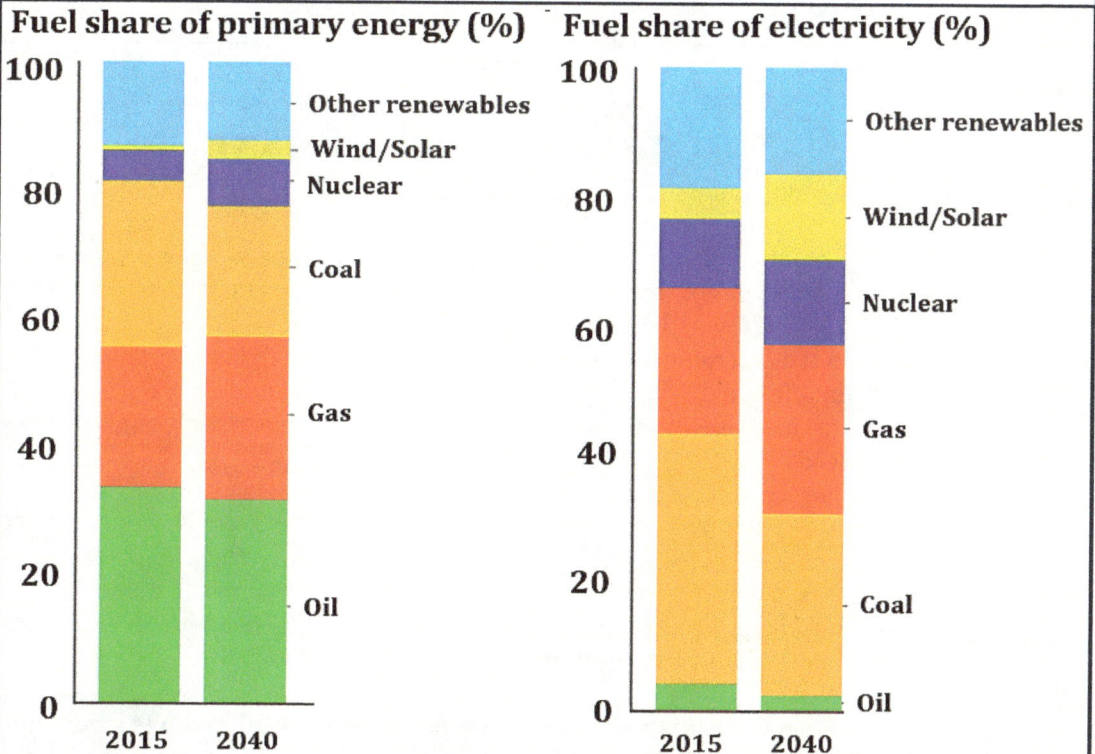

Figure 1.15b Outlook on world's primary energy use (a) by fuel, sector and region (b) by fuel for power generation *(ExxonMobil, 2017).*

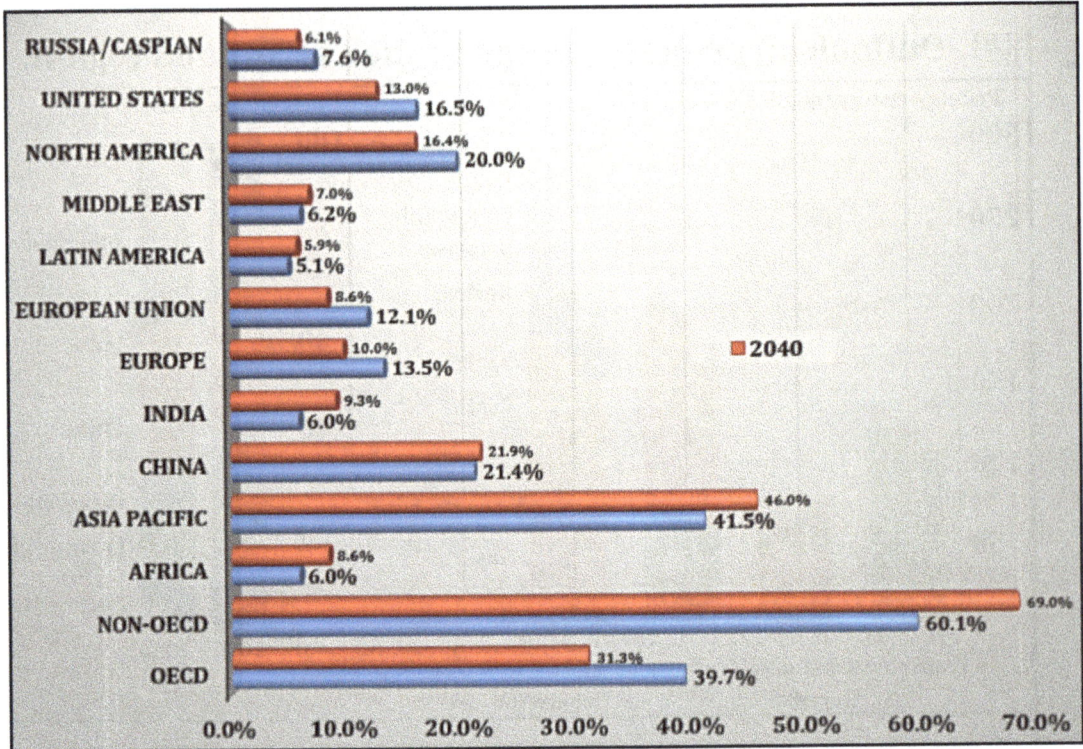

Figure 1.16 Outlook on global primary energy demand in 2040
(Data from *ExxonMobil, 2017).*

BOX 1.1
SUMMARY OF PROJECTIONS ON
GLOBAL ENERGY DEMAND

- Global economy will double by 2040, non-OECD economy will grow by 175%, rising to about 50% of the total global GDP from the current 35%. Growth in the OECD countries will be slower, about 60%. The main driver of economic growth in the emerging countries, particularly in the Asia Pacific region will be the continuing rapid urbanization and expansion of the middle class, leading to significant increases in industrial, commercial, transportation and personal energy needs.

- In spite of the relatively high economic growth, GDP per capita in non-OECD countries will still be low, around a quarter of the average for OECD countries, although there will be significant differences between countries in the region. For example, China's GDP per capita is projected to triple, reaching over $40,000 (compared with US at $80,000 and OECD Europe at $50,000).

- The world's population will grow from the current 7.3 billion to 9.1 billion by 2040, and the fastest rate will be in Africa. Working age (15-64 years) population will remain flat in OECD countries and China, while it grows at a fast rate in the emerging regions.

- The projected economic growth, coupled with population growth will drive up global energy demand by 25-35%. Growth should be around a hundred percent but for the anticipated moderating effect of increasing efficiency gains across the global economy, and decreasing energy intensity (the amount of energy used per unit of economic output). The relatively fast decline in energy intensity in the OECD countries, coupled with the slow economic growth will keep energy demand by the region relatively flat over the next two decades. Growth in global energy demand will be led by the increasing electrification of the global economy, accounting for around 55% of the total growth in energy demand. Non-OECD countries will account for about 70% of the total global energy demand in 2040. China and India alone will contribute about 45% of the total global energy demand growth, with China accounting for about 20% of the total demand, equivalent to the combined share by the United States and OECD Europe.

- Fossil fuels will continue to dominate the global energy scene, accounting for nearly 80% of the total primary energy, but there will be a significant shift in mix, with natural gas and renewables (mainly solar and wind) gaining prominence at the expense of coal, particularly for power generation, but intermittency is a major constraint which limits worldwide capacity utilization to 30% for wind and 20% for solar energy, which makes integration with fossil fuel power plants necessary for all but small installations.

- Renewables, in particular wind and solar are currently the world's fastest growing energy sources, used mainly for power generation, and represented almost two-thirds of new net electricity capacity additions in 2016. Renewables covered around 45% of the world's electricity generation growth in 2018, accounting for over 25% of global power output, second after coal. Solar PV, hydropower and wind accounted for about a third of the growth, with bioenergy accounting for the majority of the rest. Power generation from wind has been growing by an average of 12% per year over the last few years. China alone was responsible for 40% of global renewable capacity growth in 2018, and projections forecast a global increase of 43-45% over the next five years, with a total global output expansion of around 1000GW. Most of the new capacity will be filled by solar PV, with an output of around 740GW by 2022, ten times the current capacity. New solar capacity around the world grew by 50%, largely as a result of booming photovoltaic deployment, with China alone accounting for almost half of this expansion. China is also the world's largest solar PV manufacturer, accounting for around 60% of the global output. The growth of other renewable technologies such as bioenergy, concentrating solar power (CSP), and geothermal has been relatively slow, and it represented only 4% of 2016 global renewable power capacity additions. However, hydropower was still the largest renewable energy technology for power generation in 2018, accounting for about 60% of all global electricity power supply from renewables. The potential for hydropower varies across regions: while its share of renewable generation will decline in OECD countries due to decreasing number of suitable sites and increasing environmental and safety concerns in many countries in the region, deployment of small-to-large scale projects in Sub-Saharan Africa is expected to grow.

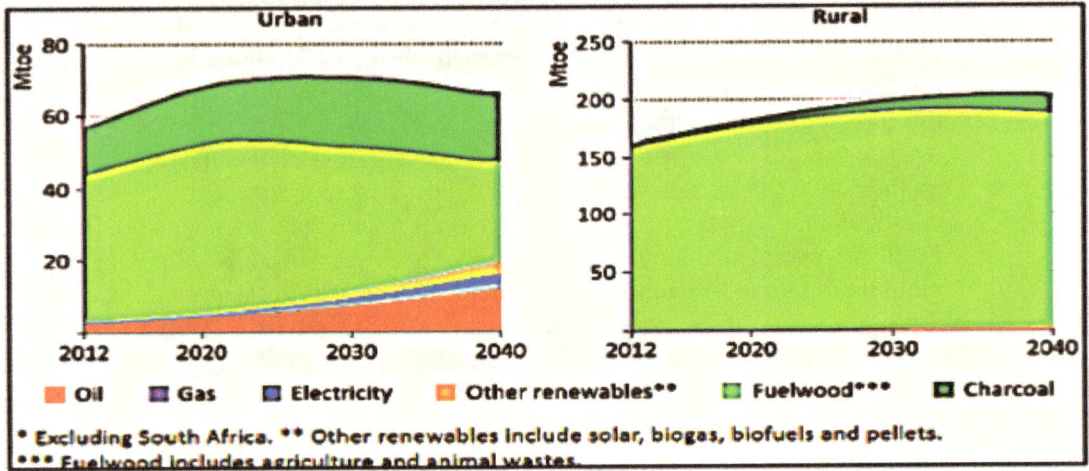

Figure 1.17 Household energy consumption for cooking by fuel in sub-Saharan Africa, taking into account policies in existence and anticipated new policies *(IEA, 2014).*

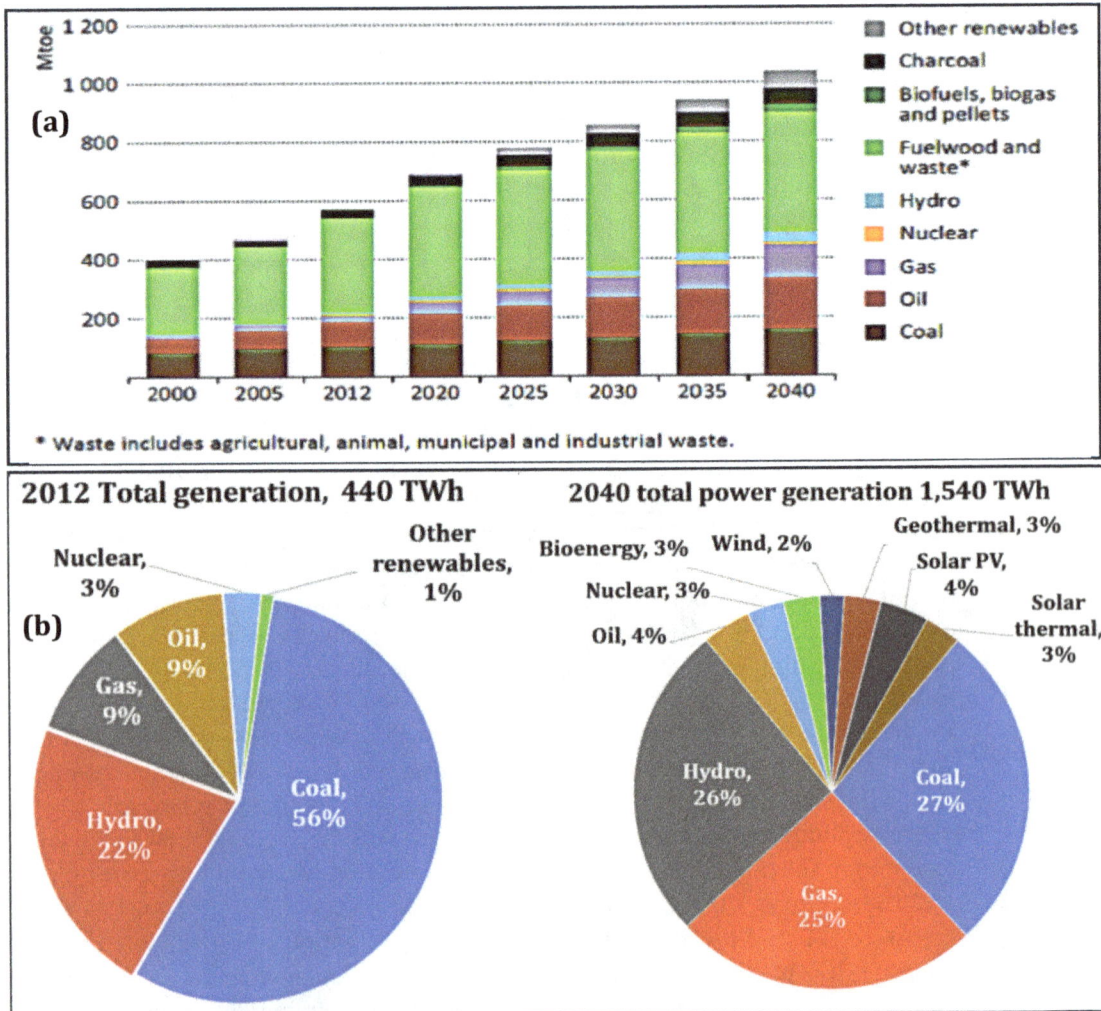

Figure 1.18a Outlook on (a) primary energy demand (b) electric power generation by fuel in sub-Saharan Africa, taking into account policies in existence and anticipated new policies *(IEA, 2014).*

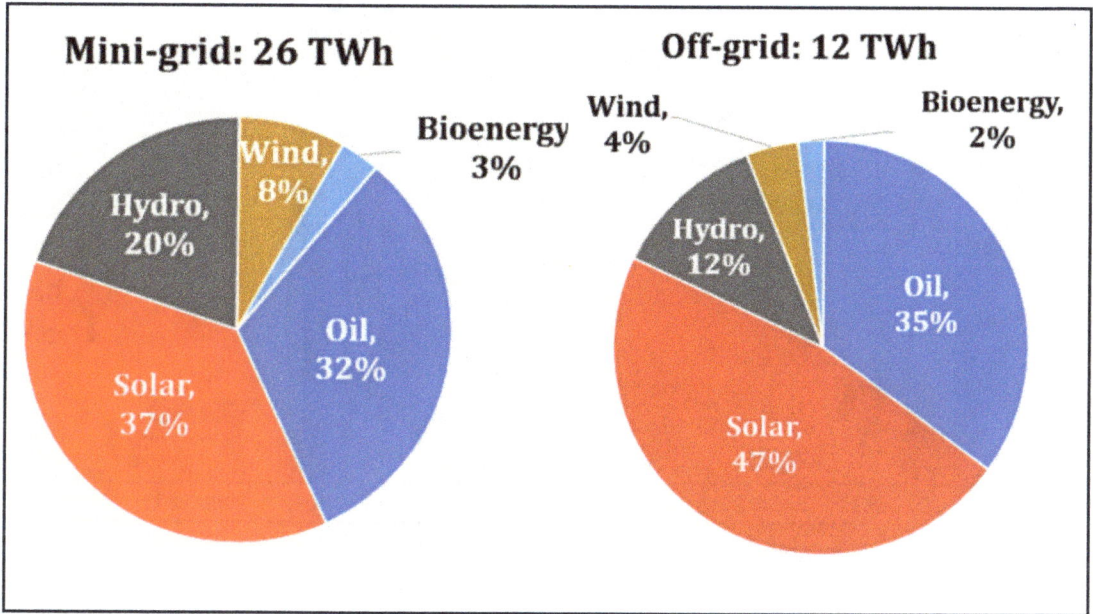

Figure 1.18b Technology mix for mini-grid and off-grid power generation in sub-Saharan Africa, taking into account policies in existence and anticipated new policies. *(IEA, 2014).*

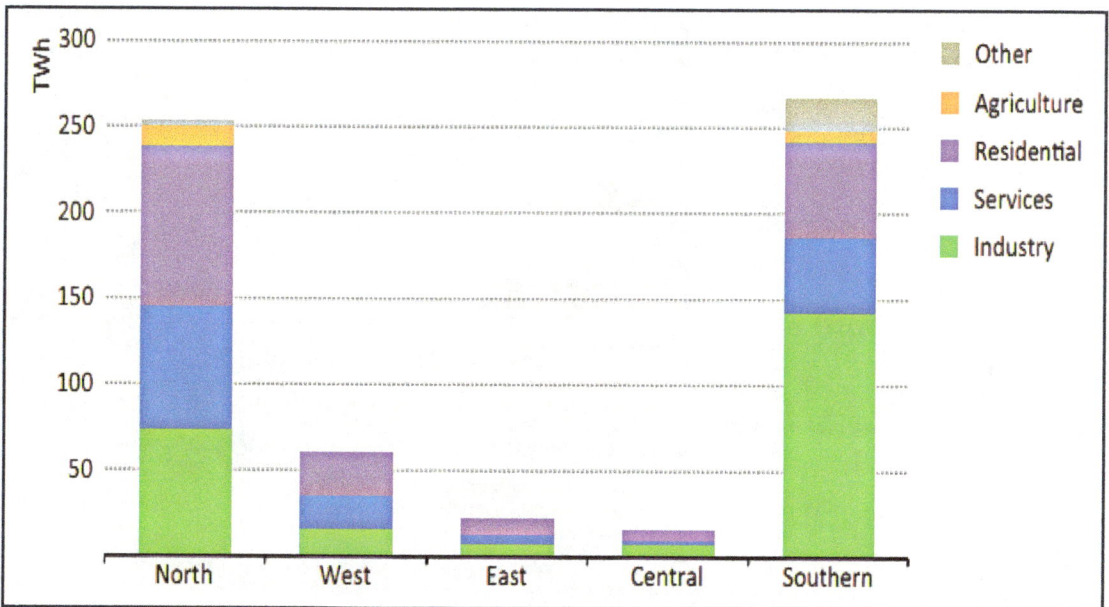

Figure 1.19a Electricity consumption in Africa by end-use sector and sub-region in 2012. *(IEA, 2014).*

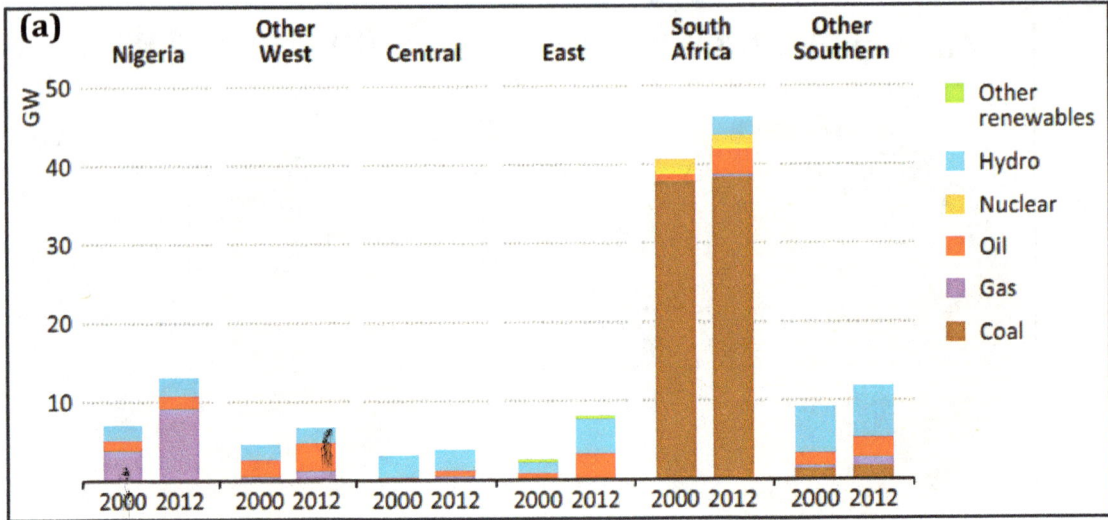

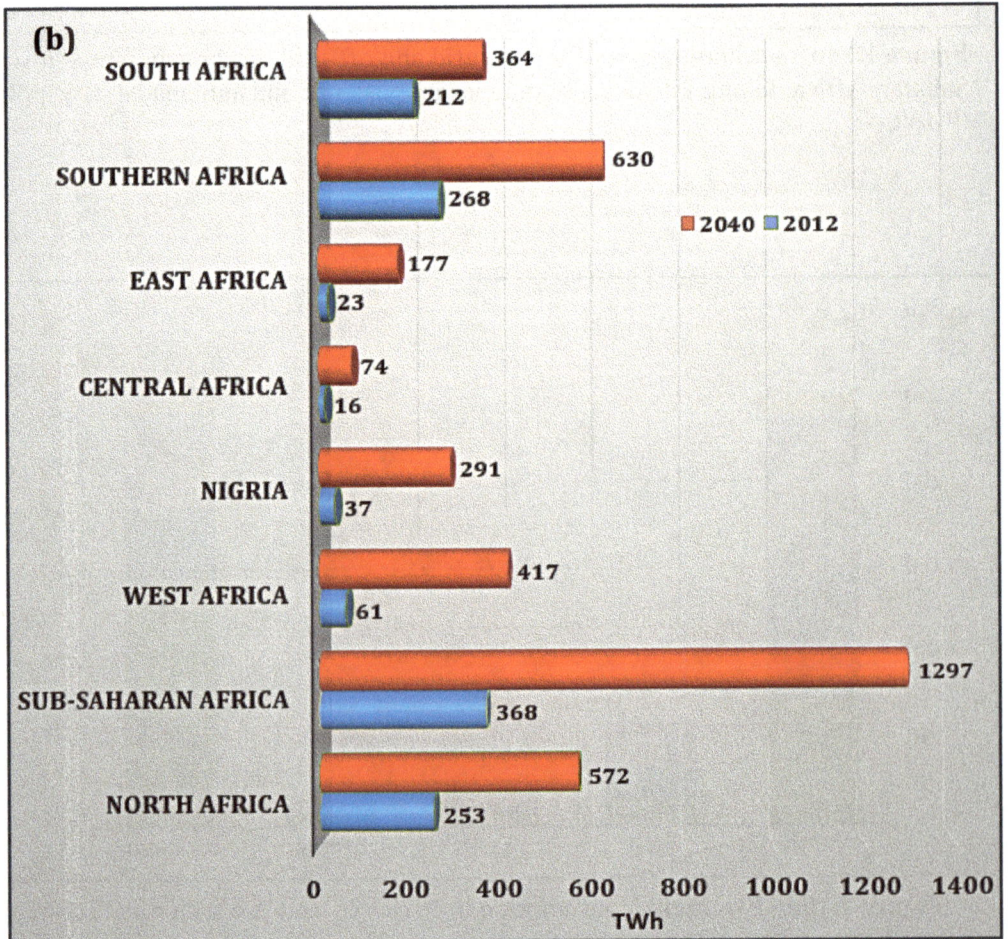

Figure 1.19b (a) Installed grid-base capacity by type and sub-region in 2012
(b) Projection of electric power demand in sub-Saharan Africa *(Data from IEA,*

1.6.3   Sub-Saharan Africa's energy outlook

Sub-Saharan Africa lags behind most other regions in terms of access to modern energy, in spite of the enormous potential energy resources. The region accounts for 13% of the world's population but less than 10% of the GDP and only 4% of the global energy demand. Energy intensity (amount of energy required per unit GDP output) is double the world average and triple the OECD average. In effect, the region uses 50% more energy than the world average for each unit of economic output. Energy consumption varies widely within the region: unlike North Africa which has achieved nearly 100% in provision of access to electricity in both urban and rural areas, sub-Saharan Africa has the lowest electricity access penetration in the world. Only about 40% of the population of the region have access to electricity, and four out of five people still rely on the traditional use of biomass, mainly fuelwood and charcoal for cooking. While there will be gradual transition to modern energy such as liquefied petroleum gas in the urban areas, projections show that the positive impact will be minimal and one-third of the population of the region (around 650 million people) will still rely on traditional biomass in 2040. Economic growth in the region has been rapid in the last two decades, rising by around 45% since 2000, and the momentum is expected to be maintained. The economy of the region will nearly quadruple in size by 2040, and energy demand will rise by around 80%. However, the population will nearly double over the same period to around 1.75 billion, a growth rate which will outpace energy growth and still leave hundreds of millions, particularly in the rural areas without access to modern energy in 2040. One of the major constraints to business development in the sub-Saharan region is inadequate electricity. With the exception of South Africa, this problem is wide spread and affects both energy resource-rich and resource-poor countries. Inadequate and erratic power supply is one of the major causes of low productivity in the region. The World Bank estimates that businesses in the region lose around 5% of annual sales on the average due to electric outages (World Bank, 2014b). Most businesses are forced to use back-up power generation to mitigate the poor public grid-based power supply, at much higher costs and higher pollution emissions. The cost of fuel for back-up generation across businesses and households is estimated at at least $5 billion a year, with Nigeria accounting for almost three-quarters of electricity supply provided by back-up generators in the region. Much of the problem of inadequate and unreliable power supply is due to lack of generating capacity, low capacity utilization (often well below 50%), limited or obsolete transmission and distribution infrastructure, and poor management. For example, Nigeria loses around 13% of generated power during transmission and distribution, compared with less than 5% for South Africa, and 45% for DR Congo. Nigeria currently has installed capacity of 12.5 GW but only 7 GW is available and the average consumption is only about 3 GW due to frequent systems collapse, poor synchrony between generation, transmission and consumption, and unreliable gas supply to the generating plants.

Sub-Saharan Africa is very rich in energy resources, more than sufficient to meet regional needs, both now and in the foreseeable future. The region holds around 7% of world conventional oil and gas reserves, and accounted for nearly 30% of global oil and gas discoveries in the last five years. Also, geophysical evidence suggests that the potential resources are much larger than the proved reserves. Substantial oil has been found in nearly twenty of the countries in the region, enough to last around 100 years at the current rate of production, gas for more than 600 years, and coal for more than 400 years. If gas flaring in the sub-Saharan region were to stop today, the life span of the natural gas resources could extend to nearly a thousand years. The region holds large deposits of uranium, and a wide range of high quality renewable energy resources, including solar, hydro, wind and geothermal. It is interesting to note that the old Belgian Congo (now Democratic Republic of Congo) holds a significant proportion of the global resources of uranium, and was the source of the uranium that powered the development of nuclear energy in both Europe and North America in the early 1940s, which eventually led to nuclear electricity in the 1950s, yet the country features among the lowest with access to electricity today (only 17% in 2017). In effect, the region is energy-poor in the midst of plenty, largely because of the lack of capability to translate theoretical potential to sustainable reality.

Sub-Saharan Africa exports about 5 million barrels a day of crude oil, but imports most of the production equipment and virtually all requirements of refined products because most of the countries lack refining infrastructure. The few that exist in Nigeria and Ghana are public-owned, old and poorly maintained, often operating at below 20% capacity. A few countries in the region - Cote d'Ivoire, Cameroon, Niger and Chad have recently benefited from Chinese investments in small refineries which are supplying most of the required oil products. Also, a private sector refinery project is under construction, in Lagos, Nigeria, with sufficient design capacity to meet all the needs of the country and several other countries in the region. It is scheduled for commissioning in 2019 and could make a significant positive impact on energy security and equity in the region.

Around 28 billion cubic meters of natural gas is estimated to have been flared in the sub-Saharan Africa region, in the process of producing oil in 2012. Apart from releasing methane, particulate matter, and other potent environmental pollutants into the atmosphere, the wasted energy could have been used to power gas turbine generating plants and increase the region's power supply by around 35%. Nigeria alone accounted for around 60% of the gas flared, and Angola, Congo and Gabon for much of the remainder, yet, all these countries are energy poor. In the same year, access to electricity in Nigeria was only 55%, Angola 30%, and Congo DR 12%. Some countries in the region have substantial reserves of natural gas and coal, and many potentially suitable sites for mini hydro-dams, wind farms, solar and geothermal energy. The technical hydropower potential in the region is estimated at around 283 GW, enough to generate around 1200 TWh a year, more than three times the current electricity consumption in the region (IEA, WEO, 2014). However, less than 10% has so far been tapped, and more than half of the remaining potential is in the Central and Eastern sub-regions which also have the least access to electricity in the region. In effect, sub-Saharan Africa is energy-poor, not because the region lacks resources, but due to a myriad of problems, including poor and unstable governance, policy deficiency, inadequate energy infrastructure, and unattractive business and investment environment.

Access to electricity varies widely within the sub-Saharan region, with Southern Africa (mainly South Africa) leading (Figure 1.19). Projections show that the total power generation of the region will quadruple by 2040, provided the region can attract the needed massive investment in generating capacity and transmission/distribution infrastructure. The fuel mix will be diverse, depending on the locally available resources but, on average, gas will lead the mix (46%), hydropower (28%) and oil (9%). The completion of the West African Power Transmission Corridor, currently under construction, coupled with improvements in the operation of the West African Gas Pipeline should greatly facilitate electricity and energy trade across West Africa. The Economic Community of West Africa (ECOWAS) is also promoting non-hydro renewables (bioenergy, solar PV, concentrated solar power, and wind), which could account for around 15% of electricity supply in the sub-region, but the contribution in the Eastern sub-region will be much higher. Several renewable power generation projects are under construction or at the planning stage, and are projected to raise power output in the sub-region from the current 8 GW to around 55 GW by 2040. Most of the increase will be powered by several large hydro-dams under construction, geothermal energy, and, to a smaller extent, other non-hydro renewables. This large projected increase will facilitate cross-border trading in electricity in the sub-region. Power generation capacity in Southern Africa will triple to 180 GW by 2040, fueled mainly by coal and natural gas. South Africa alone is projected to more than double current capacity to around 110 GW (Figure 1.20).

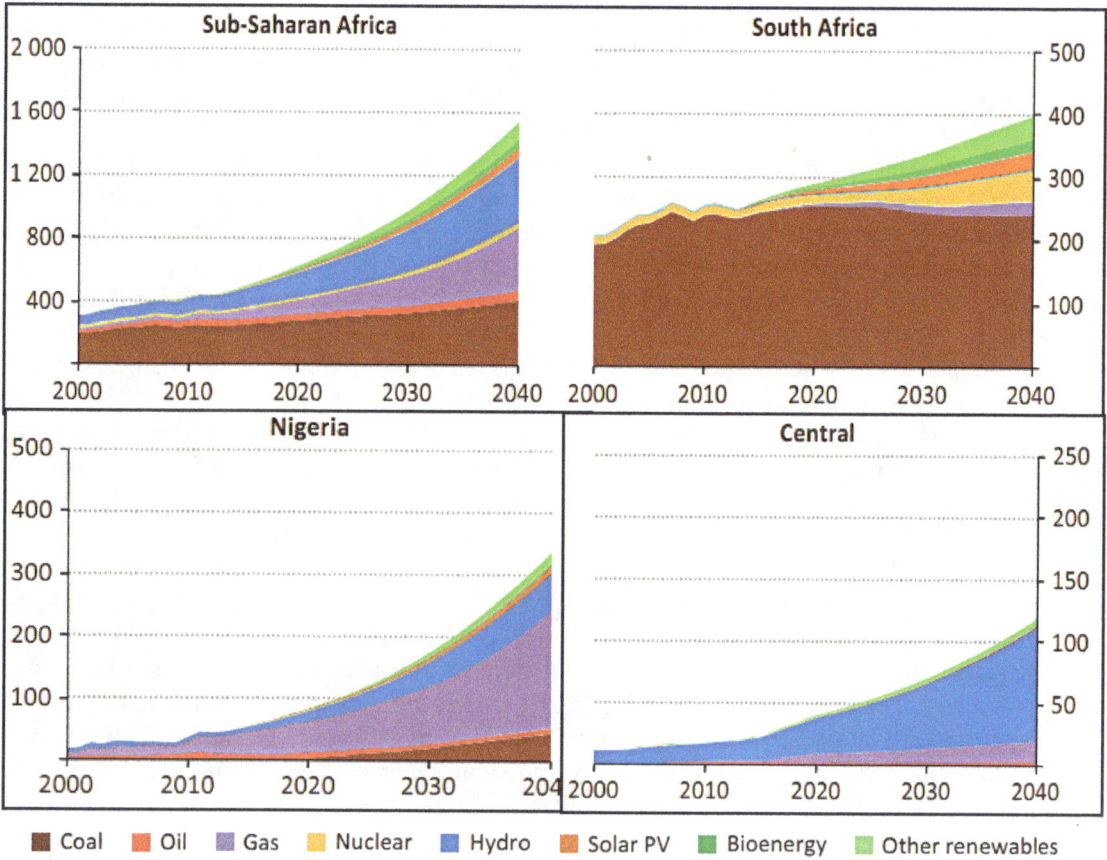

Figure 1.20 Projection of power generation capacity growth in sub-Saharan Africa by 2040 *(IEA, 2014)*.

Central Africa has the highest concentration of people without access to electricity, although there is a very large variation across the sub-region. Around 66% of people in Equatorial Guinea have access to electricity, Gabon 60%, Cameroon 54%, compared with Central African Republic (less than 3%), Chad (4%), and DR Congo (9%). Around 60 million people in DR Congo do not have access to electricity, even though the country has very large hydropower potential. The country also has one of the fastest population growth rates in the world. Grid capacity in the sub-region in 2012 was only 4 GW, equivalent to 4% of the sub-Saharan total despite the population being 12% of the total population of the region. Installed capacity is projected to increase to 36 GW in 2040, powered mainly by the expansion of the hydro-dams in DR Congo and new projects in Cameroon, complemented by natural gas in dry, low-water seasons. The proposed Grand Inga hydropower dam in DR Congo is the largest in the world (twice the size of the current largest - China's three Gorges Dam), the fourth and largest of a series of dams that have been built or are proposed for the lower end of the Congo river in DR Congo. The design capacity is 40 GW, the equivalent of twenty large nuclear power stations, around 45% of the current total power generating capacity of the sub-Saharan region. The project is the center piece of a masterplan to develop an integrated power system for the whole sub-region, supported by many regional organizations including New Partnership for Africa's Development (NEPAD), the South African Development Community (SADC), East African sub-region, supported by many regional organizations including New Partnership for Africa's Development (NEPAD), the South African Development Community (SADC), East African Power Pool (ESPP), and ESKOM, Africa's largest power utility, among others. The project plans to take advantage of low operating costs associated with large hydro-dams, but will require extensive development of a complex intra-regional power grid network, backed by competent management, to be able to move the enormous power across the sub-region.

Around 80% of people in sub-Saharan Africa rely on traditional use of solid biomass, mostly fuelwood and charcoal for cooking, and a 40% rise in demand by 2040 is projected. Much of the production/supply chain of both fuels is outside the formal economy, and the forestry stock is depleting without any systematic policies on reforestation. The fact that fuelwood is at low or no-cash cost relative to other alternative energy sources is the main reason why it will remain a fuel of choice for a large population in the sub-region in the foreseeable future, particularly in the rural areas. Although fuelwood is renewable and stock in the region is abundant, the resource is not uniformly spread across the sub-region, and is depleting in some parts due to overuse and the effect of draught. Depleting stocks of fuelwood is of lesser concern than the potentially dangerous smoke when burnt in low-efficiency stoves, often in confined areas, hence mitigation efforts by some international organizations have been focused on the development and proliferation of fuel-efficient cookstoves.

Charcoal is a popular fuel choice in urban areas because it is smokeless and delivers more energy per unit weight. However, it is expensive compared with fuelwood and, while any dry shrub is a good fuelwood, charcoal is produced from hardwood which is rapidly depleting in the sub-region. Furthermore, charcoal is produced in traditional earth-mound kilns that have a conversion efficiency of 8% to 12%. In 2012, an estimated 36 million metric tons of charcoal with market value of $11 billion were produced in the sub-region, and projections show that the charcoal market is not expected to diminish in the foreseeable future in spite of increasing costs due to exhaustion of local forest hardwood and the need to transport the product over long distances. A combination of increasing levels of consumption and higher prices is expected to push the annual charcoal market in the sub-region to around $70 billion by 2040. It seems that the only pragmatic option is to effect policies that popularize the more efficient industrial kilns that are now available. These kilns operate at around 25% efficiency and are less polluting,

In summary, a few countries in the sub-Saharan region (South Africa, Ethiopia, Gabon, Ghana and Kenya) have good prospects of achieving electricity for all by 2030, but around 75% of the 674 million globally that will still lack access to electricity, will be in the region, mostly in rural areas. The outlook on access to modern household energy is even worse, largely because the likelihood of a significant shift towards modern energy in most countries of the region in the foreseeable future is remote. This grim outlook could change if current and planned policies in the region are upgraded as summarized in Box 1.2, and a paradigm shift in modern energy procurement towards stand-alone and off-grid power supply systems is articulated.

BOX 1.2
OUTLOOK FOR MODERN ENERGY FOR ALL IN SUB-SAHARAN AFRICA BY 2030

- Political stability that transcends political cycles and ensures sustainable energy policies, and improvement in the quality of governance which has been weak across most of the region compared with other parts of the world, will be critical to the chances of achieving the international goal of modern energy for all by 2030. Poor governance implies substantial risks arising from policy and regulatory reversals and uncertainties, and can substantially deter international investments, generate distrust between countries, and retard regional cooperation.

- Development and co-ordination of strong, sustainable energy development policies at continental, regional and country levels, through existing organizations such as African Union (AU), New Partnership for Africa's Development (NEPAD), and African Development Bank (ADB) will be required. The AU/NEPAD African Action Plan and the Program for Infrastructure Development in Africa (PIDA) released recently by the AU/NEPAD/ADB partnership are a good start. The policies focus on transport, energy and information telecommunication technologies and, if implemented between 2020 and 2040 as planned, would be a major step towards relieving some of the trans-border constraints on intra-regional energy trade.

- De-emphasis of the current dominant position of state utilities which tend to be inefficient, unreliable and unable to promote and support sustainable inter-state operations will be vital. This is already happening with the operation of the West African Gas Pipeline. Furthermore, revenue collection in most countries in the sub-region, particularly for power supply is extremely weak and grossly abused, largely because most of the facilities are public-owned. Privatization will eliminate these issues and attract the much needed investments. This has happened already in the telecommunications sector in most countries in the sub-region, and the pay-as-you go revenue collection system has been very effective.

- Greater regional cooperation in the development of infrastructure linkages that support inter-country trade in energy will be important. The recent commissioning of two cross-border pipeline projects - the Mozambique-South Africa and West Africa Gas Pipeline (WAGP) - is a promising start. However there are complex obstacles - reaching agreement among multiple countries and companies, the challenge of securing adequate sources of funding, development of sustainable long-term bilateral agreements, opposition from local communities and environmentalists, security issues, etc.

- Implementation of clear, consistent and sustainable energy policies that can attract massive local, regional and international investments is crucial. Development of innovative business models that can attract venture capitalists and social entrepreneurs is vital. Extensive intra- and international support, including strong incentives for potential investors in new energy systems is also important.

- Restriction of state intervention to the development of such main energy infrastructures as grid modernization and extension, pipeline networks, etc should be a priority. The recent privatization of most of the Nigerian power plants and consumer supply sectors is a good example.

- A shift from the current thrust of providing basic electricity in rural areas, for basic lighting, cell phone charging, etc., to the provision of productive electricity that can support income generation enterprises, boost economic development, and reduce poverty is vital. Deployment of utility systems that support agriculture and food processing has strong potential for improving the economic and human development status of rural farming communities which make up the larger proportion of the population of the region. Other options include massive exploitation of renewable energy and development of on-grid, off-grid, mini grid and distributed generation systems, stand alone solar home systems that target rural communities. The varied dynamics of expanding electricity access, which also changes over time as incomes and consumption patterns change should be taken into account in determining the most cost-effective options that can be upgraded from basic to productive energy as needed.

- While the use of solid biomass for cooking is undesirable because of its contribution to environmental pollution and deforestation, the outlook is that use in the sub-region will continue to grow and the only feasible mitigation option is to minimize the health hazards by promoting the use of more efficient cookstoves and charcoal making kilns.

- Proliferation of fuel-efficient cooking stoves, production of smokeless coal briquettes, and liquefied petroleum gas (LPG) across the region could mitigate emissions from the use of charcoal significantly. Nigeria, South Africa and several other countries in the region have substantial coal and natural gas resources which could support industrial and cottage production of smokeless coal and bottled LPG across the region.

Chapter 2

World energy resources
and outlook

2.1 INTRODUCTION

Prior to the Industrial Revolution that began in the 17th century, the world depended on the natural energy resources that were readily available - traditional biomass (wood, charcoal, etc.), outcrop coal, solar, wind, geothermal energy. None of these sources was available regularly, in sufficient quantities and in a sustainable manner to power anything but small local energy requirements. The development of the first commercial coal mine in Scotland in the last half of the century provided a stable source of energy that propelled the development of a wide range of commercial-scale industries and enterprises - iron and steel, textile, commercial farming, etc. The invention of the coal-powered steam engine provided a prime mover for industrial machinery and the first mass transportation invention - the railway. Commercial coal production spread rapidly to Europe and North America, and the North American discovery and production of oil and gas in commercial quantities in early 20th century completed the energy mix that has fueled the Industrial Revolution, economic and technological development over the last two hundred years or so, in particular, mass production of primary metals, diesel, petrol and jet engines, automobiles, airplanes, polymers, computers, cell phones, etc.

2.2 WORLD ENERGY SOURCES

Primary energy refers to natural sources of energy that have not been subjected to any conversion, and the world has access to many of them, notably fossil energy (coal, oil and gas), nuclear energy, solar energy, hydro-energy, geothermal energy, wind energy, bioenergy, gravitational energy. Primary energy almost always needs to be processed through an energy conversion technology into useable forms (secondary, converted or end-use energy) required for powering machinery, automobiles, etc., the main exception being natural gas which only requires light stripping. Primary energy resources are often classified on the basis of availability into exhaustible and inexhaustible as shown in Figure 2.1. Fossil fuels dominate the primary energy scene, accounting for over 80% while renewables contribute less than 15%. Electric power is the most important in the end-use energy group and can be produced from any of the primary energy resources. Other members of the group are heat, pressurized air, refined petroleum and synthetic gaseous and liquid fuels.

2.3 WORLD ENERGY RESOURCES AND RESERVES

Providing sufficient energy is a primary prerequisite for social welfare and economic development of a society. Only four of the global primary energy resources are available everywhere, abundant, free and inexhaustible: solar, wind, gravitational and biomass energy. Hydro and geothermal are also free, but not every country has access to these resources. Furthermore, many countries which have these resources do not have the wherewithal to make them available in a continuous and sustainable manner. The global distribution of fossil and nuclear energy is inequitable and relatively few countries have the resources. Also, they are non-renewable. Actually, fossil energy is renewable but the deposits being exploited currently were formed some three hundred million years ago from dead vegetation and aquatic life remains, and new deposits are being formed all the time. However, human activities are interfering with the natural deposition and carbonification processes, and precursors which transform and progress to maturity may be relatively few. In effect, they are considered non-renewable in view of the long period between formation and maturity. For example, peat which is the first stage of carbonification of plant remains in the coal formation process is being exploited as fuel and oil spill mopping material in many parts of the world. Concerns are often expressed about the probability of exhausting the world's fossil energy resources in the next few decades but all available data show that reserves are actually increasing at a faster rate than exploitation. This is due to the fact that large areas of the world remain unexplored and many energy resources are yet to be discovered. Also, many that have been quantified cannot be exploited by currently available technologies but new, more advanced technologies are emerging continuously.

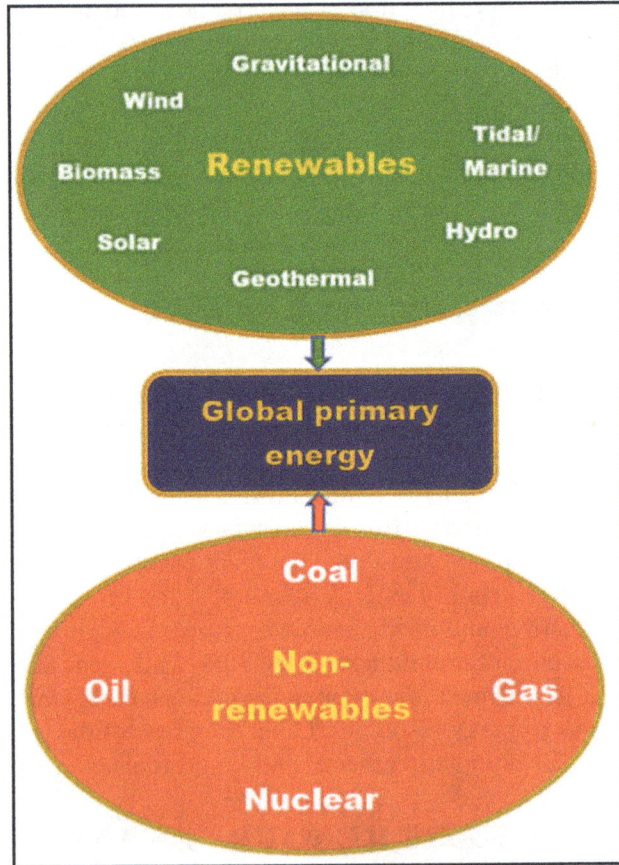

Figure 2.1 Global primary energy resources

2.4 BETWEEN RESOURCES AND RESERVES

Geological data can give valuable information on mineral deposits in the Earth's crust and under the seas. The minerals can be identified, quantities and qualities estimated, and feasibility of economic recovery assessed. Deposits so identified are *inferred resources* and are believed to be only a small fraction of *actual resources* which are yet to be identified. For example, the oceans represent about a third of the planet and hold the bulk of the global oil and gas resources. Current geophysical exploration technologies can penetrate only a few kilometers but most oceans are much deeper. Furthermore, large areas of the world remain unexplored and deposits of many minerals have not yet been discovered. Deposits that have been identified and quantified as accurately as possible using core drilling are known as *reserves in place.* A fourth category of resources known as *proved recoverable reserves* is defined by the United Nations as *"the quantity within the proved amount in place that can be recovered in the future under present and expected local economic conditions with existing available technology."* Current exploration and exploitation technologies are versatile and advanced but severely limited, and undiscovered resources are believed to be much bigger than reserves in place (Figure 2.2).

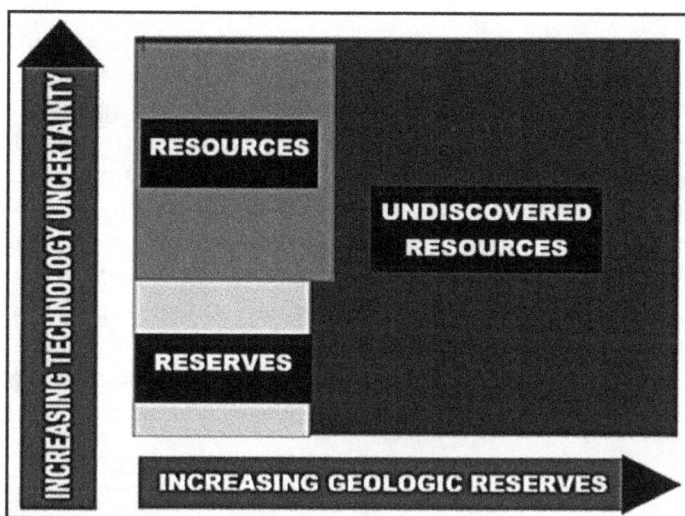

Figure 2.2 Schematic representation of world's primary energy resources and reserves. *(Afonja, 2017).*

Technological innovations are coming on stream all the time and this explains why resource estimates are rising and are being moved through reserves in place to recoverable reserves. For example, developments in hydraulic-fracturing technologies and wider applications have moved substantial oil and gas deposits from reserves in place to proved recoverable reserves. Coal reserves are also increasing because resources are being discovered, evaluated and moved to reserves.

2.5 ORIGIN OF WORLD'S PRIMARY NON-RENEWABLE ENERGY RESOURCES

The Sun's energy is probably the most important global primary energy reserve because it is abundant, free, available everywhere, and inexhaustible. If it were possible to capture and utilize a very small fraction of the Sun's energy falling on the Earth's surface in just one day, it would fill the global primary energy requirement for more than one year. Hydro-, wind, bio- and geothermal energies are also abundant and free. However, there are formidable problems with widespread capture and deployment of these energy resources. In effect, the world relies mainly on fossil energy (oil, gas and coal) and nuclear energy (Figure 2.3).

2.5.1 Coal

Coal is a complex black, brownish black, combustible, solid carbonaceous mineral formed from mainly plant debris buried underground in favorable environment hundreds of millions of years ago when the Earth was covered by steamy, swampy forests. Throughout the life span of these mostly green plants, they captured energy from the Sun, carbon dioxide from the atmosphere and water from the soil to produce through a process called *photosynthesis*, carbohydrates (complex compounds of carbon, hydrogen and oxygen) that make up plant tissues and provide food for mankind and animals. The most important element in the plant material is carbon, which gives coal most of its energy. As plants and trees died, their remains sank to the bottom of swampy areas, accumulating layer upon layer. As the layers were successively covered, their access to the air became limited and this stopped the full decomposition process, creating a soggy, dense material known as *peat*. As seas and swamps receded or dried up, these prehistoric accumulations of plant materials and peat bogs became buried, often to great depths, and went through significant physical and chemical transformations due to the combined effects of high temperatures and pressures. The original peat swamp was transformed through the *coalification* process, and passed through the progressive stages of maturation, from *lignite* through *bituminous* to *anthracite*. Intermediate stages are often classified as *sub-bituminous* and *semi-anthracite*. If the burial conditions were favorable (for example, much higher temperature and pressure when buried under rocks), *graphitization* may also have occurred and *graphite,* the most mature coal would have been the ultimate product. These natural processes are continuous but are being disrupted or truncated by human activities such as deforestation and land use.

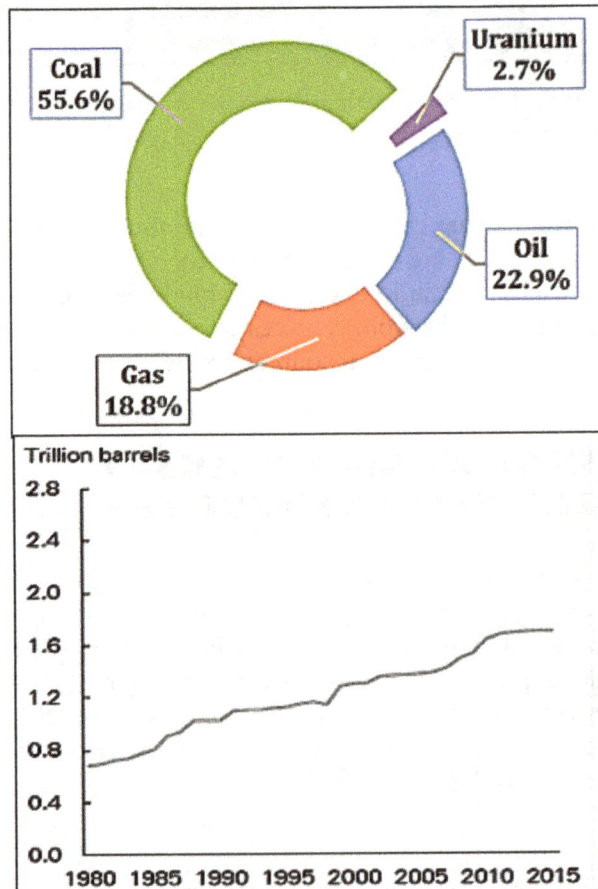

Figure 2.3   World's primary energy reserves (1360 billion metric tons of coal equivalent). Global oil reserves are rising in spite of increased exploitation. *(WNA, 2018; BP 2017).*

The environmental conditions which favor peat formation are very complex. It is estimated that about ten meters of prehistoric plant debris deposited in shallow ponds were needed to form one meter thick seam of mature coal. In effect, a one and a half meters thick coal seam which is usually considered by miners as the minimum that can be mined economically would have required the accumulation of about fifteen meters of undisturbed plant debris pond, a process which could have taken thousands of years. Furthermore, not every accumulation could have eventually resulted in the formation of coal. In fact, relatively few would have been buried under ideal conditions which include a reasonably stable water depth over a very long time. First, the peat-forming ecosystems, also known as *mire* must be deposited in swampy areas. If the water becomes too deep the plants of the swamp will drown and disintegrate. If the water cover is not maintained the plant debris will decay. Second, there must be geological activities such as earth movements and subsidence, volcanic activities, marine transgression and regression which continuously deposit earth and sediments on the mire.

Several other factors also play a vital role in peat formation and progressive transformation into high carbon material. Vital variables for peat formation include the type of deposition, the peat-forming plant communities, the nutrient supply, acidity of the environment, bacterial activity, temperature and pressure. The time frame from peat formation to full coal maturity is estimated to be around 300 million years, but many coals that are being exploited today (notably lignite and sub-bituminous coals) are in various stages of maturity. Penetration of peat into the Earth's crust proceeds gradually depending on the prevailing geological activities in the area. However, the rate of penetration varies from point to point in a particular deposit due to the variable nature and rate of deposit of the overburden, the sporadic nature of peat accumulation and biochemical activity, hence a series of horizontal seams of coal of different depths are formed. This to a large extent accounts for seam thickness variations and lateral discontinuities that are characteristic of many coal seams. Most peat mires remain at the site of formation but can be displaced by earth movements, ocean transgression and regression, etc. The ideal conditions which promote the biochemical reactions needed to initiate coal formation are highly improbable events, which explains why the conditions for forming coal have occurred only a small number of times in relatively few locations throughout the Earth's history. Modern human activities such as deforestation and exploitation of peat for energy, oil spill management, soil conditioning, etc. will likely slow down the natural process of coal formation.

2.5.2   Oil and gas

Oil and gas are believed to have been formed from algae, fish and other aquatic life that lived, died and were buried deep under oceans or rivers hundreds of millions of years ago. The remains decayed partially and were converted to oil and gas through the action of bacteria. Continuous subjection to high pressure and temperature promoted the gradual transformation of the remains into oil and gas. The reserves that are being exploited currently are believed to have been transformed over a period of around 350 million years. Thick, viscous oil is formed first but, in deeper, hotter depths, gas also forms. While most coals remain in-situ in the general location of formation, oil and gas tend to migrate and are often found together in geological traps such as porous rock formations called *caprocks* deep beneath the Earth's surface, mostly in deep ocean beds. The caprocks are dense enough to prevent the oil from seeping to the surface, although some can squeeze through the tiny pores in the rocks under the tremendous pressures that exist in great depths underground, reach the surface and become an energy resource which has been utilized by mankind from the early times. Oil and gas also migrate sideways and become trapped under dry ground, or the water dries up or migrates leaving dry overburden, which explains why oil and gas are found both under the sea and on dry ground. The formation of oil and gas (and coal also) is a continuous though exceptionally slow process but, with the exponentially increasing rate of exploitation and deforestation, mature deposits will likely become increasingly rare. Furthermore, the potential effect of ocean pollution on the process of oil and gas formation, especially on the microorganisms that break down the deposits is not clear.

2.5.3   Nuclear energy

It is possible to spit the nucleus of large atoms of some metals (for example, uranium and plutonium) by controlled bombardment with fast-moving neutrons. Enormous binding energy is released, which can be processed to generate heat, raise steam or produce electricity. The first major use of this energy source was in producing the atomic bomb that was used in the Second World War, but soon after, it found use as a prime fuel for navy submarines because only a small quantity of fuel can keep a submarine running for over a year without refueling. The technology also quickly proliferated to the power sector in the nineteen fifties, and has remained an important resource in the global primary energy mix, providing around 10% of global electricity. The 1973 global oil crisis gave prominence to the issue of energy security and had a significant effect on many countries which had relied heavily on oil importation, such as the United States, Japan and France. This experience forced such countries to invest heavily in nuclear power generation because most of the input can be sourced locally.

Uranium, the primary fuel for nuclear power generation is abundant and occurs in most rocks, although in very small concentrations of a few parts per million, and like other elements, occurs in several slightly differing forms known as *isotopes*. Uranium also occurs in sea water and can be recovered from the oceans. Natural uranium is largely a mixture of two isotopes: U-235 and U-238, the latter being the more dominant, accounting for over 99% of a typical ore. The isotope U-235 is favored for energy generation because it is easily fissionable and the rate of reaction can be controlled, but ores have to be upgraded to raise the concentration of U-235 from around 0.7% to 6-8%. Other isotopes, notably U-238 can also be fissioned when bombarded by fast neutrons, yielding significantly higher energy but the reactions are much more difficult to control. Plutonium-239, another radioactive element may be produced in traces during U-238 fission, or by processing fission waste. Plutonium 239 is highly fissionable and can sustain a nuclear chain reaction, hence the element is favored for nuclear weaponry. Uranium ore deposits have been found in many countries across many regions and proved reserves are believed to be only a small proportion of the total global resources.

2.6   RENEWABLE PRIMARY ENERGY RESOURCES

Energy from the Sun (solar energy) is probably the most dominant of the renewable energy resources. It is available everywhere in the world, though at variable times and intensities. It can be used directly as a source of heat for drying, it can also be converted into heat or electricity through solar panels and inverters. Bioenergy is also abundant, can be sourced from organic crops and waste, and used for raising steam for industrial and home heating, or for generating electric power. Biomass can also be converted to premium fuels and chemicals. Most countries have rivers that can be impounded to produce electricity and many have sea coasts where the winds or tidal waves are sufficiently strong and consistent to drive turbines for power generation. Geothermal energy comes from the hot core of the Earth and is being tapped in many countries to heat homes, pools, and generate electricity. The Earth's gravity is not a form of primary energy but is a source of potential energy that can be applied usefully. If for example a mass is lifted up it acquires potential energy due to gravity, commensurate with its height. When it is dropped, it releases the potential energy in addition to kinetic energy from acceleration due to gravity. This energy system is used often in crushing and demolition processes. As discussed above, fossil fuels and nuclear ores are also technically renewable because new deposits are being formed continuously. However, it takes several hundred million years from initial formation to maturity.

2.7   WORLD PRIMARY ENERGY RESERVES

Enormous potential primary resources are believed to exist on land and under the deep seas. The Sun's abundant energy, wind and hydro energy are largely untapped. The Earth's crust is also believed to hold abundant reserves of geothermal energy which could contribute significantly to the world's energy resources. Extensive research and development are in progress worldwide and new technologies for retrieving energy sources that are considered unrecoverable presently are evolving. Due to impressive advances in technology, proven reserves of primary energy have increased over the last two decades or so despite the increasing exploitation rate. On average, only about 50% of the original oil-in-place volumes in reservoirs are recoverable using currently available advanced technologies (IEA, 2013). New enhanced oil recovery (EOR) drilling and seismic technologies, and CO_2 injection are raising recovery factors significantly in many oil fields around the world. World reserves and resources of unconventional oil and gas are enormous and largely unexploited. These include high-viscosity oil and tar-sands, tight gas, shale gas, and coal-bed methane. New technologies are being developed for the exploitation of these vast resources, including steam processing to reduce viscosity and advanced separation processes to improve oil and gas extraction. Much of the global natural gas resources which had been considered uneconomical due to inaccessible locations are now being exploited using new, advanced drilling technologies, liquefaction processes and complex pipeline transportation.

The recoverable coal reserves using currently available advanced mining technologies are believed to be less than 5% of the resources (Table 2.1). New technologies are being developed to move more coal resources to reserves. These include advanced mining technologies to improve recovery factors of coal-in-place from operating and abandoned mines, and in-situ gasification technologies to recover and utilize underground coal seams which are considered too deep for mining, and deposits considered to be uneconomical, based on current mining techniques. Over 80% of the global primary energy reserves of primary energy are located in four regions, with Europe and Eurasia holding over 40% (Figure 2.4), although the geographical distribution varies significantly depending on the specific energy source. While only a few regions hold most of the oil and gas reserves, coal is much more widely distributed and is available in every continent of the world.

Table 2.1   World's coal reserves and resources. *(IEA, 2013).*

COAL TYPE	RESERVES		RESOURCES	
	GIGA-TONNES (GT)	TRILLION BARRELS OF OIL EQUIVALENT (BOE)	GIGA-TONNES (GT)	TRILLION BARRELS OF OIL EQUIVALENT (BOE)
Hard coal	730	3.6	18,000	88.8
Lignite	280	0.7	4,000	19.7
Total	1,010	4.3	22,000	108.5

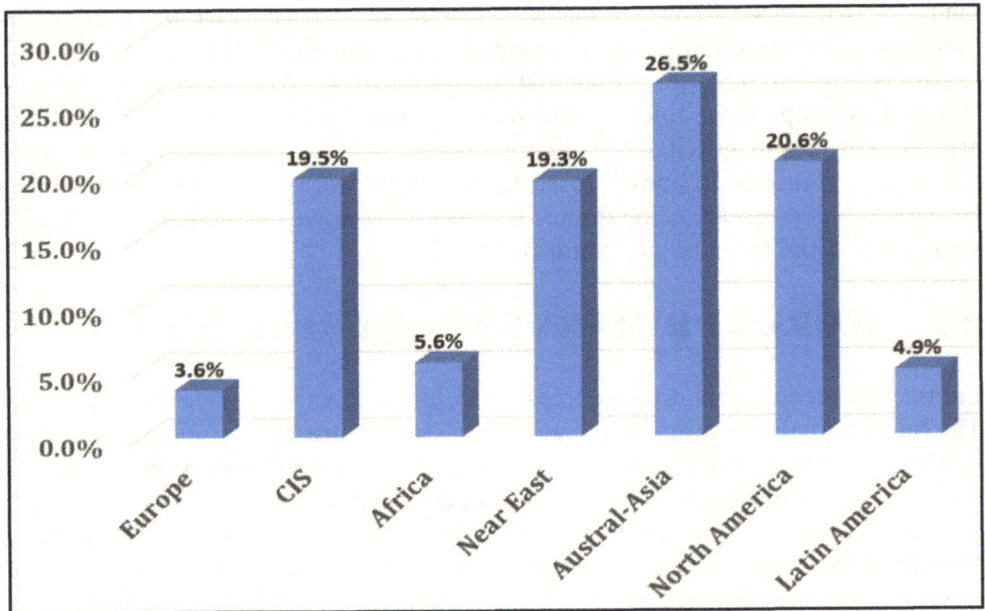

Figure 2.4 World's proven reserves of primary energy, 2011 (1,350 billion metric tons of coal equivalent). *(BP, 2014).*

As discussed earlier, reserves are energy sources that can be located and exploited using currently available technologies. It is estimated that non-renewable resources could be many times the currently proven reserves. Furthermore, in view of the fact that deposits have been formed continuously for hundreds of millions of years and the potentially disruptive of human activities intensified only in the last 200-300 years or so, it is possible that many currently immature deposits will become available to future generations. The renewables are considered inexhaustible but fossil fuel reserves could last a hundred years or more at the current rate of exploitation. Uranium could last up to a thousand years in view of the fact that very small quantities are required to generate very large energy. For example, one metric ton of natural uranium would produce about 45 million kilowatt-hours of electricity. Over 20,000 metric tons of black coal or 8.5 million cubic meters of natural gas would be required to produce the same amount of electric power. Furthermore, developments in fission and fusion reactor technologies could make nuclear fuel resources virtually inexhaustible.

2.7.1   Global primary fossil fuel reserves

Fossil fuels constitute around 97% of the proved global non-renewable energy reserves, and there are wide variations in geographical distribution (Figures 2.5 – 2.8). Various estimates indicate that proved reserves of oil and gas are adequate to meet global demand for around fifty years on the average at the current rate of exploitation. However, there are regional variations. For example, the projections for South/Central America and the Middle East for oil are 120 and 70 years respectively. Coal will last much longer, over a hundred and fifty years on the average, but again with significant regional variations. Reserves in the Middle East will last only fifty years or so while projections for North America and Europe/Eurasia are in the range 300 – 350 years. It should be noted however that proved reserves may be a small fraction of actual resources for reasons given in section 2.4.

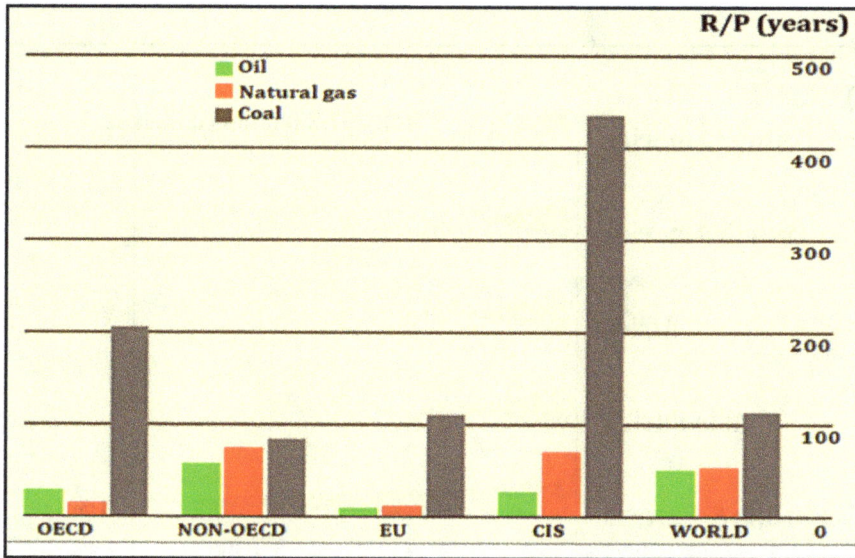

Figure 2.5   Fossil fuel reserves-to-production (R/P) ratios, 2015. *(BP, 2016)*

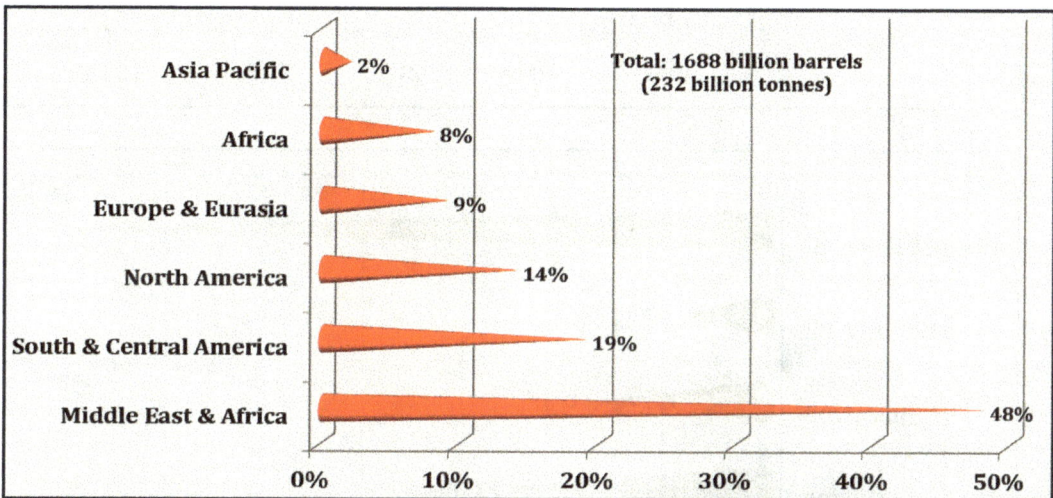

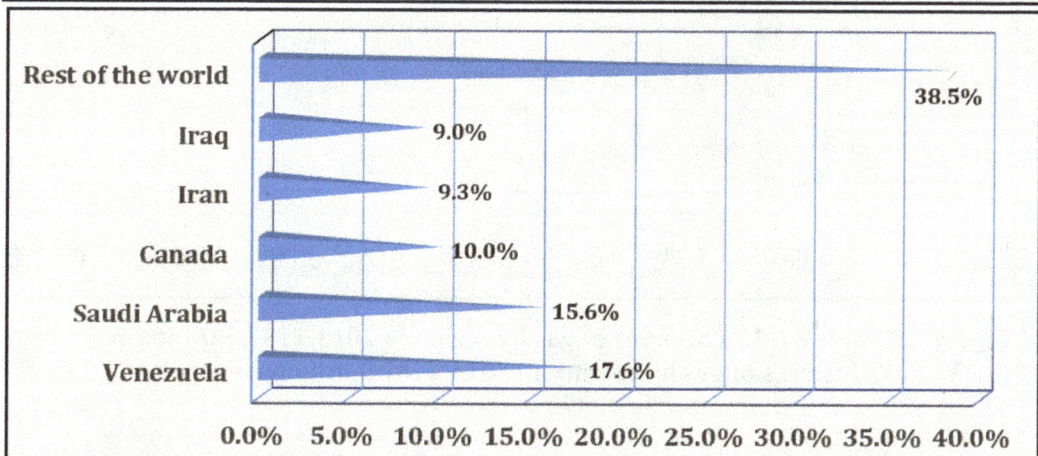

Figure 2.6   World's reserves of oil by region and country in 2013 (1687.9 thousand million barrels, 238.2 thousand million metric tons). *(BP, 2014).*

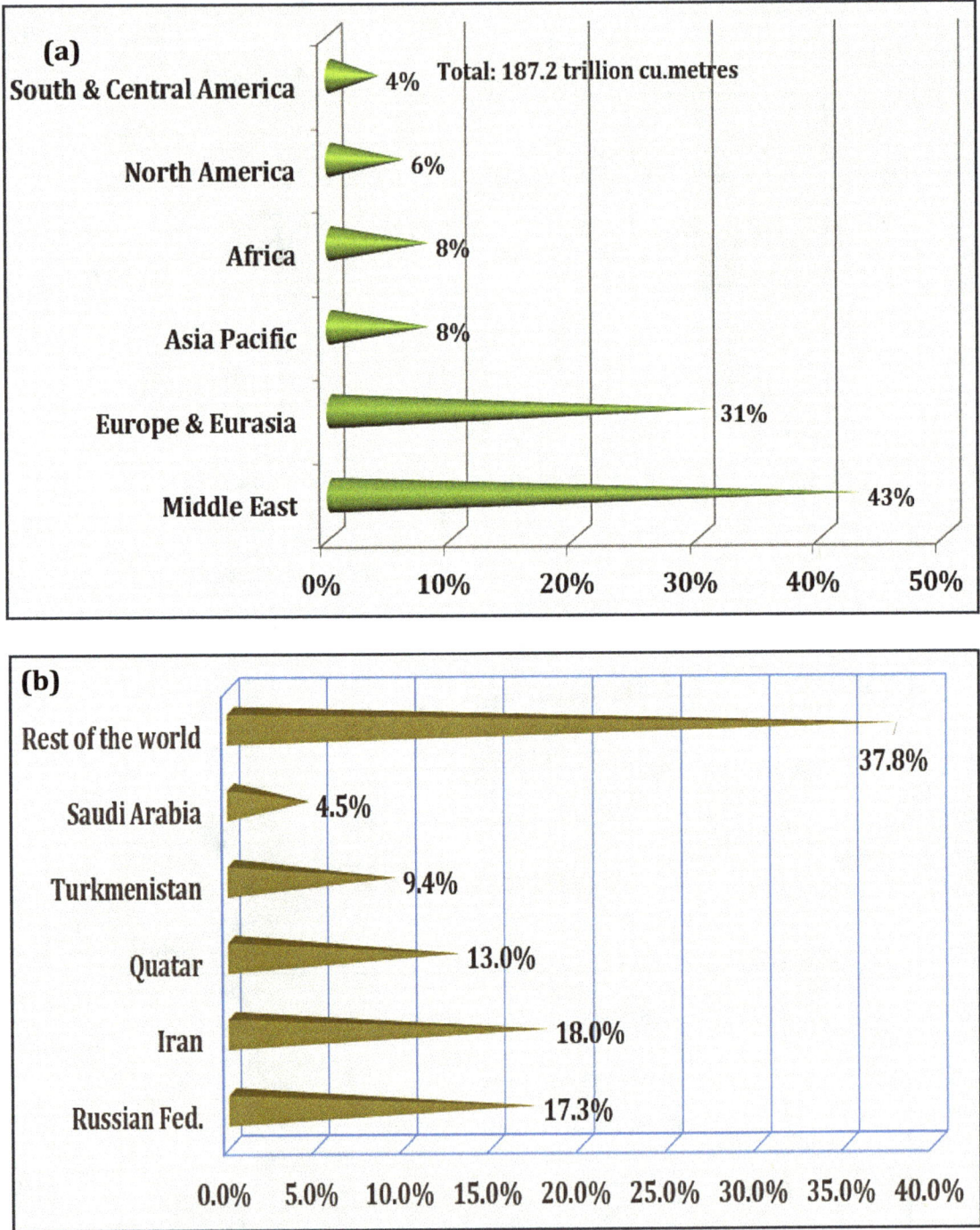

(a)
South & Central America — 4%   Total: 187.2 trillion cu.metres
North America — 6%
Africa — 8%
Asia Pacific — 8%
Europe & Eurasia — 31%
Middle East — 43%

0%   10%   20%   30%   40%   50%

(b)
Rest of the world — 37.8%
Saudi Arabia — 4.5%
Turkmenistan — 9.4%
Quatar — 13.0%
Iran — 18.0%
Russian Fed. — 17.3%

0.0%   5.0%   10.0%   15.0%   20.0%   25.0%   30.0%   35.0%   40.0%

Figure 2.7   (a) World's reserves of gas by region in 2013 (187.3 trillion cubic meters)
(b) World's reserves of gas by country in 2013 (187.3 trillion cubic meters). (*BP, 2014*).

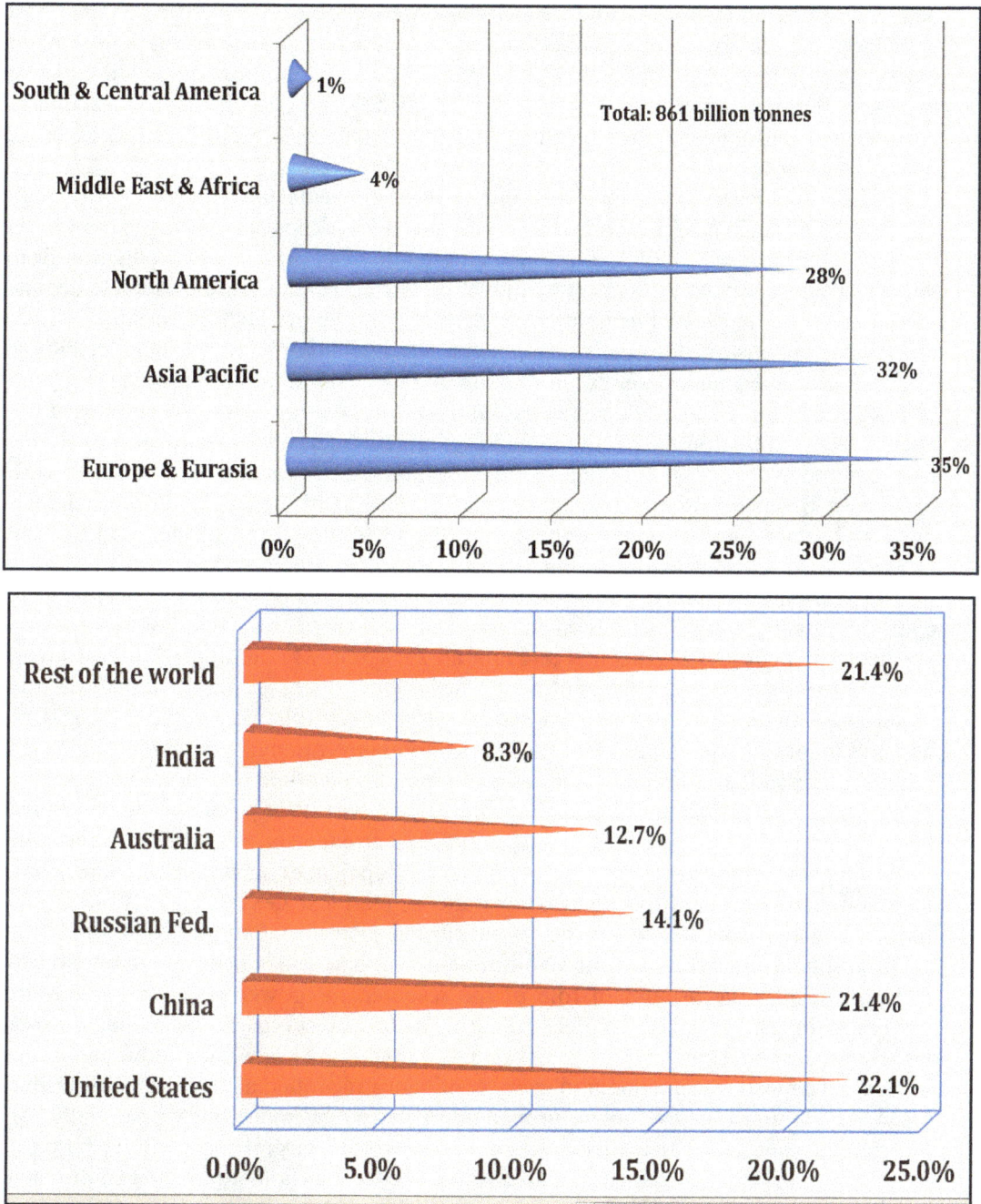

Figure 2.8 World's reserves of coal by region and country in 2013. *(BP, 2014).*

Most of the world's oil and gas reserves are off-shore, the bulk of it at depths well beyond the reach of current drilling technologies. Furthermore, on average, only about 50% of the original oil-in-place volumes in reservoirs are recoverable using currently available advanced technologies (IEA, 2013), but new enhanced oil recovery (EOR) drilling and seismic technologies, and CO_2 injection are raising recovery factors significantly in many oil fields around the world. The recoverable coal reserves using currently available advanced mining technologies are less than 5% of the resources. Furthermore mines previously considered depleted are being re-opened and deposits that were previously regarded as inaccessible or uneconomical are now being gasified *in-situ* to generate heat and power. This explains why reserves are increasing by the year as new exploration and exploitation technologies become available. For example, shale oil is believed to hold around four times as much oil as conventional oil deposits, and coalbed methane resources are estimated to be at par with natural gas resources.

Substantial reserves of shale oil and coalbed methane have been identified in many countries in all regions of the world. However, only a few deposits are being exploited currently because appropriate technologies are just emerging. Also, oil and gas production from shale deposits is more expensive than conventional processes, and investments tend to decline when global oil prices are low. Furthermore, there are many unpredictable variables which make reserve projections of fossil fuels difficult and unreliable. For example, new technologies are improving efficiencies of energy production and use, and therefore reducing the amount of primary fuel required to produce a quantum of end-use energy. Also, the amount of delivered energy required to produce a unit amount of work (for example, energy use per GDP or auto fuel required to cover 100 kilometers) is decreasing. Other unpredictable variables are the rate of growth of energy demand in emerging countries, turbulence in the global oil market dynamics, emergence of new technologies, the very large areas of land and oceans of the world that are yet to be explored, etc.

Five countries held about 62% of the global reserves of oil in 2017. However, many countries remain largely unexplored and may hold substantial reserves. For example, attention has been turning to Africa recently and around 35% of newly discovered reserves worldwide in the last five years are on the continent. Furthermore, there are large unquantified reserves of unconventional fuels - shale, oil sands, extra heavy oil, natural bitumen and methane deposits in many parts of the world. All of the non-conventional fossil fuels (shale, heavy oil, bitumen, tar sand) are potential sources of oil and, if taken into account, total reserves could be four times larger than the current conventional reserves (WEC, 2013). Natural gas resources are estimated to be more than oil resources and, by current estimates, five countries hold about 62% of the total global reserves. Interest in the development of natural gas resources has grown in recent years because of its increasing status as the most flexible and efficient fossil fuel for power generation. It also has the lowest carbon footprint of all the fossil fuels.

Investment in coal exploration and mine development has been low compared with oil and gas, despite the prominent role of the resource for power generation, due largely to economic forces. In the last two decades, proven global reserves have declined by about 14% largely because investment in prospecting has all but dried up, while production has gone up 62%. Coal deposits are abundant in every continent and around a hundred countries including many developing countries have proven reserves. Many more have potentially large but unexplored resources. The United States holds the world's largest coal resource base estimated at 22% of the global reserves and, with four other countries hold about 79% of the total world reserves. China led the other four countries as the world's largest producer of coal, and was also the world's largest consumer in 2016, accounting for nearly half of the global consumption. Coal will remain prominent in the global primary energy scenario for decades to come, primarily because there is no viable substitute in many applications, notably iron and steel and cement production but also because, for many emerging countries, it is the only locally available primary energy source that can readily provide access to energy for their teaming populations. However, increasingly stringent emission regulations in many developed countries and dropping prices of natural gas and renewable energy are forcing existing coal plants to switch to gas or close down, but many still depend heavily on coal for power generation – United States and Germany, 43% each; Australia, 69%; Poland, 86%; China,

81%; South Africa, 94%. Furthermore, the import-export market for hard coal for steam raising and coking coal for steelmaking is strong and likely to remain so in the foreseeable future.

2.7.2 Nuclear energy

Uranium, the primary source of nuclear energy occurs naturally as uranium oxide in the Earth's crust, in rocks, in granite which makes up 60% of the Earth's crust rivers and seas, and in some coal deposits. The concentration is usually very low, mostly between 0.01% and 0.05%. However, some deposits have sufficiently high concentration and can be processed to obtain nuclear fuel. These are classified as uranium ores. These deposits may be no more than about 100 meters deep and can be recovered by opencast mining. Deeper deposits are exploited by underground mining (WNO, 2017a). Many ores have low concentrations of around 0.1% or less, but are still good for processing into energy although such ores need to be upgraded by enrichment processes to 3 - 7% U-235 by weight. Uranium constitutes less than 3% of the total world energy reserves (in metric tons of coal equivalent) (Figure 2.3). Over half (52.8%) of the reserves are located in Austral-Asia, 16.7% each in North America and CIS, 8.3% in Latin America, and 5.5% in Africa (Figure 2.9). Another recent estimate by OECD/IAEA (2016) which gives much higher reserve values categorizes resources and reserves in terms of value and major locations of deposits as shown in Table 2.2 and Figure 2.9b. Total reserves (in situ and recoverable) was estimated at about 38 million metric tons but undiscovered and unproven resources are believed to be much higher, perhaps a hundred times for once-through fuel cycles and several thousand times for breeder reactors which are capable of producing more fissile fuel than they consume.

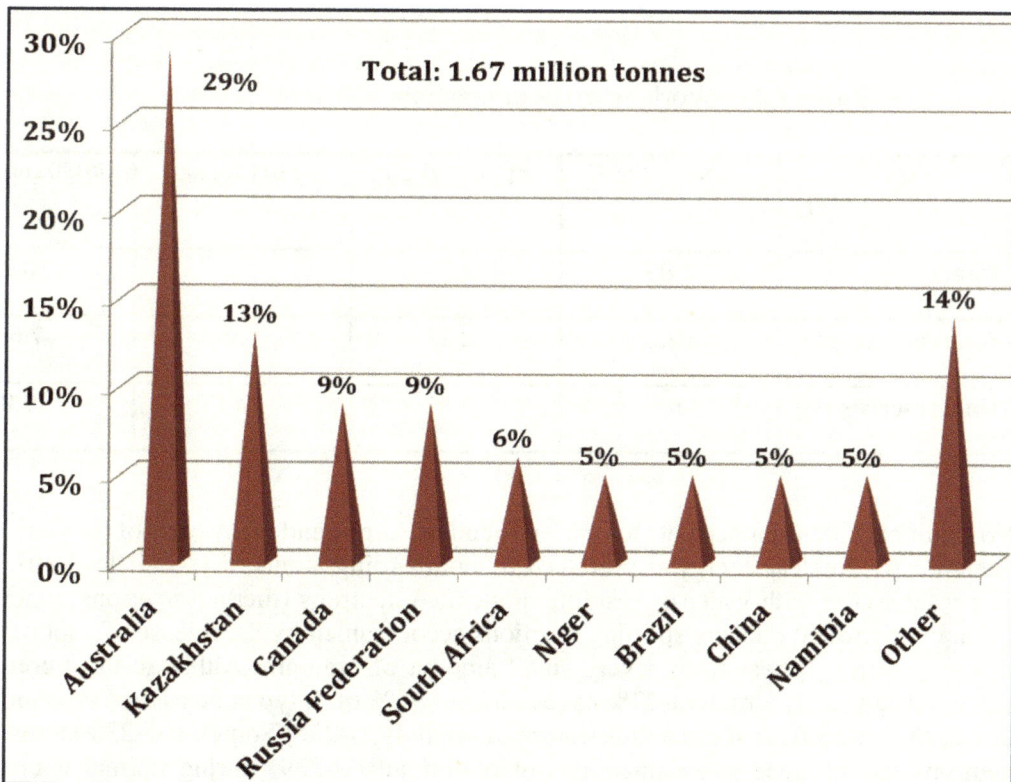

Figure 2.9a Global proved reserves of nuclear fuel. *(Data from world-nuclear.org, 2018).*

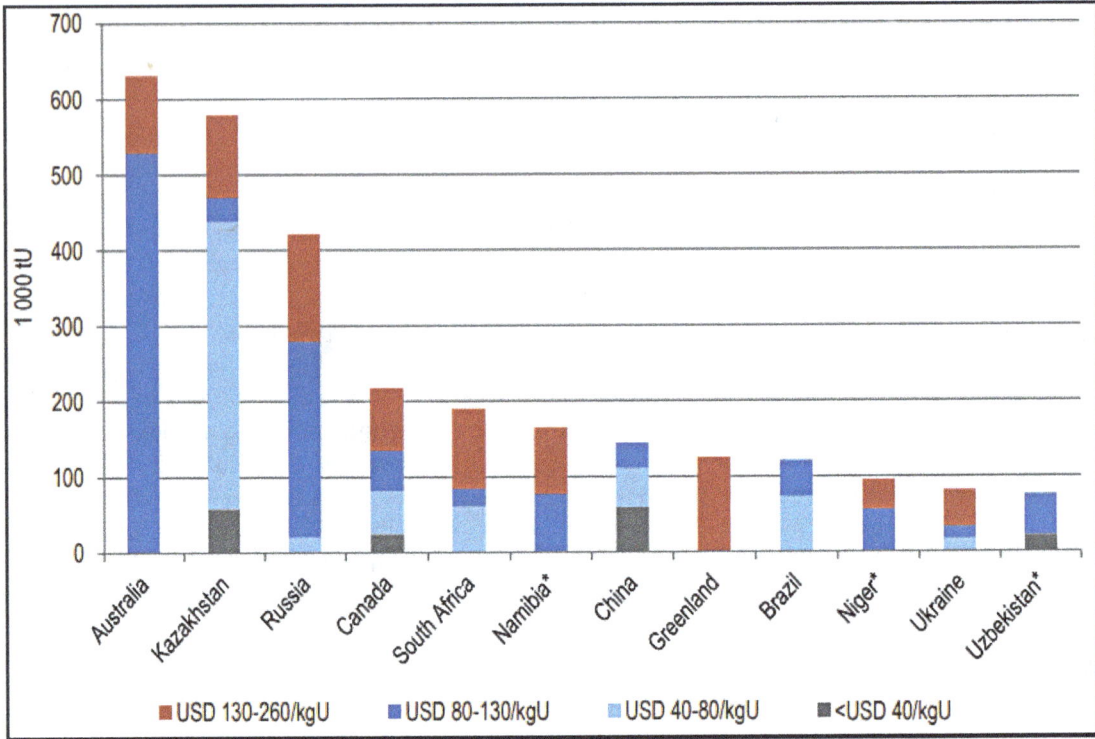

Figure 2.9b   Global proved reserves of nuclear fuel. *(OECD/IAEA, 2016).*

Table 2.2   World reserves of uranium. *(OECD/IAEA 2016)*

Reserves (million tonnes)	<USD40/kgU	<USD80/kgU	<USD130/kgU	<USD260/kgU
Total in situ	0.85	2.7	7.7	10.2
Total recoverable	0.65	2.1	5.7	7.6
Total reserves	1.5	4.8	13.4	17.8

No significant uranium deposits have been found in Europe and many parts of the world remain unexplored. Nuclear energy is produced by bombarding atoms of U-235 low-enriched ore in a reactor core with water or graphite-moderated neutrons (thermal neutrons) that set up a chain of controlled nucleus-splitting reactions accompanied by the release of a lot of binding energy in form of heat from a very small amount of uranium. Although most commercial reactors use U-235, Uranium-238 makes up over 90% of a typical enriched uranium ore. In effect, the waste from the reaction comprises mainly U-238. Some of U-238 atoms capture neutrons and produce small quantities of plutonium (U-239) during normal nuclear plant operations, and about half of this is also fissioned, providing about a third to over half of the reactor's energy output, depending on the reactor type and design. A typical spent fuel would have about 1% U-235, around the same level of plutonium, and about 95% of U-238. A new generation of reactors known as fast breeder reactors use fast neutrons which are not so good for fissioning U-235, but are better absorbed by U-238 which is more abundant in the ore. Apart from the capability of generating up to 30% more energy than normal reactors, operating conditions of fast breeder reactors can be controlled to produce plutonium-rich spent fuel. In effect, the spent fuel of a normal thermal nuclear reactor which still contains around 96% of its original uranium, is a valuable feed fuel for fast breed reactors which are expected to be in common use in the near future. This is why U-238 (which is also the most abundant type in natural uranium ore) is considered a fuel of the future. Spent nuclear fuel can be reprocessed to extract U-238 for fueling fast breeder reactors, or to extract plutonium, the prime fuel for nuclear weaponry.

Fast breeder reactors produce more energy than they use, and wide deployment can extend the life of global resources of uranium ore indefinitely. However, after decades of development, there is at present no commercial fast breeder reactor anywhere in the world, although there are research units in some countries. Of particular concern is the fact that spent fuel of breeder reactors can be enriched to produce nuclear weapons-grade plutonium, which is at cross purposes with the current global efforts to control the spread of nuclear arsenal. Also, reprocessing will create highly radioactive waste and potentially high radiation exposures. For these reasons, most of the leading countries involved in the research and development (U.S., U.K., Germany, France) have shut down their research reactors, but some countries, notably India, Russia, Japan and China still have operational fast breeder reactor programs.

2.7.3 Renewable energy reserves

Renewable energy is the group of energy sources that derive from resources which are naturally replenished – solar, biomass, hydro, wind, tidal, geothermal. Technically, as discussed earlier, fossil fuels which are classified as non-renewable also fall in this category, the difference being the timescale. While it would take 350 to 400 million years to replenish the fossil fuel resources currently being exploited, energy is available on a continuous basis from the Sun, the wind, the tides, and a third of the Earth's surface is covered by water which could be harnessed for hydropower or tidal energy. Also, while fossil fuel resources are concentrated in relatively few geographical locations, renewable energy resources are available worldwide. Bioenergy is a group of fuels of biological origin. The group includes wood which is perhaps the oldest source of energy used by mankind apart from solar energy. Other fuels in the group are charcoal, organic waste from forestry and agriculture, municipal solid waste, animal manure, human waste, and biofuels derived from crops including soybean, sugarcane and corn.

Biomass contains stored energy from the Sun in plants in form of carbon produced through photosynthesis. Animals and humans that feed on plants also excrete waste that is rich in carbon. The energy can be released by combustion to generate heat and electricity, or processed biochemically to obtain biofuels. Biomass resources can be widely regenerated on an annual timescale and some countries, notably Brazil and the United States are doing it on a large scale. Geological activities in the Earth's crust which generate enormous geothermal energy occur on a continuous basis and are becoming a significant source of heat and electricity in some countries. The Sun's energy that falls on the Earth's surface daily is enormous and it is estimated that capturing less than 0.02% will provide enough energy to meet the world's total annual primary energy requirements (Tester *et al.*, 2005). Solar and wind power are the two renewables expected to lead the decarbonization of global energy use, in particular, the power sector, but there are formidable challenges: solar and wind are variable sources of energy, and their output depends on real-time availability of wind and sunlight respectively, making it very difficult to maintain the necessary balance of electricity supply and consumption at all times. This makes their output variable over time, hence the classification as variable renewable energy (VRE). It is also not possible to perfectly predict resource availability with any degree of precision ahead of time. Integrating a VRE source into a conventional power system has been a major challenge because the system itself faces variable power demand that cannot be predicted with perfect accuracy, and possible unexpected outages. Variability of demand has always been an issue with electric power supply, since load variability within the day between peak and minimum demand can be very substantial, often by a factor of two. On the other hand the supply side is full of uncertainties, the most serious being unexpected failure of plants or other system components, and unpredictable fluctuation in real-time load demand. This is complicated by the fact that there are as yet no technologies for storage of electricity in large quantities economically. However, emergence of increasingly sophisticated control systems is making it possible to integrate VRE systems with as low as 5% to 10% generation into conventional power systems. The share of renewables in the global energy demand mix is currently less than 10%, used mainly in power generation, heat generation and transportation. However, growth rate is very fast and an increase of about 20% is projected over the next five years (IEA, 2018). The fastest growth rate is in the electricity generation sector, expected to grow from the current 24% to around 30% in 2023, meeting more than 70% of the global electricity generation growth in the period, led by solar PV and followed by wind, hydropower and bioenergy. However, in terms of total contribution, hydroenergy will remain the largest renewable source, meeting 16% of global electricity demand, followed by wind (6%), solar PV (4%) and bioenergy (3%) in 2023 (Figure 2.10). A similar growth rate in renewable heat consumption (buildings and industry) is expected over the next five years with a share of about 12% of the heating sector demand. The contribution of renewables to the transport sector, mainly from biofuels is the lowest, around 3.5% currently, and projections indicate that contribution will still be less than 4% in 2023, but growth of renewable electricity consumption in transportation is very rapid, although starting from a very low base, and contribution is expected to grow by about 65% over then next few years.

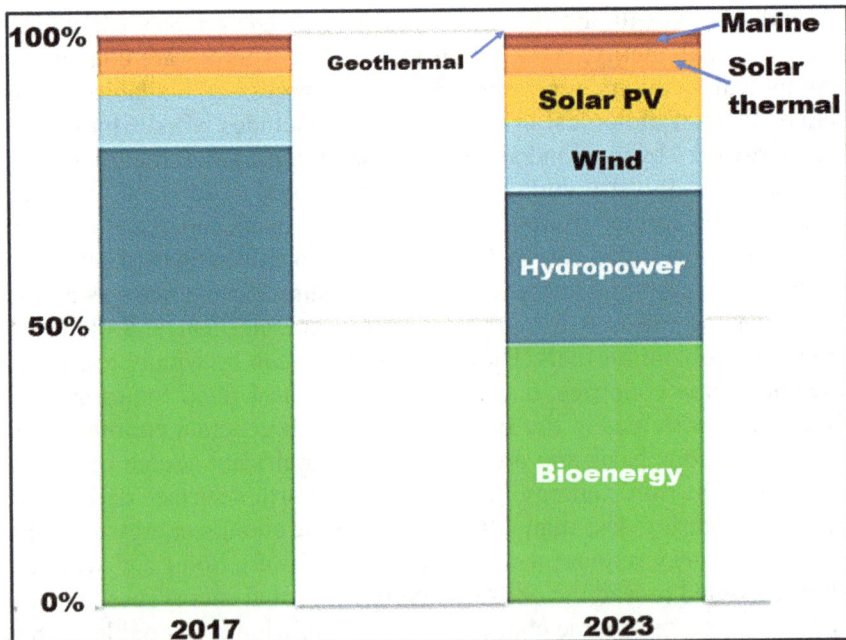

Figure 2.10 Projection of renewable energy mix by technology, 2017-23 *(IEA, 2019e).*

2.7.3.1   *Bioenergy*

Traditionally, bioenergy has been produced by direct combustion of biomass (organic matter). Using such fuels remains the primary energy source for many people in the developing world but, in view of the inefficient combustion leading to harmful emissions with serious health implications, this traditional biomass use is unsustainable. However, there are many pathways to the conversion of organic biomass to solid, liquid and gaseous forms that can more conveniently provide useful, environment-friendly energy, and possibly replace fossil fuels in many applications. A wide range of technologies have now become available for converting biomass to biofuels which are more efficient, more environmentally sustainable, and can be utilized or transported more conveniently (Figure 2.11). Also, a wide range of biomass feedstocks can be used as sources of bioenergy. These include organic waste, municipal waste, agro-waste, crops grown specifically for energy, such as corn, soybean, wheat, sugar and vegetable oils, and nonfood crops such as ligno-cellulosic plants and grasses.

Most biofuels are made by fermentation of sugar derived from wheat, corn, sugar cane, sugar beets, molasses, etc., using microorganisms and enzymes. Starch (from corn, potato, fruit waste), vegetable oils and animal fats are also fermented to produce alcohols, mostly ethanol but propanol, butanol and biodiesel are also produced. While butanol can be used in gasoline engines directly as a replacement for gasoline, ethanol is blended with gasoline, and diesel is also blended with biodiesel. Traditional use remains the largest end use of biomass and biowaste, but modern use is growing, particularly in the heat sector. Biogas (mainly methane, also known as landfill gas) that forms when garbage, agricultural waste and human waste decompose is becoming important as a source of household energy, and small digesters are in use especially in developing countries. Biofuels hold great promise as an effective option for decarbonizing the heat and transportation sectors of the global economy which account for 80% of the total final energy consumption: gasoline blended with bio-ethanol and diesel containing up to 20% biodiesel are already available in fuel stations in some countries; solid biofuels are used increasingly in generation of heat for the housing and industrial sectors.

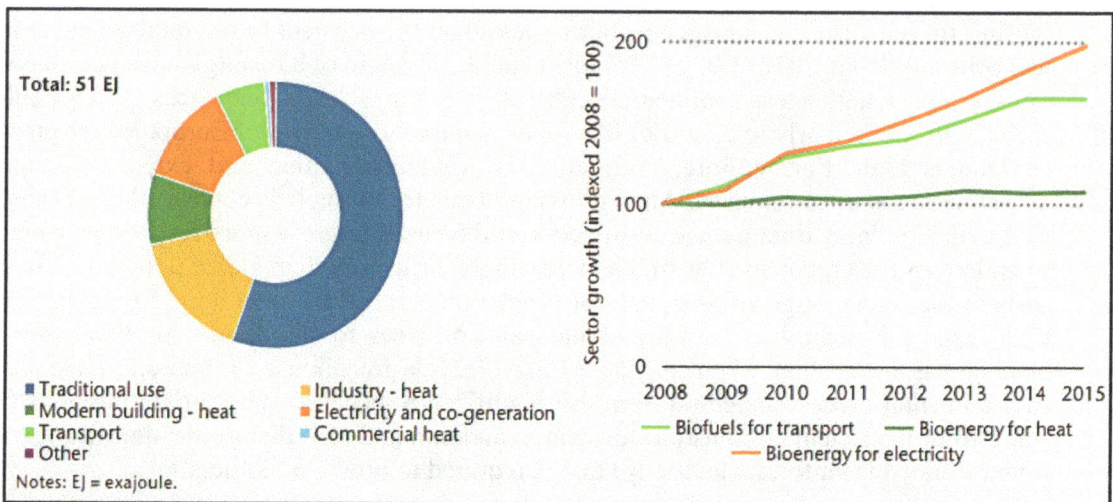

Figure 2.11a   End-use of biomass energy pathways: from biomass to final energy use *(IEA, 2017j)*.

FEEDSTOCK	PRODUCTION PROCESS	PRODUCT	END USE
Oil crops (e.g. palm, canola, sunflower)	Transesterification	Biodiesel	Biofuels for Transport
	Fermentation	Bioethanol	
Sugar and starch crops (e.g. sugar cane, corn, cereals)	Advances biofuel processes	Renewable diesel	
		Cellulosic ethanol	
	Chipping	Other advanced biofuels	
Lignocellulosic biomass from forestry, agriculture and other industries (e.g. forestry residues, straw, bagasse)	Pelletisation	Woodchips	Combustion for □ Electricity □ Heat □ Co-generation
		Pellets	
	Pyrolisis	Pyrolysis oil	
	Gasification	Bio-synthetic gas (Syngas)	
Biomass from waste (e.g. biomass fraction of MSW, wet wastes from agriculture)	Sorting, separating and fuel preparation	Refuse Derived Fuel (RDF)	
	Anaerobic digestion	Biogas	Biomass-based materials and products

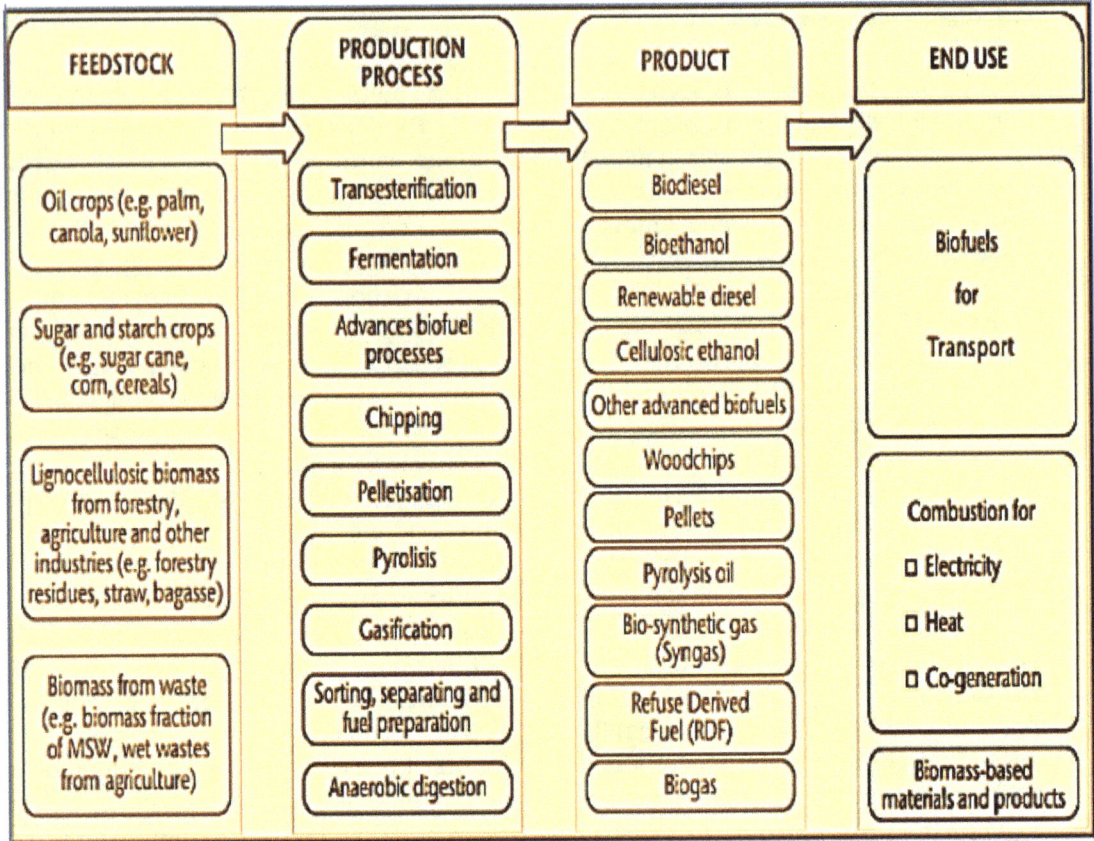

Figure 2.11b Potential bioenergy pathways: from biomass to final energy use *(IEA, 2017j)*.

Sustainable Development Aerofuel (SDA) promoted by the International Energy Agency under its Sustainable Development Scenario (SDS) projects biofuels reaching around 10 per cent of aviation fuel demand by 2030, and close to 20 per cent by 2040. However, aero biofuel use is starting from a very low base, contributing less than 0.1 per cent to the total global aviation fuel consumption in 2018. The development and deployment of bioenergy have been slow due to two major challenges: commercial cultivation requires large land areas which could be problematic particularly in countries like Japan and many countries of Europe where premium on land is high. Furthermore, cultivation is weather-sensitive and extensive irrigation infrastructure is often required. Also, growing plants for biofuels is controversial because the land irrigation infrastructure and fertilizers could be used to grow more food crops which are considered more urgent in view of the increasingly virile growth in global population. In some parts of the world, large areas of natural vegetation and forests have been cut down to grow sugar cane for ethanol and soybeans and palm oil trees for biodiesel. Another issue with bioenergy is the fact that, when considered on a lifecycle greenhouse gas emissions, the positive environmental credentials could diminish significantly because substantial emissions could come from inefficient combustion for heat generation and also due to the significant energy (often sourced from fossil electricity) that is required to produce biofuels.

2.7.3.2   Solar energy

The Sun's radiation about 60% of which reaches the Earth is the most abundant primary energy source both in intensity and geographical availability. The annual potential of solar energy is several times larger than the total world primary energy consumption (UNDP, 2000). Solar energy use has been boosted by significant developments in solar PV technology, notably, sharp cost reductions and strong policy support. Power generation has been the main beneficiary of solar energy, and China leads the world, with India following closely. In 2016, solar PV power generation capacity grew by 50%, with China alone accounting for almost half of this expansion. The record low generation cost per kilowatt-hour is becoming increasingly comparable or lower than generation cost of new, modern gas and coal power plants. Solar energy deployment has the greatest potential for decarbonizing power generation, because of the declining costs of solar PV, but also because of the very wide variety of deployment options, from small units that power cell phones, to utility-scale stand-alone household units, to large power plants. The global PV generating capacity has grown five-fold in the last five years, due largely to dramatic cost reductions of PVs, around 50% over the period. Variability is the main challenge and capacity utilization could be as low as 15-20%, but rapid developments in battery storage technologies and advanced control systems that help integrate solar PV power with power grids will help to a large extent in mitigating this problem.

2.7.3.3 *Wind energy*

Wind energy has been used for centuries in driving corn mills, water pumps, etc. Like solar energy, wind power is available virtually everywhere on the Earth. In fact, wind energy derives from the uneven heating of the atmosphere by the Sun. The speed of movement of the wind is controlled by the terrestrial thermal balance and distribution, the Earth's terrain, the Earth's rotation, vegetation, oceans and seas, etc. Wind speed is usually strongest over open oceans which cover over 70% of the planet because of fewer obstructions, and also because they pick up energy from warm waters. Strong wind passing through a propeller can generate considerable thrust, very much like in the aircraft, and can be used to drive turbine-alternator systems to generate electricity. If it were possible to harness just 5% of the available potential wind energy, it would provide electric power equivalent to the current global total primary energy needs. However, suitable sites are relatively few and variability is a significant challenge. Also, plants operating in the best locations often return no more than 15-30% efficiency. Wind speed is a critical feature of wind-powered systems because it determines deliverable energy, hence terrains with as few obstructions as possible are the best sites, and there are many all over the world. The major challenge is the intermittent nature which makes it hard to synchronize availability with demand for electricity. Also, good sites are often located in remote areas far from power demand centers. Furthermore, because wind farms require substantial land areas and are noisy, resistance is building up in some developed countries which account for most of the investments in wind energy. Compared with fossil fuel-powered generating plants, capital investment is significantly higher, but a life-cycle cost analysis shows that wind-powered plants are cheaper because of the lower operating costs, since wind is free and there is no fuel to purchase. For this reason, wind energy has high value as a secure source.

2.7.3.4 *Hydropower*

Hydro-energy is a renewable resource because water is the primary fuel, and the global water cycle is controlled by the Sun. Virtually every country has potentially suitable hydro dam sites and, although capital investments can be very high, the technology produces electricity at a cost that is lower than any other power generation technology, renewable or otherwise, and produces no direct emissions. Also, impounded reservoirs offer a variety of recreational opportunities, notably, fishing and recreation. Hydropower is the leading source of renewable electricity generation globally, currently supplying over 70% of renewable electricity and 16.4% of the world's electricity from all sources. There has been a major upsurge in hydropower development globally in the past decade or so during which total installed capacity has grown by around 40%. The rise has been concentrated in emerging countries where they not only provide electricity, but also water services and energy security. China has the world's highest installed hydro-dam capacity, three times the capacity of the next country: the United States. Although worldwide undeveloped potential of hydropower is huge, fewer new dams are being built in the developed countries because of environmental concerns. However, development in emerging regions remains strong, energized by the United Nations' sponsorship of hundreds of mini-hydro-dams for developing countries.

2.7.3.5   *Geothermal energy*

The Earth's crust stores an enormous amount of energy, believed to have originated largely from radioactive decay of minerals but also during the Earth's formation. The core mantle boundary which is about ten kilometers deep is believed to be about 4,000°C, hot enough to melt rocks and set up a geothermal gradient between the core and the Earth's surface. High pressures force molten rocks towards the Earth's surface, and eruptions of volcanoes and hot lava occur frequently in some regions of the world. Underground water that comes in contact with the heat becomes heated and penetrates to the Earth's surface as warm springs, and when hot enough, they heat homes, swimming pools, etc. In some locations the temperature could be as high as 380°C, enough to produce superheated steam which can be captured for electric power generation. It is estimated that the upper ten kilometers of the Earth's crust contains 50,000 times as much energy as found in all the World's fossil fuels. Geothermal energy is abundant in areas located around the boundaries of tectonic plates which make up the Earth's outermost shell also known as the lithosphere. Countries which are rich in geothermal energy include North and Latin American countries, African countries, Japan, Australia and Russia. Geothermal energy has been used since the ancient times as source of domestic hot water and space heating, and, currently, is being used extensively for space and industrial process heating, heating of pools and greenhouses, and water treatment in around 70 countries. However, focus has shifted in the last decade or two to electric power generation and commercial power plants are operating in around 25 countries of the world with a total geothermal generating capacity of 14 Gigawatts in 2017, and is projected to increase to about 17 Gigawatts by 2023, with most of the additions from Indonesia, Turkey, Kenya and the Philippines. The United States leads the world, accounting for around 30% of the global capacity while many other countries, in particular those located along the Great Rift Valley of Africa (Kenya, Ethiopia, Uganda), and others located around the East Mediterranean also have operating geothermal power plants. The potential for geothermal energy is believed to be enormous and it is estimated that only about 7% has been tapped so far (GEA 2015).

The world's largest geothermal plant (850 megawatts) is located in California, USA, and hundreds of new plants are being developed around the world. El Salvador and Iceland currently derive nearly 30% of their electricity from geothermal energy and Kenya plans to source over half of its electric power requirements from geothermal energy within the next decade (Brown, 2009). New technologies are being developed which involve drilling to the hot rock layer several kilometers below the Earth's surface, fracturing the rock and pumping water into the crack to generate superheated steam for driving steam turbines in power generating plants above ground. It is believed that when this new technology becomes fully developed, geothermal energy could supply the entire electricity requirements of many countries, including the United States. The technology roadmap for geothermal heat and power developed by the

International Energy Agency offers a strategic plan to maximize deployment of geothermal energy use and projects that 1,400 TWh of electricity per year could come from geothermal power by 2050, up from 85 TWh at present. The carbon footprint of geothermal energy is very low, around 5% of that of a coal-fired power plant hence the technology has considerable potential in decarbonizing energy.

2.7.4 New renewable energy reserves: hydrogen and nuclear fission

Hydrogen is a versatile energy resource which can be produced from many sources including sea water, coal, natural gas, coalbed methane, and biomass. It can be burned in oxygen to produce energy that powers gas turbines for electric power generation, and in the propulsion of spacecraft. It is used to power fuel cells and also has the potential to be mass-produced and commercialized for passenger vehicles and aircraft. Hydrogen energy is considered renewable because it can be produced from water and biomass. It is also considered a zero-emission fuel because the only products of combustion are heat and water. However, production of hydrogen from any source is considered inefficient because more energy is always required to produce the gas than can be retrieved from it in end use (Zehner, 2012). Furthermore, electricity required to produce the gas will likely come from generating plants powered by fossil fuels, renewables or nuclear energy, hence hydrogen has a significant carbon footprint on a life-cycle basis. Also, handling hydrogen is extremely dangerous and this could remain a significant constraint to widespread use as an energy source. There are many precursors to hydrogen production but the most common is natural gas which is reformed by reacting with high pressure steam at high temperature in the presence of a catalyst, in accordance with Equation 2.1. The carbon monoxide is reacted with high pressure steam at lower temperature to produce more hydrogen in accordance with Equation 2.2.

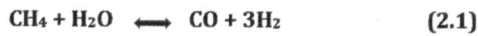

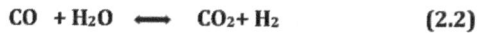

$$CH_4 + H_2O \rightleftharpoons CO + 3H_2 \qquad (2.1)$$
$$CO + H_2O \rightleftharpoons CO_2 + H_2 \qquad (2.2)$$

Synthetic gas (syngas) produced from coal or biomass contains mainly hydrogen and carbon monoxide and can also be steam-reformed to produce pure hydrogen (Equation 2.2). Research on new sources of hydrogen is intensive and the focus is on the electrolysis of water. The process involves application of an electric current to separate water into its components: hydrogen and oxygen. This process has numerous advantages: Water is abundant, electricity can be produced from wind or solar powered generators and the hydrogen produced can be liquefied, stored or transported for use as required, thus resolving the issue of variability of electric supply. Liquid hydrogen is a very versatile fuel and has powered space rockets for decades; many power plants feature hydrogen-powered gas turbine; and electric vehicles with hydrogen-fueled solar cells are already on the roads in some developed countries.

Another new energy technology which simulates the continuous process of energy production by the Sun is under development and may become important in the near future. Considerable research is on-going on simulation and exploitation of the processes by which the Sun and the stars generate energy. In contrast with nuclear fission which involves splitting atoms to release enormous energy, the Sun generates energy by fusing two hydrogen atoms under enormous pressure and high temperature to form helium thereby releasing a large amount of binding energy. Hydrogen is the primary constituent of the Sun and the prevailing internal temperatures and pressures are sufficiently high to sustain the reactions on a continuous basis. Nuclear fusion reactors which are currently under development derive hydrogen from sea water and other sources, and can produce enormous net energy that can be harnessed for electric power generation. The high temperatures and pressures required to sustain the reaction and availability of appropriate materials are major constraints and, although there are many research fusion reactors in the world, none has reached a commercial stage. Nuclear fission is often classified as renewable energy, first because it can be produced from hydrogen which comes from water, a renewable resource, but also because, unlike all other primary energy sources, it produces more energy than it consumes.

2.8   WORLD ENERGY RESOURCES OUTLOOK

As discussed in Section 2.4, it is difficult to project global energy resources and reserves because inferred resources and in situ reserves based on geological and geophysical data are believed to be much greater than proved reserves which can be exploited using currently available technologies. Furthermore, it is believed that many fossil fuel resources have not yet been located because large areas of the world remain unexplored. For example, much of the global oil resources is believed to be buried deep under the oceans, currently inaccessible using available instrumental and physical drilling techniques. This explains why global reserves are being reviewed upwards almost yearly as new prospecting and drilling technologies become available, and virgin areas are being investigated. Many parts of the developing world remain unexplored and, considering that about a third of new proven global oil reserves in the last five years have been found in sub-Saharan Africa and investments in coal prospecting have declined significantly, it is expected that proven fossil fuel reserves will continue to grow in the foreseeable future. Fluctuations in the global energy market and economy are also important unpredictable variables which can accelerate or depress growth rate of energy demand and alter the global energy mix. For instance, declining prices in the global oil or natural gas market always tend to depress investment in renewable energy. In effect, the data presented in Figure 2.5 showing a life span of 50-120 years for fossil fuels, based on current resource/production ratios should be regarded as tentative.

Uranium constitutes less than 3% of the total world energy reserves (in metric tons of coal equivalent), with over half located in Australasia. In view of the fact that the energy content of one metric ton of the fuel is equivalent to around 20,000 metric tons of coal, and with a growth rate of over 10% in the last five years, global reserves are very substantial. Current reserves are projected to last about a hundred years at the current rate of use, but geological and geophysical data indicate that global resources could be ten times more than current proved reserves, and commercialization of fast breeder reactors which are currently under development could extend the life span of the resource 100-fold or more. This type of reactor can be designed to operate in closed circuit with its reprocessing plant, or in a flexible configuration that can use both natural or depleted uranium. In effect, each metric ton of ore yields vastly more energy than in a conventional reactor. Global resources and reserves of renewable energy are inexhaustible, and could eventually supply the total annual global energy requirements, provided the numerous constraints to development and deployment can be surmounted. This will be discussed in some depth in the next chapter.

Chapter 3
World's energy utilization

3.1 INTRODUCTION

Energy is the prime mover of economic development and availability in sustainable and acceptable forms will to a large extent control the pace of world economic and human development in the foreseeable future. The world is endowed with abundant energy resources: fossil fuels (wood, coal, oil gas), renewable energy (solar, wind, water, biomass, geothermal), and nuclear energy. The problem is how to manage these resources to provide a balanced mix to satisfy the strongly growing world energy requirements in a sustainable manner that does not continue to degrade the environment. The inequitable geographical distribution of fossil fuel resources has made energy supply, particularly oil a potent variable in geopolitics. Furthermore, it has made agreement on environmental issues associated with energy use very problematic, since coal, the most potent pollutant is also the most widely available, and often the only major source of primary energy in many developing countries and some developed countries as well. Energy demand is growing in all countries of the world, the primary drivers being the rapid growths in the global economy and population, and most countries are faced with the complex problem of providing adequate energy in a secure and environmentally sustainable manner. The global energy sector is being transformed rapidly: new technologies in all areas of exploration, production and utilization are making a significant impact in all sectors of energy technology, from exploration to production and end-use. New exploration technologies are moving resources to reserves, and deployment of new technologies in power generation, industrial processes, transportation and building is raising utilization efficiencies and lowering pollution emissions.

3.2 GLOBAL PRIMARY ENERGY DEMAND

Economic and human development depend critically on access to a range of energy choices. The global economy is expected to at least double over the next twenty years and much of the projected growth is driven by emerging economies, with China and India accounting for around half of the increase. Growth in world economy requires more energy and about 30-35% growth is projected over the next two decades. Virtually all the growth in demand is expected to come from the developing world, again with China and India accounting for over half of the increase. On the contrary, growth rate in demand by the developed economies has been slowing down and the trend is expected to continue in response to the adoption of more efficient technologies, and movement away from high energy intensity production (primary metals, chemicals, cement, etc.), a sub-sector now dominated by the developing countries. For decades fossil fuels have supplied most of the global demand for primary energy, accounting for over 81% of the total demand in 2016 (Figures 3.1 & 3.2). The fossil fuel share in 1973 was 86.7%, a drop of only 5% in nearly five decades. Global climate change mitigation effort to decarbonize energy is intensifying, but all projections published recently indicate that there will be little change in the dominance of fossil fuels over the next two decades, still accounting for nearly 80% of the total primary energy supplies in 2040 (Figures 3.3 to 3.8). However, there will be a significant adjustment in the fuel mix and non-fossil fuels (renewables, nuclear and hydro energy) will provide around half of the increase in global energy demand by the end of the projection period, but the total contribution of non-fossil energy to the global primary energy use will still be less than 20%. Also, there will be a significant shift in production of energy-related pollutions to the emerging nations many of which rely heavily on coal use.

The renewable sector has grown rapidly in terms of investments and emergence of new technologies, and is playing a more prominent role in heat and power generation. All recent projections show a rapid growth in renewable energy use across all regions, but the total contribution to global primary energy over the next two decades will still be less than 20%. All the outlooks show continuing modest growth in oil, with gas growing more rapidly than oil and coal. There are some significant differences among the outlooks, especially in fuel mix projections, which reflect differences on key assumptions, such as: the availability and cost of oil and gas supplies; the speed of deployment of new technologies; the pace of structural change in China; and the impact of energy and environmental policies on such issues as the speed of transition to lower carbon economies. However, It should be noted that all projections on energy resources, demand and supply are subject to many unpredictable variables, notably the dynamics of the global economy, global oil availability and pricing, growth rate and pattern of the transportation sector, and the emergence of new technologies. For example, the decline in natural gas prices over the last few years due to increasingly rapid deployment of hydraulic-fracturing technologies has stimulated investments in gas-fired power plants and slowed down investments in renewable energy development (IEA 2019). On the other hand, declining prices of natural gas are having negative impact on current operation and future investments in relatively expensive hydraulic fracking technologies. This explains the significant variations in nine different projections which were analyzed and compared by BP (2017) (Figure 3.9).

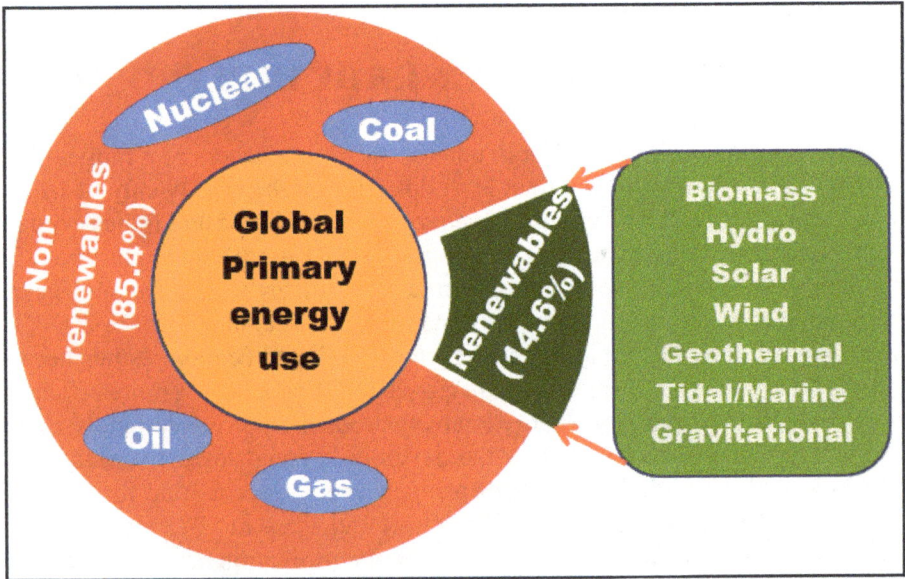

Figure 3.1a World total primary energy mix in 2016 (Mtoe). *(IEA, 2018)*

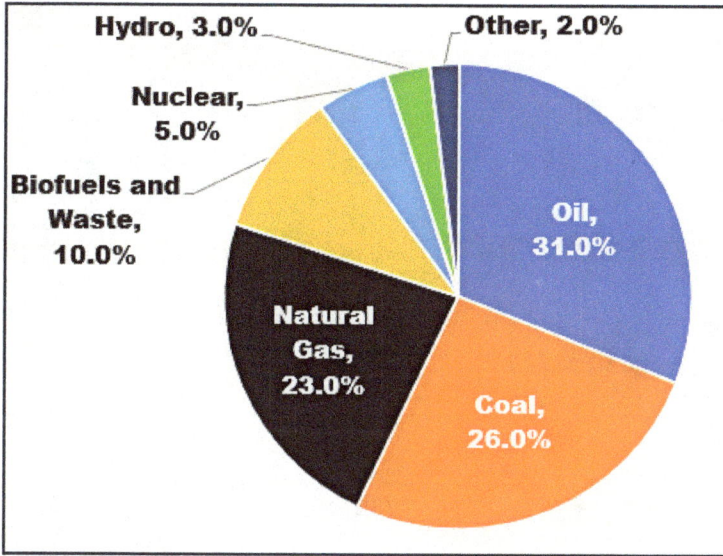

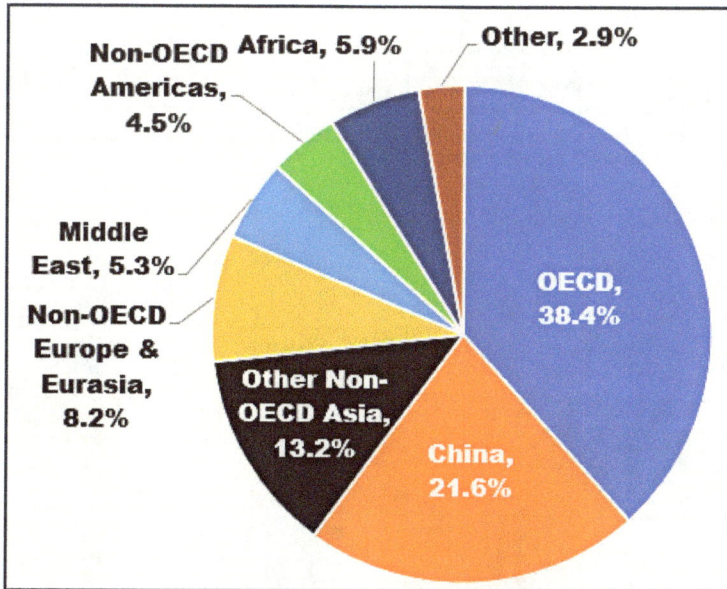

Figure 3.1b World total primary energy supply by fuel and region in 2017.
(IEA, 2019f).

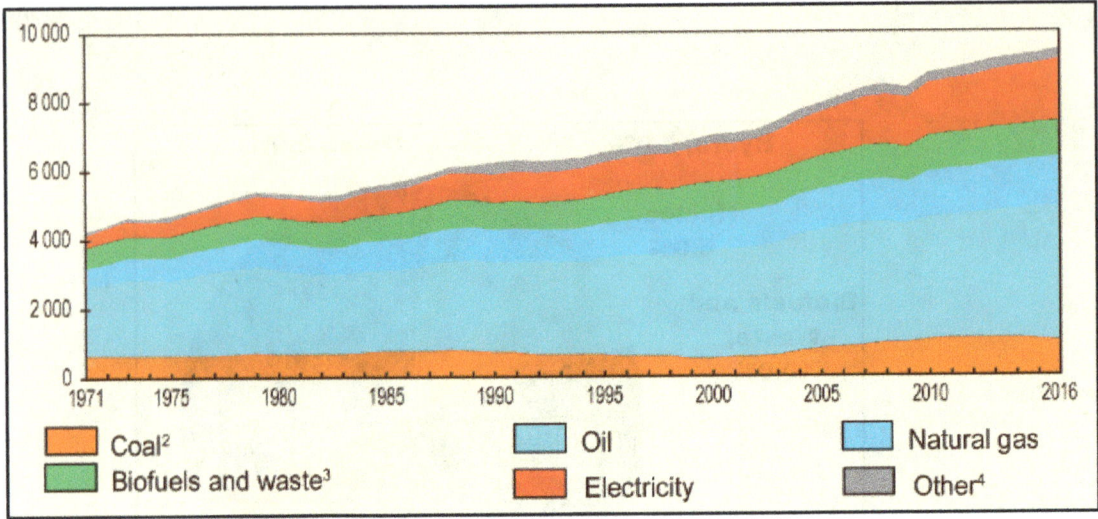

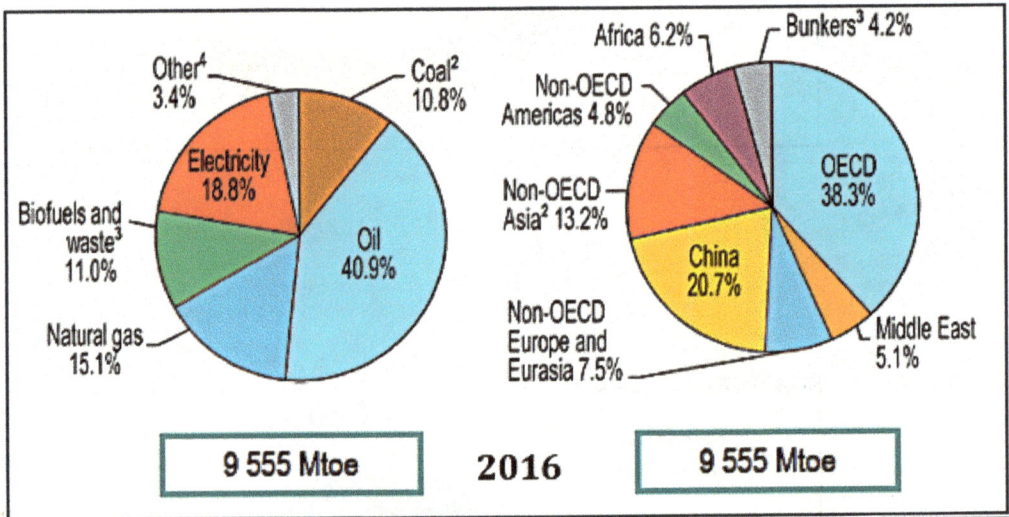

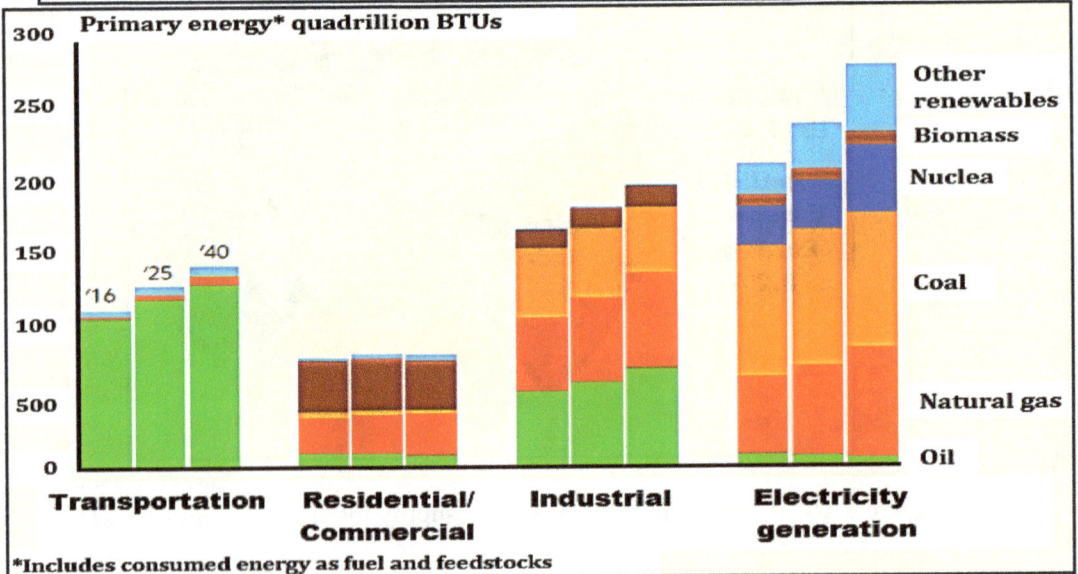

Figure 3.2   World total final energy consumption (TFC) by fuel, region and economic sector in 2016 (Mtoe). *(IEA, 2018).*

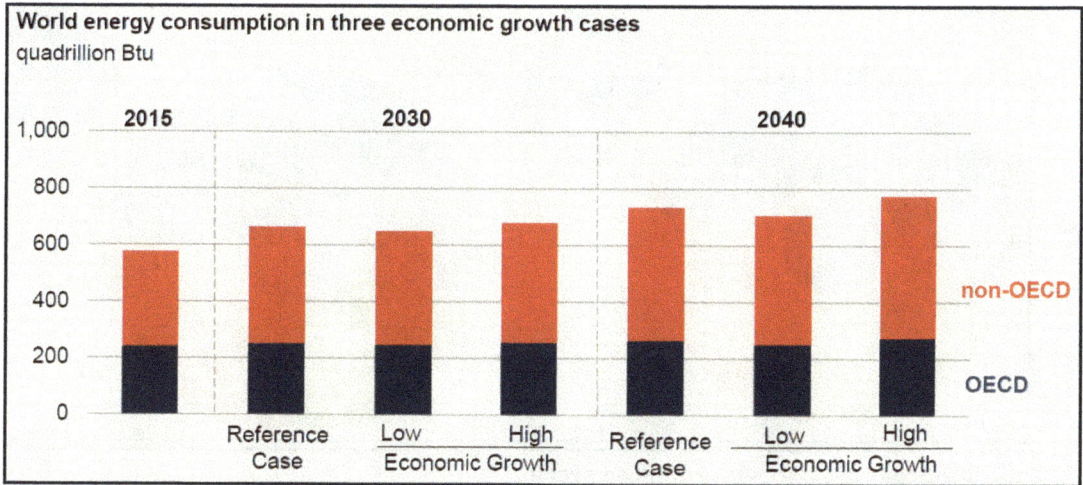

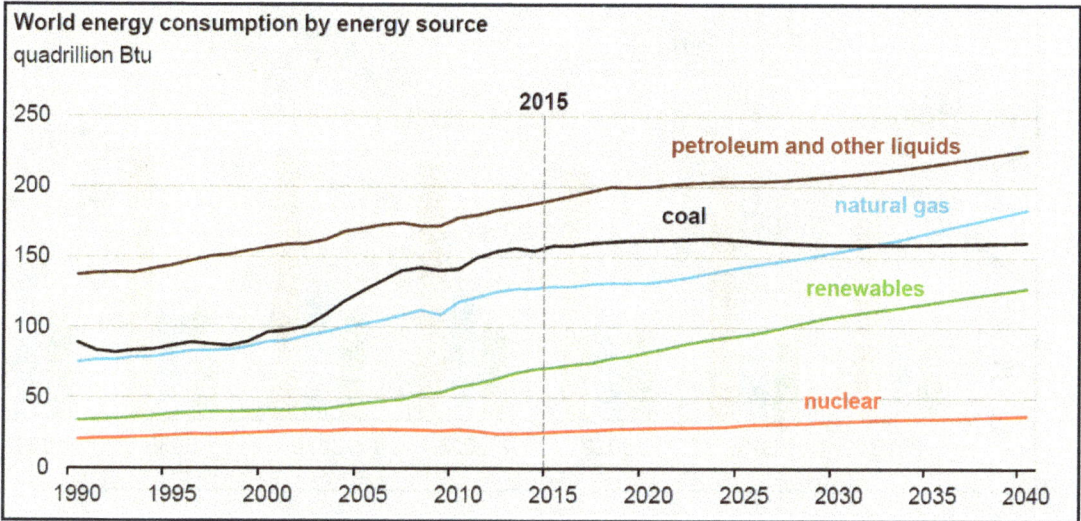

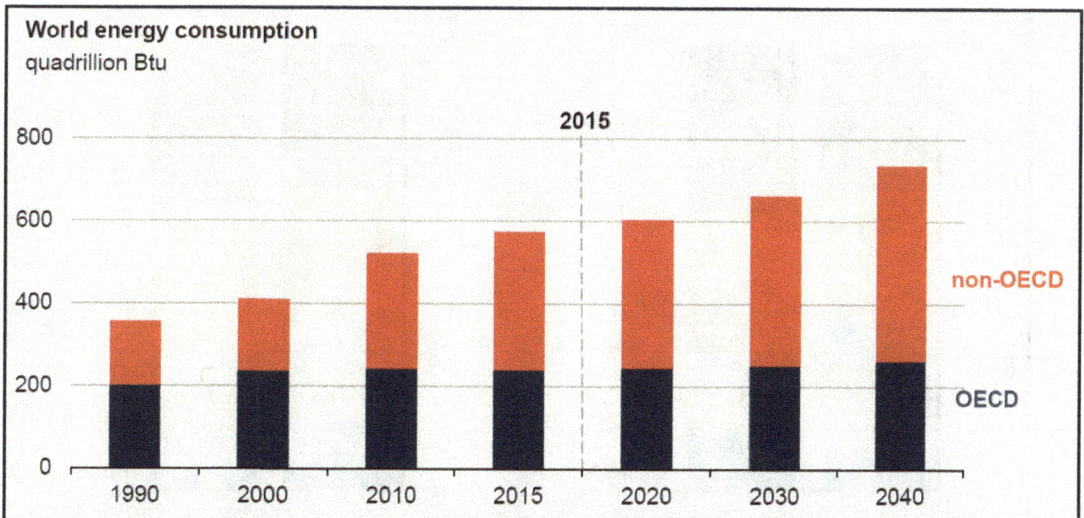

Figure 3.3 Projections of global primary energy demand by economic scenario, fuel and region region. *(EIA, 2017).*

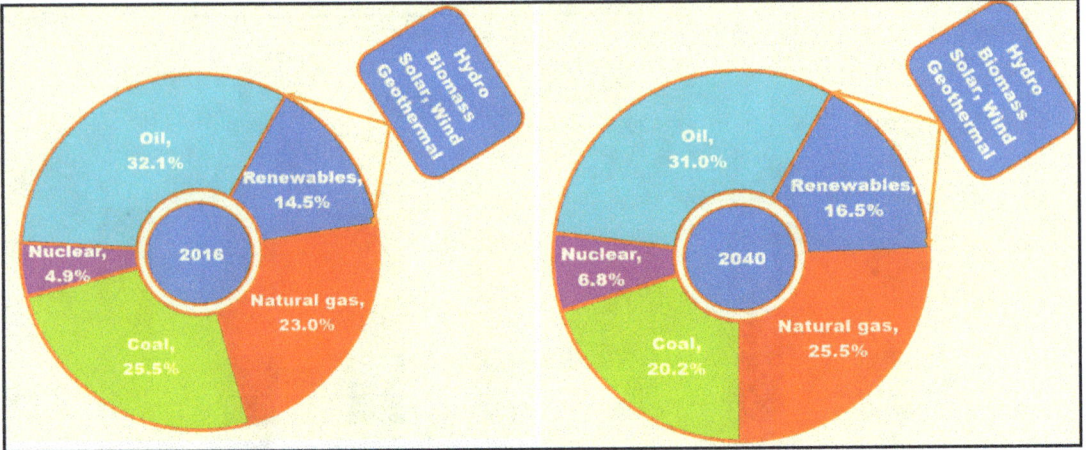

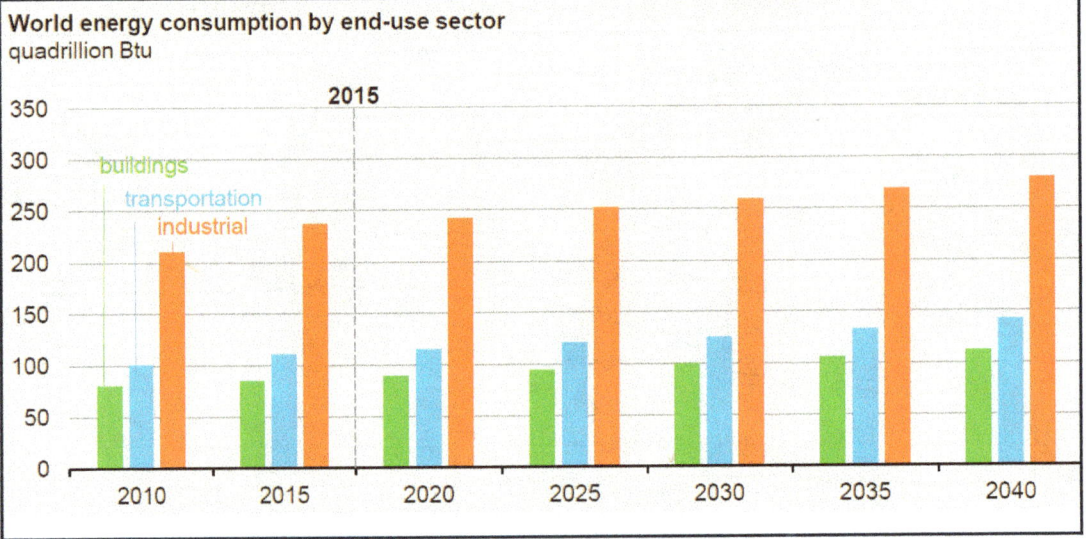

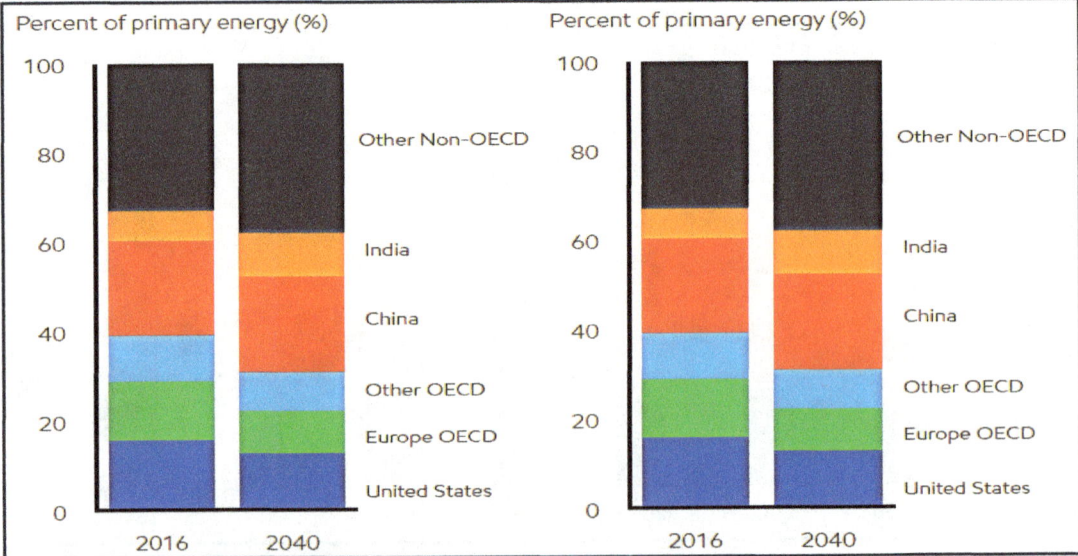

Figure 3.4   Projections of global primary energy demand by sector and source. *(EIA, 2017).*

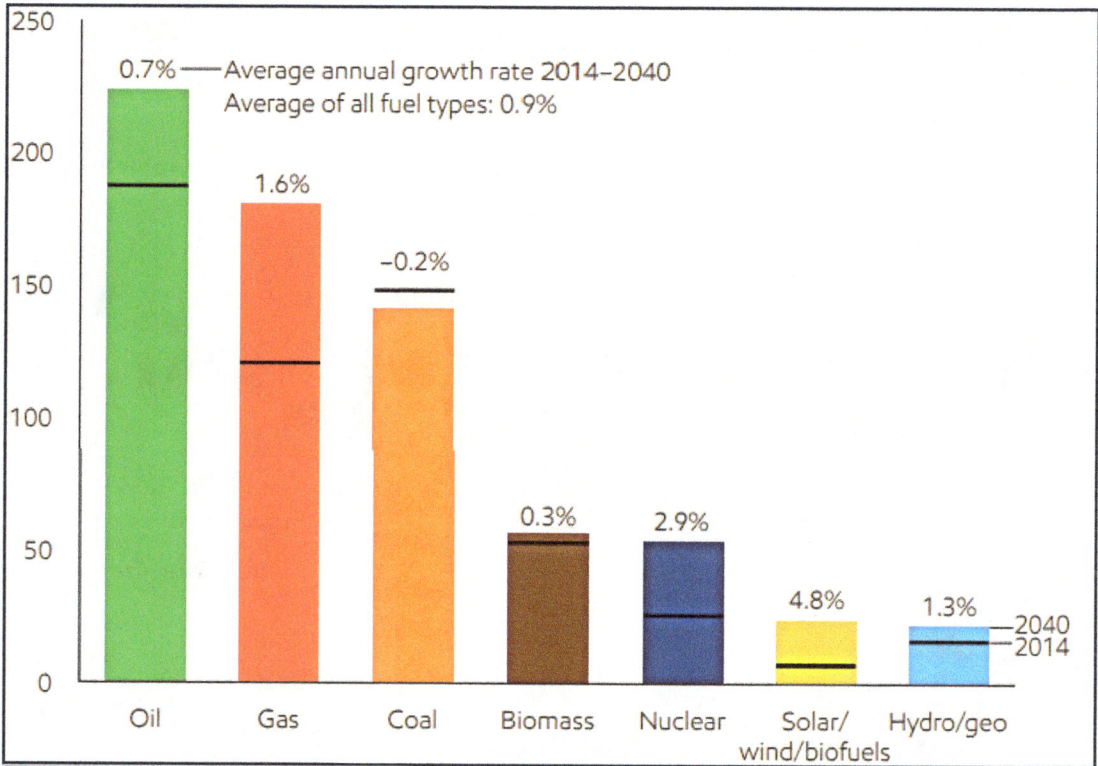

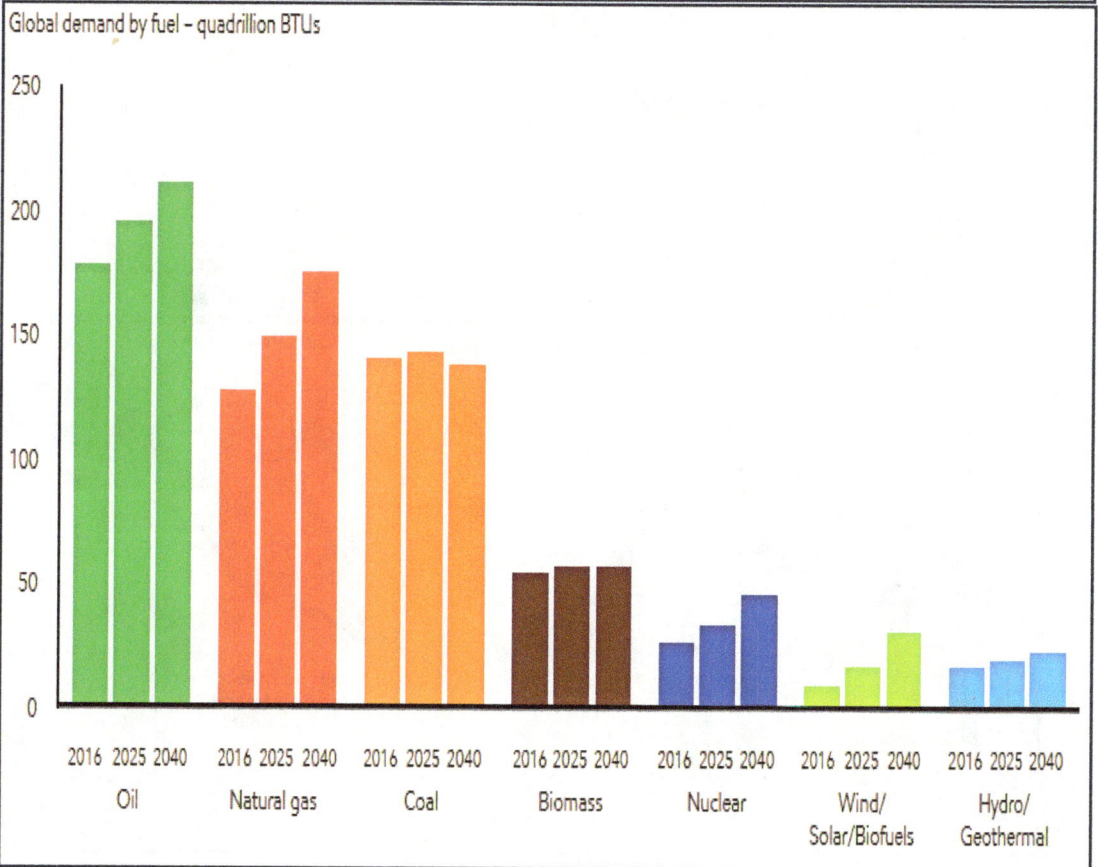

Figure 3.5 Projections of primary energy consumption (TPC) by fuel. *(ExxonMobil, 2018).*

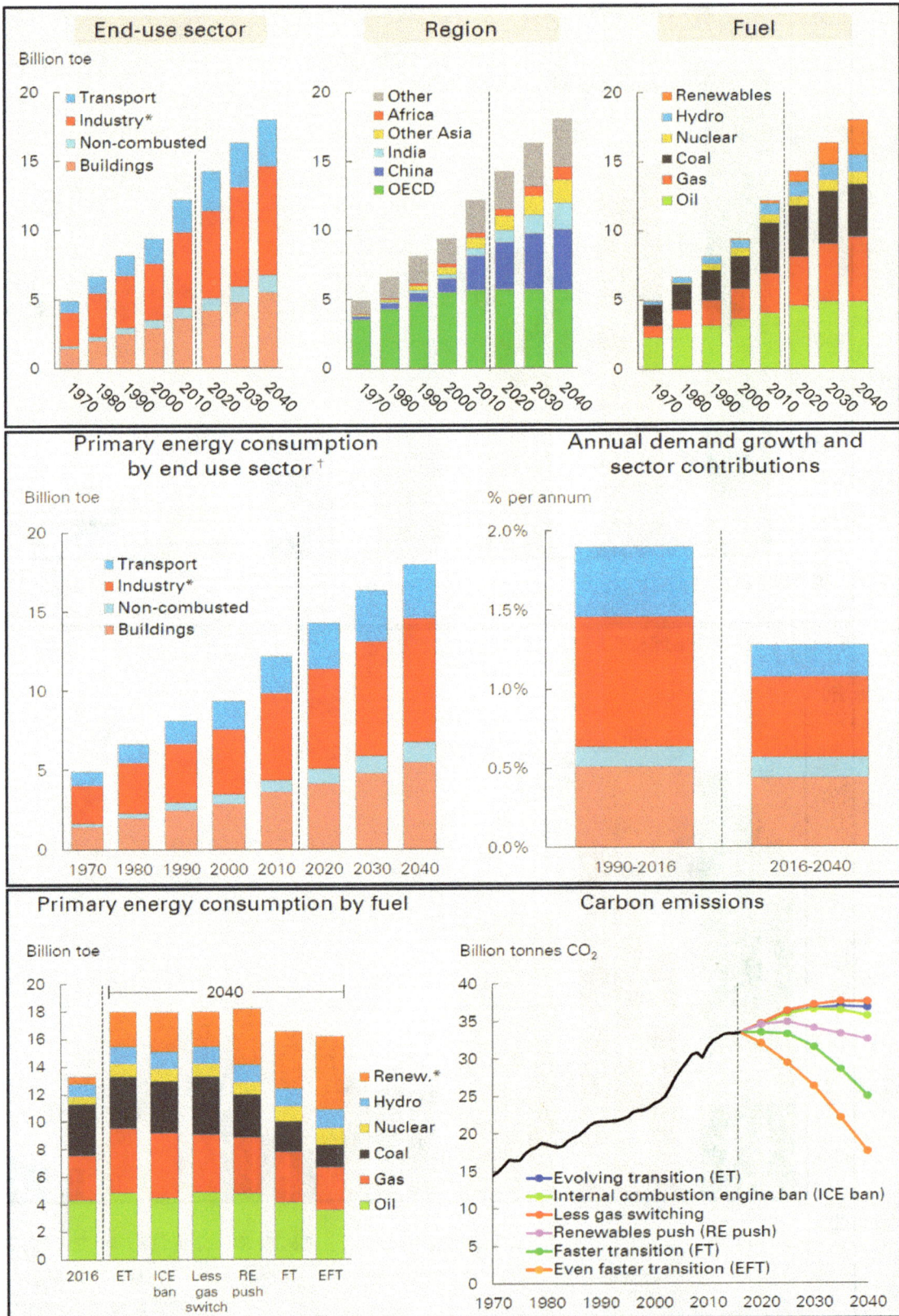

Figure 3.6 Projections of global primary energy demand by end-use, region and fuel mix *(BP, 2018).*

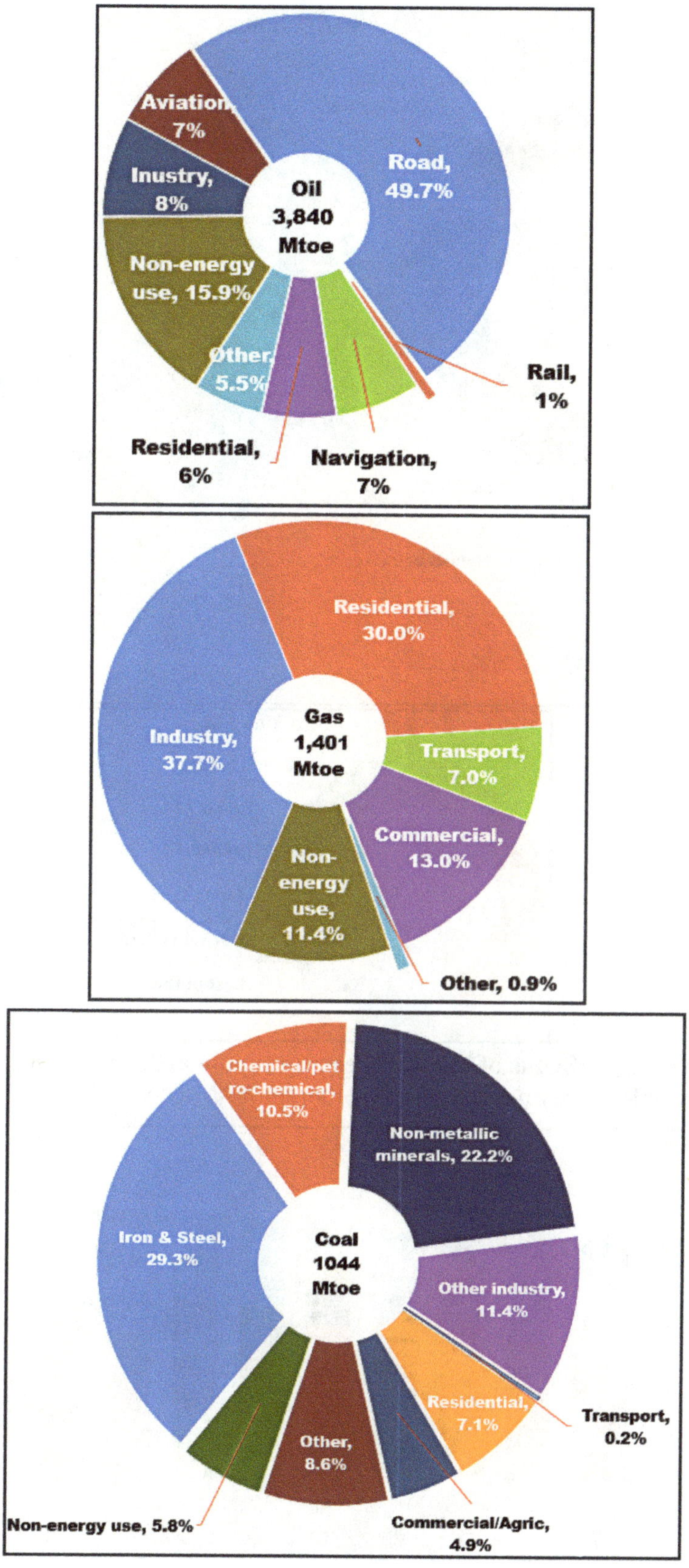

Figure 3.7 Current primary energy use by sector (TPEC) *(IEA, 2017).*

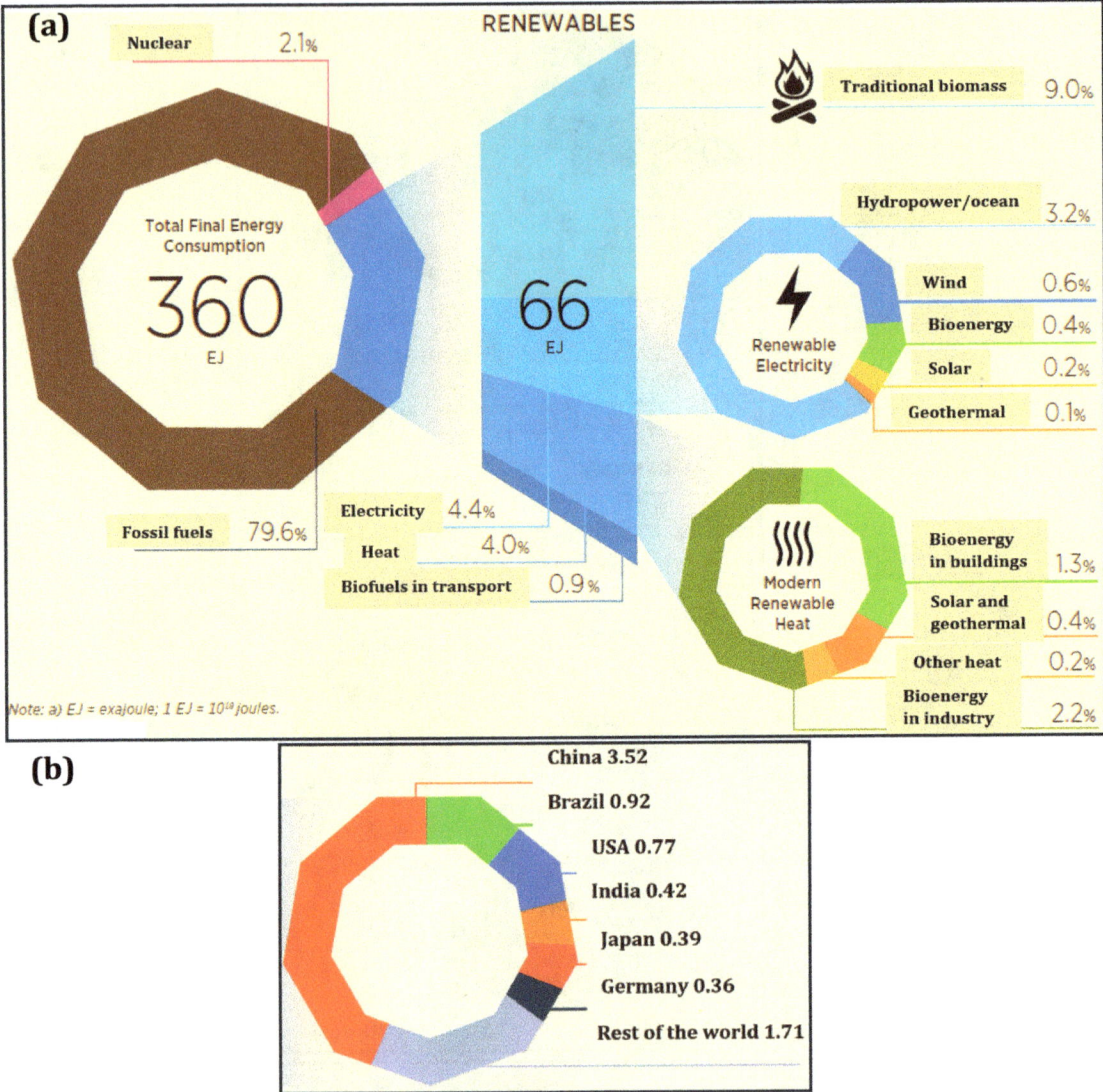

(a)

RENEWABLES

Nuclear 2.1%

Total Final Energy Consumption

360 EJ

Fossil fuels 79.6%

66 EJ

Electricity 4.4%
Heat 4.0%
Biofuels in transport 0.9%

Traditional biomass 9.0%

Renewable Electricity

Hydropower/ocean 3.2%
Wind 0.6%
Bioenergy 0.4%
Solar 0.2%
Geothermal 0.1%

Modern Renewable Heat

Bioenergy in buildings 1.3%
Solar and geothermal 0.4%
Other heat 0.2%
Bioenergy in industry 2.2%

Note: a) EJ = exajoule; 1 EJ = 10^{18} joules.

(b)

China 3.52
Brazil 0.92
USA 0.77
India 0.42
Japan 0.39
Germany 0.36
Rest of the world 1.71

Figure 3.8 (a) Status of the world energy supply in 2015 (b) Six leading countries in renewable energy use *(IRENA, 2016b, 2017).*

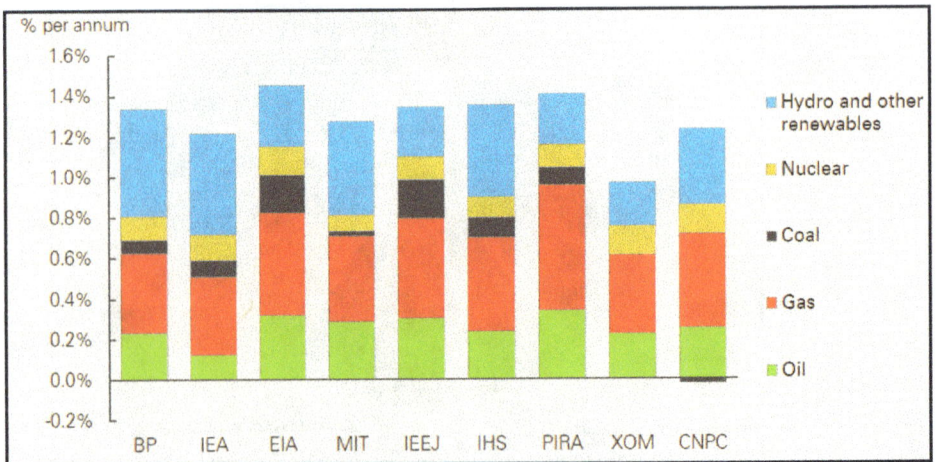

Figure 3.9 Comparison of projections from many sources *(BP, 2017).*

3.2.1   Fossil fuels

Fossil fuels (oil, natural gas and coal) have supplied the bulk of global primary for nearly a century, accounting for over 81% of the global energy demand (roughly the same for the last four decades), and 70% of the growth around the world in 2017. Demand grew 2.1%, more than twice the growth rate in 2016. For decades, efforts have been made to extend the global energy mix to include less carbon-intensive energy - renewables and nuclear - but the rate of deployment has been slow for many reasons. All recent projections show that the situation will not change significantly over the next two decades or so, although renewables, in particular, solar and wind will play a much more prominent role in power generation. Fossil fuels drive most of the important sectors of the economy, just as the economy drives fossil fuel use, and recent projections indicate that the fuels will remain dominant in the foreseeable future. Oil is the world's leading fuel, accounting for around 32% of total global primary energy consumption in 2016, and the transportation sector is the primary consumer. About 90% of the fuel demand by the sector is filled by oil, and the prospects of viable substitutes over the next few decades are not very bright. Natural gas and coal are used mainly for power generation but gas is also used extensively in industrial processes, commercial and public services, domestic heating and cooking, while coal is the primary fuel for iron and steel and cement production. All the fossil fuels are also precursors to many non-energy products - chemicals, plastics, fertilizers, etc.

Coal has been the major fuel for power generation in decades but it is also the most carbon-intensive fossil fuel. Increasing efforts to decarbonize fuel use by increasing the contribution of renewables, emerging technologies for producing cheaper gas, and strong policy disincentive are beginning to reduce the dominance of coal in power generation overall but many countries (mostly emerging nations) are increasingly meeting their growing energy demands with cheap, readily available coal. On the contrary, developed economies are slowing down on the use of coal and saving fuel by improving efficiencies across fossil fuel production and use in industry and transportation. Many coal-fired power plants are being de-commissioned or converted to more efficient, increasingly competitive, and less carbon-intensive gas-fired plants. Coal is predominantly an indigenous fuel, mined and used in the same country, and only hard coals are traded internationally. Despite international action to reduce dependence on coal for heat and power generation, consumption has remained around 40% for the last 40 years, fueling 80 – 98% power generation even in some developed countries (Figure 3.10). Although the share of coal in primary energy demand and in electricity generation slowly continues to decrease, it still remains the largest source of electricity and the second-largest source of primary energy. China leads the world in coal-fired power output, followed by USA (Figure 3.11).

In spite of the very poor environmental credentials, coal has remained prominent in the global primary energy mix for five decades, accounting for the second largest share after oil, and the largest source of electricity. However, the role of coal has continued slowly over the last few years. After several consecutive years of slower growth rate, coal share increased again over the last two years by around 1% per year, although significantly lower than the typical 4.5% growth rate per year from 2000-10. The rise over the last two years is believed to have been partly triggered by the unusually cold weather in Europe and North America over the period, but also due to increasing shift of demand and use to Asia Pacific that accounted for around 40% of the global power generation in 2017, with coal as the primary fuel. The region also produced over 60% of the total global output of coal in 2017.

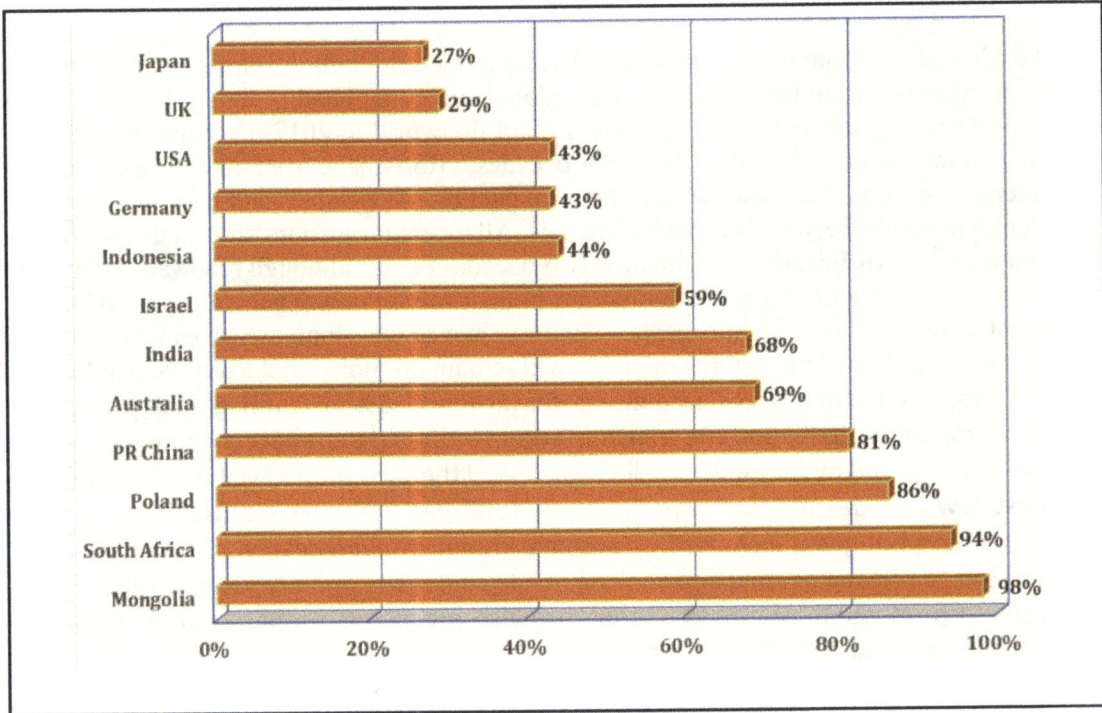

Figure 3.10 Twelve countries with the highest dependence on coal for power generation *(IEA, 2013b)*.

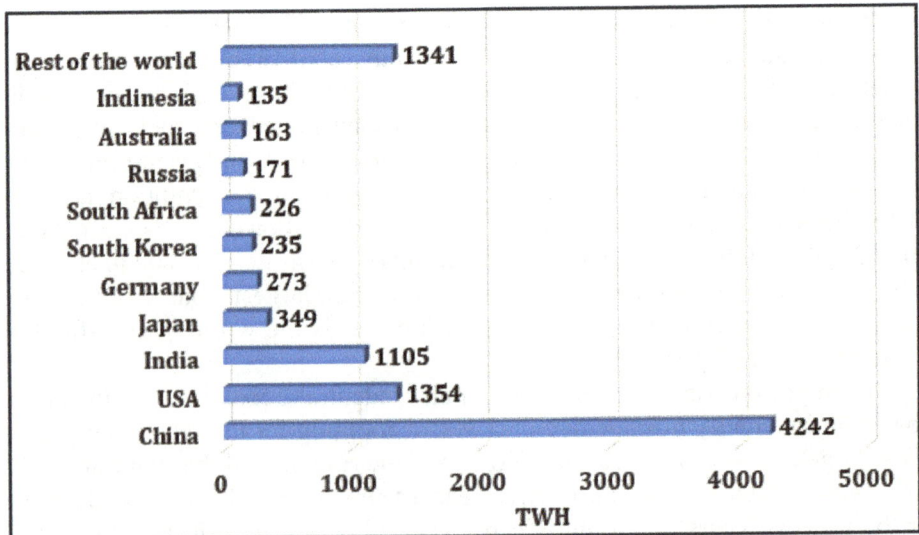

Figure 3.11 Country ranking, coal-fired power generation *(data from IEA, 2018)*.

Technologies for gasifying coal for power generation have been available for decades but have gained impetus only in the last decade or so as an effective way of reducing pollution, and many new coal-fired power plants across all regions will likely adopt this technology. Coal is still the primary energy for the production of iron and steel, and 15-20% of the annual global consumption is accounted for by the sector. There is currently no feasible replacement for coal in the blast furnace and coal-fired direct reduction process, both of which produce around 75% of global steel requirements, but an alternative gas-based direct reduction technology is gaining ground, especially in developing countries which have access to cheap local gas resources. The main reason for the continued dominance of fossil fuels well into the foreseeable future is the lack of viable replacements in many applications. In spite of massive investments in renewable energy sources and the projected future rapid growth, the main contribution to global primary energy demand over the next twenty years or so will be less than 20%, and power generation will account for most of the use, accounting for 24-35%.

3.2.2 Nuclear energy

Before 1954 when the first nuclear electricity plant was commissioned, uranium was used primarily for weaponry. Over the next fifty years or so, the number of installations for power generation grew very rapidly and nuclear plants were supplying nearly 20% of the global electricity in the late nineteen eighties. Development was bolstered by the oil crisis of the early 1970s which brought energy security to the forefront of global concerns. For three decades, the OECD dominated global nuclear power generation, accounting for nearly 93% of the total nuclear electricity generation in 2015 (Figure 3.12). The United States leads the world in terms of nuclear power output, accounting for nearly a third, but the proportion in the domestic power generation fuel mix was less than 20% in 2016. In contrast, nearly 80% of the total power output in France in the same year was from nuclear plants (Figure 3.13). However, two serious accidents (Three Mile Island in 1979, Chernobyl in 1986) marked the beginning of a steady decline which reached its lowest after another accident in Fukushima, Japan in 2011. Sixteen countries depend on nuclear energy for at least one quarter of their electricity. Nuclear power share of global power generation decreased for the tenth year in a row to about 11% in 2016, yet still accounting for nearly a third of the world's low carbon electricity production. In spite of policy-driven rapid expansion of renewable use in power generation, fossil fuels, especially coal have remained dominant in global electric power generation. Although new renewables (wind, solar, geothermal power) have surpassed nuclear power in total installed capacity, their share of actual electricity generation is less than one third of that produced by nuclear power because of their intermittency (IAEA, 2017).

3.2.3 Renewable energy

Renewable energies are natural resources that can be converted to useful energy - solar, wind, biomass, tidal, geothermal, hydropower. They are the fastest growing primary energy sources used mainly for steam raising and power generation, meeting a quarter of global energy demand growth in 2017, with wind energy accounting for 36% of the growth in renewables-based power output (IEA, 2018). China and the United States led this highest growth ever, contributing around 50% of the increase in renewables-based electricity generation, followed by the European Union, India and Japan. In spite of the unprecedented growth rate in the last few years, renewables accounted for less than 15% of the global primary energy supply in 2016, shared equally between modern renewables and traditional biomass, and accounted for only

24% of global power generation in 2018 (IEA, 2019k). Most recent projections show that renewables will still contribute no more than 17-25% to electric power generation by 2040, although there will be significant variations between countries, with China increasing renewable share by close to 40%. Hydroelectricity has dominated the renewable energy power generation for decades but its contribution is projected to reduce significantly due to increased use of solar and bioenergy and environmental concerns in the developed countries (Figures 3.14 - 3.17). Renewable energy will dominate the growth in global primary energy demand over the next two decades, with its share increasing from about 4% currently to around 15% by 2040.

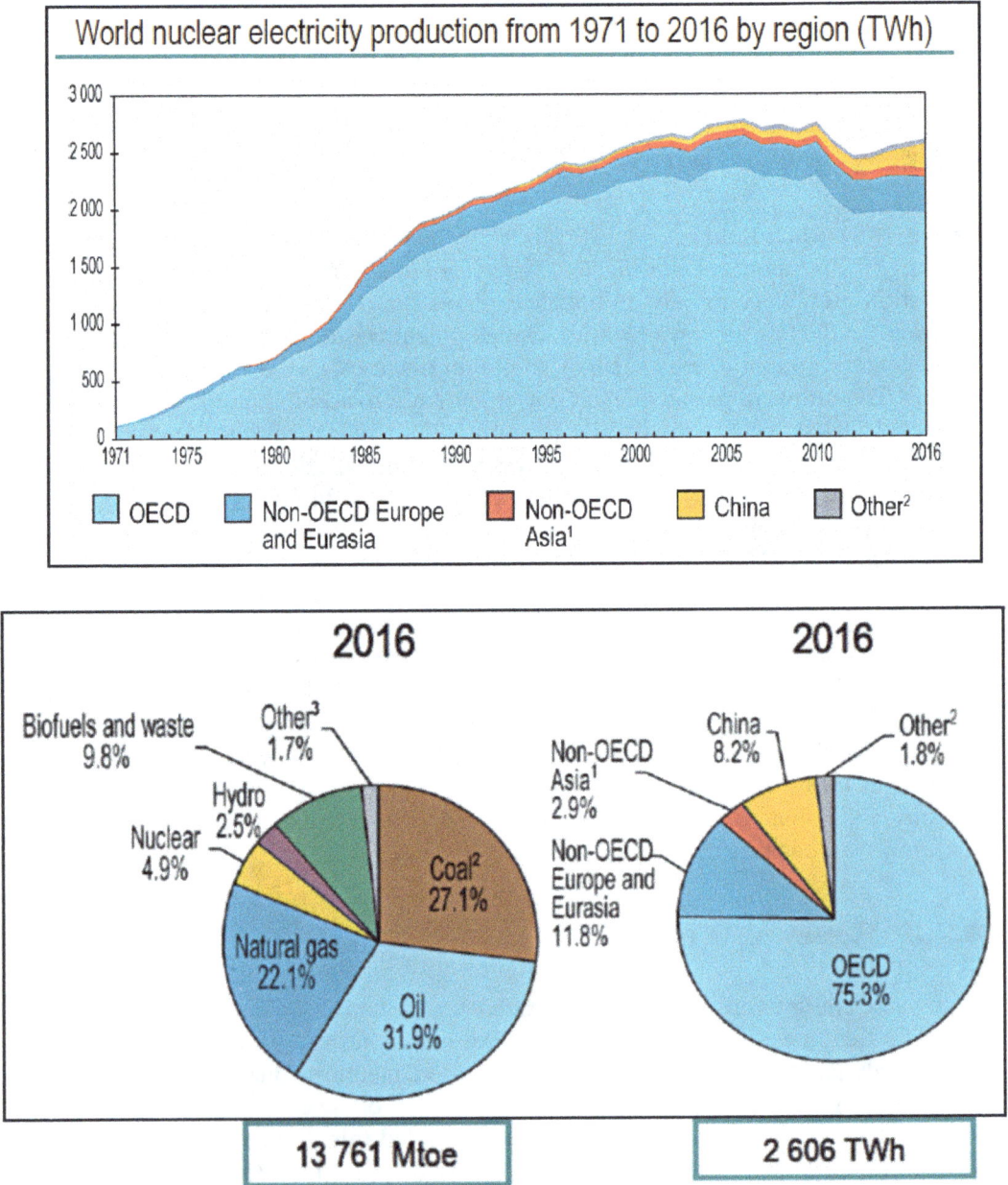

Figure 3.12 World nuclear electricity production from 1971 to 2015 by region; 2016 by fuel and region. (TWh) *(IEA, 2017a, 2018).*

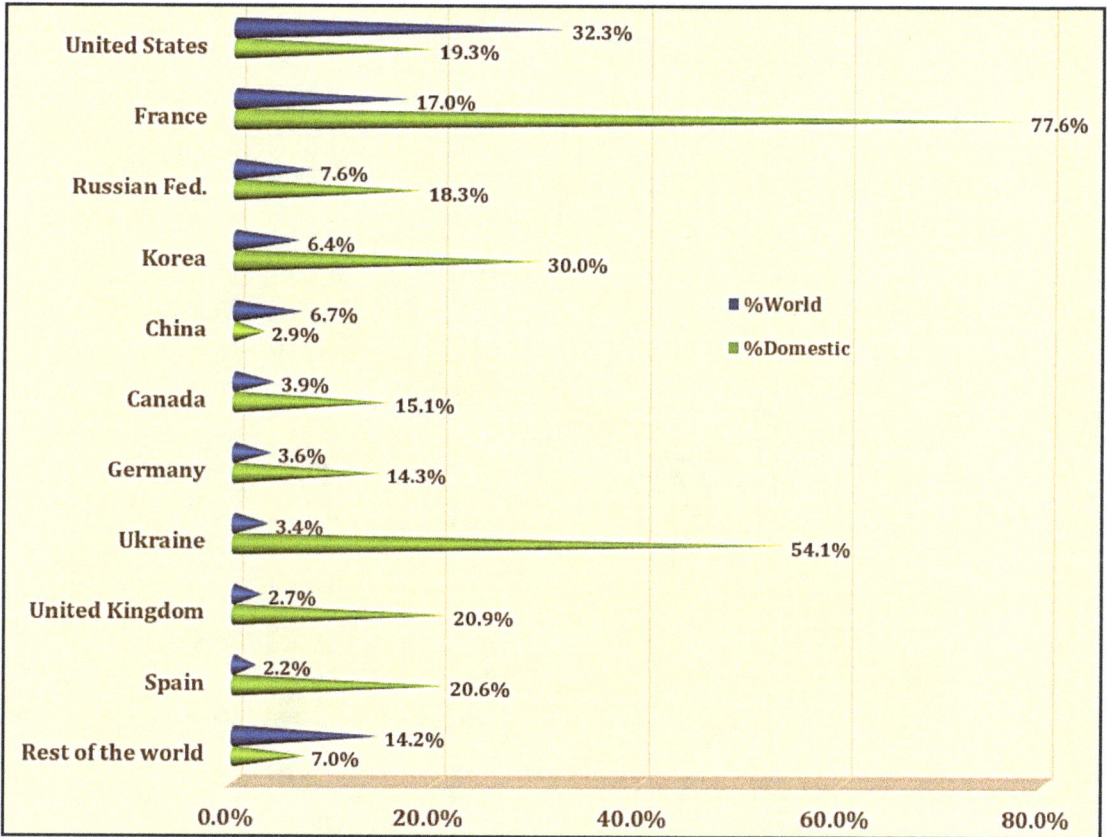

Figure 3.13   World's top ten producers of nuclear electricity in 2016.
(IAEA, 2017; IEA, 2018b; 2017a).

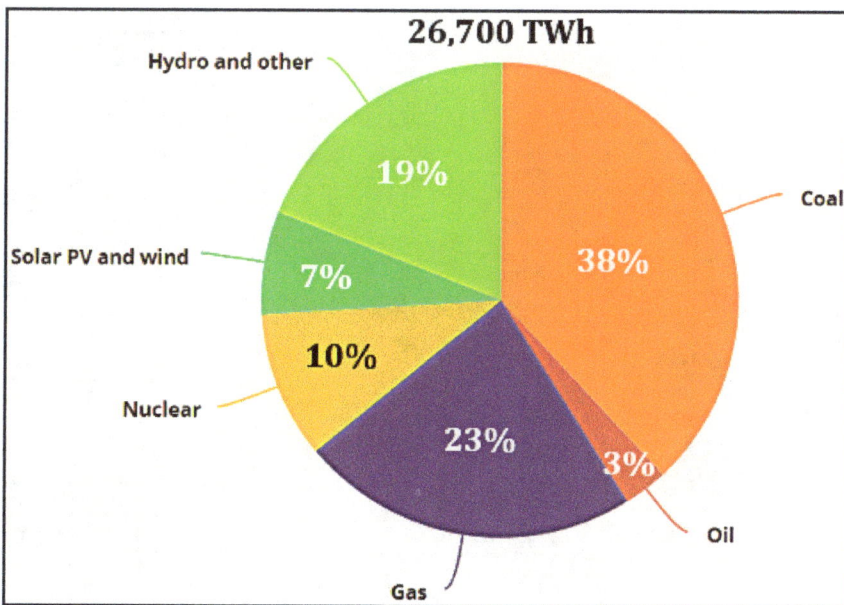

Figure 3.14a   Global electricity generation by source *(IEA, 2019k).*

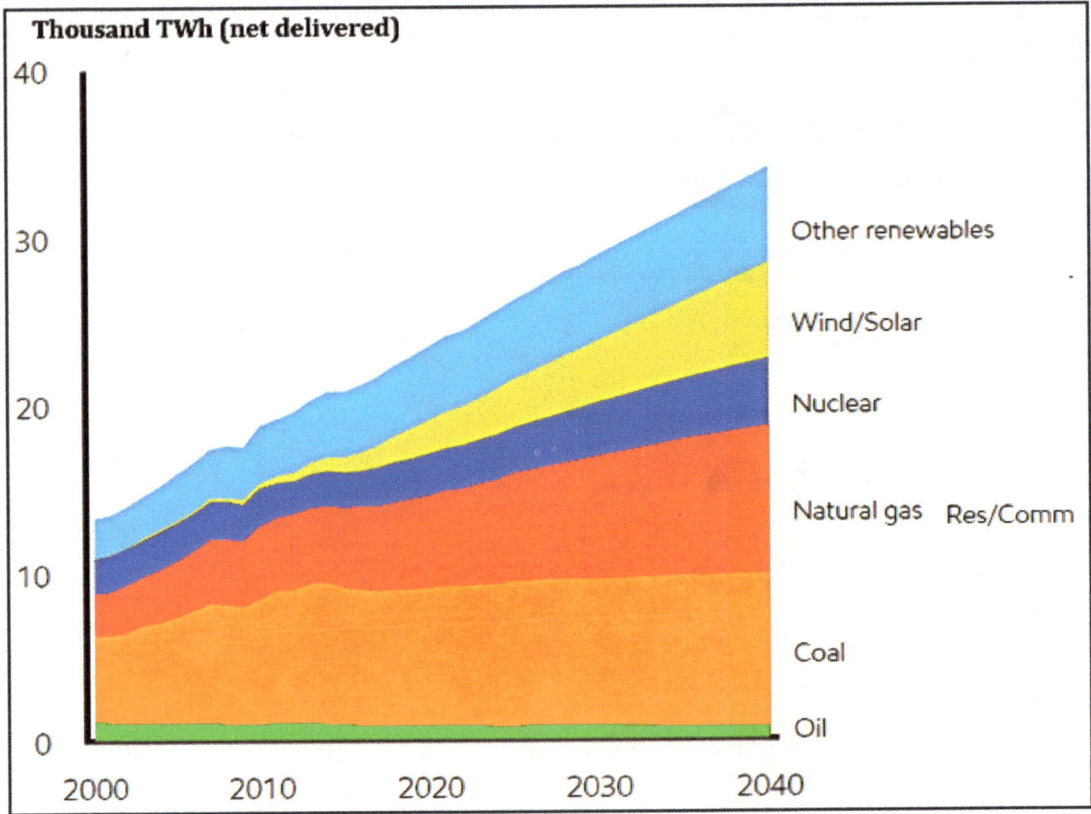

Figure 3.14b World electricity generation by source *(ExxonMobil, 2018).*

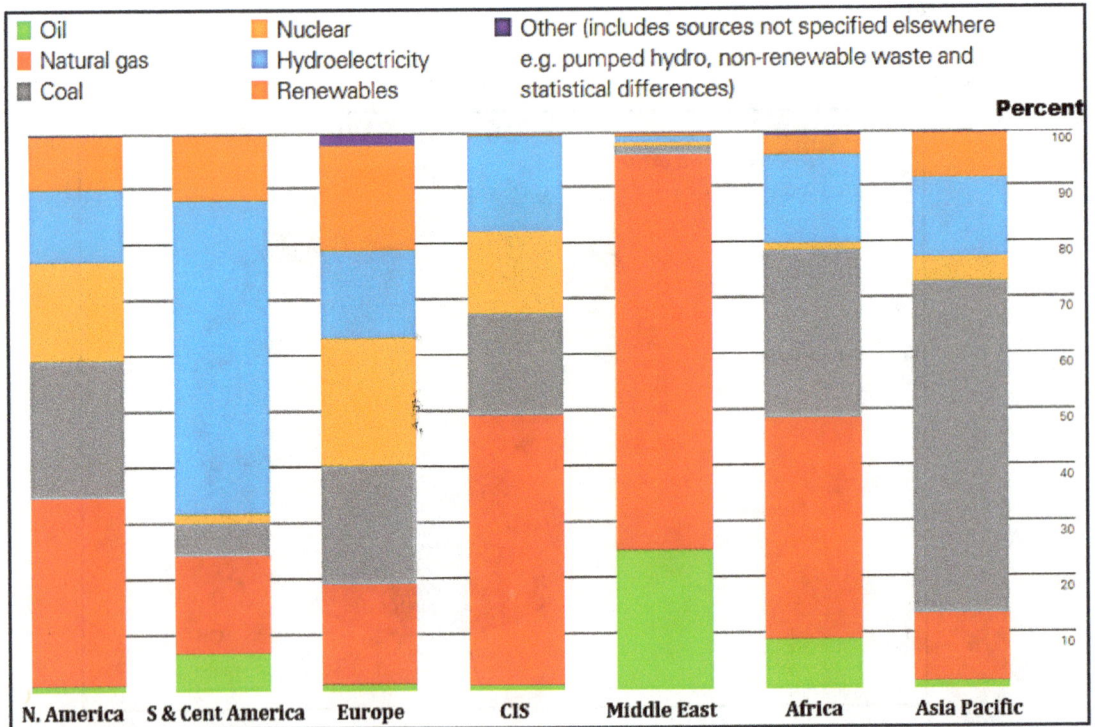

Figure 3.14c Regional electricity generation by fuel in 2018 *(BP, 2019b).*

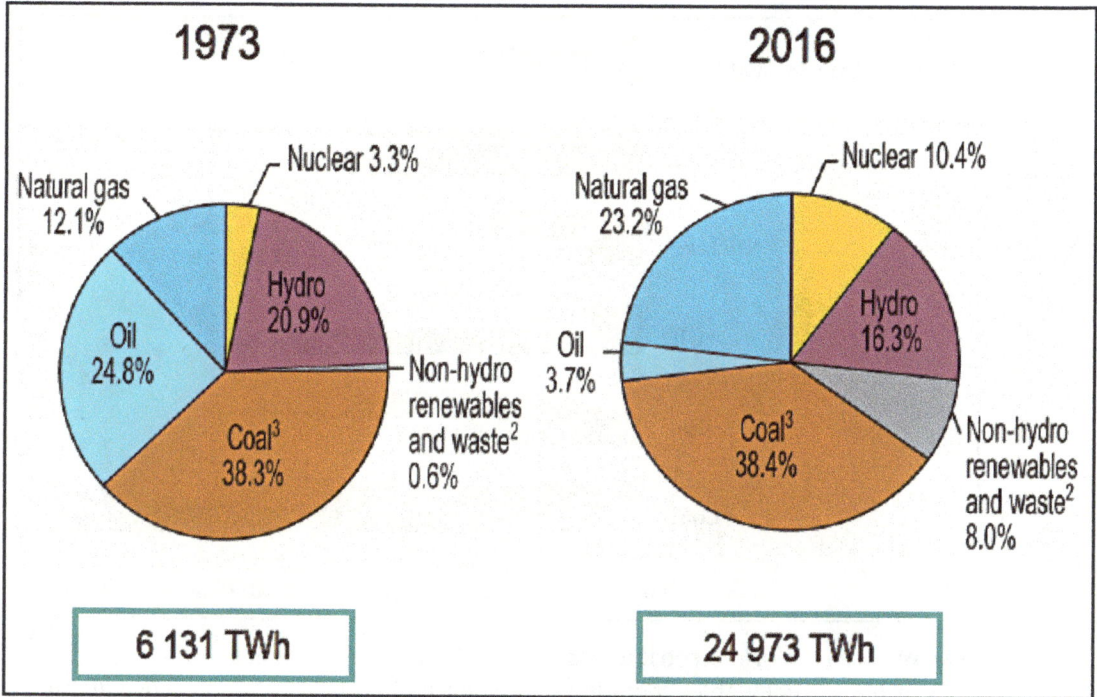

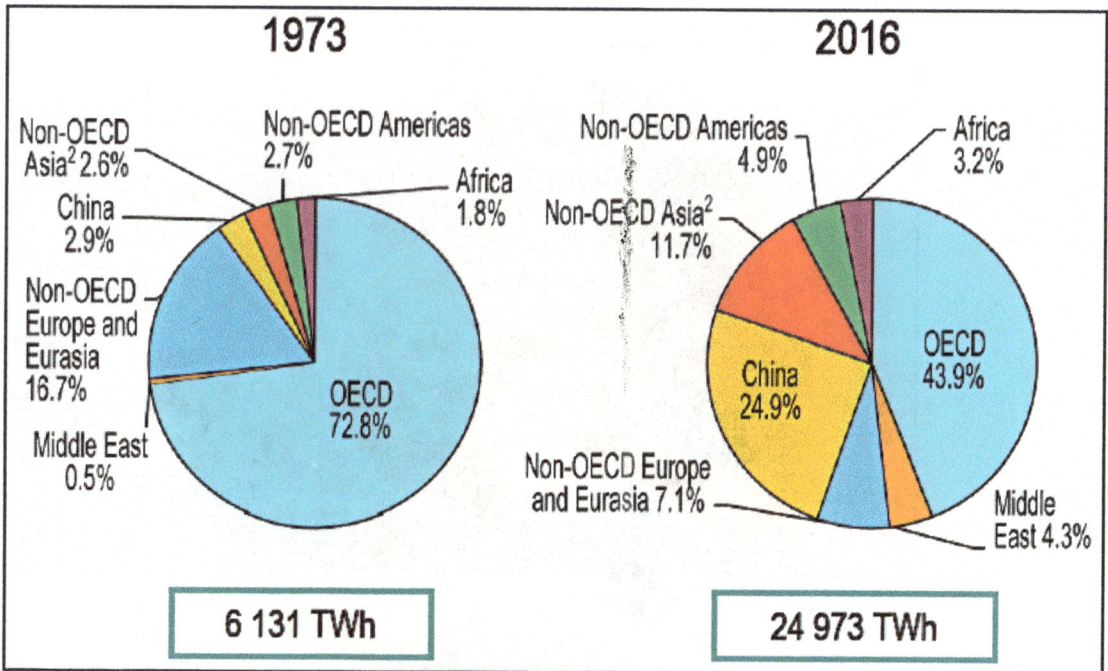

Figure 3.15 World electricity generation by fuel and region *(IEA 2017a, 2019f).*

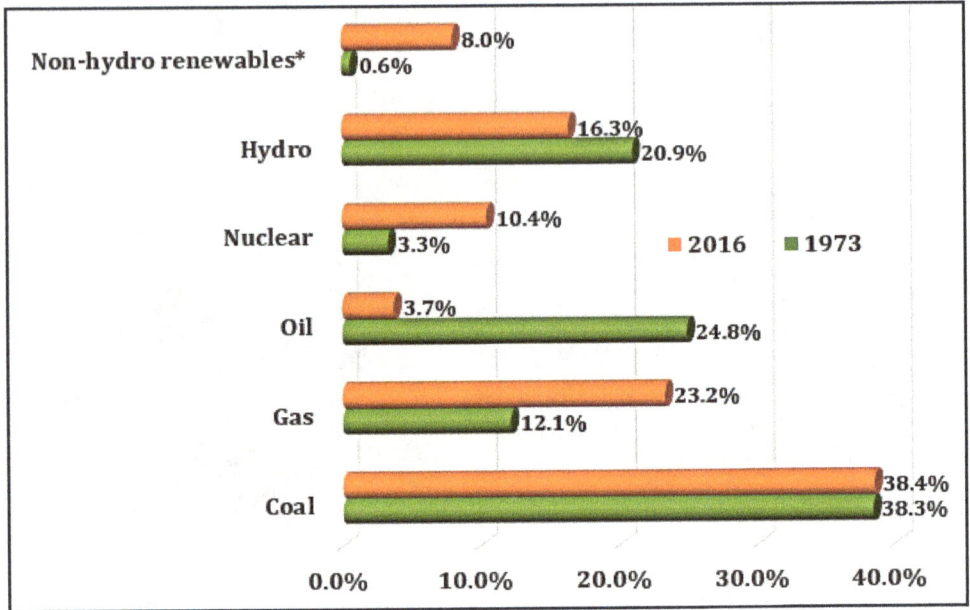

Figure 3.16 World electricity generation by fuel and region
* Non-hydro renewables include wind, solar and biomass. *(Data from IEA 2017a).*

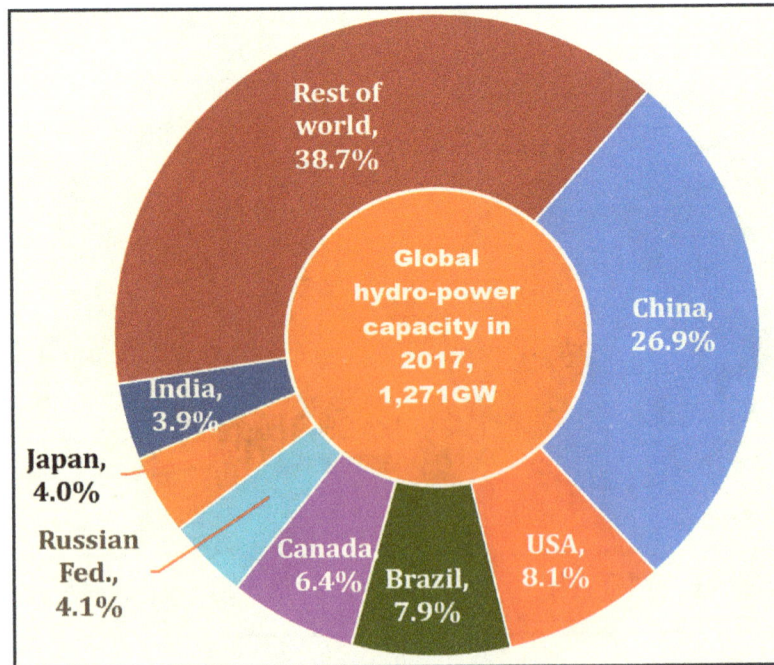

Figure 3.17 World hydroelectricity generation *(Data from IEA 2017a).*

3.2.3.1   *Solar energy*

Solar energy is the most abundant global energy resource and is available everywhere. It can be converted to heat for use in steam raising, power generation and many other applications. About 60% of the total energy emitted by the Sun reaches the Earth's surface, if just 0.1% of this energy can be converted into heat at an efficiency as low as 10% it could supply four times the global electric power demand. Solar radiation also has indirect impact on other forms of renewable energy – wind, biomass, marine, hydro energy, etc. Solar energy may be concentrated by collectors and used to heat water and produce steam for power generation (solar thermal systems), or converted directly into electricity by photovoltaic cells (solar PV systems). Rapid developments in solar PV and concentration technologies are making it possible to deploy solar electricity plants even in locations where radiation intensity is low or variable. Solar radiation is available in the appropriate intensity and for adequate duration in relatively few global locations, mostly in regions hosting emerging countries that do not have the wherewithal to harness and deploy solar-powered power generation on any massive scale. One major exception is Morocco where the world's largest solar thermal power plant was commissioned recently. China, Japan, United States, and Germany also have extensive solar power projects. Most of the growth in solar power deployment over the next two decades will be stand-alone and rooftop systems for small establishments, rural communities and homes across all regions of the world. The contribution of solar energy to global power generation is still very low, only 1.8% in 2017. However deployment of solar power is growing rapidly and projections indicate the total contribution to global power generation could double by 2023 (IEA, 2019e). Although the cost of PV cells has dropped drastically in recent years, variability and low load factor (around 20%) remain formidable issues.

3.2.3.2   *Wind power*

Wind is available everywhere in the world and if just 1% could be effectively utilized, it could provide electricity equivalent to the capacity of all global electric power generating plants in the world today. However, the intensity in many locations varies widely. Furthermore, there are relatively few locations where the wind is sufficiently strong and available for enough time to power electric generation plants. Even then, the load factor is low, 15-40% compared with 75-90% for thermal plants. Wind power is useful only when the intensity and consistency are sufficient to drive a wind turbine, and the best locations are offshore mostly in remote locations where obstructions are minimal. Wind power contributed 4.4% to the total global electric power generation in 2018, although there were significant variations between countries. For many countries, wind power has become a pillar in their strategies to phase out fossil and nuclear energy in power generation. For example, wind power produced 43% of Denmark's and 13% of Germany's electricity, and an increasing number of countries - Ireland, Portugal, Spain, Sweden, Uruguay - have reached double digit wind power share. Around 3,000 wind-powered generating plants are operating offshore in 15 countries and many more are planned or under construction. Seven countries accounted for three quarters of the total global wind power generation capacity in 2017, with China alone accounting for one-third. Like solar energy, wind energy is making a significant contribution to supply of secure, reliable electricity in many countries, particularly, OECD countries. However, the contribution to the total global final energy supply is less than 1%.

3.2.3.3 *Hydropower*

Hydropower is available in over a hundred countries and is supplying a significant amount of electric power throughout the world. Hydropower currently accounts for around 70% of the total contribution of renewables to the total global final energy use. The global average of contribution to electric power generation is around 15% but several countries depend on the resource for over 50% of local power demand. Just five countries - China, United States, Brazil, Canada and India accounted for about half of the total global capacity in 2016 (Figure 3.17). Drivers for hydropower's growth include a general increase in demand not just for electricity, but also for qualities such as reliable, clean and affordable power as countries seek to meet the carbon reduction goals set out in the Paris Agreement. Hydro-dams in emerging countries also serve as municipal water sources and recreational facilities. There has also been a significant increase in pumped storage capacity in the last few years, a growing recognition of hydropower's role in enhancing the use of variables renewables (solar and wind) in more predictable power generation - surplus energy from these sources are used to pump water to higher ground, and the water is released as in a normal hydro-dam to generate power when needed. Hydropower has great potential in developing countries and the United Nations is pioneering the installation of hundreds of small hydro projects in many emerging countries. However, there is increasing resistance to proliferation in developed countries due to environmental concerns.

3.2.3.4 *Bioenergy*

Bioenergy refers to energy generated from various feedstocks of biological origin. Fuelwood, charcoal and agricultural residues often referred to as traditional biomass have been used to generate energy from early times. However, although there has been a significant mix diversification in the last two decades or so, with the wide availability of industrial biowaste and crops which are grown specifically for conversion to energy - sugarcane, corn, soybean, sorghum, etc., - traditional biomass still dominates (Figure 3.18). Biomass can be converted into a wide range of energy sources including heat for power generation and heating, liquid and gaseous fuels. The major drawbacks that have prevented wider use are the logistics for collecting and processing biowaste and the need for massive land areas and sufficient water for the cultivation of feedstock which are also important human and animal food items. Brazil and the US lead the world in the production and utilization of biofuels from starch or sugar derived from corn, soybean, sugar cane, cassava, etc., while the main focus of Europe is on the production of biodiesel from oils and plant/animal fats (soy, rapeseed, palm oil, mustard, sunflower oils etc.) by catalyzed transesterification. The process of producing fuel from oils and animal fats involves reacting an alcohol with an ester to produce a different alcohol through an exchange of organic carbonyl carbon groups in the feedstock with organic groups of alcohol. Biodiesel produced by this process is more environment-friendly and produces 60% less pollution compared to fossil-sourced diesel oil.

3.2.3.5 *Geothermal energy*

The potential of geothermal energy in the global primary energy mix is enormous. It is estimated that stored thermal energy down to 3 km within continental crust is around 43×10^6 EJ, which is considerably greater than the world's total primary energy consumption (WEC, 2016a). However, the contribution to the global total primary energy consumption is very small and power generation from geothermal energy accounted for less than 1% of the world's output in 2015. The largest operating geothermal powerplant in the world is located in the United States, Turkey leads the world in new plant projects and accounts for half of the new global capacity additions, although Indonesia has the highest global planned capacity additions. New plants are also under construction in many countries (Figure 3.18b). Apart from power generation, geothermal energy is also used directly for domestic and industrial heating.

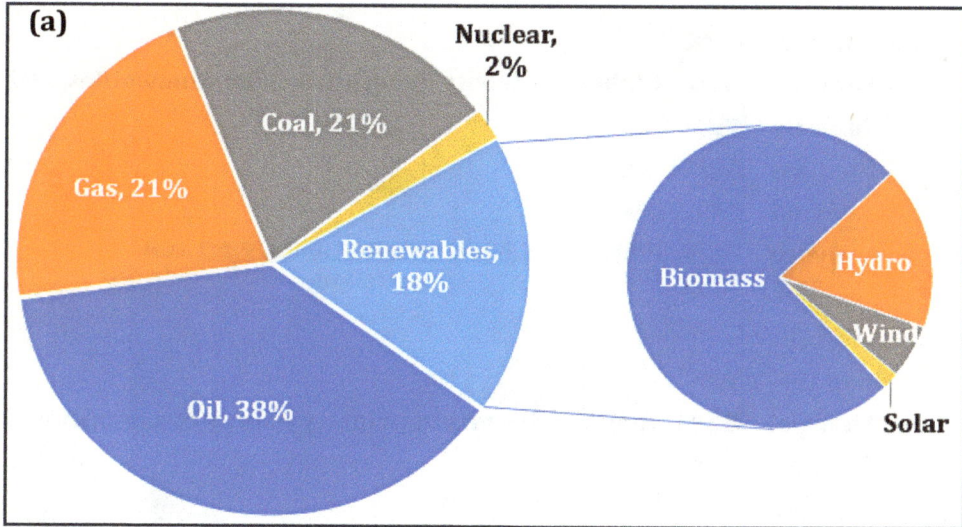

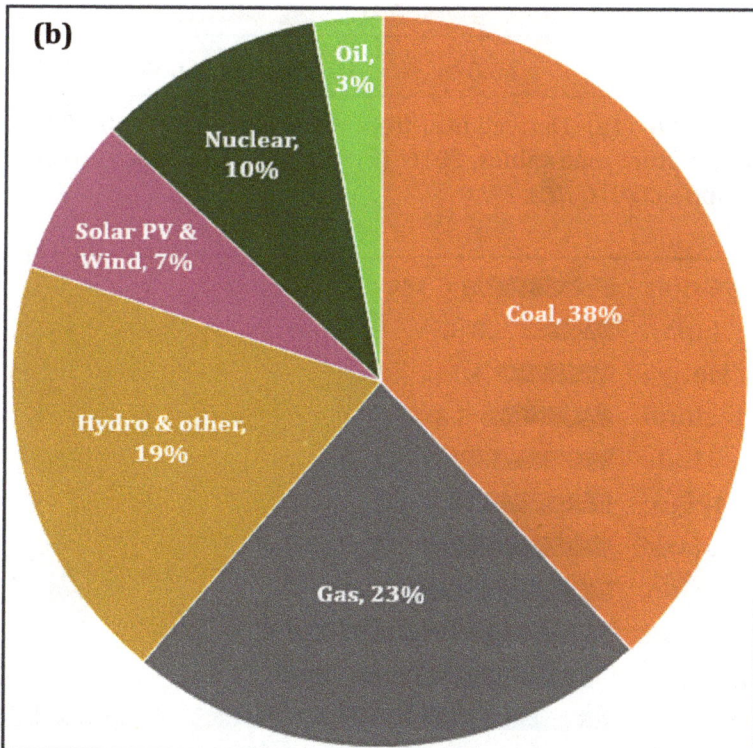

Figure 3.18a (a) Gross global primary energy consumption, 2018
(b) Global electricity generation fuel mix, 2018 (IEA,2019i)

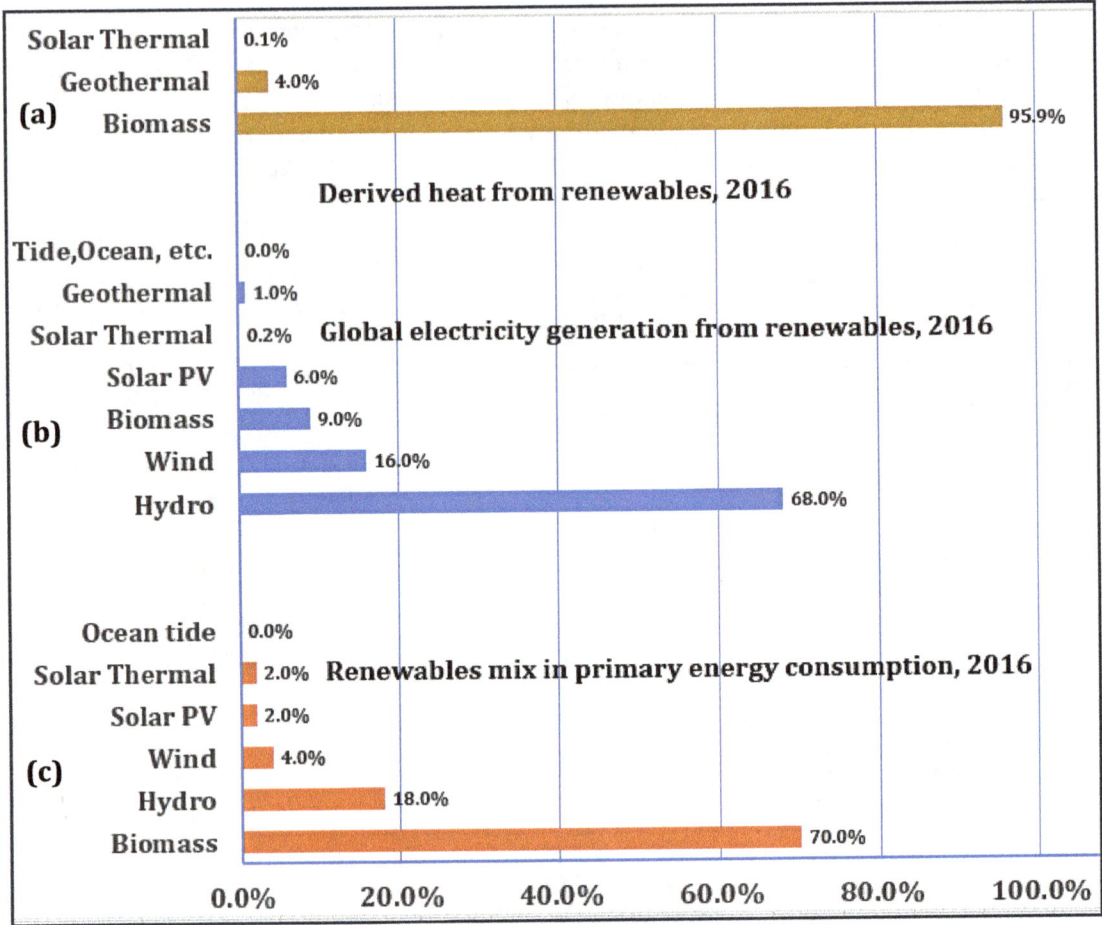

Figure 3.18b   (a) Derived heat from renewables, 2016 (b) Global electricity generation from renewables, 2016 (c)   Renewables' mix of total primary energy consumption, 2016 *(IEA, 2018)*

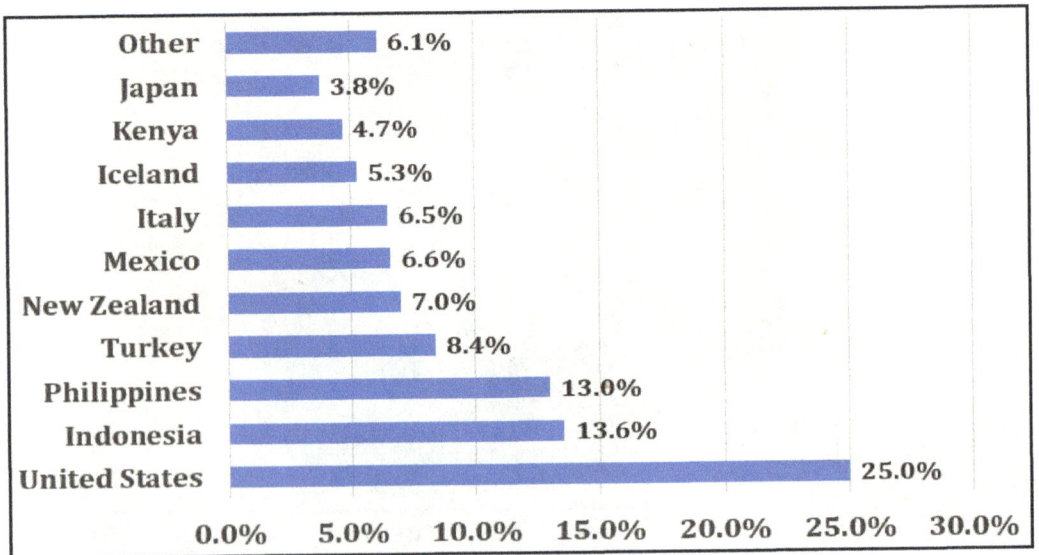

Figure 3.18c Top ten countries with highest geothermal capacities in 2016 (data from thinkenergy.com)

3.2.4   Unconventional energy

There are other primary energy sources apart from the traditional ones discussed above. These include shale oil, gas and bitumen, coal/biomass-based gaseous fuels, coal-bed methane, coal/gas-based liquid fuels, and hydrogen often classified as unconventional energy. Exploitation of these resources is increasing due to the development of new technologies and the impact on the global primary energy reserves, demand and supply will increase significantly over the next two decades. The highest growth will be in unconventional natural gas which is expected to account for around 60% of the projected rise in global natural gas supply, about a third of the total natural gas requirements, and nearly 90% of North America gas production by 2040. Shale oil is being processed into petroleum products, shale (natural) gas is being produced by hydraulic fracking, and exploitation of coal bed methane for power generation is increasing.

3.2.4.1   *Shale oil and gas*

Substantial gas and some oil resources are trapped in petroleum-bearing formations of low permeability: in rocks (caprocks), shale or tight sands deep beneath the Earth and seabeds, and are not recoverable by conventional techniques, especially when the gas may be dispersed in the rock rather than occurring in a concentrated underground reservoir. Fine-grained sedimentary rock may contain significant amounts of a solid mixture of organic chemical compounds (kerogen) from which liquid (shale oil) and gaseous (shale gas) fuels can be extracted. Unconventional oil recovery accounts for about 30% of the global recoverable oil reserves and oil shale contains three to four times as much oil as conventional crude oil reserves which are projected at around 1.2 trillion barrels (WEC, 2016a).

About 600 shale oil deposits have been identified in 33 countries on all continents of the world but less than fifty have been proved to be economically viable. Current estimates put the global resources of shale oil (also known at tight oil or shale-hosted oil) at around 6,050 billion barrels, around four times the size of the world's conventional crude oil resources. The United States alone holds about 80% of the total global reserves, with China, Russia, Israel, Jordan and DR Congo holding most of the balance. It is believed that current estimates for oil shale resources are rather conservative, as many deposits have not been adequately explored and many others have not even been identified. It should be noted also that only a small fraction of the resources has been proven and moved to reserves, that is, shale oil that is economically extractable using currently available technologies. There is little doubt that new technologies will improve access in the near future.

Global resources of shale gas are estimated at around 0.2 trillion cubic meters, enough to fill the global demand for natural gas for over sixty years at the current rate of consumption. Again, many potential resources are either not yet identified or evaluated. Six countries (U.S., China, Argentina, Algeria, Canada and Mexico) hold around two-thirds of the technically recoverable global shale gas resources but only three countries - the U.S., Australia and China - are actively exploiting shale gas deposits. Production of shale oil and gas requires special technology known as hydraulic fracturing. Vertical wells are drilled to depths of up to several thousand meters and may include horizontal sections. Shale rocks are fractured by pumping high-pressure fluid, typically water, proppant (sand, ceramic pellets, other small incompressible particles) and proprietary chemical additives. The high-pressure fluid fractures the rocks while the proppants hold the fractures open to allow gas or oil flow. Hydraulic fracturing technology was developed in the late 1940s and has been in use since the 1980s, initially for fracturing coal seams to release methane, primarily for safety reasons, since coalbed methane is both poisonous and a major cause of coal mine explosions (Figure 3.19).

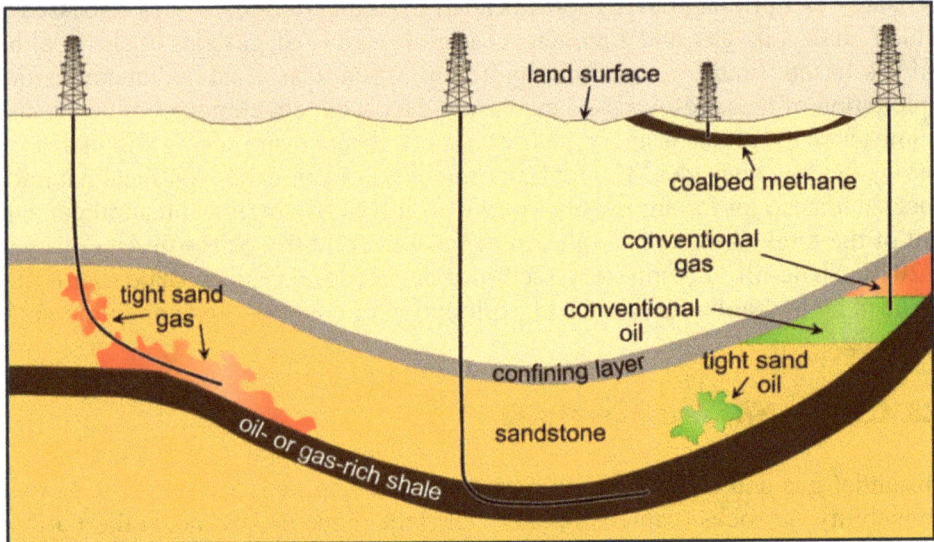

Figure 3.19 Unconventional production of coalbed methane, shale oil and gas by hydraulic fracturing. *(Wikipedia, 2018).*

The fracking technology has been further developed and deployed primarily in the United States and Canada, and can produce large volumes of shale oil and gas at relatively low cost. Two other countries - China and Argentina complete the list of the only four countries in the world currently exploiting shale oil and gas commercially. In 2017, shale oil provided 50% of total U.S. crude oil production and shale gas provided 60% of total natural gas production. The rapid development of fracturing technology is projected to transform the U.S. from a net importer to a net exporter of energy by the early 2020s. The country is currently the world's largest producer and consumer of natural gas globally. Production of unconventional oil and gas raises new environmental problems. Fluids employed in cracking up caprocks and coalbeds are potential groundwater contaminants and fracking is believed to induce geological instability which may increase the incidence or intensity of earth tremors and earthquakes.

3.2.4.2   Bitumen/Oil sand/tarsand

Bitumen (oil sand/tar sand) refers to naturally occurring thick oil trapped in loose sandstone and clay. The texture varies from highly viscous oil to solid mass, depending on the depth of the deposit. Deposits have been found in many countries all over the world and while some are less than a hundred meters deep, most deposits are located deeper underground. Global bitumen resources are estimated at more than 2 trillion barrels although much of it is yet to move to reserves. Canada holds the largest deposit in the world and utilizes the most advanced production technologies which include complex mining and in-situ oil recovery. Currently, tar sands represent about 40% of the country's oil production. Venezuela is the only other country in the world currently exploiting bitumen in commercial quantities, although on a much smaller scale. Complex technologies are required to separate the oil from the sand and clay, and refine it to the quality required for processing by regular refineries.

3.2.4.3   Coal-bed methane (CBM)

Methane is the major constituent of natural gas and is normally associated with petroleum deposits. However, the gas also occurs in association with coal deposits, often referred to as unconventional natural gas. During coalification, large quantities of methane-rich gas are generated and stored within the coal on internal surfaces. Methane occurs wherever large coal deposits are found and is released into the atmosphere regularly during coal mining for safety reasons. Some coalbeds are vented through boreholes before mining, but in the majority, the methane is diluted to below its explosive limit of around 5%. The gas is not only poisonous but also potentially explosive, and a cause of many mining accidents. Also, methane is a potent anthropogenic gas and releasing the gas into the atmosphere causes major pollution problems. Some mines burn the methane on site but this also releases carbon dioxide (another potent anthropogenic gas) into the atmosphere. In recent years, coalbed methane has gained prominence as a potential energy resource.

Because coal has such a large internal surface area, it can store very large volumes of methane-rich gas; six or seven times as much gas as a conventional natural gas reservoir of equal rock volume can hold (USGS, 2000). In addition, much of the coal, and thus much of the methane, lies at shallow depths, making wells easy to drill and inexpensive to complete. With greater depth, increased pressure closes fractures (cleats) in the coal, which reduces permeability and the ability of the gas to move through and out of the coal. Exploration costs for coal-bed methane are low, and the wells are cost-effective to drill. Unlike natural gas from most petroleum deposits, coalbed methane contains very little of the heavier hydrocarbons such as propane and butane which need to be removed from associated natural gas before use. In a conventional oil or gas reservoir, for example, gas lies on top of oil which, in turn, lies on top of water. An oil or gas well draws only from the petroleum that is extracted without producing a large volume of water, but water permeates coal beds, and its pressure traps methane within the coal. The methane is in a near-liquid state, lining the inside of pores within the coal (called the matrix). The open fractures in the coal (called the cleats) can also contain free gas or can be saturated with water. CBM can be recovered from underground coal before, during, or after mining operations. It can also be extracted from unmineable coal seams that are too deep, too thin, or of poor or inconsistent quality.

Coalbed methane resources are commensurate with the world's conventional gas resources. Methane concentration in the mixture of coalbed natural gases reaches 95 to 98% but carbon dioxide and ethane may also be present in significant quantities reaching around 50% and 10% respectively. Although coal deposits abound all over the world and all contain methane, not all deposits are suitable for coalbed methane exploitation and not every type of coal deposit is suitable for methane production. Thus, long-flaming brown coal fields are featured with low methane content and anthracite coal is characterized with high gas content but, it cannot be recovered economically due to high density and very low permeability of the deposit. The coals that fall somewhere in between the brown coals and the anthracite coals (bituminous) are attributable to the most favorable ones for methane production. Another important determinant of the economic potential of coalbed methane is the cost of competing primary energy resources, in particular, natural gas. Active drilling for coal bed methane began in the 1980s, mostly in the U.S. but spread rapidly to many other countries and the global CBM market is expanding rapidly. North America holds the largest share, followed by Europe and Asia Pacific. However, many emerging countries are also developing and exploiting resources, in particular, China, India and Indonesia. Application areas are diverse - power generation, heat for industrial processes, transportation, and steam-raising for commercial and residential use.

3.2.4.4   *Synthetic gas and liquid fuels*

Coal, and biomass can be converted to more efficient, lower carbon fuel including synthetic oil gas, synthetic natural gas, hydrogen, gasoline, diesel oil, and even ethanol which is mixed with gasoline. Also, natural gas can be steam-reformed to produce cleaner, higher-value hydrogen fuel. The technologies are well developed and have been around for decades. While syngas produced from coal, natural gas and biomass is fueling many power plants all over the world, proliferation of coal/gas/biomass-to-fuels has been slow because the economics is closely tied to global oil pricing which has been unstable for many years. Conversion of coal to transportation fuels has been the main source of liquid energy in South Africa for many decades and China also operates a commercial plant, with many more being planned or under construction. The main advantage of coal-to-fuels from the environmental point of view is that most of the pollutants emitted at the gasification stage can be captured at source, thus greatly reducing the life-cycle carbon footprint of the synthetic fuels.

3.2.4.5   *Flared fuels*

Combustible gas/liquid fuel is released during normal or unplanned operation in many industrial processes - oil and gas extraction, chemical plants, coal mining, landfills - and is usually burned off. The flaring process releases anthropogenic gases and particulate matter into the atmosphere and is a major environmental issue. Flaring in the oil and gas industry is perhaps the most prominent in terms of atmospheric pollution but also because enormous potentially valuable fuel is being wasted. The World Bank reports that between 150 to 170 billion cubic meters (BCM) of gases are flared or vented annually with no energy benefit and significant environmental damage, an amount valued at about $ 30.6 billion, equivalent to one-quarter of the United States' or 30% of the European Union's gas consumption annually. Russia leads the world in gas flaring and, with nine other countries, (mostly developing countries with low access to electricity), account for around three-quarters of the total global flaring (Figure 3.20).

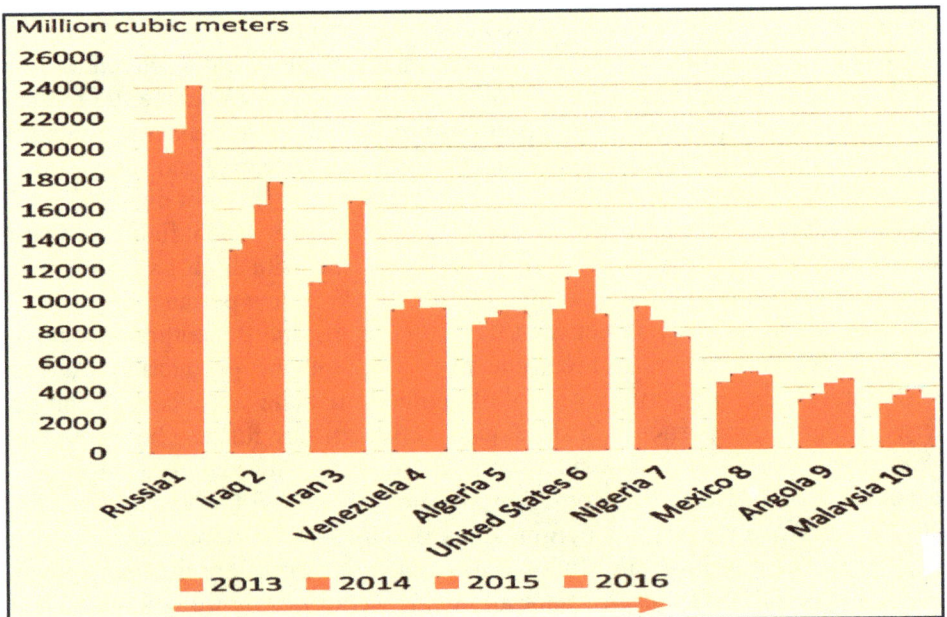

Figure 3.20   Top ten flaring countries, *2013-2016 (World Bank, 2018).*

The current amount of flaring releases around 350-400 million metric tons of $CO_2.eq$ into the atmosphere, in addition to toxic particulate matter. This level of emissions, combined with a similar levels in midstream operations, (leakages, routine venting during handling and transportation), account for around 20% of total greenhouse gas emissions from natural gas systems (the balance of 80% is accounted to combustion processes). Flaring gas wastes a valuable energy resource that could be used to support economic growth and progress, as well as contributes to environmental pollution by releasing millions of tonnes of $CO_2.eq$ to the atmosphere. This has prompted a World Bank initiative of "Zero Routine Flaring by 2030." For decades, global efforts to reduce flaring had not been very successful primarily because global prices of conventional oil and gas were high and there was little incentive to invest in expensive associated gas gathering and utilization infrastructure. However, the drastic fall of crude oil prices in the last decade, coupled with increasingly stringent environmental regulations, have repositioned associated fuel as a potentially valuable primary fuel. It is interesting to note that a significant amount of flaring is practiced in Africa which also has one of the lowest rankings in global modern energy access. It is estimated that flared gas by oil and gas companies in Africa could supply around 50% of the continent's electricity needs (Andersen *et al*., 2012).

3.2.5 Hydrogen

Hydrogen is the simplest element in the chemical chart and the most abundant in the universe. Because the gas is very reactive, it is never found in pure state but is always combined with other elements. Water is a combination of hydrogen and oxygen and many organic compounds contain hydrocarbons which are compounds of hydrogen and carbon - gasoline, natural gas, coal, methanol, propane, etc. Hydrogen is considered the fuel of the future because of its low carbon intensity and high versatility. Apart from the fact that the gas can be produced from many sources including fossil fuels, biofuels and water, it can provide enormous energy with little or no pollution residue (the main residue is almost always clean water). However, virtually every hydrogen production process requires substantial amounts of energy, and the total lifecycle carbon footprint can be very substantial depending on the source of energy, especially electricity. Areas of use include powering gas turbines in power generation, fuel cells which are powering engines, motors and appliances all over the world. Liquid hydrogen fuel cells have powered the electrical systems of space shuttles since the 1970s, and the potentials in transportation are high, with hydrogen-powered cars already on the market in some developed countries.

3.3 PRIMARY ENERGY USE

Primary energy is used in a variety of ways: Oil is transformed into a variety of fuels by refining; coal, natural gas, biomass, nuclear power are transformed into electricity by power plants; coal, natural gas and biomass are used directly in heating; coal, natural gas and biomass are transformed into biofuels; and coal, natural gas, oil are feedstocks for the production of a wide variety of chemicals and petrochemicals. The final consumers are industry, transportation, buildings, and a few others. Global primary energy demand is driven largely by economic growth, which in turn is driven by the rising population and increasing prosperity, particularly in the developing world. However, the overall growth in energy demand is moderated significantly by declining energy intensity (energy used per unit of GDP).

3.3.1   Total primary energy supply (TPES)

Fossil fuel share of the global total primary energy supply (TPES) in 2017 was around 80%, with oil accounting for the largest share of 31% (Figure 3.1). Non-OECD countries accounted for around two-thirds of the demand, with Asia accounting for over one third, and China alone consuming 22%. Oil is refined into gasoline, diesel oil, jet fuel, etc., and used mainly by the transportation sector; coal, gas, nuclear fuel and renewables are used mainly for power generation; the industrial sector uses coal, gas and electric power as energy sources, and coal, oil and gas as feedstocks for the chemical and petrochemical sub-sector. Primary energy demand is projected to grow by about 35% by 2040, broadly based across all end-use sectors, with the industrial sector accounting for around half of the overall increase. Nearly all of the growth in energy demand will come from fast-growing developing economies, with China, India and other emerging Asia accounting for around two-thirds of the growth. Renewables will grow faster than any other primary fuel source, accounting for about 40% of the expected total growth in primary energy demand over the next two decades. Around two-thirds of the increase will be used to generate power to meet the exponentially increasing demand for access to power in the developing world. Over 90% of the primary energy end-use is for combustion to produce heat, power, and fuel internal combustion engines which drive automobiles, airplanes, ships, etc., the remaining (around 6%) feeding mostly the petrochemicals industry. However, non-combusted use is growing rapidly and is expected to become the largest source of fossil fuel demand growth in the next two decades.

3.3.2   Total primary energy consumption (TPEC)

Less than 70% of the global total primary energy supply actually is available as end-use energy, the balance being lost to inefficiency across the whole spectrum of energy production and utilization. Global primary energy consumption (TPEC) is mixed and includes both primary and transformed fuels - refined transportation fuels, natural gas, oil, coal, biofuels. Primary energy is used in four key sectors of the global economy: electric power generation, industry, transport, and buildings. Oil accounted for the largest share (40.9%) of TPEC in 2016, followed by electricity (18.8%) (Figures 3.2).

3.4   PRIMARY ENERGY FOR POWER GENERATION

Electric power is not a primary energy because it is produced by converting a primary energy source, and any of all available primary sources is potentially suitable for conversion. Electricity can be generated from any of the primary fuels but fossil fuels still account for by far the largest share of the total global electric power generation. However share of fossil fuels (mainly oil and coal) will drop significantly over the next two decades, due to substitution with natural gas and increased use of renewables. Electricity is the prime energy source for industry, commerce and the social sector - homes, hospitals, recreation centers, etc. Electricity powers all aspects of human development - factories that make consumer goods, commercial buildings, heat, light, air conditioning, the Internet and everything that connects to it. Electric power generation currently accounts for about 42% of the global primary energy use, and the sector is projected to account for around 70% of the growth in primary energy consumption over the next two decades, with comparable rise in both the industrializing and mature economies. Presently, electric power accounts for nearly 40% of total global delivered (end-use) energy and demand is expected to rise as new technologies evolve, many more countries industrialize, and more people in emerging nations gain access to electricity. Industry (which includes agriculture and other minor sectors) and the buildings sectors consume the highest proportion of final energy use (including electricity) and will account for the majority of the growth over the next two decades. In 2018, fossil fuels provided 66-70% of the primary energy used for power generation, and demand by the non-OECD countries has risen dramatically over the last few decades, accounting for around 55% in 2015 (Figure 3.14). Several projections indicate that demand will rise by 60-70% by 2040 and, although increases are expected from all parts of the world, the bulk of the rise (85%) will come from non-OECD countries. Also, fuel mix for power generation will change significantly globally as well as across regions. Industry accounts for about half of global electricity usage while the other half comes from commercial and residential usage. All reviewed projections agree that the share of coal in power generation will decline from about 40% to around 30% by 2040, due to increasing shares of natural gas, renewables and nuclear energy (Figures 3.21-3.25).

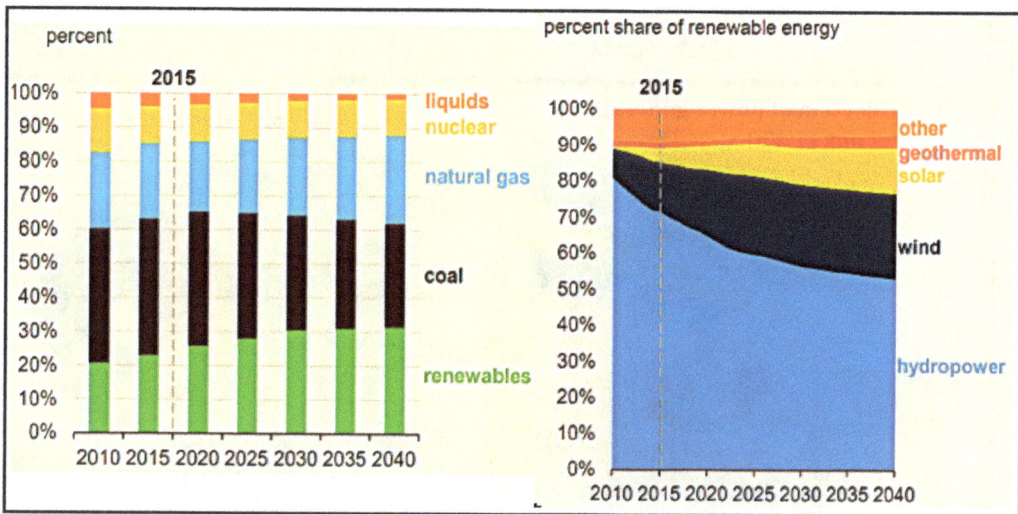

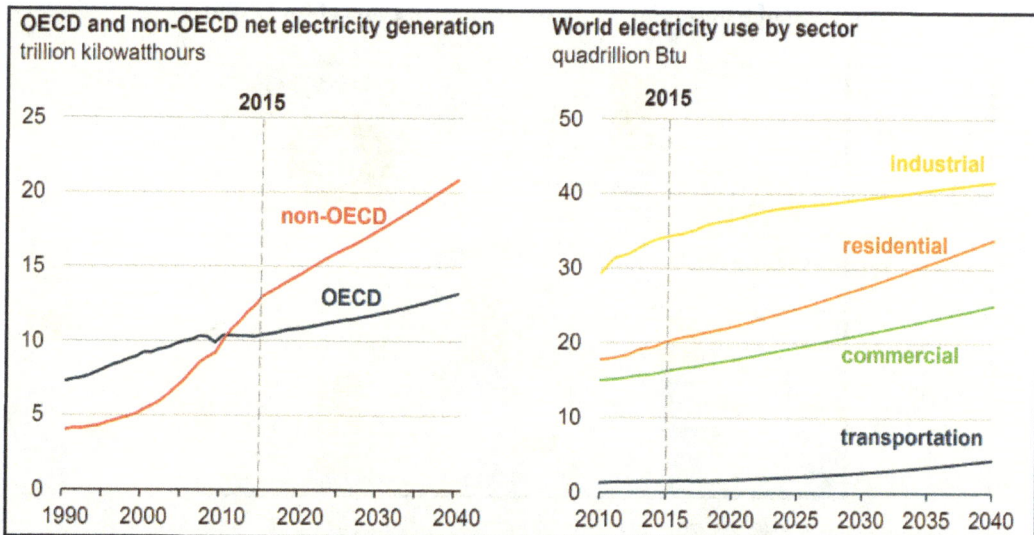

Figure 3.21 Projections of electric power generation by fuel mix, region and sector *(EIA, 2017).*

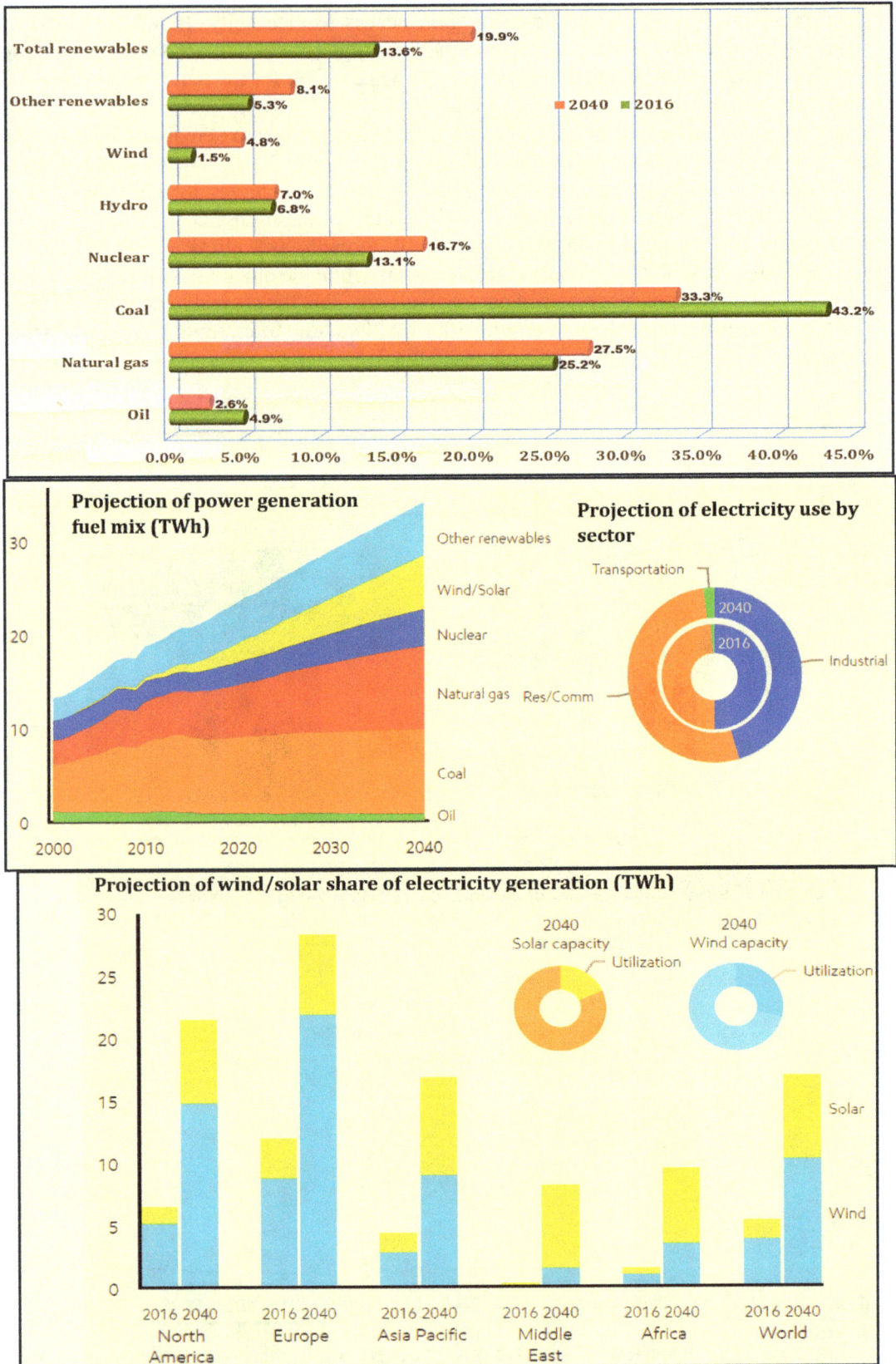

Figure 3.22 Projected energy for power generation by mix and region/country.
(ExxonMobil, 2018).

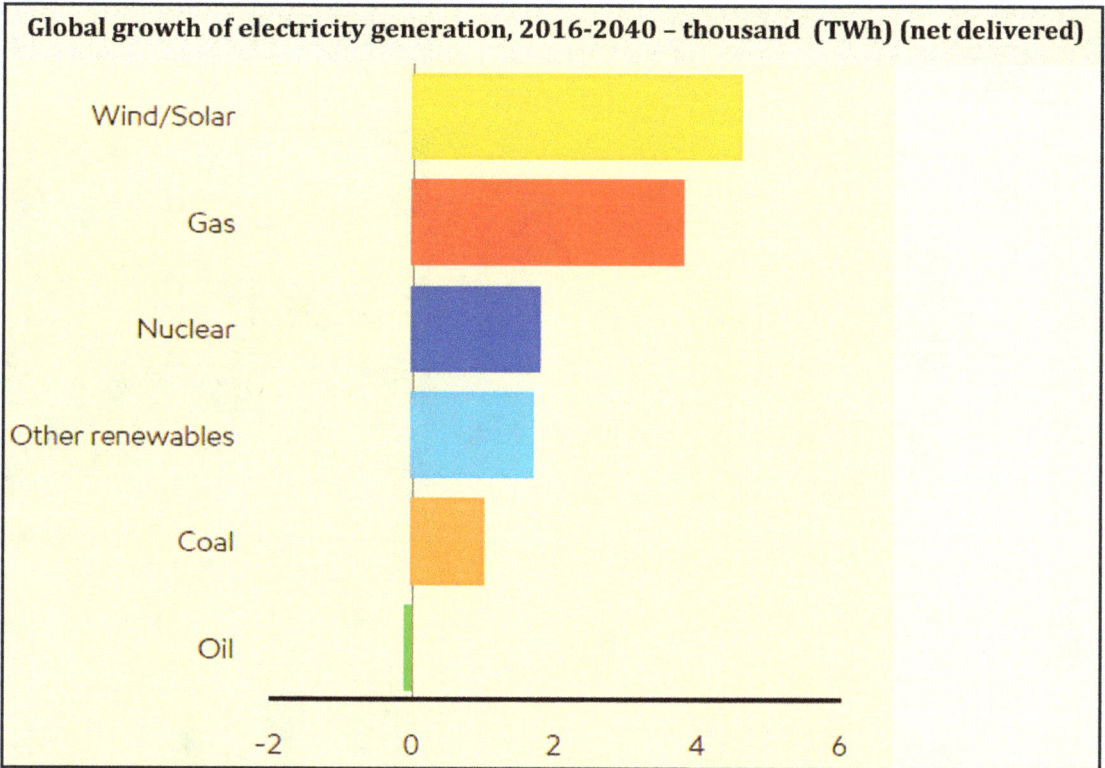

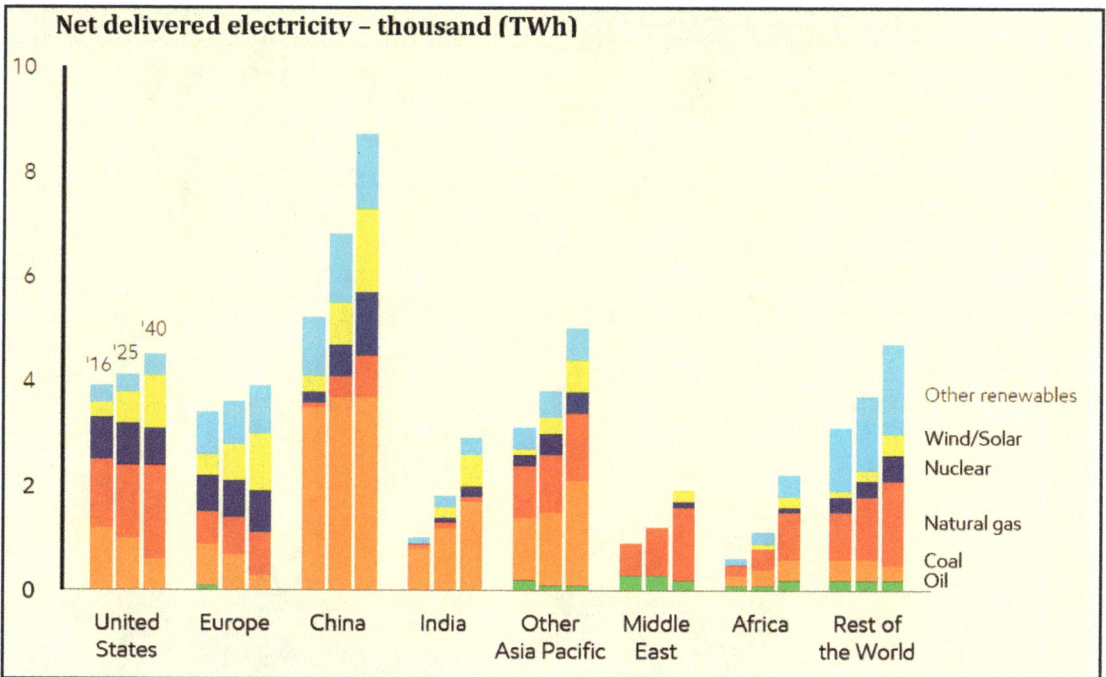

Figure 3.23   Projected energy for power generation by mix and region/country *(ExxonMobil, 2018).*

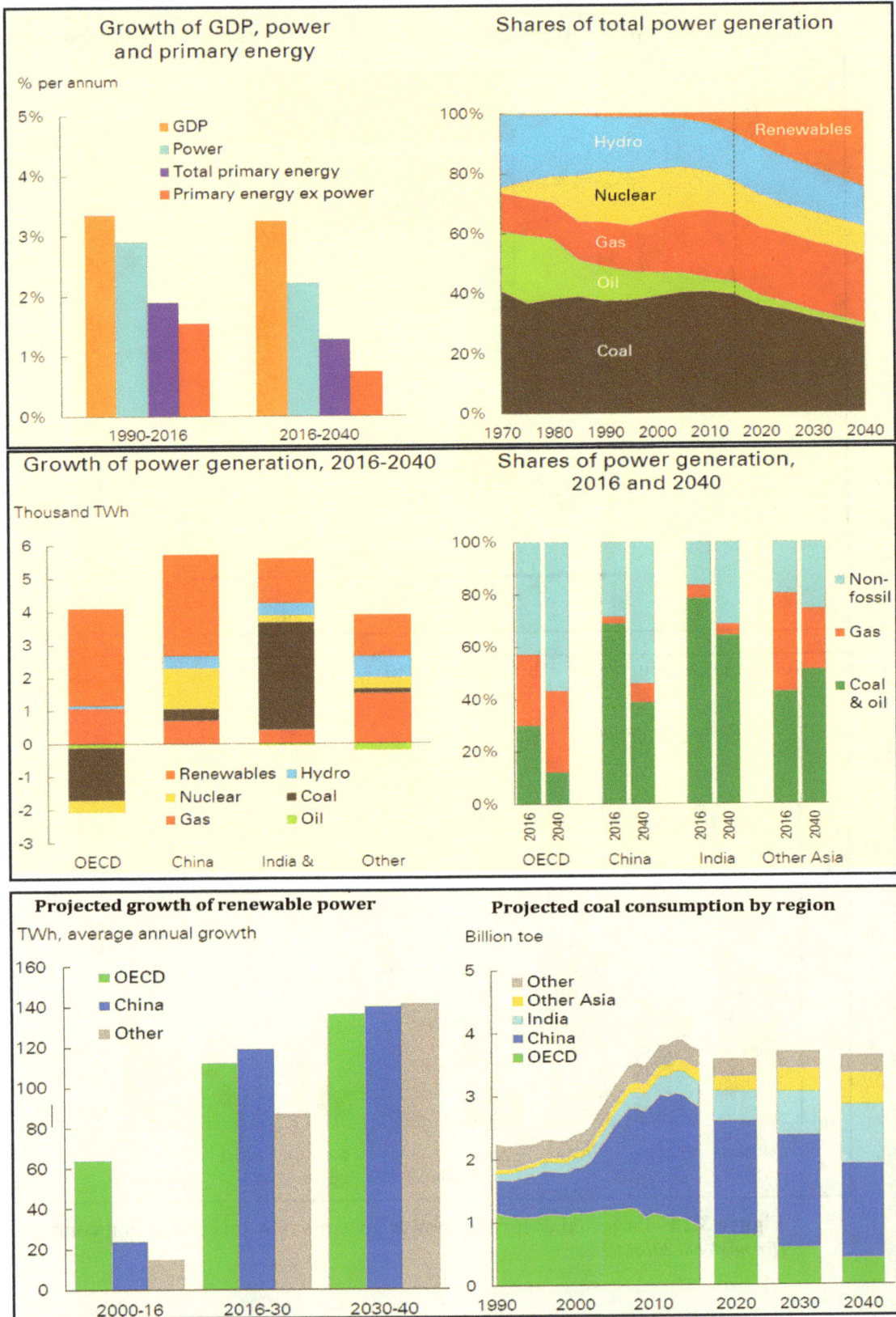

Figure 3.24   Projected energy for power generation by mix and region/country *(BP, 2018).*

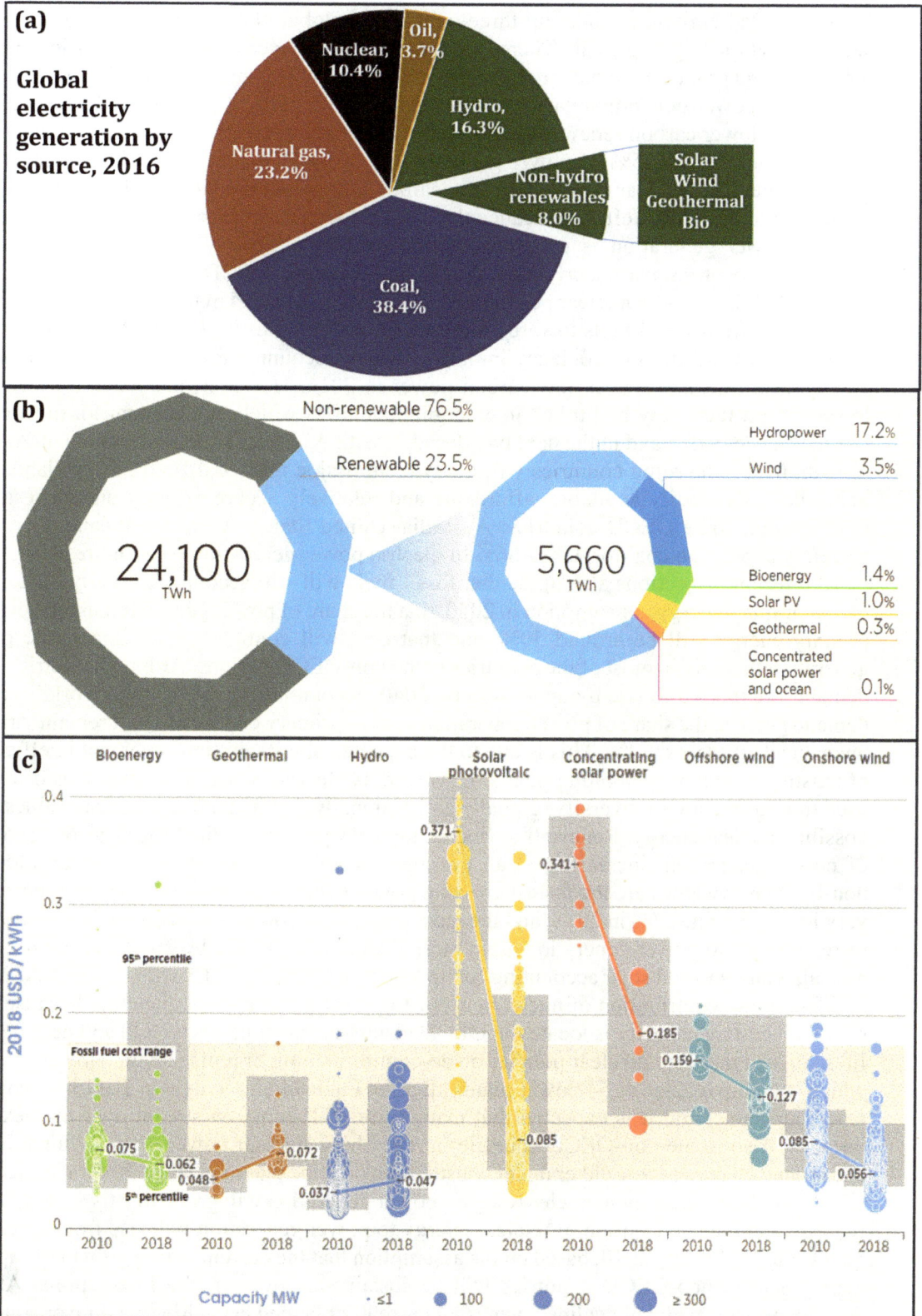

Figure 3.25 (a) Renewable share of global electricity generation in 2016; (b) Renewable share of global electricity generation by type in 2015; (c) Global weighted average levelized cost of electricity for utility-scale power, 2010-2018 *(IEA, 2018; IRENA, 2017a & b).*

Energy-related emissions make up three-quarters of global GHG emissions, and the power sector shows the largest growth. There is a consensus on the increasing urgency to decarbonize the sector, and promoting greater use of renewables and nuclear energy is considered a priority. Renewable power generation capacity has been growing particularly in OECD countries in an effort to lower carbon renewables, and some optimistic projections expect that renewables could account for 25-35% of power generation fuels by 2040, especially if stronger policy support emerges. Nuclear power, as a low carbon energy source has significant potential to contribute to the global effort to reduce the carbon footprint of electricity, but the future of nuclear power generation is uncertain. While nuclear electricity is expected to grow in non-OECD countries, particularly Asia, growth will decline in OECD countries which currently host most of the world's nuclear power capacity. There has been a major global shift of energy investments from fossil fuels towards low carbon technologies in the past few years, mostly in renewables and energy efficiency initiatives which accounted for over 95% of total global energy sector investments in low-carbon power in 2016. However, the scope for switch to lower carbon fuels may be limited in emerging countries which will account for most of the rise in electricity demand in the next two decades, with Asia accounting for around 60% of the growth. Many emerging countries will likely commission new coal-fired power plants, coal being the only locally available, affordable and relatively secure primary energy resource. Furthermore, some OECD countries: Australia, United States, Australia, Poland, Germany, currently depend on coal for 45% - 86% of electric power generation (see Figure 3.10).

All recent projections also agree that fossil fuels will still account for nearly 80% of the global primary energy consumption in 2040, that the share of power generation in total end-use primary energy will be around 50%, and that coal will continue to dominate the power generation fuel mix in most of the countries in the non-OECD regions. Although electric power demand is expected to rise by up to 70% by 2040, accounting for 55% of the world's energy demand growth, the share of power generation in total primary energy use will remain virtually unchanged, at around 42%. This is due to the expected significant improvement in efficiency of existing and new generating plants. However, as electricity use rises, the types of energy used to generate it will diversify, globally and regionally, led by natural gas, renewables, and, possibly, nuclear energy. Renewables (including hydropower) are the fastest-growing sources of power generation, increasing at an average rate of 2.8% per year. However, although non-hydro renewables are the fastest growing power generation fuels, they are starting from a very low base of just 7% in 2018, and the most optimistic projections estimate the contribution of renewables to power generation at no more than about 20-35% by 2040, around the same as coal, with wind and solar accounting for up to a third (IEA, 2017; EIA, 2017; IRENA, 2017).

The future contribution of nuclear energy to power generation is uncertain because most of the generating capacity is located in OECD countries, most plants are old and nearing their lifespan, and there are no clear policies on de-commissioning or replacement. However, Japan which shut down 52 of its 54 power plants after the Fukushima accident in 2011 has started to reactivate some, and it is expected that many more will come on stream over the next two decades. Also, some non-OECD countries already have nuclear power plants (China, India), and around 60 reactors are either under construction or at the planning stage. Several projections show that global capacity in nuclear electric power generation will grow, mostly in non-OECD regions, and the contribution of nuclear energy to power generation will rise from its current 10-11% to 17-18% by 2040, based on the assumption that the current strong growth of capacity deployment in non-OECD countries will be sustained. However, the International Atomic Agency projects a global decline, even if all currently planned capacities become operational, unless many more countries especially in the OECD region decide to include nuclear electricity in their fuel mixes.

Production of biofuels (ethanol, biodiesel) from sugarcane, corn, soybean, sorghum, and other organic matter has been rising, although sustainability has been questioned in recent years in connection with food-versus-fuel priority in land use. The products are used primarily in transportation, although a small proportion (biowaste) is also used for power generation, accounting for around 2% of global electricity output. The development of renewable energy, with the exception of hydro-energy has been slow despite the extensive availability of renewable resources, due to the intermittent nature of these resources and dependence on weather by most of them. For example, draught affects water levels for hydropower generation and cultivation of crops for biomass energy. The bulk of renewable energy, in particular, biomass, solar and wind (variable renewable energy, VRE) is used for electric power generation and the intermittent nature makes it difficult to integrate the power produced with national electricity grids. Also, the very low utilization capacities of solar and wind energy (20% and 30% respectively) are major constraints to their share in power generation, which was only about 5% in 2016, projected to grow to about 10-12% in 2040, while the total contribution of all renewables including hydropower will rise to only about 25%. In the last two decades or so, efforts have been made to develop bi-fuel hybrid power plants which combine wind, solar, biomass or tidal power with a fossil fuel. Some hybrid solar-gas plants are already in commercial operation in some developed economies, in particular, the United States. Renewable energy is deployed whenever available and the intermittent gaps are filled by natural gas. Developments in pumped storage are also creating new options of using solar/wind electricity when available to pump water to higher ground for later use in hydropower generation.

Technology advances, particularly those that unlock efficiencies, and falling costs in recent years have been driving the adoption of renewable energy, particularly in the developing world, with the power sector leading the way. Weighted average levelized cost of PV solar and off-shore wind power generation now fall within the fossil fuel range, and projections show that both will become even more competitive over the next few years (Figure 3.25). Global investment has also been increasing, much of it in the developing world. In 2015, developing countries attracted the majority of renewable energy investments, with China alone accounting for about one-third of the global total (IRENA, 2017). Most of the investments in recent years have been in PV solar and wind, which together accounted for about 90% of total global investments. Solar PV is particularly suitable for utility power generation, serving small off-grid communities, commercial units, or household power generation, enabling consumers to produce power for their own needs and feed surplus energy into the grid. Utility scale installations are becoming increasingly competitive with new fossil-fuel power generation, even without subsidies. Renewables increased by 4% in 2018, accounting for almost one-quarter of global energy demand growth. The power sector led the gains, with renewables-based electricity generation increasing at its fastest pace this decade. Solar PV, hydropower, and wind each accounted for about a third of the growth, with bioenergy accounting for most of the rest. Renewables covered almost 45% of the world's electricity generation growth, now accounting for over 25% of global power output. Solar PV, hydropower, and wind each accounted for about a third of the growth, with bioenergy accounting for the majority of the rest. Taken together, renewables were responsible for almost 45% of the world's increase in electricity generation. They now account for almost 25% of global power output, second after coal. China accounted for over 40% of the growth in renewable-based electricity generation, followed by Europe, which accounted for 25%. The United States and India combined contributed another 13% (IEA, 2019c).

Many developed countries are also providing attractive incentives to promote investments in solar and wind power generation, with very positive results. Renewables accounted for about 12% of the U.S. total energy consumption in 2017, with hydropower contributing only 2%.

A combination of reductions in technology costs and implementation of policies that encourage the use of renewables at the state level (renewable portfolio standards) and at the federal level (production and investment tax credits) has been driving down the costs of renewables technologies, in particular, wind and solar photovoltaic, supporting their expanded adoption. Wind and solar generation will lead the growth in power generation fuel use by the country over the next two to three decades, accounting for about 64% of the total electric generation growth. The increase in wind and solar generation will lead the growth, accounting for around 94% of the total growth in renewable energy use (EIA, 2018). Many other countries in Europe also have strong policies that are promoting the use of renewables. It should be noted however that growth of renewable use is very sensitive to the cost of alternative fuels, in particular, natural gas. Lower natural gas prices always impact negatively on renewable power investments, as has happened in the last two years or so. Furthermore, although growth in solar and wind energy deployment has been very strong and is projected to rise even faster in the coming years, both are starting from a very low base and contributed only 7% to the global power generation in 2018.

3.5   END-USE (DELIVERED) ENERGY CONSUMPTION

Three sectors (industry, transport and building) use both primary and converted fuels. Industry and the building sectors use a variety of fuels, mainly coal, natural gas, and electricity while most of the products of oil - gasoline, diesel, jet fuel - are used by the transport sector. Natural gas and electricity account for less than 5% of the transport sector energy use. However, industry also uses oil, coal and natural gas as raw materials for chemicals and petrochemicals production, and bioenergy features significantly in the household energy mix (Figures 3.3 – 3.6).

3.5.1   Industry sector

Modern economy depends critically on the industrial sector (including infrastructure and agriculture) for survival and growth. The sector is the largest user of energy globally, accounting for around half of the primary energy demand and 50% of electric power consumption. Areas of utilization include mining, manufacture of machinery, automobiles, production of consumer goods, primary metals, chemicals, paper, cement, metals polymers, agriculture, pharmaceuticals, food, etc. Industry uses energy both as a fuel and as feedstock for the production of chemicals, petrochemicals, lubricants, waxes, asphalt, etc. Oil, natural gas and electricity dominate the industrial fuel mix, and will contribute about one-third of industrial energy growth over the next two decades. The projected rise in oil share is due mainly to its use as a chemical feedstock. Coal plays a key role in the production of iron and steel, cement and other non-metallic manufacturing, heat and steam raising, and the emerging economies will account for most of the growth in demand. Non-OECD regions currently consume seven times as much coal as OECD regions, and this scenario is expected to continue in view of the continuing movement of heavy industrial manufacturing to the emerging nations. Fuel switching opportunities are relatively rare for these industries, and, in any case, many of the host countries have no access to any other cost-effective fuel.

With the expected growth in the global economy, human development and consumer demand, energy demand by the industrial sector is expected to grow by about 25-30% by 2040, but much of the growth will be in the emerging countries in view of the expected rapid growth in industrialization, especially the heavy industries such as iron and steel, chemicals, petrochemicals and cement production. Demand for fossil fuels as feedstocks for chemicals production has been rising. Chemicals are the building blocks for a very wide variety of consumer products: fertilizers, adhesives, cosmetics, textiles, plastics, paints, medical devices, plastics pharmaceuticals, cars, computers and many other industrial and home goods. The primary products industries are energy-intensive and offer relatively small scope for efficiency improvement (Figures 3.26-3.28). Energy demand projection in the industrial sector has been based on the assumption that the current global upward trend in industrial manufacturing efficiencies and adoption of new technologies which translate to lower energy intensity, will be sustained. While this is likely in OECD countries and China where energy intensity in the industrial sector has improved markedly over the last two decades, other emerging economies may not be able to afford the added cost of deploying newer, more energy efficient technologies.

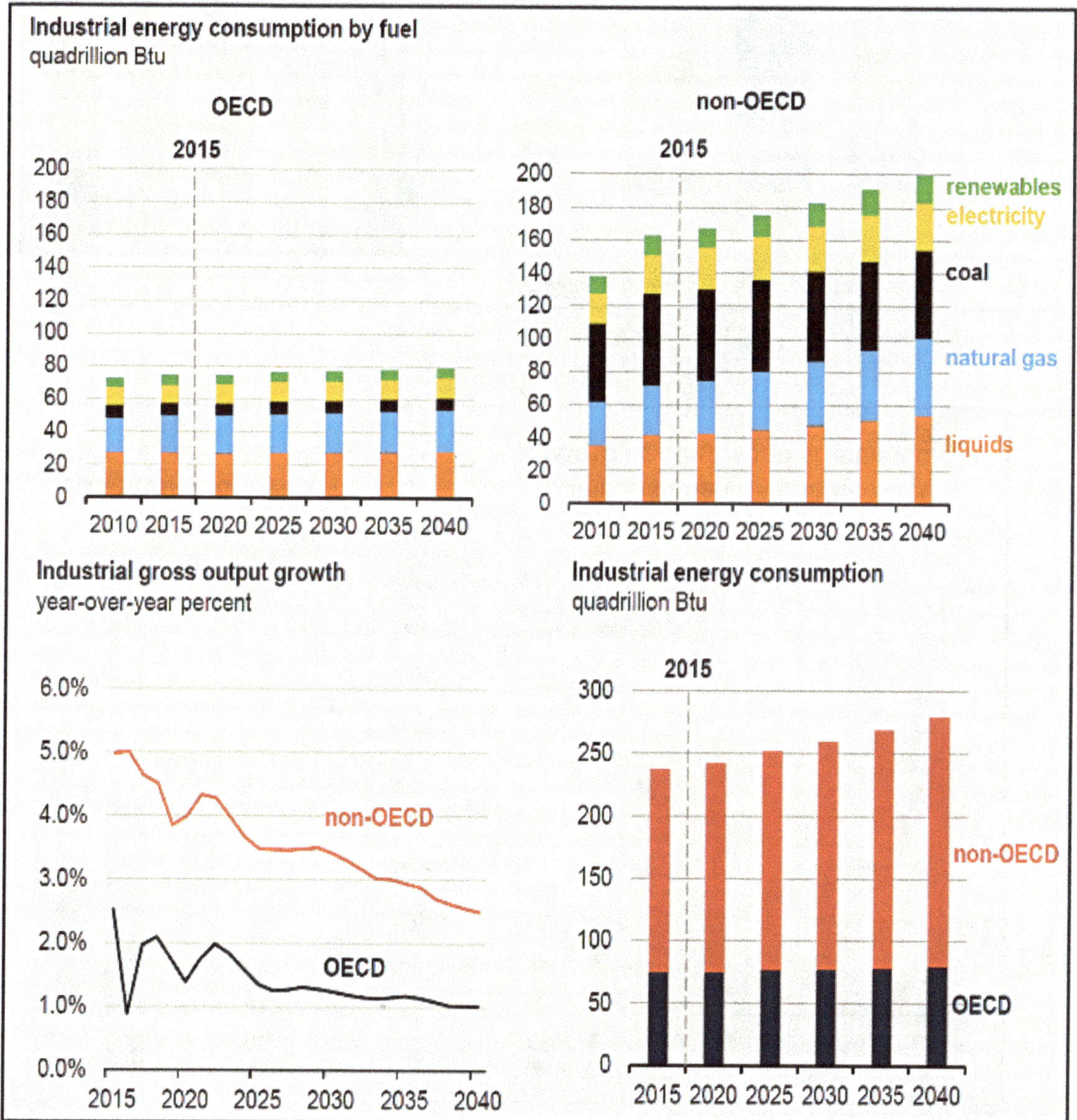

Figure 3.26 Projections of industry energy demand and mix *(EIA, 2017).*

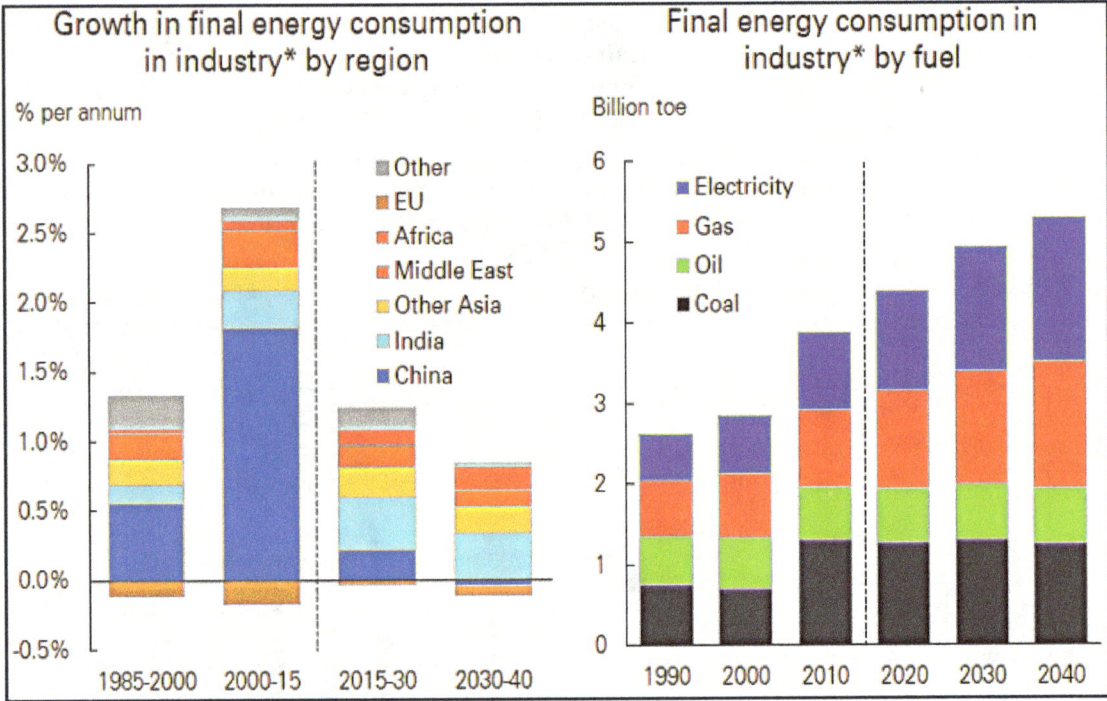

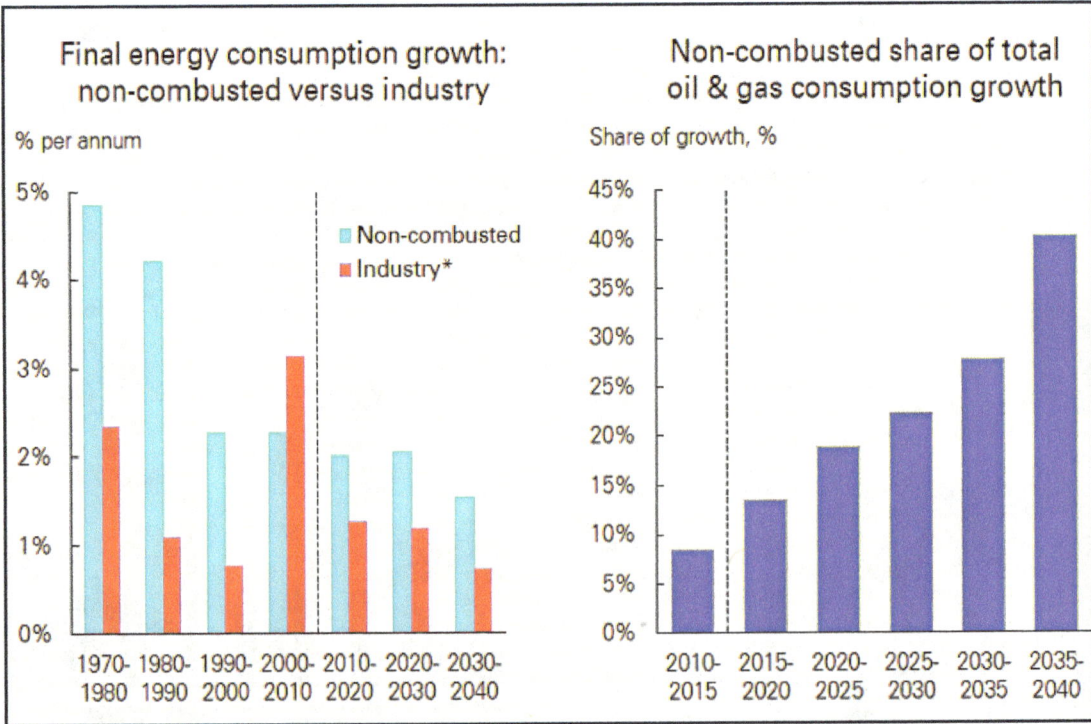

Figure 3.27 Projections of industry energy demand and mix , excludes non-combusted fuels *(BP, 2018).*

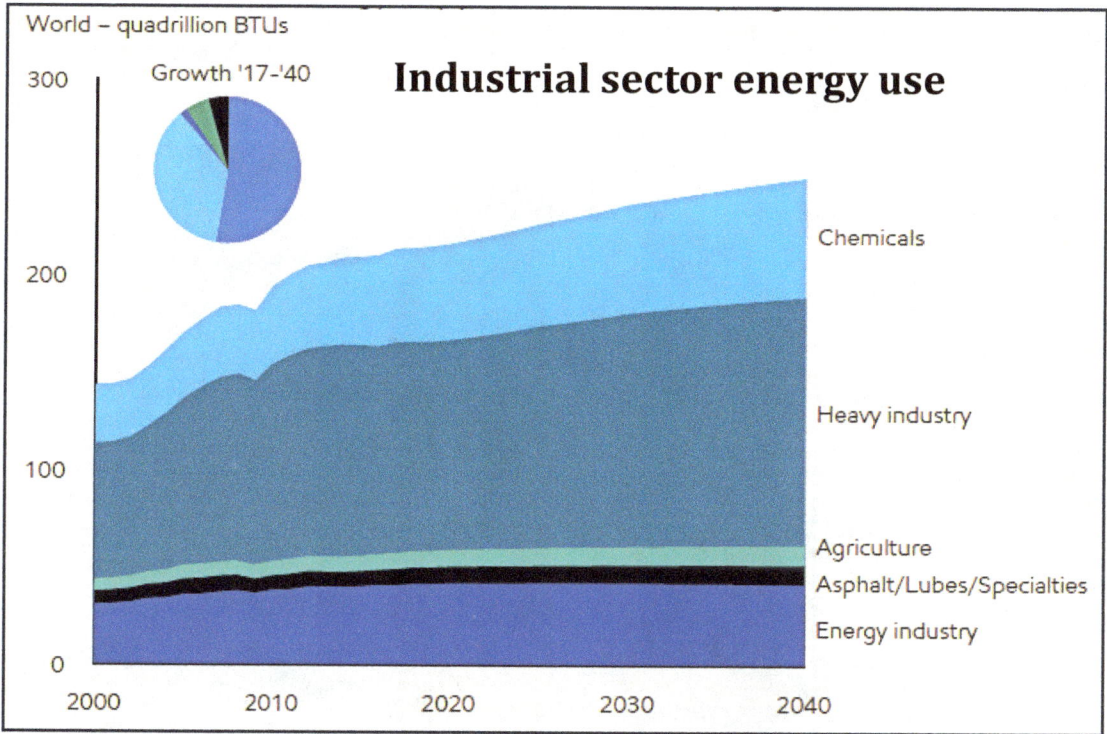

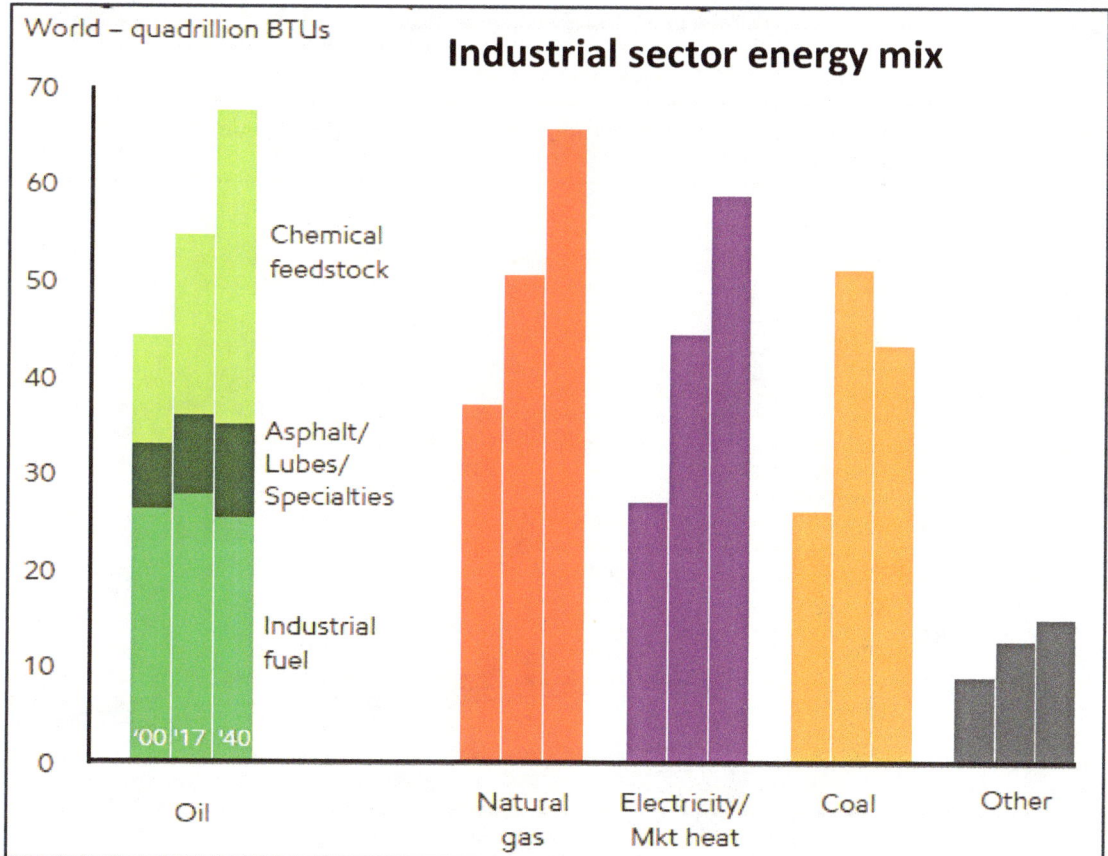

Figure 3.28a Projections of industry energy demand and mix *(ExxonMobil, 2019).*

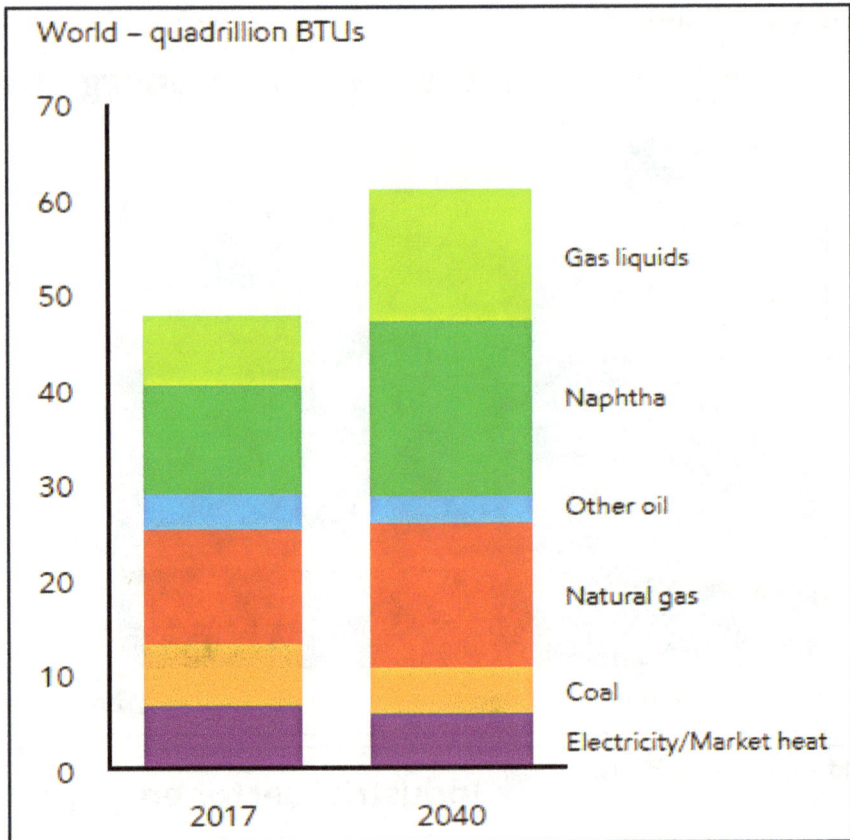

Figure 3.28b Projections of industry energy use for chemicals production *(ExxonMobil, 2019).*

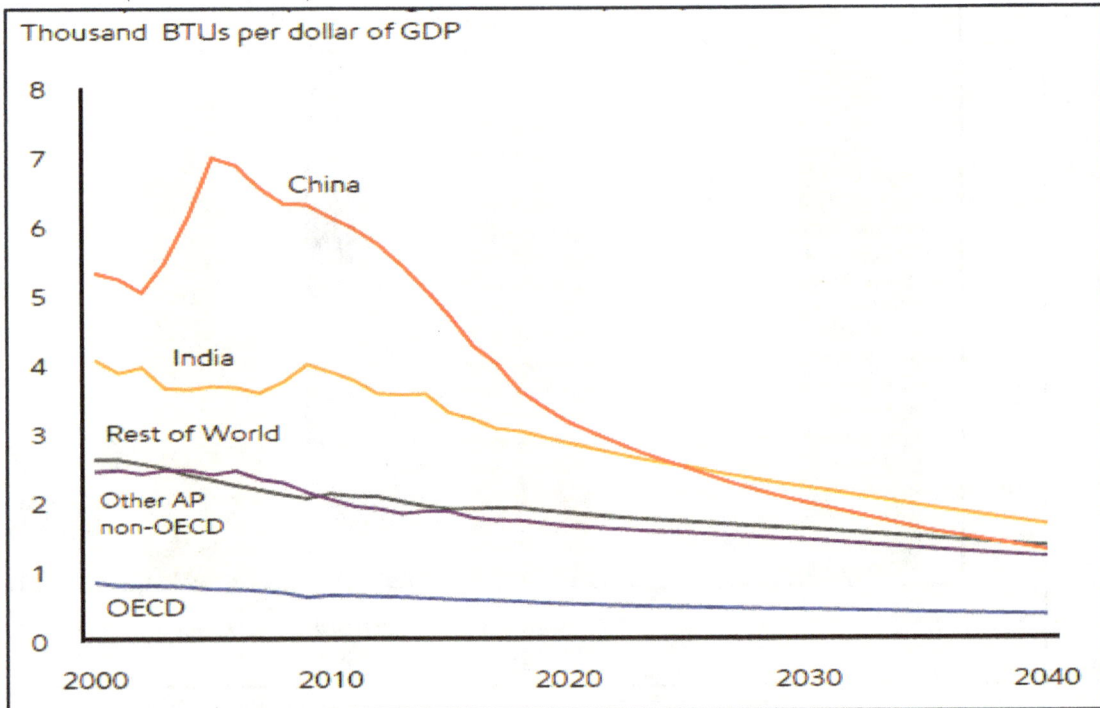

Figure 3.28c Projections of industry energy intensity *(ExxonMobil, 2019).*

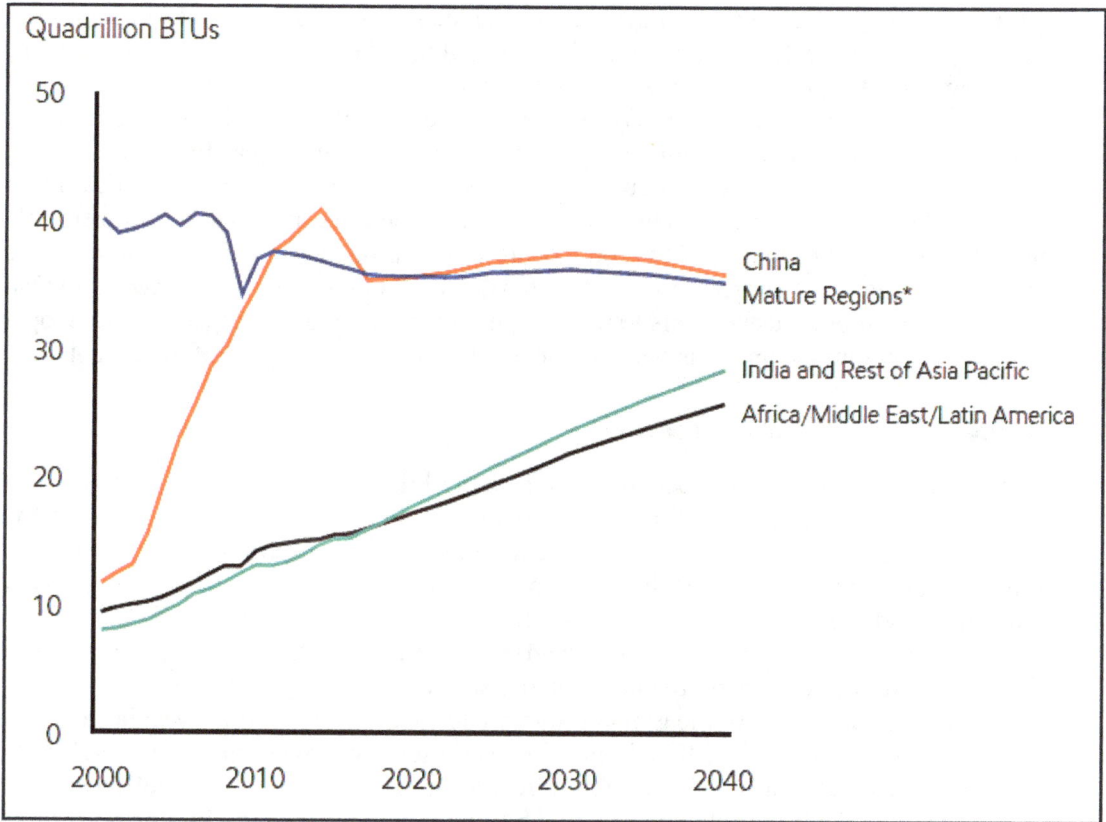

Figure 3.28d Migration of heavy industry to emerging markets *(ExxonMobil, 2018).*

The overall projected rise in energy demand by the industrial sector by 2040 is around 30% but there will be significant variations in the sub-sectors. While the growth in the iron and steel and cement industries will be about 25%, energy demand by the chemical industry sector will be double, around 50%, largely because of the strong growth in demand for products in the emerging regions. A significant shift in the fuel mix is expected by 2040, particularly in the heavy industry sub-sector. Coal use will fall from over 20% to around 15%, natural gas will rise to about 25% and oil use for energy will decline. China accounted for the bulk of the growth in industrial energy consumption in the last two decades but a significant shift in the regional dynamics is expected, and India will account for the highest growth rate over the next two decades. About 10% of current oil and gas demand by industry is used in non-energy applications, mainly as feedstock for the chemical, petrochemical, and other industries. Around 6% of the global demand of coal, 16% of oil and 12.5% of natural gas were used as raw materials for non-energy products in 2016 - chemicals, fertilizers, polymers, waxes, lubricants, bitumen, etc. The demand for these products has been rising, consistent with rising prosperity across all regions. The chemicals sector (which uses energy both as fuel and feedstock) is emerging as a major energy consumer in the industrial sector of the economy, particularly in many emerging markets where demand for chemical products is growing exponentially, outpacing GDP growth.

Growth of non-combusted fossil fuel demand has been very strong, and is projected to rise almost twice as fast the rate for other sub-sectors of industry, rising by around 45% from current level by 2040. Oil will account for nearly two-thirds of the growth in the non-combusted use of energy, with natural gas providing much of the remainder. The chemicals/petrochemicals industry is also the fastest growing energy user in the industrial sector, and has been migrating to many emerging nations with competitive advantage like low-cost feedstocks or increasing local demand. Developing Asia Pacific will account for the largest growth, driven by population size and strong economic growth. By 2040, growth of the sub-sector is expected to more than double in the Middle East, stimulated by access to low-cost oil and gas feedstock, while China is also expanding its chemicals industry which uses coal both as fuel and chemical feedstock. Most of the products of the chemicals industries - naphtha, methanol, fertilizers, plastics, pharmaceuticals, and many other products are traded globally as intermediate products. For example, China supplies around 40% of global demand for methanol, produced from coal. The chemicals sector of industry offers limited scope for efficiency gains, especially when operating in emerging countries, and has become the most polluting sub-sector of the global industry.

3.5.2 Transportation sector

The transport sector currently accounts for 21% of global primary energy demand. The global car fleet in 2015 was about 0.9 billion and is expected to double to around 2 billion by 2040 (BP 2018). Almost all the growth is in emerging markets where rapid economic growth, urbanization and rising incomes boost car ownership. Car ownership in non-OECD countries will triple over the same period, from 0.4 billion to 1.2 billion. A similar growth pattern in the commercial vehicle sector is also projected. Oil will remain the world's primary fuel in the foreseeable future, driven by demand for transportation and by the chemicals industry. Oil accounted for about 32% of the global primary energy in 2015 and there will be no significant change in 2040 - around 32%. About 94% was used for transportation in 2018 but some reduction is expected in 2040 (possibly down to 85%) due to increasing efficiency and the expected rise in the use of electric cars and biofuel-generated power (Figures 3.29 to 3.33).

The demand for diesel is expected to grow by about 30% by 2040 to meet the projected increase in trucking and marine transportation, while jet fuel will rise by about 50%. Fossil fuel sourced petrol, diesel, gas, and aviation fuel remain dominant in the transportation sector but increased conversion of biomass to gasoline and diesel will help to reduce demand to some extent over the next two decades or so. Fuel efficiency of transportation vehicles has been rising and the trend is expected to continue. The average kilometers per liter of fuel is expected to double by 2040, and this should reduce demand for oil in 2040. However, the population of vehicles is growing exponentially and is expected to double by 2035-2040, and the bulk of the growth will be in the emerging world where there is little policy-inspired fuel efficiency drive. This is expected to push up the demand for transportation fuel by about 30%.

The main components of the oil demand are gasoline, diesel oil and aviation fuel, gasoline being in the highest demand. Diesel accounts for around 35% of the total energy used for transportation but demand is growing rapidly and is expected to surpass gasoline by 2040. The main driver is the growth in heavy-duty truck transportation which accounts for about 80% of the total demand for diesel oil. Global energy demand for heavy-duty vehicles is expected to increase by about 45% by 2040 and about 85% of the growth will be in emerging economies where economic activities are increasing most rapidly, but growth in the developed countries will also be strong because of the rising trend in e-commerce. Light-duty vehicles and marine transportation account for the remaining 20%. Clearly, diesel oil will remain predominant in the heavy-duty transportation sector in the foreseeable future. Some heavy trucks and buses are designed to use compressed natural gas but the share of the fuel is only about 2% of the total transportation fuel demand, although this is likely to rise to about 5% by 2040. The main problem is the cost of gas-powered trucks which is significantly higher than the cost of a comparable diesel truck but, depending on local availability and pricing of gas, there could be significant savings in fuel cost. Some city bus and heavy truck fleets are running already on liquefied natural gas in some OECD countries.

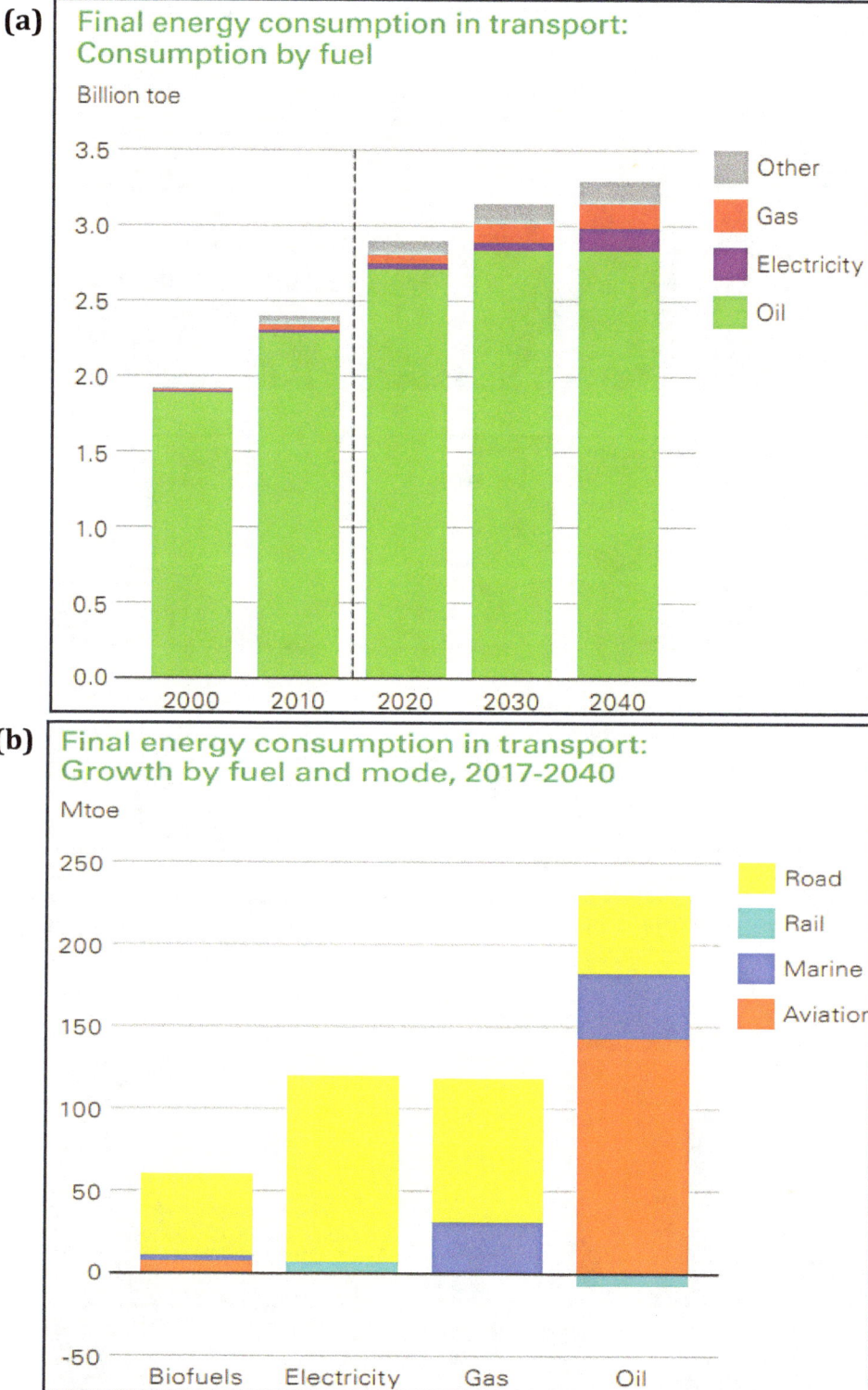

Figure 3.29 (a) Global transportation energy consumption by fuel (b) Final transportation energy consumption by fuel and mode *(BP, 2019).*

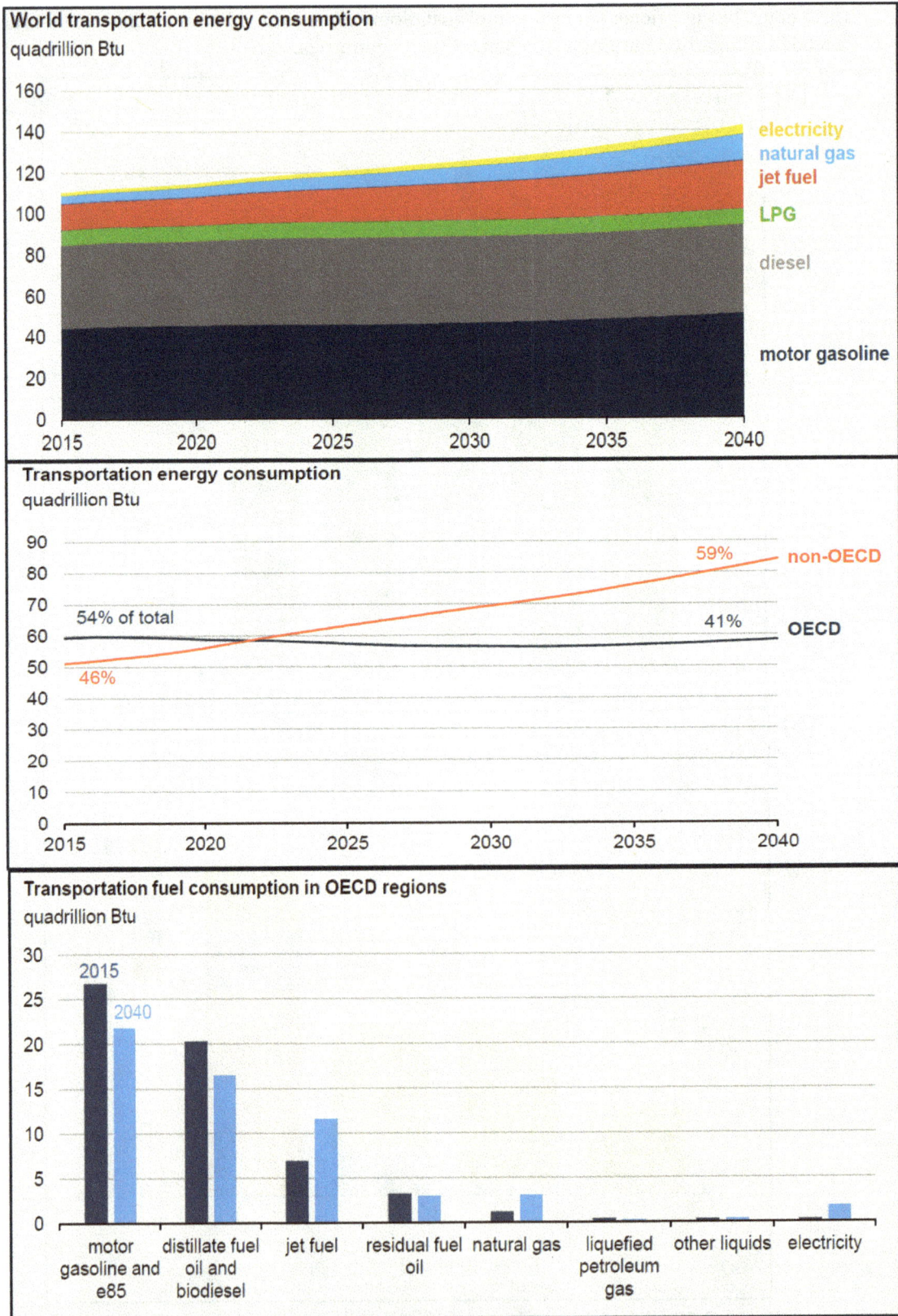

Figure 3.30 Global transportation energy consumption by region and fuel *(EIA, 2017).*

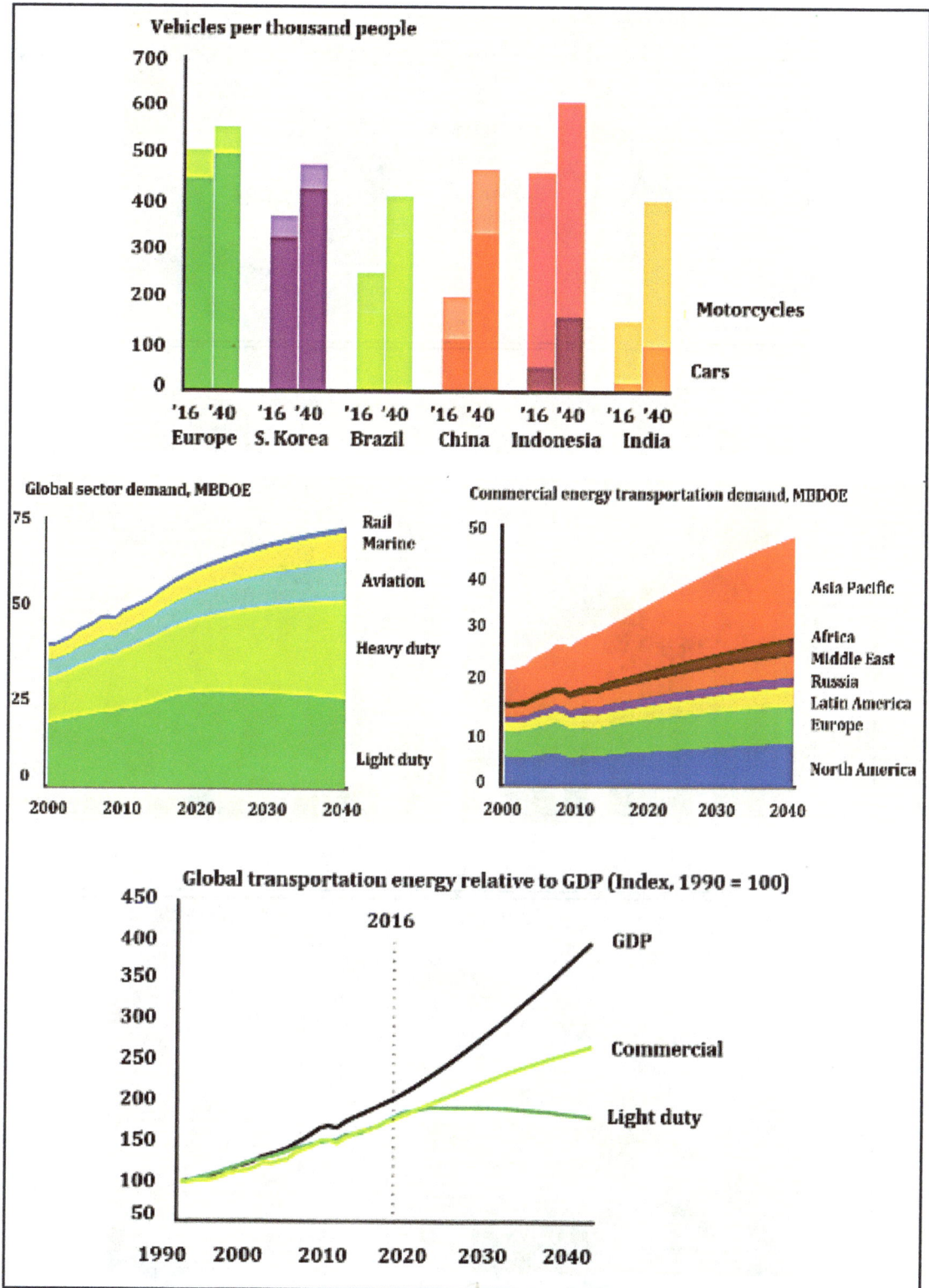

Figure 3.31 Global vehicle population by type and region, and transportation energy demand by region and sector. (MBDOE = Million oil-equivalent barrels per day). *(ExxonMobil, 2018).*

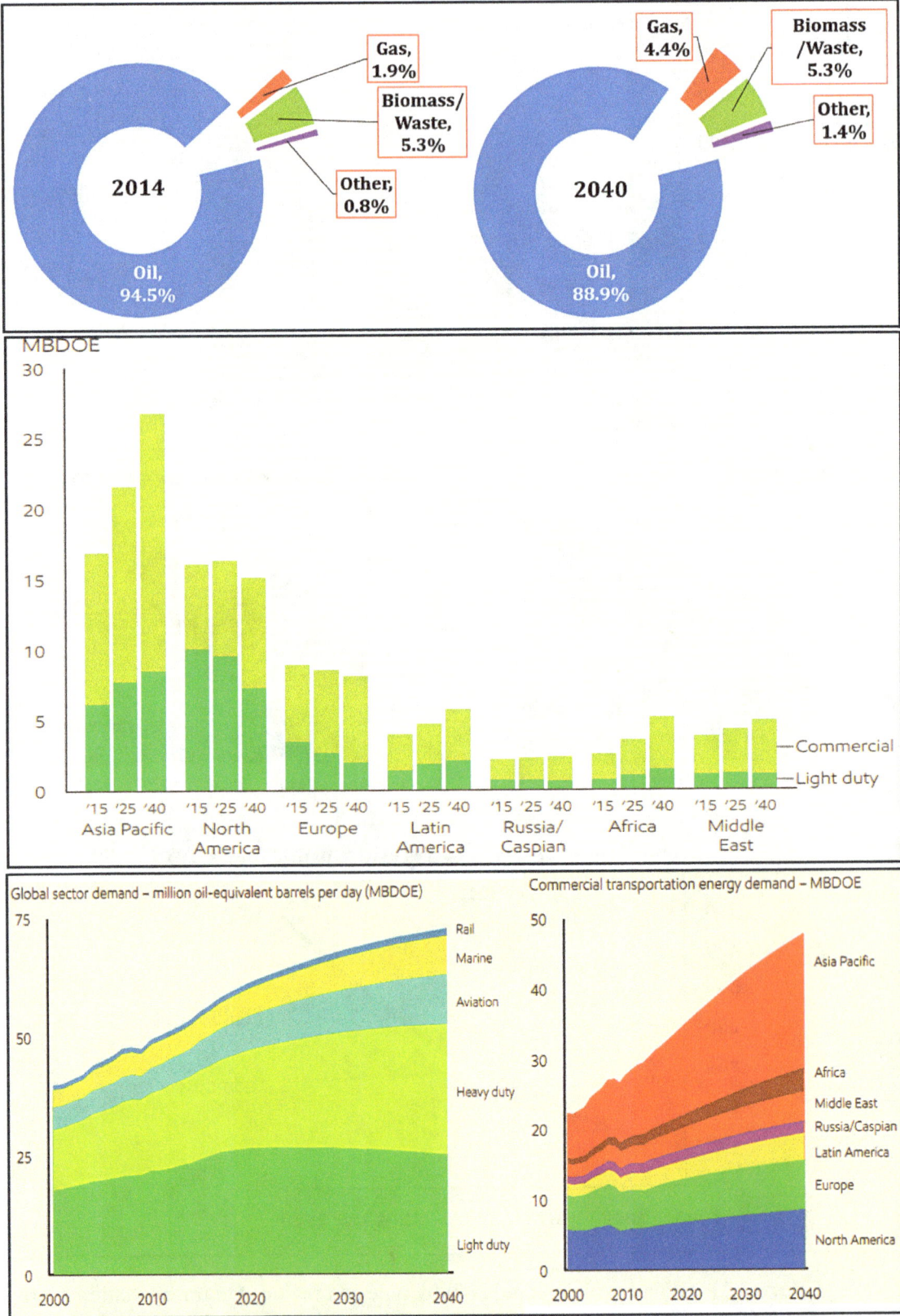

Figure 3.32 Transportation fuel demand by fuel/sector/region. (MBDOE = Million oil-equivalent barrels per day). *(ExxonMobil, 2017/2018).*

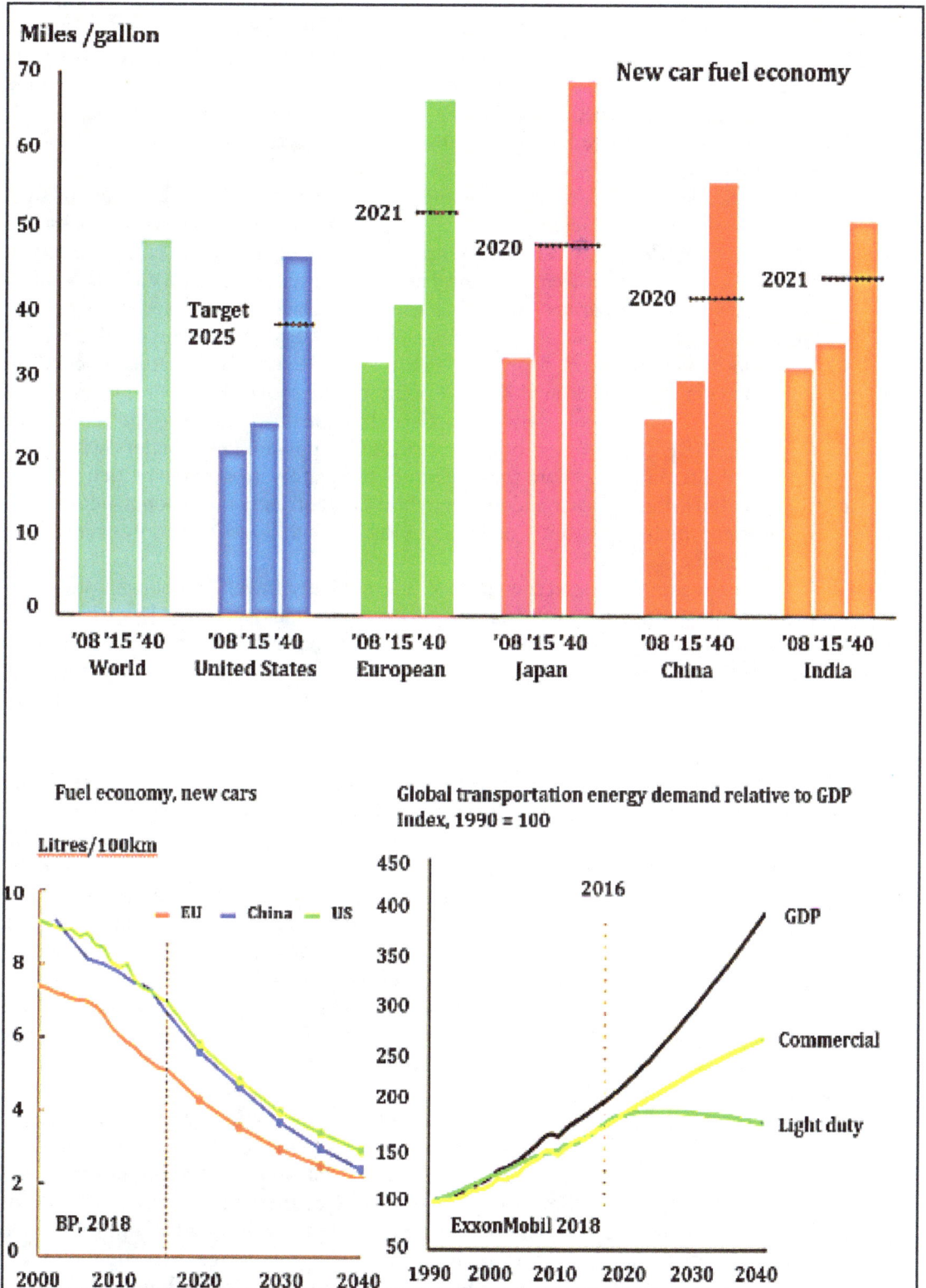

Figure 3.33 Global growth in vehicle fuel efficiency *(ExxonMobil, 2016/2017/2018).*

Fuel use by the air planes, ships and trains is growing rapidly from about 50% of the amount used by light-duty vehicles in 2014 to about 85% in 2040 (ExxonMobil, 2016). Over 90% of the demand by the sector will be met by oil in spite of increasing fuel efficiency, largely because demand for jet fuel is growing rapidly in response to the increasing air travel worldwide, and there is as yet no viable substitute, although jet fuel from biomass is beginning to emerge as jet fuel mix in some developed markets. However, biofuel contribution to jet fuel use in 2018 was less than 0.1%

The electric car market has been growing, be it slowly over the past decade or so. There are three types of electric vehicles: Battery-electric (BEV) which is powered solely by a battery charged from mains electricity; Plug-in Hybrid (PHEV) which features both - battery and an internal combustion engine powered by petrol or diesel; Extended range EV (E-REV), a version of PHEVs, powered by a battery but with a petrol or diesel powered generator on board. A battery-electric vehicle runs purely on mains-charged battery and has a typical range of up to 150 kilometers between charges, depending on driving style and use of heating/air conditioning; the PHEVs have a range of around 45 kilometers on pure electricity but once the battery is depleted, the vehicle can switch to hybrid mode and run on gasoline/diesel, delivering 400-450 kilometers; the E-REV is a variant of the PHEV, and is powered by a battery with a petrol or diesel powered alternator on board that recharges the battery continuously, delivering a total range of up to 450 kilometers. Conventional hybrids are currently more popular compared with plug-in vehicles which rely entirely on batteries that have to be recharged frequently, but they are more expensive.

The main advantages of electric cars are the relatively lower fuel consumption (around 30% fuel economy compared with gasoline powered vehicles) and very low carbon footprint because water is the only tailpipe emission. Also, EVs are particularly well-suited to intra-city driving, one of the main sources of ambient pollution. Apart from being environmental-friendly, plug-in cars offer a number of other benefits compared to conventional vehicles. Depending on the relative costs of gasoline and electricity, specific energy cost (energy cost per kilometer) for BEVs could be only a quarter of the cost of a similar internal combustion engine (ICE) vehicle, especially when owners have access to low off-peak electricity tariffs; there are fewer mechanical components in an electric vehicle when compared with conventional vehicles, which often results in lower servicing and maintenance costs; and there are many policy incentives which include subsidies on initial purchase cost of eligible EVs and cost of installation of charge points; lower taxes; and even free parking in some urban areas. However, there are also several issues with electric vehicles, associated mainly with batteries and access to charging bays. Batteries of electric cars are huge and expensive, it takes 6-8 hours to fully charge a lead-acid battery which delivers only a few hundred kilometers between charges; there are few commercial battery charging bays compared with conventional gasoline stations; and, crucially, consumer interest is low and government policies supporting EV ownership are generally weak in many countries. However, several more efficient batteries with fast recharge cycles are already emerging.

The penetration of electric vehicles into the global auto market has been very slow. There were about 5 million electric vehicles in the global fleet (about 0.2% of the total) in 2018, with 45% in China, 24% in Europe and 22% in the United States. Several recent projections on EV market penetration show that the global population of electric vehicles compared with conventional autos will remain insignificant in the foreseeable future. The most optimistic outlooks project that electric vehicle population concentrated within passenger cars, light duty vehicles and public buses will grow to about 350 million (of which around 300 million are passenger cars) accounting for about 15% of the total global vehicle fleet. One optimistic projection expects that around 14 million EVs (around 26% of the projected global sales of new vehicles) could be sold in 2020 mainly in China, Japan, United States, and Western Europe. About 3 million will be fully electric cars and range extenders, while the balance of 11 million will be hybrids (BCG, 2018). The Global electric vehicle outlook published recently by the International Energy Agency (IEA, 2019) shows that the global electric car fleet in 2018 was about 5 million, up by around 2 million from the previous year, rising to between 130-250 million by 2030 (Figure 3.35). The projection was based on historical analysis and all electric vehicle development initiatives. Electric cars and light commercial vehicles dominated the market but there has also been substantial growth in the global stock of two-wheelers, electric buses and medium electric truck sales. China remained the world's largest electric car market, accounting for nearly half of the total stock, followed by Europe and the United States. This growth rate is phenomenal but the total EV population was still less that 3% of the global vehicle fleet. A combination of technological innovations particularly in battery performance and cost, and strong policy support instruments could push EV vehicle sales to around 23 million in 2030 and the total global stock to 130 - 150 million, but this would still be less than 10% of the projected 1.8 - 2 billion vehicle population. A more optimistic EV30@30 initiative (with a combined pledge of 30% market share of EV vehicles by 2030 projects sales and stock at 43 million and 250 million respectively. BP (2019) projects a steady rise to around 350 million by 2040, with roughly equal proportions of PHEVs and BEVs, but electric vehicles will still account for much less than a fifth of the expected 2 billion cars. Another projection by ExxonMobil (2018) is less optimistic and expects a total market penetration to be about 160 million in 2040, with BEVs accounting for around 60% (Figure 3.34). The significant difference between the projections of market penetration of electric vehicles in these two analyses arises from many uncertainties about the likely growth rate of the market. It is unclear how quickly emerging technologies will resolve the formidable problems of battery cost and paucity of private/public low-speed/high-speed charging bays, and battery swapping bays. Also, a faster pace of fuel efficiency improvement of conventional engines could slow down EV market penetration. A typical internal combustion engine car will consume around 3 liters per 100 km, in 2040 compared with 5 liters today and 7 liters in 2000 (BP, 2018). Furthermore, the prevailing slow-down in global economies and drop in oil prices over the last few years could slow down market penetration of EVs. Only about 2-4% of global electricity demand will come from the transportation sector in 2040 and oil will still account for around 85% of total fuel demand by the sector, down from the current 94-95%. Natural gas, electricity, and a mix of other types of fuels are each projected to account for around 5% of transport fuel by 2040 (BP, 2018).

The mostly inefficient two-wheeler population in China still dominates the global EV market but the estimated average global emissions of 41 million tonnes CO_2-eq in 2018 was nearly 50% lower compared to an equivalent internal combustion engine fleet. However, the lower carbon footprint (on well-to-wheel basis) ultimately depends on the power mix: while the carbon advantage would be significant in Europe due to major progress in decarbonizing power generation, the positive impact of EVs on the environment could be minimal in China which hosts around half of the global population of EVs but generates most of its electricity (around 65%) from coal, the most highly-polluting fossil fuel. Furthermore, a substantial proportion of EVs would be hybrids which still use fossil fuels. For example, although more than two-thirds of EVs sold globally in 2018 were battery electric vehicles (BEV) almost half of the electric vehicles sold in the United States were Plug-in hybrids (PHEV), driven largely by the downward trend in the price of petrol. Also, about 70% of electric vehicles sold in Japan in the same year were PHEVs. There are several major constraints to the growth of the EV market, notably consumer apathy, inadequate and shifting policy support, and lack of adequate number of fast charging bays. Although there were as many battery charging infrastructures as there were EVs in 2018, most were private, low-charging systems which require 6 - 8 hours recharge time. This effectively excludes the large number of people who live in apartments or homes without a garage or front parking lot, and accentuates fears of being stranded on trips due to discharged batteries.

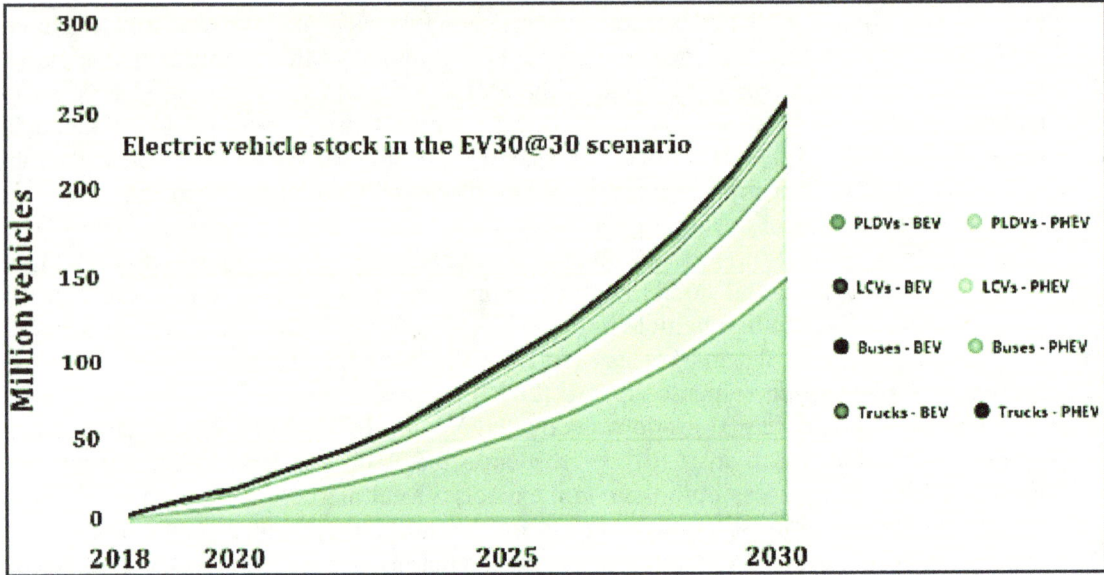

Figure 3.34a   Outlook on electric vehicle stock in 2030 in the EV30@30 scenario (30% by 2030) *IEA, 2019b)*

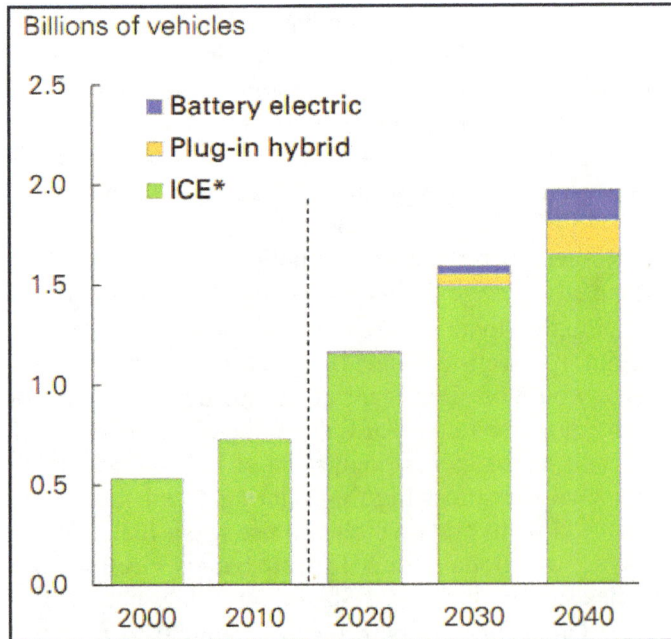

Figure 3.34b   Projections of global passenger car population type *(ExxonMobil, 2018).*

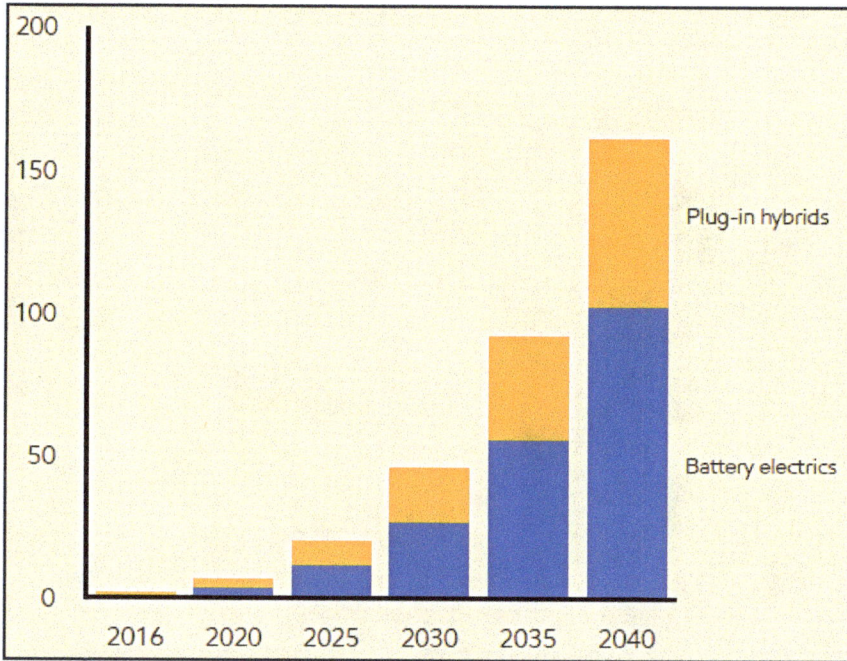

Figure 3.34c   Projections of global passenger car population type
(BP, 2018).

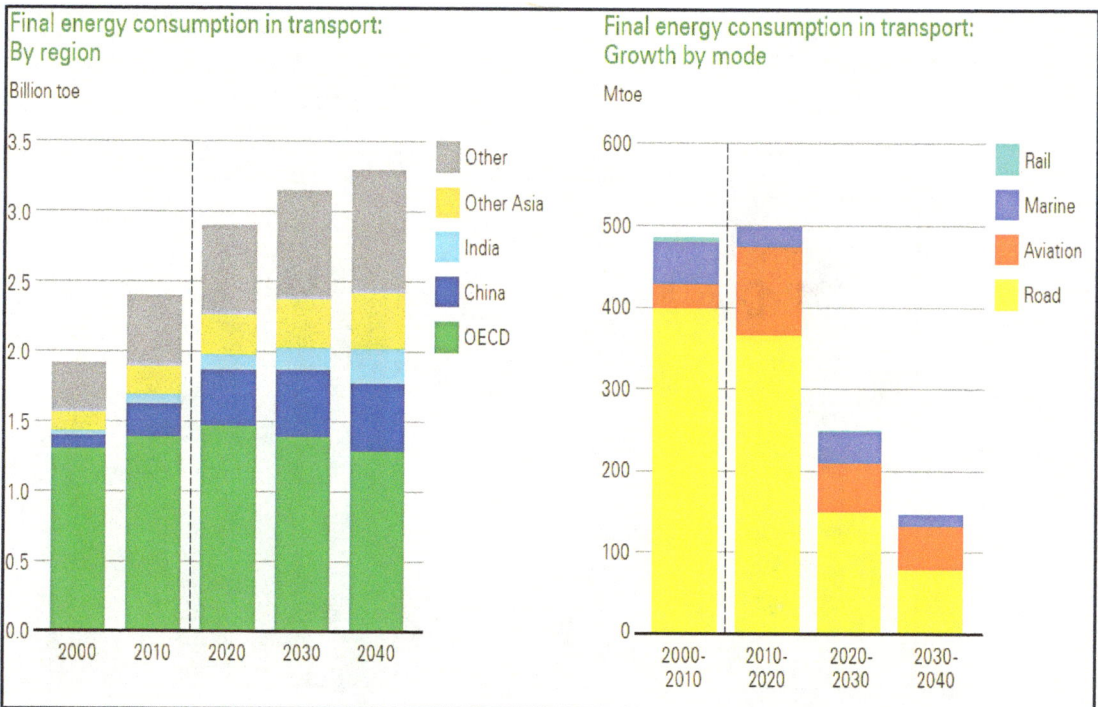

Figure 3.34d   Projections of transportation energy by region and sub-sector
(BP, 2019).

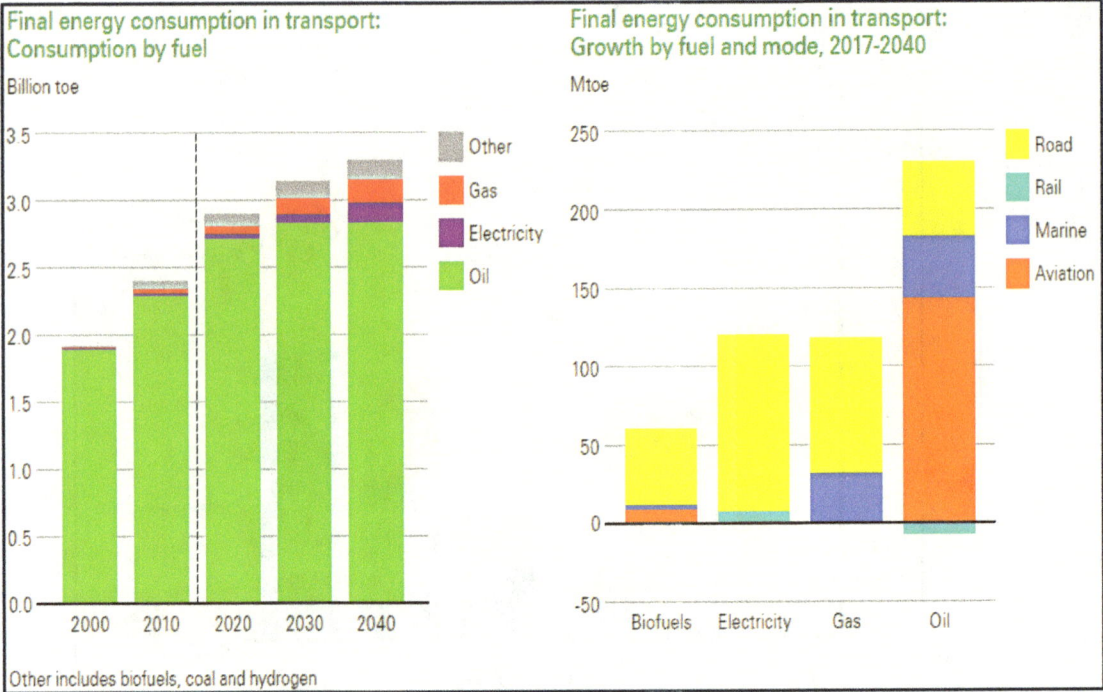

Figure 3.34e Projections of transportation energy by fuel and sub-sector *(BP, 2019).*

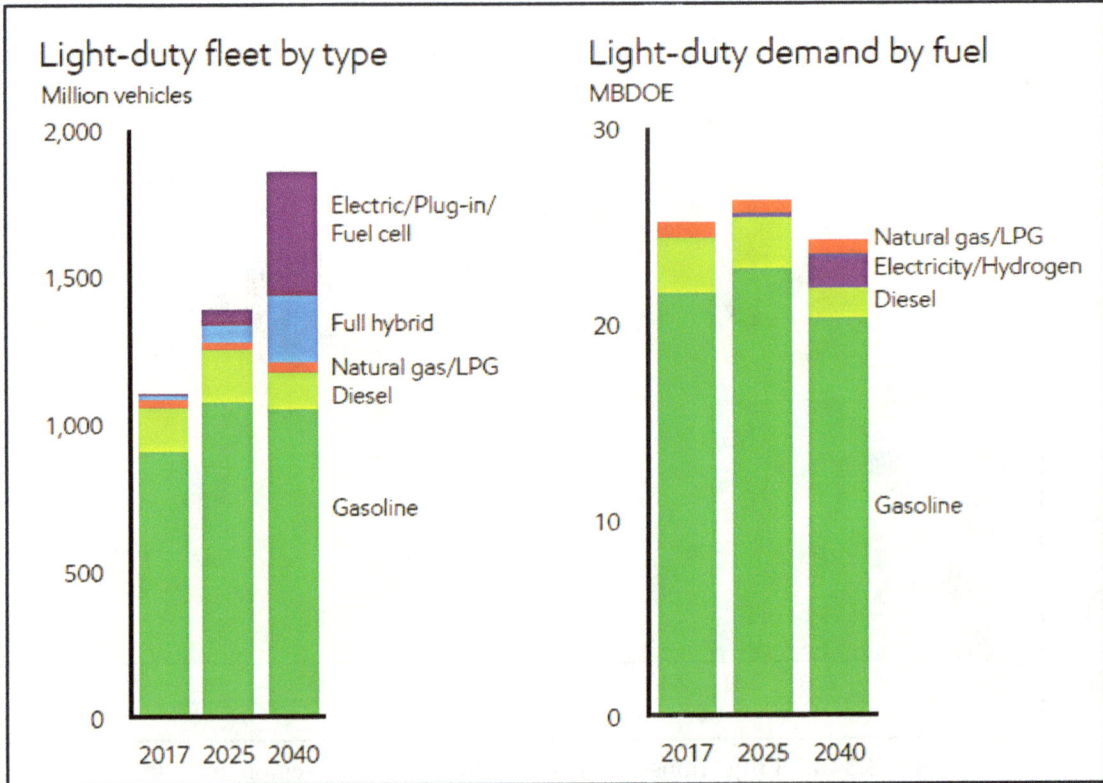

Figure 3.34f Projections of light-duty transportation energy by vehicle and fuel type *(ExxonMobil, 2019)*

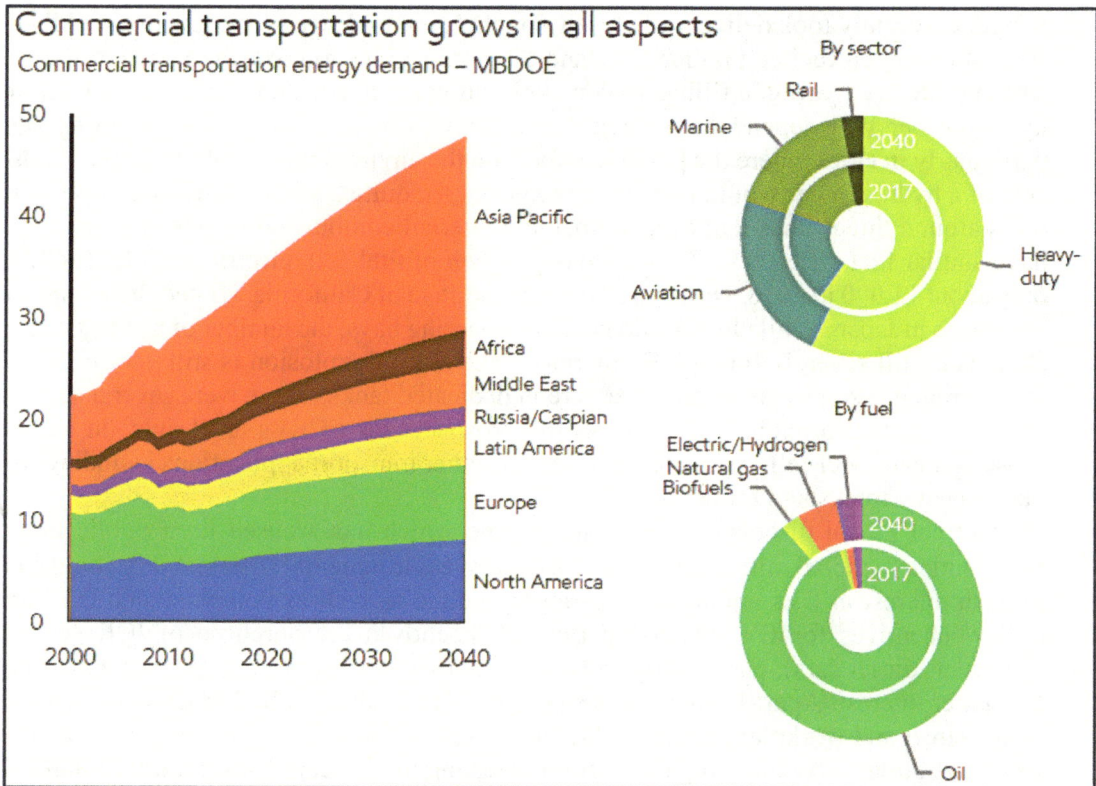

Figure 3.34g Projections of commercial transportation energy by region and type
(ExxonMobil, 2019)

The hydrogen-powered car (FCEV)is a new variant of the electric vehicle which is becoming a reality, with several automobile manufacturers bringing out models. The car is powered by a fuel cell which generates electricity from hydrogen that can be produced from many sources. The main advantage is that there is no direct ambient tailpipe pollution which is a major source of city smog and risks to human health and other ecosystems since the only emissions are heat and water. Also, hydrogen vehicles do not require battery charging, and hydrogen refilling stations could produce hydrogen from water on site, using solar power when available, and store as liquid hydrogen, thus eliminating the need for bulk transportation of highly inflammable liquid hydrogen. Furthermore, any pollution associated with hydrogen production (which can be substantial) can be contained at source. One major disadvantage is that hydrogen production requires enormous electricity, hence it is environmentally sustainable only if the electricity comes, not from fossil fuels but from renewables or nuclear energy. Solar energy is particularly promising because it can be used to produce liquid hydrogen when available and the product can be stored or transported easily like conventional liquid fuel. This takes care of the prominent disadvantage of solar power intermittency. It also means that both the electricity and liquid hydrogen can be produced on site. Hydrogen can be produced from many sources: fossil fuels (natural gas, coal, oil), biomass, water. Water is the most environment-friendly source. It is abundant, renewable, and cheap, but production is power intensive.

A typical hydrogen car currently on the market delivers 400-450 kilometers on a tankful of liquid hydrogen, comparable with a regular petrol/diesel car, and higher than most mainstream electric cars. Some developed countries are already promoting hydrogen cars and developing hydrogen filling stations. The London Metropolitan Police Service in the United

Kingdom recently took delivery of eleven Toyota Mirai hydrogen cars. It is the world's largest fleet of hydrogen fuel cell police cars, and the Force plans to grow the fleet to 550 by 2020. The current five hydrogen filling stations in London are adequate for now, but success of the hydrogen car will depend on an aggressive policy-driven proliferation of filling stations, particularly in cities where the positive impact on the environment would be highest. Also, the cost of a hydrogen car would be a major deterrent for quite a while - a hydrogen-powered car costs around three times that of a comparable internal combustion engine car, although they have similar fuel efficiency. The global population of fuel cell electric vehicles (FCEV) was only about 11,000 in 2018, with half of them in the state of California, United States and another 25% or so in Japan. While FCEVs do not need charging bays, the number of hydrogen refueling stations is still severely limited. Furthermore, the risk of explosion is still very high because the hydrogen is stored under high pressure in dedicated tanks - there were several accidents in Europe in recent months. Nevertheless the prospects for FCEVs are bright: in addition to passenger cars, fuel cells are featured in various other transport applications including fuel cell buses, heavy trucks and trains.

Current global discourse on climate change which has aroused keen public interest in contributing to global mitigation efforts on energy-environmental sustainability could continue to push interest in EVs among the car-buying public as well as policymakers. Furthermore, policy incentives for EV ownership (and disincentives for purchase of ICE vehicles, in particular, diesel cars), particularly in North America and Europe are beginning to stimulate consumer interest. Apart from subsidies on cost of purchase of eligible EVs and provision of home, street and workplace charging facilities, conventional ICEs are facing stiff tax regimes which are already discouraging many from investing in new cars. However, there has been no dramatic rise in the sale of electric vehicles either, indicating that people may be confused and are simply delaying investment in new vehicles for now.

3.5.3 **Building sector (Residential and Commercial)**

Building sector primary energy demand which currently accounts for about 29% of the global total is projected to grow to nearly a third by 2040, faster than the growth in demand by the industrial and transport sectors. The total number of households worldwide will increase by around 40% by 2040 and over 90% of the growth in building energy use will be in the emerging world. The relatively warm climate in most of these regions means a low demand for space heating while the majority of household energy demand will be driven by the need for space cooling. Also, the growing prosperity and urbanization in these regions will drive up ownership of electrical appliances. There will be rapid growth in urbanization, and upward social mobility in many developing countries, and building sector energy demand will rise by about 40% while it will flatten or decline in the developed world (Figures 3.35-3.37). The energy use per household has been declining and the trend is expected to continue to 2040. This is due to the increasing energy-efficient homes, appliances, etc. Household energy demand will rise by only a quarter, all of it accounted for by the emerging countries. The increase should be around 100% but for the mitigation of rising energy efficiencies in building construction, fittings and appliances across all regions of the world. About 90% of this demand growth will be met by electricity which will emerge as the preferred household energy. Led by the growing economies of the non-OECD nations, average worldwide household electricity use will rise by about 30% by 2040, again nearly all of it in emerging countries where increase is expected to double. Residential, commercial and public places account for about 15% of the global primary energy use and half of the global electricity demand for heating, lighting, refrigeration, air conditioning, cooking etc. Biomass currently dominates the energy mix in the sector, accounting for about 40% followed closely by electricity. The bulk of the biomass consumption currently is in the emerging world and comprises mainly wood and charcoal. The share is expected to decline to about 30% due to rapid urbanization and more people entering the middle class and gaining access to more modern fuels - electricity, natural gas, liquefied petroleum gas (LPG), solar energy.

China and Africa will each account for 30% of the increase in commercial and residential energy demand. Residential electricity use will increase by 75% and account for over 40% of the total end-use energy share of the sector. Most of the expected steep rise will be accounted for by non-OECD countries where demand will rise by around 150%. Increase in India and Africa is expected to be even higher, around 250%, but use per household will remain low. Use of natural gas will rise by 25%, demand for biomass should peak around 2020 and then start to decline, but oil and coal use have been declining and the trend is expected to be sustained.

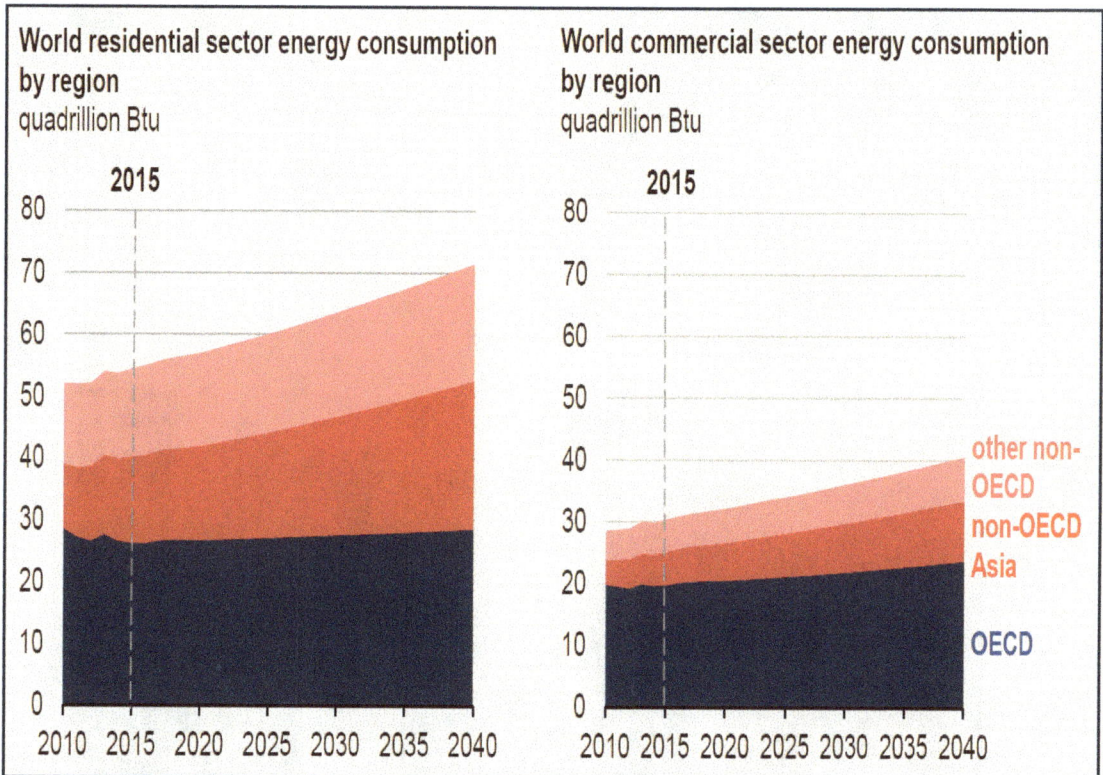

Figure 3.35 Projections of energy use by building/commercial sector by region *(EIA, 2017).*

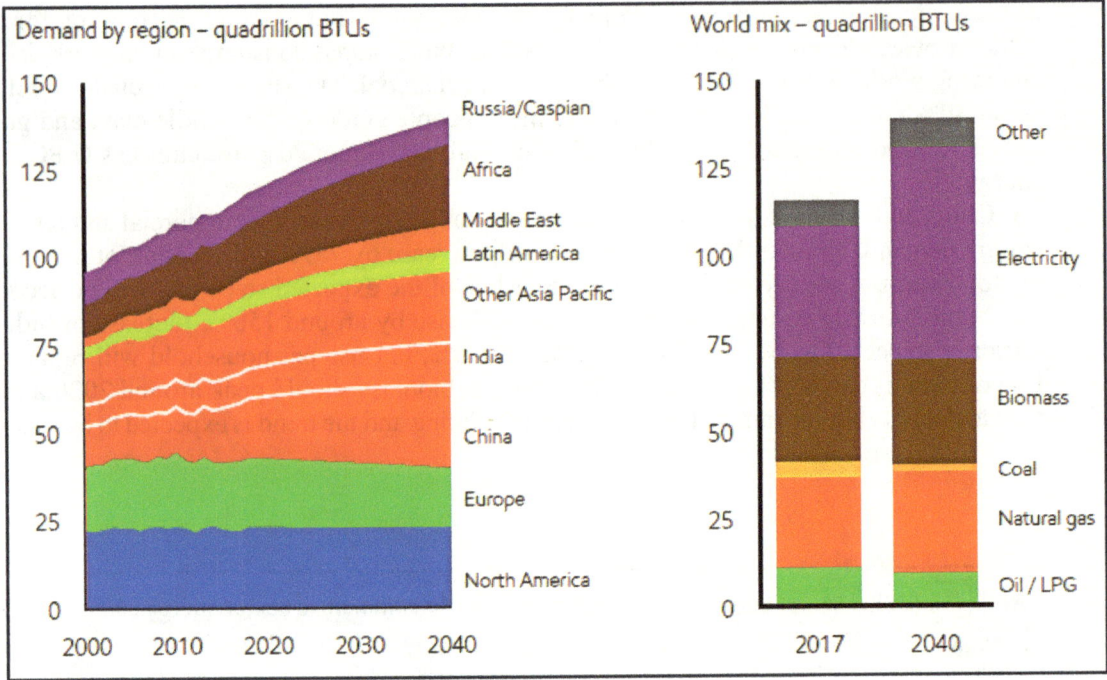

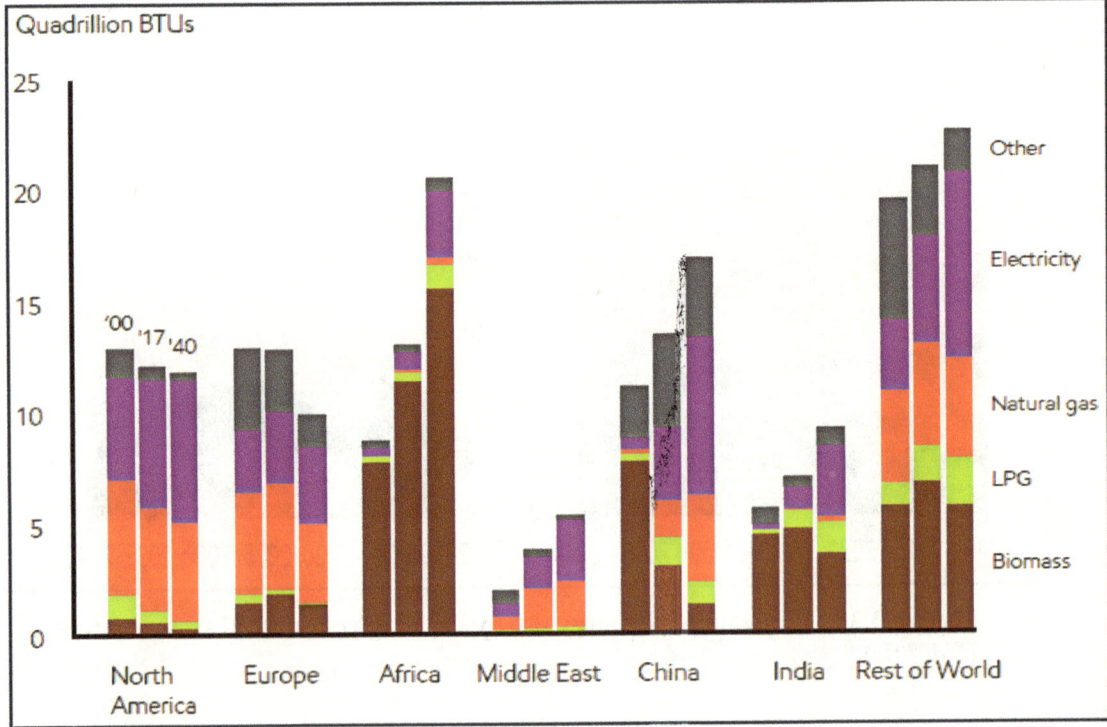

Figure 3.36 Projections of energy use by building/commercial sector by fuel, country and region *(ExxonMobil, 2019).*

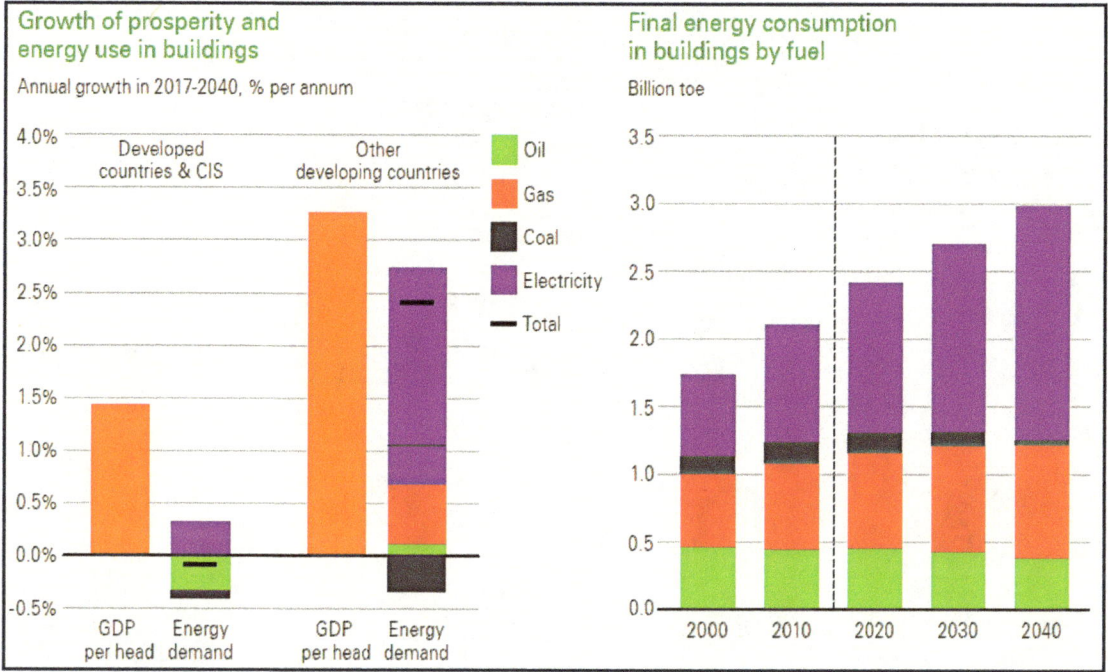

Figure 3.37 Projections of energy use by building/commercial sector by fuel, country and region *(BP, 2019).*

Chapter 4
Effect of fossil energy use on the environment

4.1 INTRODUCTION

The Earth and its environment have co-habited for millions of years and the way they interact largely determines the sustainability of life on earth. Physico-chemical activities in the atmosphere moderate the Sun's energy that reaches the Earth's surface, without which life would not be sustainable; the Earth's rotation around the Sun determines climate; the winds, move global surface energy around, picking up moisture from the oceans and dumping rain on land, largely controlling the weather. Many natural processes like the Earth's rotation, gravity, internal structure, and other unknown phenomena can alter both the weather and climate on Earth. Extensive archeo-geological evidence shows that desert areas of the world were at one time fertile land, and many land areas were once occupied by oceans. Advances in atmospheric science have identified many of the variables which control these natural phenomena, and concluded that many human activities, particularly in the last two centuries or so can interfere with the natural processes, with undesirable consequences, and environmental pollution is the main focus.

Environmental pollution is the introduction into the atmosphere of contaminants which can alter significantly the natural balance of the environment, with potentially serious consequences. Pollution can be in many forms, including particulates, chemical substances, or in energy form - heat, noise, light, radioactivity, etc. There are many naturally occurring contaminants but the environment has in place a system of processing, which is crucial to the sustenance of life on Earth. Energy is a primary input into most human activities, the fundamental driver of economic and human development, and all processes of energy production and utilization have mostly negative as well as some positive impact on the environment. The major areas of pollution are land, water and air. Around 68% of the global anthropogenic emissions (environmental pollutions that emanate from human activities) come from primary energy production and utilization, although there are significant country and regional variations (IEA, 2017a). Carbon dioxide makes up about 90% of energy-related pollutions, others being methane (9%), nitrous oxide (1%), and particulate matter.

Virtually all energy production and utilization processes generate pollution gases and particulates. Oil drilling and production, gas harvesting and coal mining processes have negative effects on the environment because mining and winning degrade landscapes and release particulate and gaseous pollutants, in particular, methane into the atmosphere. Conversion and utilization of energy also result in the release of anthropogenic pollutants. However, the greatest intensity arises from energy utilization, in particular electric power generation and transportation. Energy demand has been increasing for decades, propelled by worldwide economic growth and development. Global primary energy supply (TPES) increased by almost 150% between 1971 and 2015, powered by fossil fuels which accounted for 86% of TPES in 1971, and has dropped only a few percentage points to 82% in 2015 (80% in 2018), due largely to increasing use of nuclear power and renewables. The dominance of fossil fuels in the global primary energy mix has played a key role in the upward trend in CO_2eq emissions, in particular, since the Industrial revolution. In 2015, global CO_2eq emissions reached 32.3 Gt CO_2eq, compared to near zero in 1871 (Figure 4.1). By 2018, the level had risen to 36.2 Gt of which 91% was energy-related.

The Intergovernmental Panel on Climate Change Report (Fifth Assessment Report, Working Group 1, (2013, 2018) provided scientific evidence of human influence on the climate system and identified the use of energy as by far the largest source of anthropogenic greenhouse gases, with smaller shares from agriculture and livestock production, and non-energy related industrial processes. Atmospheric pollution caused by energy utilization is of major global concern and has been linked to various environmental problems which are believed to have increased significantly with the exponential increase in fossil fuel utilization over the last century or so. Negative consequences of fossil fuel utilization include global warming, extreme weather, ocean flooding, desertification and a wide range of human ailments. There are no physical atmospheric boundaries between countries and regions and consequences of activities in one location can resonate over long distances, hence the current efforts by the United Nations to galvanize concerted global effort to resolve the problem of environmental pollution.

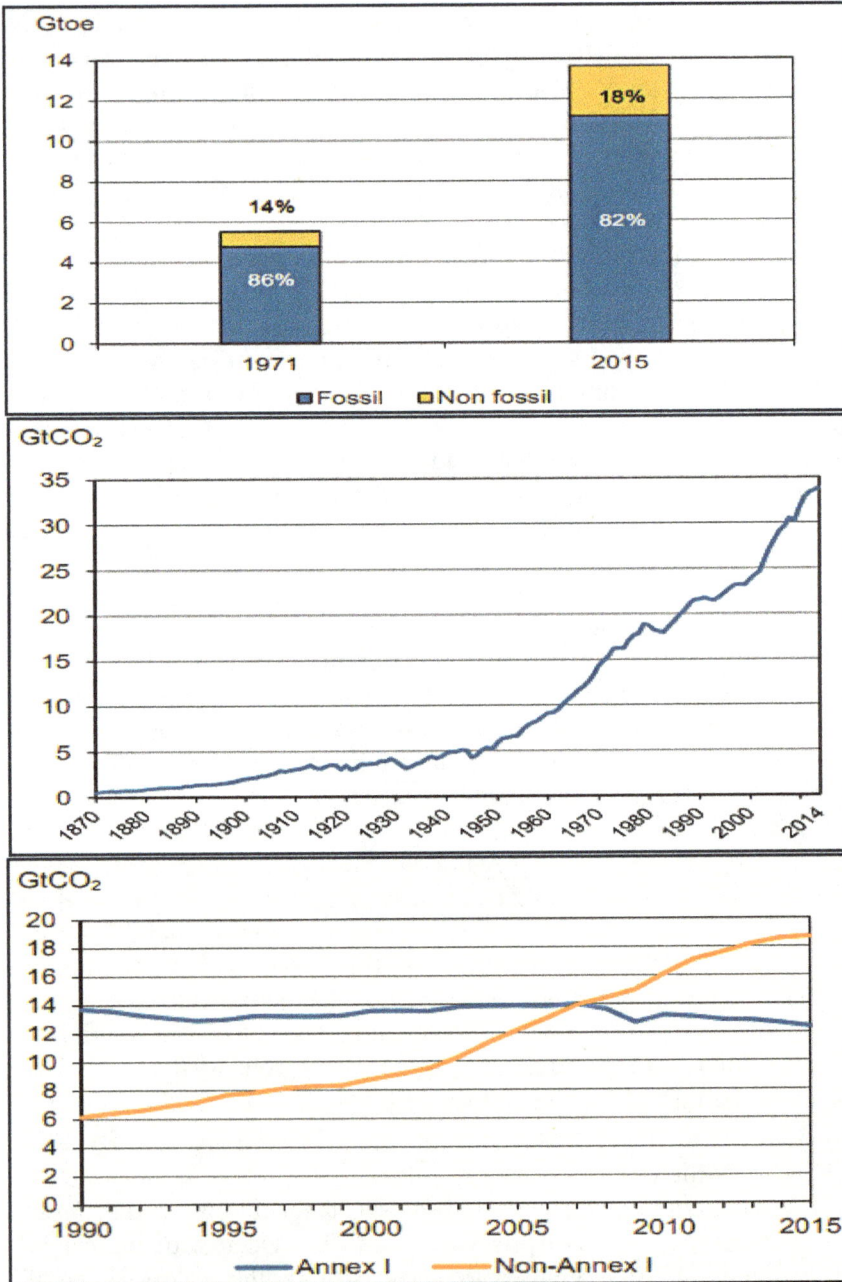

Figure 4.1 Global total primary energy supply (TPES) and energy-related CO_2eq emissions *(IEA, 2017e)*. Note: Annex 1 comprises mostly developed countries, and Non-Annex 1 are mostly developing countries.

4.2 THE EARTH'S ENVIRONMENT

The Earth-environment system (EES) comprises the Earth and the environment that surrounds it. The system is both the source and sink of all human activities. Oxygen that sustains life is derived from the Earth's atmosphere and the expelled carbon dioxide, a primary compound for plant photosynthesis goes back into the atmosphere. Food, minerals and energy are derived from the Earth's crust and its environment, and the waste that results from utilization is recycled through the system. The Earth's environment mantle is a natural system that includes the Earth and the ecological units, and natural resources that surround it. These include all living (mankind, animals, birds, fish, microorganisms), all non-living (soils, vegetation, oceans, rocks, etc.) and the atmosphere (air, climate, energy and magnetic radiation).

The Earth is central to a global system commonly defined in terms of four major spheres: the *lithosphere, hydrosphere, atmosphere*, and *biosphere* (Figure 4.2). The hydrosphere may be sub-divided into fresh and frozen water, hence the common fifth member: the *cryosphere*. The lithosphere comprises the Earth's crust (from the surface to a depth of about 100 km), and the upper part of the solid mantle that extends to a depth of about 3000 km. The hydrosphere comprises water, liquid or frozen, that exists under or over the surface of planet Earth - oceans, lakes, streams, glaciers, ground waters. The lower atmosphere (troposphere) comprises layers of gases that extend from the Earth's surface to about 20 km into space, and hosts most of the natural activities that determine the local weather - wind, clouds, precipitation, etc. There several other layers above the troposphere which have different characteristics and have different influences on the Earth's natural environment. The biosphere refers to parts of the land, sea, and atmosphere that host all living organisms, microorganisms and plants.

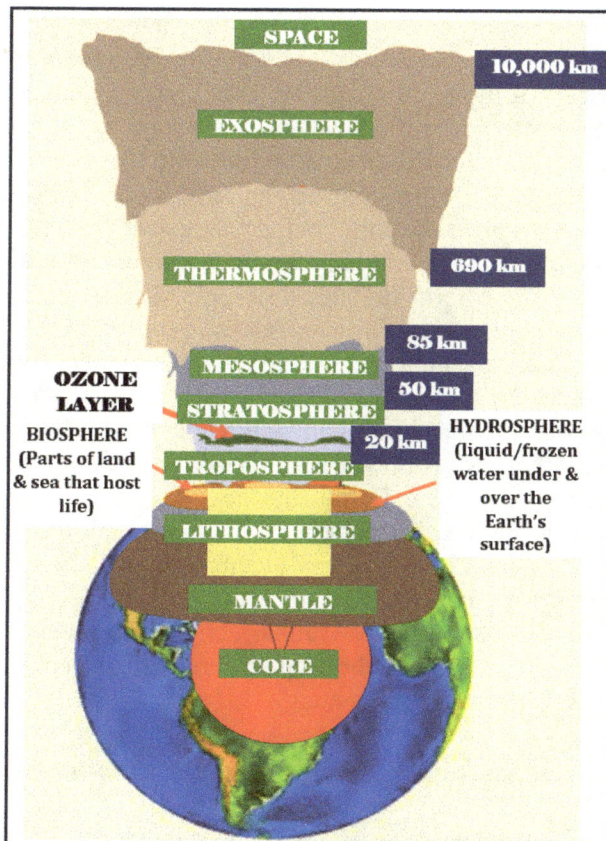

Figure 4.2 The Earth and its environment *(Afonja, 2017).*

The atmosphere (in particular, the first two layers) sustains life on Earth through the supply of oxygen needed by most organisms for respiration, carbon dioxide required by vegetation (plants, algae etc.) for photosynthesis, the regulation of the Earth's temperature, and the control of the potentially damaging effects of the Sun's ultraviolet radiation. Activities in the troposphere have the most immediate impact on the living world, particularly on the weather and human health because it is the layer in direct contact, but events in the next layer (stratosphere) have more long-term impact, mostly on climate.

4.3   THE EARTH'S NATURAL AND ANTHROPOGENIC ENVIRONMENTS

The four major spheres that make up the ecological system are characterized by intricate intra and interactions, all interdependent and interconnected. The interaction between these spheres to a large extent determines the climate. The functions of the spheres are controlled by natural regulatory processes and the boundaries between them may be clear or ill defined. While the boundaries between organisms and vegetation are fairly well defined there are no clear boundaries between air, water, climate, energy, radiation, etc. The environment that is controlled by natural forces which have the capacity to dilute, absorb or dissipate pollutants is known as *natural environment.* The natural environment is imperfect in many ways. Extreme weather, flooding, volcanoes, earthquakes are natural occurrences which have always been part of the ecosystem.

4.3.1   The tropospheric layer (lower atmosphere)

The first fifteen to twenty kilometers of the atmosphere, known as the troposphere or lower atmosphere is the part of the atmosphere that is in direct contact with life on Earth. It is also the zone that largely determines the weather; it is the source of oxygen that sustains human and life and the depository for exhaled carbon dioxide that is crucial to the survival of plant and vegetation on Earth. The layer controls the direction of the winds at any given time, and the water cycle which involves evaporation from the Earth's surface, seeding, precipitation and condensation in the troposphere resulting in clouds, rain, snow, slits etc. It also controls the movement of winds which form thunderstorms, hurricanes and tornadoes, heat waves and season patterns around the world. While the relative proportions of the constituent gases of the troposphere remain fairly constant, the water vapor is variable, depending on evaporation and transpiration processes on the Earth's surface which is the source. The evaporation process plays a prominent role in controlling temperatures on the Earth's surface because the energy that sustains the endothermic processes is sourced from the solar energy and thermal radiation re-emitted or reflected from the Earth's surface. The natural activities controlled by the troposphere determine the *weather* pattern on the Earth and, ultimately the *climate* (weather pattern over many decades). In effect, any events that alter the natural balance of the tropospheric activities will have an immediate, direct impact on life on Earth, as well as well as the weather, and ultimately, the climate. Human activities produce a wide range of chemical compounds that can accumulate in the troposphere and affect global weather, human and animal health, aquatic life, vegetation, etc.

4.3.2   The stratospheric layer (middle atmosphere)

Many gases present in the Earth's *stratosphere* (middle atmosphere, between about 20 km and

40 km or so above the Earth's surface) are capable of trapping, holding and releasing the Sun's thermal rays. They are known as *greenhouse gases* (GHGs). Although they are present in very minor quantities relative to nitrogen and oxygen, they play a crucial role in the control of the amount of heat that reaches the Earth's surface, and therefore the Earth's average surface temperature. The main natural GHGs are water vapor, carbon dioxide, methane, and nitrous oxide, but several others [ozone, chlorofluorocarbons (CFCs) hydrofluorocarbons (including HCFCs and HFCs)] are also active. Natural carbon dioxide is a product of respiration and methane is produced primarily by anaerobic (oxygen-deficient) decomposition of organic matter in biological systems. The compound is also produced by termites which harbor bacteria that are capable of breaking down organic matter in wetlands and swamps.

Agricultural activities (biogenic sources) such as wet rice cultivation, soil fertilization, and vegetarian livestock production release significant quantities of methane into the atmosphere. For example, the digestive process of a cow emits about a quarter of a kilogram of methane into the atmosphere daily. Significant methane emissions also emanate from decaying vegetation, especially in landfills, hydro dams, artificial lakes and reservoirs. Water vapor is the largest contributor to the natural greenhouse effect and plays an essential role in the Earth's climate. However, the amount of water vapor in the atmosphere at any given time is controlled mostly by air temperature, rather than emissions, and therefore the maximum varies across the globe. The atmosphere can retain around 7% more moisture for every one degree centigrade rise in temperature, hence, a column of air in the tropics may hold ten to twenty times more moisture than a similar column in the polar regions. For these reasons, water vapor is considered as a *feedback* agent rather than a *forcing* to climate change, and therefore usually not included in lists of anthropogenic greenhouse gases.

4.3.3   The Greenhouse Effect

About two-thirds of the solar short wave radiation reaching the Earth is absorbed and the balance is reflected into space as infrared long-wave radiation, mostly in the daytime. Greenhouse gases present in the stratosphere absorb part of this long-wave radiation and reflect it in all directions. Part of this radiation is directed towards the Earth's surface, warming it. The intensity of the downward radiation will depend on the atmosphere's temperature and on the concentration of the greenhouse gases in the stratosphere. When the Earth's surface temperature drops, heat is radiated from the gas layer back to Earth. In effect, the gases form a heat blanket in the stratosphere which helps to control the temperature on the Earth's surface, keeping it at a life-supporting average of 13 °C. This natural absorption-emission-absorption cycle is known as the *greenhouse effect* (GHE). Without this natural activity, the temperature of the Earth's surface would be around minus 18°C and would be uninhabitable. An increase in the concentration of any of the greenhouse gases enhances its ability to capture and reflect heat energy back to the Earth's surface, thereby increasing temperatures (Figure 4.3)

4.3.4   Stratospheric ozone layer

Natural ozone is concentrated in a thin layer mostly in the lower portion of the stratosphere, from approximately 15 to 30 km above Earth's surface (Figure 4.4). The concentration is highest (2-8 parts per million) in the 20-40 km altitude range, although the thickness of the layer varies seasonally and geographically. Ozone in the Earth's stratosphere is created by the Sun's ultraviolet rays (with wavelengths shorter than 240 nm) which supply energy for the dissociation of some of the oxygen molecules constantly being introduced into the atmosphere by photosynthesis. The molecules split into two highly unstable oxygen atoms which react with oxygen molecules to form ozone, (an exothermic reaction) with the release of heat energy. The net effect of the above two reactions is the formation of two molecules of ozone from three molecules of oxygen, accompanied by the conversion of light energy to heat energy. The ozone formed is unstable and absorbs the Sun's ultraviolet rays which causes it to decompose into oxygen molecules and atoms, with the release of heat energy.

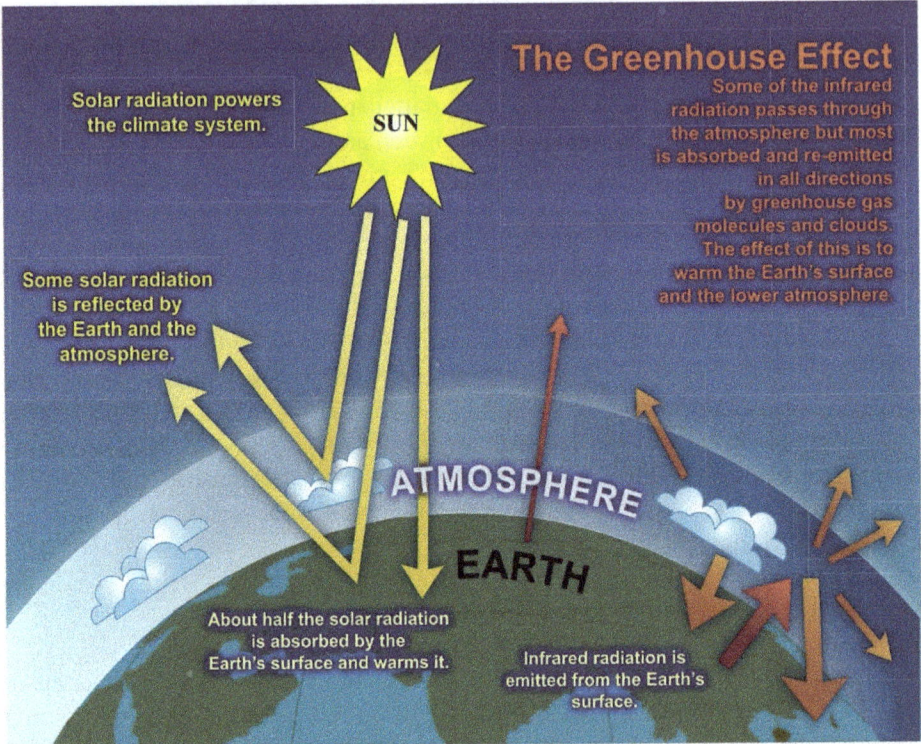

Figure 4.3   The Greenhouse Effect. *(geogaphyiseasy.worpress.com).*

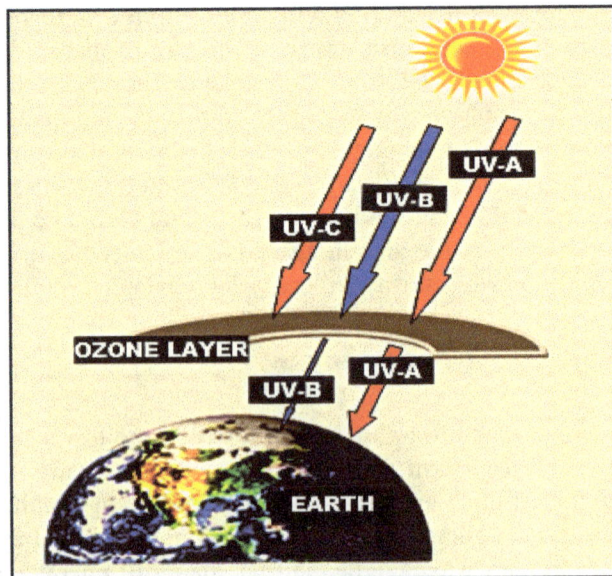

Figure 4.4   Role of the ozone layer in controlling the Sun's ultraviolet rays that reach the Earth *(Afonja, 2017)*.

This continuous set of reversible reactions involving the breakdown and formation of ozone with the accompanying absorption and dissipation of energy is known as the *ozone-oxygen cycle*. These reactions, also known as Chapman cycle, summarized in Equations 4.1 to 4.6 play a critical role in screening off potentially harmful radiations from the Sun from the Earth's surface.

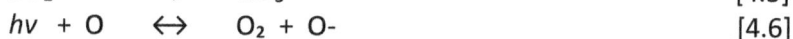

$$hv + O_2 \rightarrow 2O\text{-} \qquad\qquad [4.1]$$
$$O_2 + O\text{-} \rightarrow O_3 \qquad\qquad [4.2]$$
$$O_3 + O\text{-} \rightarrow 2O_2 \qquad\qquad [4.3]$$
$$O_3 + hv \rightarrow O_2 + O\text{-} \qquad\qquad [4.4]$$
$$3O_2 \rightarrow 2O_3 \qquad\qquad [4.5]$$
$$hv + O \leftrightarrow O_2 + O\text{-} \qquad\qquad [4.6]$$

There is no net ozone depletion because the process produces atomic oxygen that reacts with molecular oxygen to form another ozone molecule. A natural ozone layer is thus created in the stratosphere between 10 to 50 km above the Earth's surface, although most of the compound (around 90%) is concentrated between 20 to 40 km. However, the natural ozone-oxygen equilibrium is disrupted by the presence of certain radicals that contain chlorine, bromine, nitrogen, hydrogen or oxygen atoms. Even though ozone concentration in the ozone layer is very small, it is vitally important to life because it absorbs biologically harmful ultraviolet radiation coming from the Sun and prevents up to 99% of the Sun's medium frequency ultraviolet UV-B (400-315 nm) and UV-C (280-100 nm) which are biologically harmful from reaching the Earth's surface, by using them to energize the chemical reactions discussed above. The most dangerous component of the Sun's rays (UV-C) is screened out completely, but some UV-B gets through, which is beneficial because it energizes the production of Vitamin D, an important vitamin required for the regulation of the amount of calcium phosphate needed for healthy human bones and teeth. However, excessive exposure is the main cause of sunburn, and has been linked to many harmful effects, including skin cancers, genetic damage, cataracts, and harm to some crops and marine life. Ozone is transparent to most of UV-A which is the visible, longer wavelength radiation from the Sun that reaches the surface of the Earth as light. In effect, the stratospheric ozone layer in which ozone concentration is fairly constant because the gas has the capability to continuously regenerate itself, acts as a heat shield which screens out dangerous ultraviolet radiation from the Sun and protects life on the Earth.

4.4   ENVIRONMENTAL POLLUTION

The natural processes of controlling the Earth's average surface temperature can be influenced or affected by human interference through actions which generate different pollutants that can potentially cause undesirable effects on life on earth. An environment contaminated and affected significantly by human actions is no longer a natural but an *anthropogenic environment*. The greenhouse effect and ozone layer are nature's mechanisms for controlling the temperature on the surface of the Earth, and screening out components of the Sun's radiation which are potentially injurious to life. Human activities have always had a negative effect on the Earth-Environment system but the rate of damage was such that they were neutralized by natural system processes. However, the Industrial Revolution of the late 19th century greatly accelerated industrial and commercial development, the prime mover of which was energy.

Coal was the main source of energy and mines sprang up quickly in many parts of the world. Coal fueled iron and steel production, steam engines which were the prime movers, the electric power generation plants, and the rail system which became the major form of transportation. The development of diesel and petrol engines towards the end of the century was made possible by production of oil and gave birth to auto and air transportation industries in the first two decades of the 20th century. Since then, the demand for coal and other fossil fuels (oil and gas) has grown exponentially, with profoundly negative impact on the Earth-Environment system. Combustion of fossil fuels produces greenhouse gases which have greatly increased the natural concentration in the stratosphere, causing more infrared radiation from the Earth to be trapped and reflected back to the Earth's surface. As a result, global temperatures have been increasing steadily over the past five decades, with potentially negative effects on the ecosystem. This phenomenon is known as *global warming*, and heat-trapping greenhouse gases in the atmosphere which emanate from human activities are known as *anthropogenic greenhouse gases*.

Human activities are also generating organic compounds known collectively as halocarbons. Many of the halocarbons are also powerful greenhouse gases, and increased concentration in the atmosphere disturbs the natural balance and contributes to global warming. However, their net greenhouse effect in the atmosphere is reduced because they react with ozone which is also a greenhouse gas, and cause depletion, which in turn reduces the effectiveness of the gas as the Earth's shield from the Sun's harmful ultraviolet radiation. One CFC molecule can destroy about a hundred thousand ozone molecules and reduce the ozone concentration in the atmosphere, causing 'ozone holes' through which more harmful Sun's ultraviolet rays reach the Earth's surface and cause ecological damage. Halocarbon gases are very stable and are believed to be capable of remaining in the atmosphere for hundreds of years. In the last six decades or so gases used in refrigeration and air conditioning have become potent contributors to enhanced GHGs, accounting for over 20% of the total.

4.5   EFFECTS OF ANTHROPOGENIC POLLUTION

Anthropogenic pollution affects the environment in many ways: greenhouse gases released into the atmosphere enhance the natural concentration in the stratosphere, causing gradual global warming; the potential consequences of global warming include extreme weather and climate change; particulate and chemical aerosols from pollution concentrate mainly in the troposphere, interfering with the natural cloud systems, causing smog hanging over cities in all regions of the world, health issues, and damage to the ecosystem. In the last few decades, different methods have been developed for describing and quantifying the potential damage to the environment caused by different pollutants, and individual human contribution. These include radiative forcing and global warming potentials of greenhouse gases, and carbon intensity/carbon footprint of fuels and end-use energy.

4.5.1   Radiative forcing

The Earth's climate is largely determined by the radiant energy (sunlight) received from the Sun. The Earth may absorb all the energy received, or some of it in which case the balance is reflected back into space. The balance between the absorbed and reflected solar energy determines the average global temperature which in turn largely determines the climate. Radiative forcing (RF) is a quantitative measure of the difference between energy received by the Earth from the Sun and the amount radiated back into space (net change in energy balance in response to some external perturbation). Positive RF means that the Earth receives more energy than it radiates and negative RF implies that the Earth loses more energy into space than it receives. Net energy gain (positive RF) will cause global warming while negative RF means that the Earth cools. The RF concept is valuable for comparing the influence on global mean surface temperature (GMST) of most agents affecting the Earth's radiation balance. Human activities have changed and continue to change the Earth's surface and atmospheric composition. Some of these changes have direct or indirect effects on the energy balance of the Earth and are thus drivers of climate change.

Greenhouse gases are present in the atmosphere in very minor quantities compared with oxygen and nitrogen which make up over 99% of the gases in the atmosphere. However their regulatory role is critical, and there are many natural events that can move RF either way. However, any human activities that increase the concentrations will also cause (mostly positive) radiative forcing. The concentration of these gases in the atmosphere has increased exponentially over the last hundred years or so due to human activities (anthropogenic emissions), in particular, use of fossil energy, and is believed to be the main cause of positive radiative forcing and gradual increase in the average global temperature over the last two hundred years or so, since the exploitation of fossil fuel resources started to intensify.

4.5.2   Global warming potentials and atmospheric lifetimes

Not all of greenhouse gases make an equal contribution to the greenhouse effect. For example, one molecule of methane (CH_4) has 20 times the impact of a molecule of carbon dioxide for a 100-year time scale; nitrous oxide (N_2O) 300 times; ground-level ozone 2,000 times; and a chlorofluorocarbon molecule has from 13,000 to 20,000 times the impact of a molecule of carbon dioxide. Greenhouse gases may remain in the atmosphere for very short periods or for as long as 150 years. *Global Warming Potentials* are a quantified measure of the globally averaged relative radiative forcing impacts of a particular greenhouse gas with reference to carbon dioxide.

Global Warming Potential is a cumulative radiative forcing, (both direct and indirect effects) integrated over a period of time from the emission of a unit mass of gas relative to some reference gas which is carbon dioxide (IPCC 1996, 2007). Carbon dioxide being by far the largest anthropogenic gas emission in terms of quantity (around 90%), is assigned a unit global warming potential (GWP) over all time, and serves as the baseline. In effect, GWP of any other greenhouse gas is a measure of how well the gas absorbs reflected solar energy from the Earth, preventing it from immediately escaping into the atmosphere, compared with carbon dioxide. The higher the GWP, the more positive the radiative forcing (RF), and the more warming the gas causes.

Carbon dioxide equivalent (CO_2eq) is a method of placing emissions of various radiative forcing agents on a common footing by accounting for their effects on the environment. It describes, for a given mixture and amount of greenhouse gases, the amount of CO_2eq that would have the same global warming ability, when measured over a specified time period, usually 100 years. Most quoted values represent a basket of greenhouse gases listed in the Annex A to the Kyoto Protocol which was the first major global mitigation action on anthropogenic greenhouse gas emissions. Each of the greenhouse gases can remain in the atmosphere for different amounts of time, ranging from a few years to thousands of years. Methane remains in the atmosphere for 10-12 years before it degrades, nitrous oxide has a lifespan of 100 -120 years, and carbon tetrafluoride (tetrafluoromethane, [CF_4]) may remain in the atmosphere for up to 50,000 years. All these gases remain in the atmosphere long enough to become well mixed, in effect, the concentration in any atmospheric location remains approximately constant regardless of the source of the emissions. A gas which has a high radiative forcing but a short life compared with carbon dioxide could have a high GWP on a 20-year time scale or a low value on a 100-year time scale. On the other hand, a gas that has longer atmospheric life than the reference gas will have higher GWP with the time scale. (IPCC, 2014) (Table 4.1).

Table 4.1 Global warming potentials (GWP) and atmospheric lifetimes (yrs) of select greenhouse gases *(Extracted from IPCC, 2007).*

Gas		Atmospheric Lifetime (yrs)	GWP (100 yrs)	GWP (20 yrs)
Carbon dioxide (CO_2)		50-200	1	1
Methane (CH_4)*		12±3	21	56
Nitrous oxide (N_2O)		120	310	280
Hydrofluorocarbons				
	HFC-23	264	11,700	9,100
	HFC-125	32.6	2,800	4,600
	HFC-134a	14.6	1,300	3,400
	HFC-143a	48.3	3,800	5,000
	HFC-152a	1.5	140	460
	HFC-227a	36.5	2,900	4,300
	HFC-232fa	209	6,300	5,100
HFC4310mee		17.1	1,300	3,000
CF_4		50,000	6,500	4,400
C_2F_6		10,000	9,200	6,200
C_4F_{10}		2,600	7,000	4,800
C_6F_{14}		3,200	7,400	5,000
Sluphur hexafluoride (SF_6)		3,200	23,900	16,300
* The methane GWP includes the direct effects and those indirect effects due to the production of tropospheric ozone and stratospheric water vapour				

Direct effects known as *direct radiative forcing* occur when the gas itself is a greenhouse gas and *indirect radiative forcing* refers to situations when chemical transformations involving the original gas produce a gas or gases that are greenhouse gases, or when a gas influences other radiatively important processes such as the atmospheric lifetimes of other gases. For example, methane has both direct and indirect effects, an atmospheric lifetime of 12 ± 3 years and a GWP of 56 over 20 years, 21 over 100 years and 7.6 over 500 years. The decrease in GWP at longer times is because methane is degraded to water and carbon dioxide through chemical reactions in the atmosphere. The effect of radiative forcing varies substantially in space and time, and therefore the influence on climate response may also vary. For example, carbon dioxide has the largest forcing in the subtropics, decreasing towards the poles, with the largest forcing in warm and dry regions and smaller values in moist regions and in high-altitude regions (Taylor et al., 2011; IPCC, 2013).

4.5.3 Carbon intensity

Carbon Intensity (also called carbon intensity index or emission intensity, or specific emissions index) is the amount of metric tons of carbon dioxide or equivalent produced per unit of output or an activity, for example per Megajoule of energy produced, per capita or per GDP, or even per human activity - like driving a hundred kilometers. Population and GDP are the major determinants of primary energy consumption and therefore a country's energy-related emissions, but two factors may mitigate carbon intensity, namely *energy intensity* and *fuel mix*. Energy intensity/GDP gives some indication of a country's level of economic development and the extent of adoption of energy-efficient technologies, but simply moving away from high energy intensity production, like is happening in many developed economies can also lower energy intensity. On the other hand, emerging countries which are increasingly hosting primary production such as primary metals, polymers, cement, chemicals, will have high energy intensities largely because of the inherent energy-intensive nature of these production processes, but also because they often cannot afford to adopt the most efficient technologies.

Emissions intensity also depends on fuel mix (carbon content of energy consumed). Coal has the highest carbon content, followed by oil. Natural gas has the lowest carbon content of fossil fuels, while most renewables and nuclear energy have relatively low carbon content. The developed countries are decarbonizing energy by moving away from coal and substituting less carbon-intensive natural gas and renewables, mainly for power generation. On the contrary, coal is the only readily available and affordable fuel for power generation, steam raising, home heating, etc. in many emerging countries, hence carbon intensity will be relatively high. Carbon intensity is useful in comparing the effect of greenhouse gases from various sources on the environment. Values of carbon intensity can be calculated for electric power plants per unit of energy produced using different fuels, for heat generating units per joule of heat, and also for various production processes (Figure 4.5).

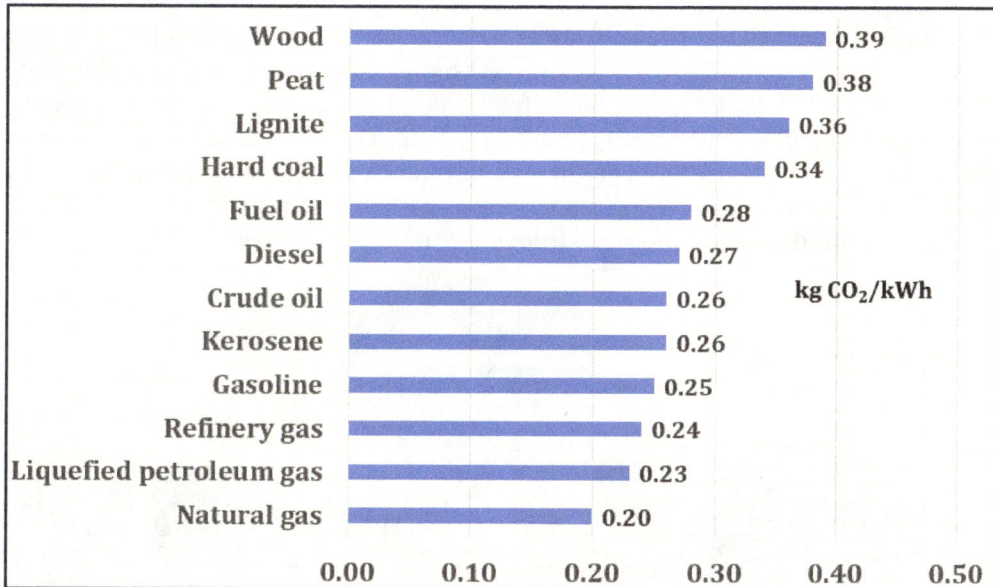

Figure 4.5   Specific carbon dioxide emissions of various fuels
(*Data from volker-quaschning.de/datserv/CO2-spez/index_e.php*).

4.5.4   Carbon footprint/carbon dioxide equivalent

Carbon footprint literally means an estimate of the amount of direct or indirect contribution of greenhouse gas emissions from any source over a time frame (usually one year), for example, events, products, organizations or persons. A comprehensive, quantitative assessment based on this definition would not be possible since every person breathes oxygen in and carbon dioxide out and there are many other natural sources of greenhouse gases. A carbon footprint is measured in metric tons of carbon dioxide equivalent (tCO_2eq).

Although accurate calculation is problematic, many approximate methods have been developed and carbon footprint calculation has become a powerful tool for assessing the impact of human (including personal) behavior on global warming. It should be noted that carbon represents all greenhouse gases which are now added as carbon dioxide equivalents. The foods and goods that humans buy everyday, polymer shopping bags, travels all have carbon footprints which must be accounted for in calculating personal contribution. For example, ten liters of petrol or diesel burnt in a car, or of oil used in home heating contribute 23-27 kg of carbon dioxide to the atmosphere; the manufacture of three empty one-liter plastic bottles of water/soft drink or twenty plastic shopping bags releases about 1 kg of CO_2eq into the atmosphere. In effect, reducing personal carbon footprint, like rationalizing energy use, cutting out unnecessary driving, opting for energy-efficient vehicles, opting for energy-efficient homes, fittings and appliances, or supporting district recycling efforts, are effective ways of reducing personal carbon footprint, and a personal contribution to global efforts at mitigating energy-related emissions. The average carbon footprint varies widely by country: 15.3 ($kTCO_2$/capita/yr), Bahrain, 24; Curacao, 46.8; United Kingdom, 5.7; China, 7.7; Somalia, 0.1; Nigeria, 0.5; Qatar, 37.1; Palau, 64.9 (Wikepedia, 2019).

4.5.5   Life-cycle analy,sis of greenhouse gases

Total Life-Cycle Analysis (LCA) is an accounting process that considers the environmental impact of all stages of a particular process or product chain. For example, LCA for an energy resource involves consideration of emissions at every stage from production to utilization, from material and fuel mining, manufacturing of components, construction, installation, de-commissioning, to waste management. A typical cradle-to-grave LCA of fossil fuels which begins with extraction of natural gas or coal and ends with electricity delivered to the consumer is shown in Figure 4.6. Lifecycle analysis of carbon intensity is a very useful method of comparing anthropogenic emissions from generating plants using different fuels (Figure 4.7).

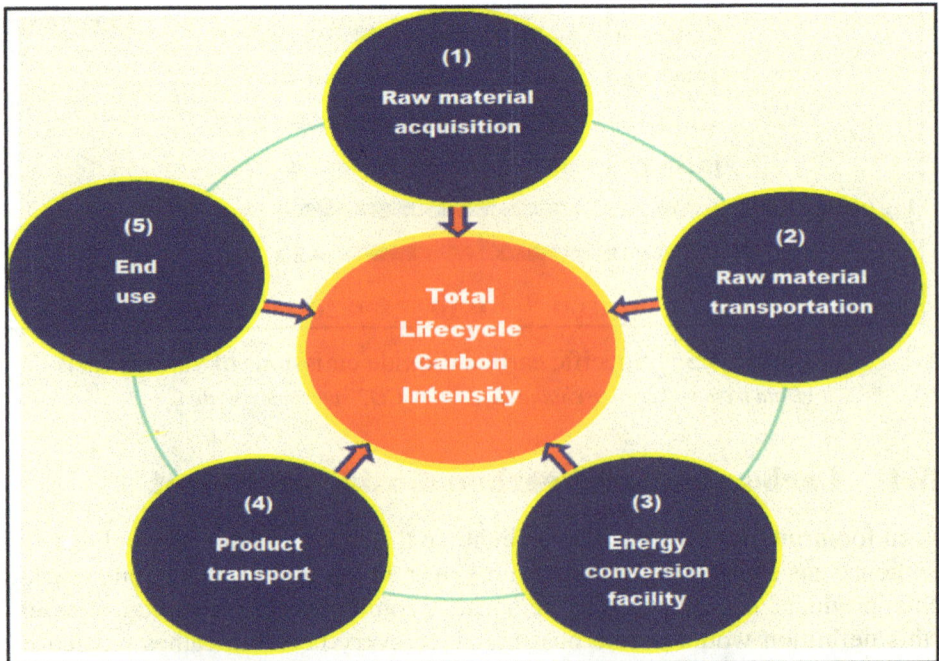

Figure 4.6 Cradle-to-Grave Life Cycle Analysis of fossil fuels.
(Adapted from NETL-Life Cycle GHG Perspective Report, 2014).

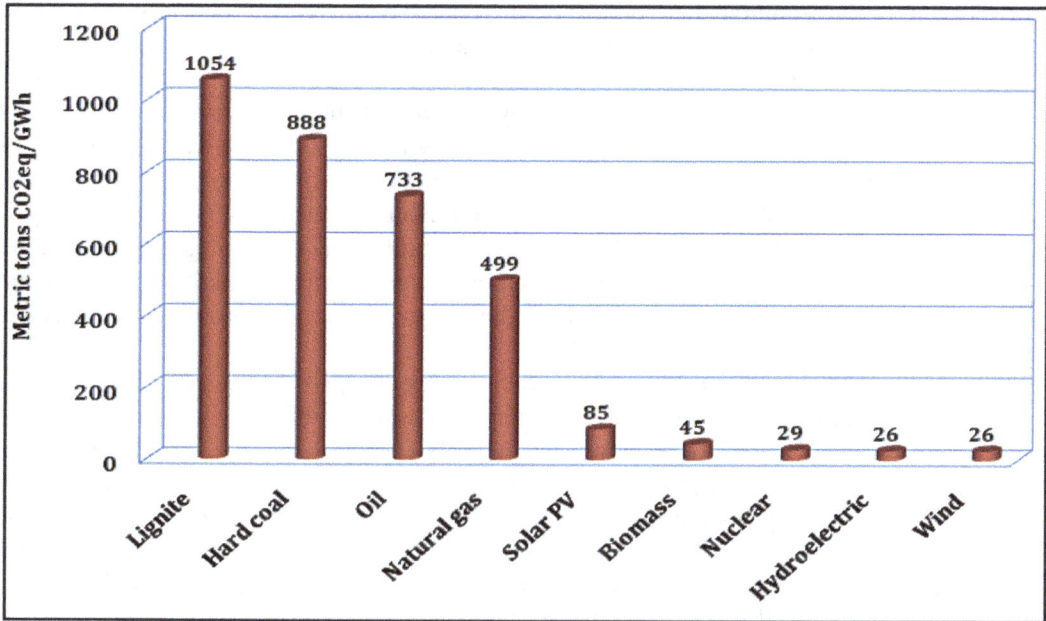

Figure 4.7   Lifecycle analysis of GHG emissions by fuel in electric power generation. *(Data from WNA Report, 2011).*

Calculations of the total life cycle emissions of a fuel take into account emissions at the mining stage, transportation, processing, ultimate utilization and recycled or discarded solid or heat by-products. Emissions from the production of materials of construction such as steel and cement used in levee construction, materials used in the transportation and distribution of natural gas and fuel oil are also taken into account. For example, solar energy is considered a zero emission energy resource in the traditional evaluation process. However, Life-cycle-analysis presents a different picture: the overall carbon footprint is significant because enormous energy (usually supplied by coal-fired plants) is required for the production of silicon used in the manufacture of photovoltaic solar cells, collectors, associated equipment, building and infrastructure, etc. In effect, there is nothing like near-total generation of electricity from zero carbon sources (solar, hydroelectric, nuclear, wind) as claimed by some developed countries - all energy sources have carbon footprint. It is also a common assumption that gas-fired power generating plants emit around 50% lower environmental pollutants compared with coal. However, total life-cycle analysis shows that, unless methane emissions at the gas production stage, especially from gas flaring, processing and transportation, can be kept below 2-3%, there is no relative advantage, since methane is a significantly more potent anthropogenic gas compared to carbon dioxide which is the predominant pollutant in coal-fired plants (Nalbandian-Sugden, 2015). Life cycle analysis also involves the quantification of all emissions in the chain, including emissions saved (displacement emission) where for example the by-products of one process are used to replace products of another process, thereby avoiding the emissions associated with those replaced products. Steel, aluminium, copper and indeed most metals are 100% recyclable. The energy required in processing recycled steel is about 30% less than required for processing virgin steel raw materials. Recycled aluminium only consumes about 10% of the energy required to produce the material from bauxite ore and recycling of lead, copper, tin, etc. also saves significant energy and associated emissions. Furthermore emissions from mining and processing ores for recycled metals will be avoided.

It is commonly assumed that replacement of steel with lighter materials such as plastic or aluminium makes an automobile lighter and therefore more fuel efficient. While this may be true, the analysis fails to take into account the fact that the production processes of most of these lighter materials are much more energy and carbon-intensive than steel. In effect, in order to determine the contribution of any process to environmental degradation, the total life cycle should be evaluated. This would enable society to make informed choices on the environmental impact of products and processes, and personal behavior, like supporting recycling. Fly ash, usually a waste product from coal combustion is now used increasingly to replace cement in concrete mixtures, thereby reducing the cement required and saving the emissions that would have been discharged into the atmosphere during its production, and the emissions associated with the energy saved.

Life-cycle analysis for conventional steelmaking comprises emissions from the coke oven, sinter plant, blast furnace, oxygen converter, slab caster, rolling mills, oxygen and electricity production and utilization. Utilizing waste process energy from a typical integrated steel plant to generate electricity can displace about 8 GJ of grid electricity which in turn displaces the combustion of the equivalent of 22 GJ of feed coal in a coal-fired electricity power plant, and cuts process emissions significantly. These should be taken into account in a comprehensive LCA. Life cycle analysis can significantly change the widely held views about the relative emissions from fossil energy resources. About 97% of coal emissions are generated at the combustion stage, the balance at the production and cleaning stages. By comparison, only about 60% of emissions from natural gas occur at the burner tip. However, considerable emissions, mainly methane occur at the well head (about 15%), pipe line leakages during transportation, and at other points in the cycle. Conventional emission analysis indicates that the contribution of coal is about 30% higher than that of natural gas. However, life cycle analysis shows no significant difference if the coal-fired plant is of comparable efficiency. In summary, the true assessment of a product should not only take account of direct emissions from the manufacturing and transportation to retailers but must also consider a host of indirect emissions such as those caused by the production of the raw materials used in the production of the good.

4.6   EFFECTS OF POLLUTION FROM FOSSIL ENERGY USE

Human activities especially since the beginning of the Industrial Revolution have consistently undermined the natural balance in the carbon cycle as designed by nature, with various activities that produce anthropogenic gases, dust and other chemical compounds which end up in the atmosphere (Figure 4.8). Release of carbon dioxide into the atmosphere from fossil fuel and land use perturbs the natural carbon cycle. Over half of the anthropogenic emissions is removed from the atmosphere by carbon sinks in terrestrial ecosystems through enhanced photosynthesis, but also in the oceans, with potentially negative consequences. Increases in the natural concentration of the greenhouse gases greatly disturb nature's control system which enables the sustenance of life on the planet, and could have a profound effect on climate and ecosystems. Higher concentrations of GHGs will reflect more heat back to the Earth. Global average temperature is rising, causing ice caps to melt, and oceanic flooding. Increased concentrations of dust and smog alter the natural response of clouds to the Sun's radiation, interfering with the natural processes of moisture aspiration and transpiration, altering local weather and global climate, and also causing health problems for human beings and other life on Earth.

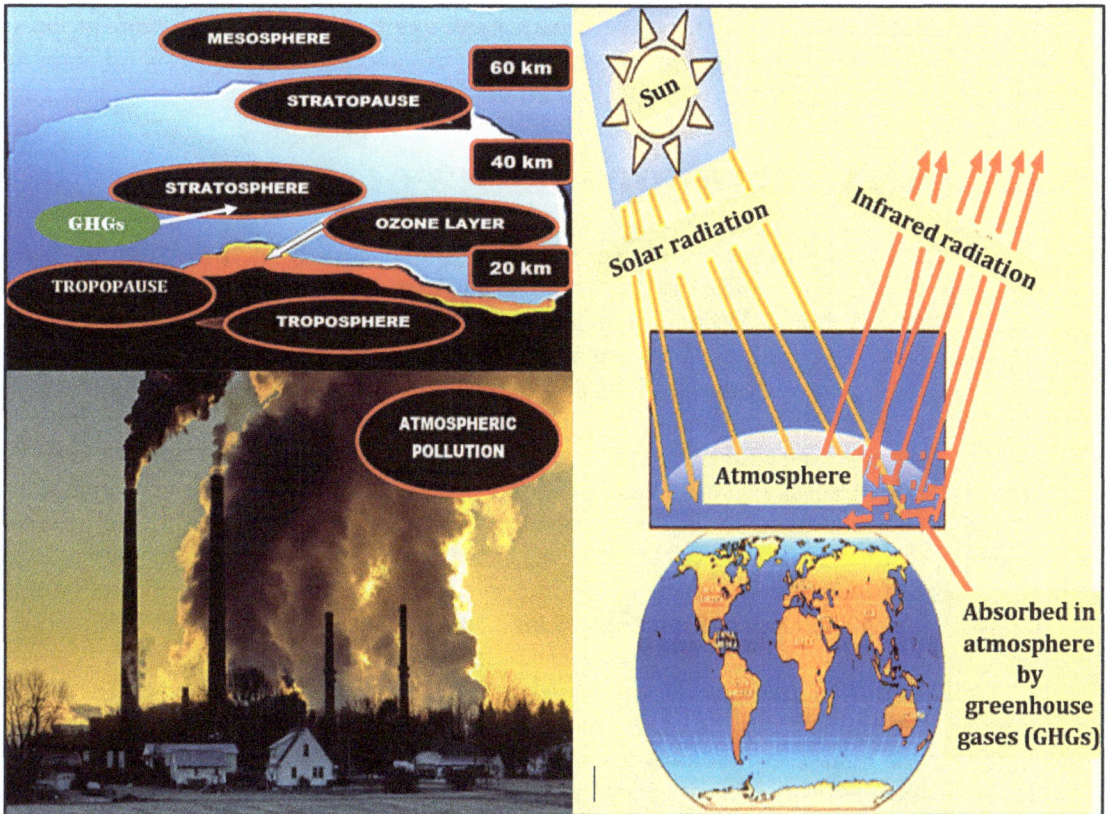

Figure 4.8 (a) Schematics of the Earth an its atmospheres (b) Greenhouse effect *(Afonja, 2017).*

The natural atmospheric ozone layer also known as 'good ozone' is an effective shield which prevents most of the harmful ultraviolet rays from the Sun (especially the potentially lethal UV-B radiation) from reaching the Earth and causing health problems including skin cancer, eye diseases, immune deficiency, and respiratory diseases. UV-B radiation affects the physiological and developmental processes of plants and could cause fundamental changes in the biochemical cycles of plants, and plant diseases. UV-B radiation causes damage to the natural developmental processes of marine life and reduces survival rates of a very important group of marine organisms known as phytoplankton. These organisms play a vital role in sustaining aquatic food pool, as well as the formation of oil and gas under the seabeds. Ozone is also released into the lower atmosphere from fossil fuel combustion and many industrial applications. This type is often called 'bad ozone' because it is a potential source of smog which is a health and safety hazard, particularly in urban areas.

Human activities - fossil energy production, power generation, use of automobiles, forest fires, biomass burning, use of basic biofuels, etc. release potent black carbon, volatile organic compounds, and other aerosols, mostly into the lower atmosphere (troposphere). Although most of these substances have a short life in the atmosphere, the potentially negative effects on humans and the ecosystem are extensive. Apart from causing smog over cities (which reduce visibility and interfere with the natural cloud activities), they have been linked to various diseases, premature deaths, and observed issues with agriculture and aquatic life. Synthetic as well as naturally occurring biopolymers degrade and become brittle when exposed to UV-B radiation, and special additives used to stabilize and protect these materials from radiation damage (such as bisphenols) are also potentially injurious to human health.

Management of polymer waste has become a major environmental issue, a large proportion of discarded polymers is ending up in oceans, and traces are beginning to show up in the human food chain.

4.7   TOTAL ANTHROPOGENIC POLLUTION

Global anthropogenic emissions have increased gradually since the Industrial Revolution of the 1800s and have intensified in the last hundred years or so, driven largely by fast economic and population growth. Although energy production and use account for nearly 70% of the total global anthropogenic pollution, there are many other sources (Figure 4.9). Carbon dioxide is emitted in forestry and other land use (FOLU) and many industrial processes; methane comes from animal and rice production; and most of the halocarbons come from refrigeration/air conditioning and household goods industries. Forestry and land use produce about 11% of carbon dioxide emissions while methane comes partly from oil, gas and coal production but also from agricultural activities that involve composting, fermentation, landfilling and animal breeding. Nitrous oxide is produced from use of fertilizers and fossil fuel combustion. Industrial processes, refrigeration and many consumer products are the main sources of anthropogenic F-gases, which include hydrofluorocarbons (HFCs), perfluorocarbons (PFCs), and sulfur hexafluoride (SF_6).

Emissions from the energy sector have increased exponentially - CO_2eq emissions (carbon dioxide, methane and nitrous oxide) have more than doubled in the last forty years. Carbon dioxide is the main constituent of energy-related pollutants, accounting for around 90% and, although the gas is the least potent in terms of atmospheric damage, it has become the benchmark for the evaluation of the anthropogenic effect of all other gaseous pollutants, expressed in carbon dioxide equivalent (CO_2eq) (Figure 4.9). Over 99% of energy-related emissions come from the production and utilization of fossil fuels (coal, oil and gas), with coal accounting for around 45%. Oil was the greatest pollutant in the 1970s but has been overtaken by coal. The regional distribution of emissions has also changed significantly. In the early part of the 20th century, virtually all emissions originated from the United States and Europe, today together they account for less than 30%. Just two countries (the U.S.A. and China) accounted for around 45% of the global CO_2eq emissions in 2016, estimated at 32.3 billion metric tons (Figures 4.10 - 4.12) (IEA, 2017). Carbon dioxide emissions stagnated for two consecutive years up to 2016 due primarily to strong efficiency improvements and low-carbon technology deployment, but then started rising again in 2017, reaching a record 33.1 Gt CO_2 in 2018, believed to be a result of the unusually hot and cold weather in some parts of the world. Although emissions from all fossil fuels increased, the power sector accounted for nearly two-thirds of emissions growth and coal use in power generation accounted for around a third of the total global energy-related emissions in 2018, mostly in Asia. China, India and the United States accounted for 85% of the net increase in emissions over the last two years (IEA, 2019c).

Different sectors of the economy generate different types and levels of pollution emissions. The energy supply sector (energy extraction, conversion, storage, transmission, and distribution processes that deliver final (end-use) energy to the end-use sectors - industry, transport, building, agriculture, forestry - is the largest contributor to global greenhouse gas emissions and accounted for about 42% of total energy-related anthropogenic GHG emissions in 2016, the largest single source. Industry contributed around 21% which came from fossil fuels used for in-plant power generation, metallurgical, chemical and mineral production, as well as feedstocks for chemicals and petrochemicals production (Figures 4.13-4.15). The building sector was the least polluting, contributing only about 6%. Much of the increase in energy-related anthropogenic pollution in the last two decades or so has been driven by large increases in the production of energy-intensive materials such as steel, cement, chemicals and petrochemicals, and all these processes depend heavily on fossil fuels and electricity. Most of these heavy industries are now in emerging and developing countries, many of which have also expanded very rapidly fossil-fueled power generation capacities. All recent projections indicate that these countries will account for nearly all of the increase in primary energy consumption (and related carbon dioxide emissions) over the next two decades. Production is often inefficient and energy intensity is high. Emissions by region and economic sector are presented in Figure 4.16. Clearly, there are significant regional differences: while Asia accounted for 53% in 2015, Africa contributed only 4%.

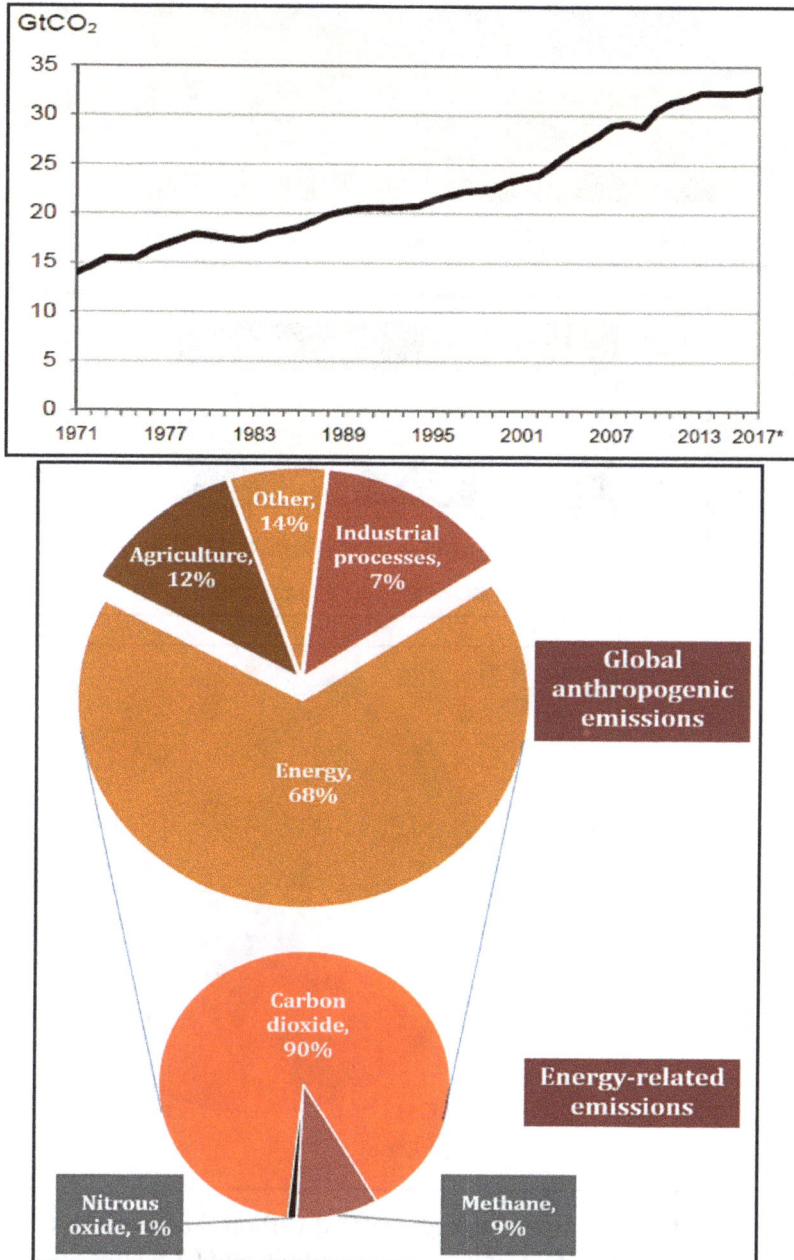

Figure 4.9 Growth trend and estimated shares of global anthropogenic greenhouse gas (GHG) emissions *(Data from IEA, 2017e. 2018; IPCC, 2014).*

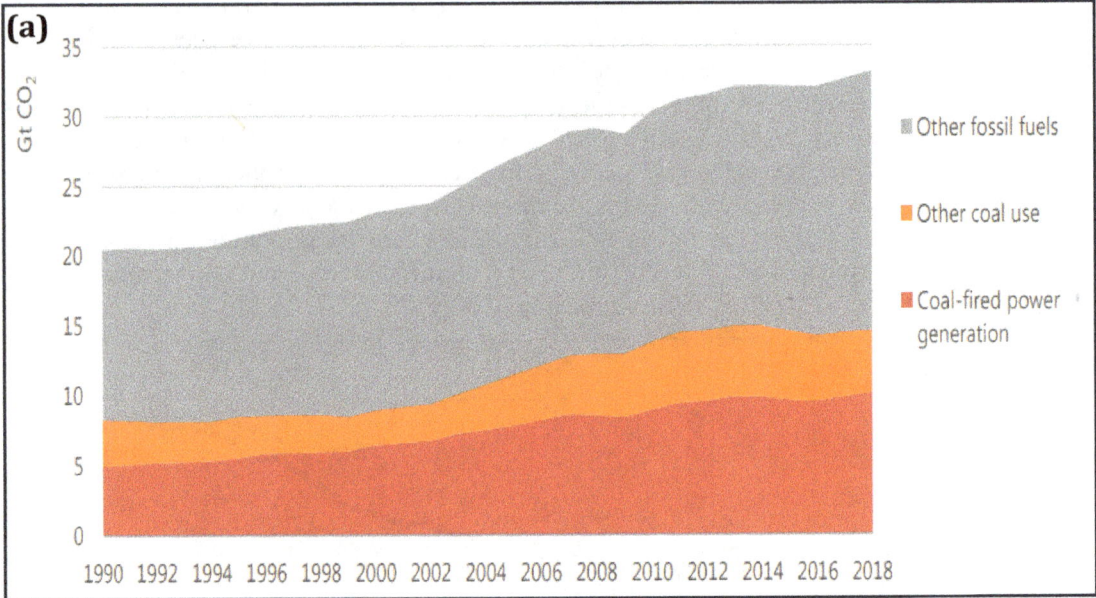

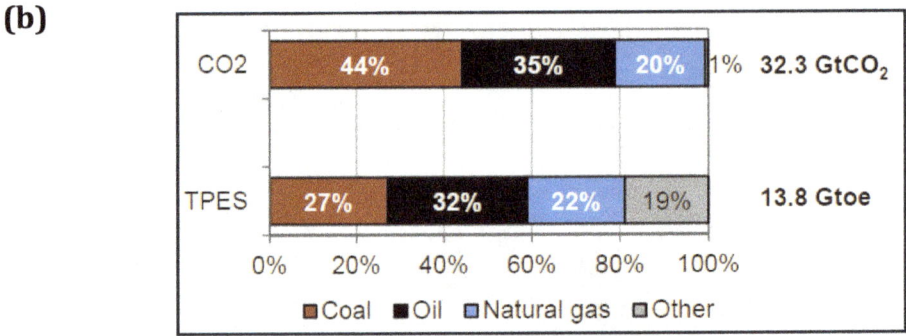

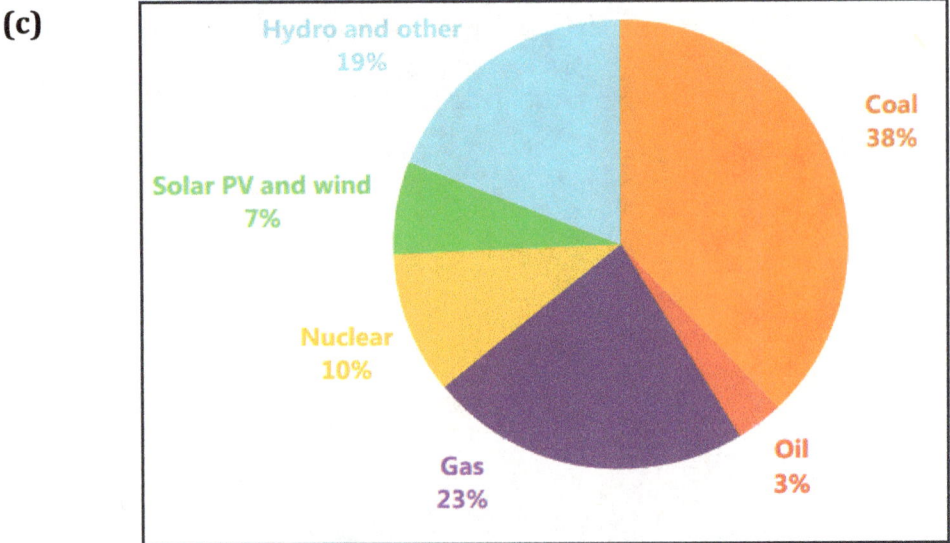

Figure 4.10 (a) Global energy-related carbon dioxide emissions by source, 1990-2018 (b) Total primary energy supply and CO_2 emissions, 2016 (c) Electricity generation mix, 2018 *(IEA, 2018b, 2018c)*.

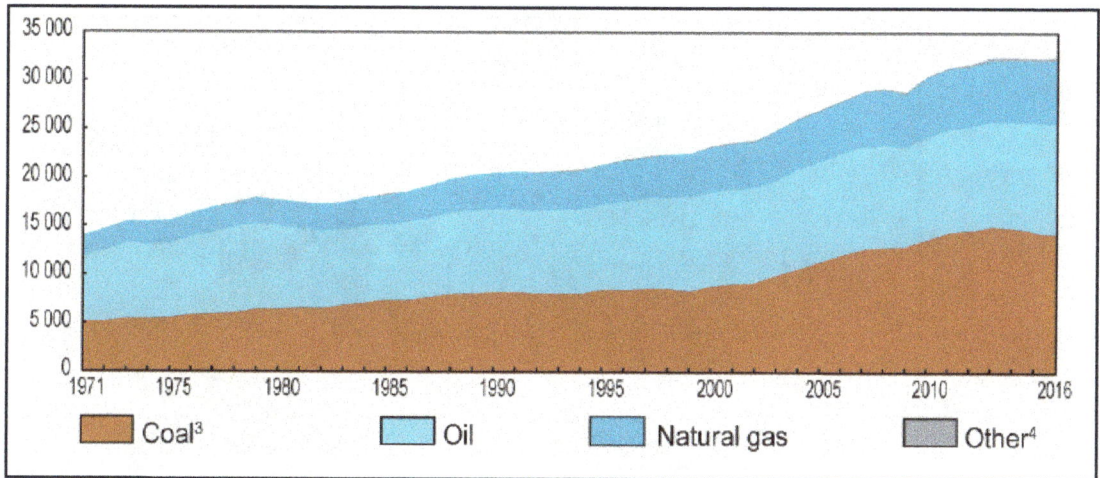

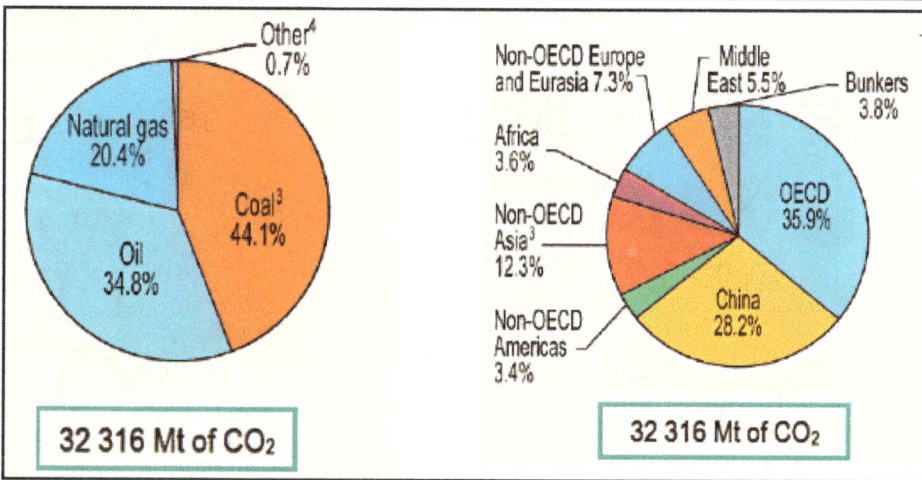

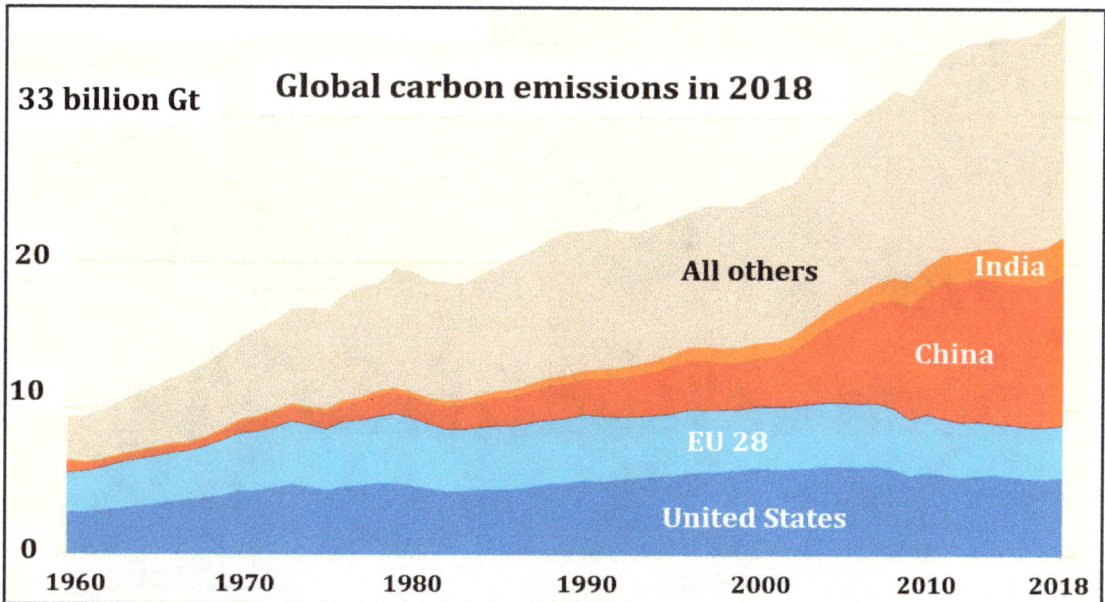

Figure 4.11 World CO_2 emissions from fuel combustion from 1971 to 2018 by fuel, country and region*(IEA, 2018c; 2019; the guardian.com).*

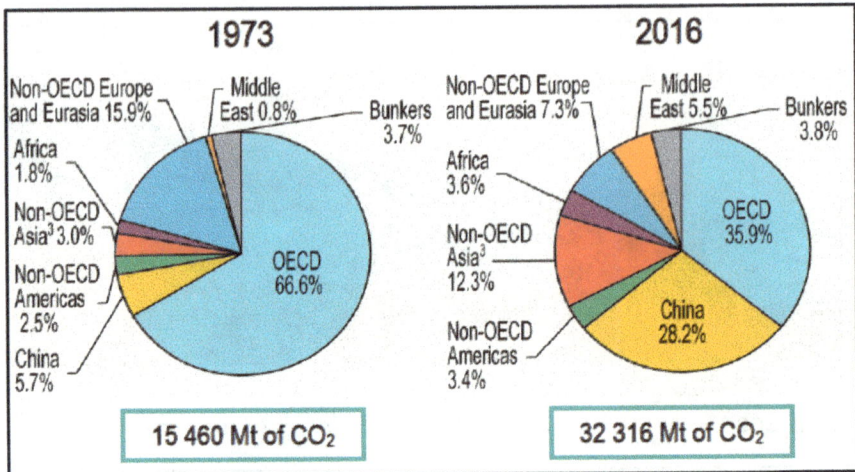

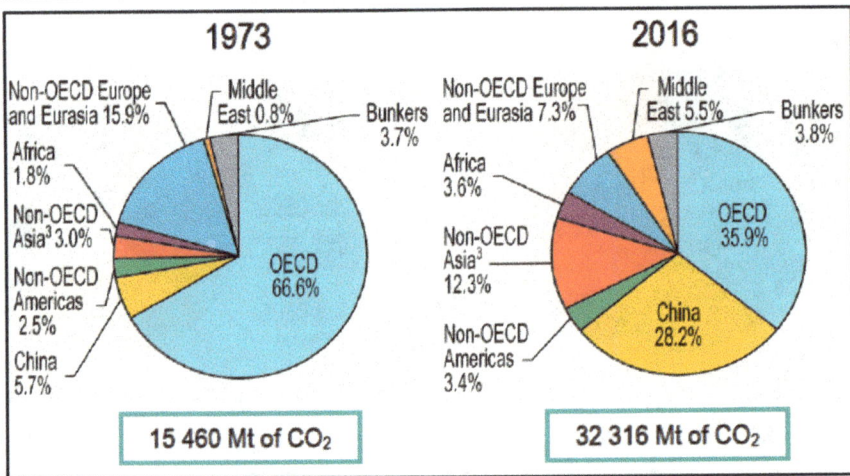

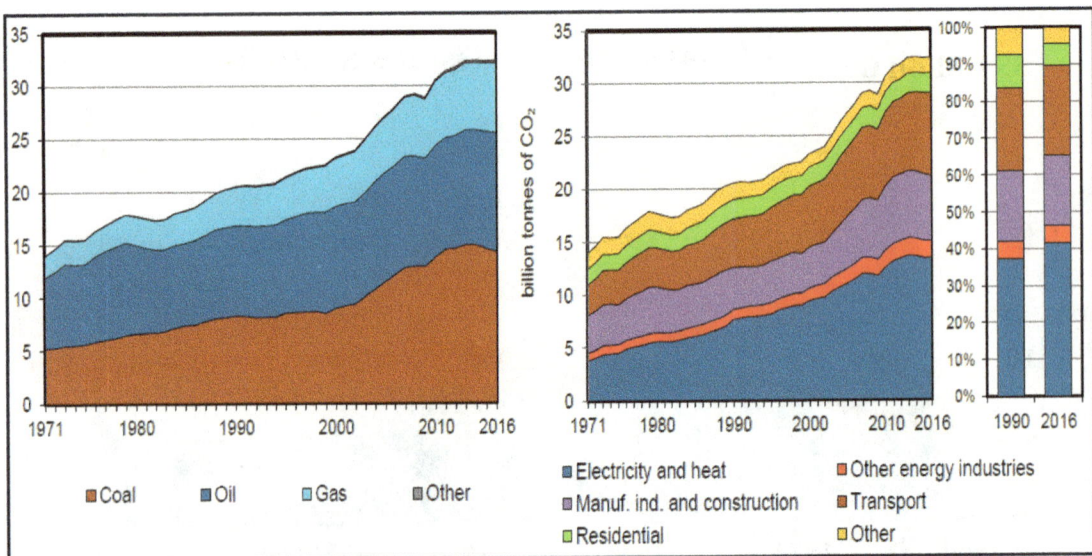

Figure 4.12 World CO_2 emissions from fuel combustion from 1971 to 2016 by fuel, sector, region and country *(IEA, 2017a; IEA, 2017e, 2018b).*

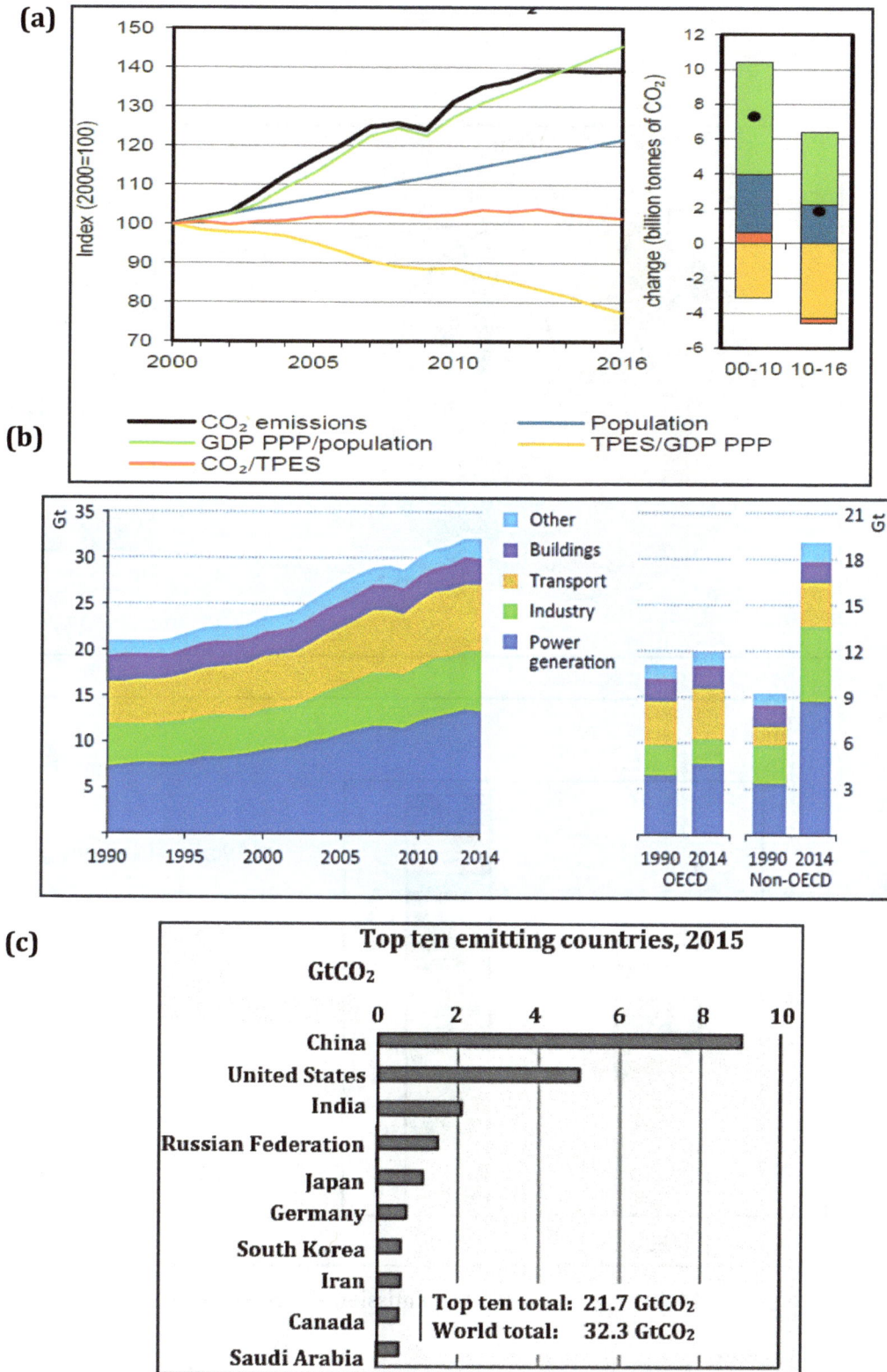

Figure 4.13 World CO_2 emissions (a) total GHG emissions (b & c) energy-related by sector (c) Top ten emitting countries *(IPCC, 2014a; IEA, 2017e, 2018).*

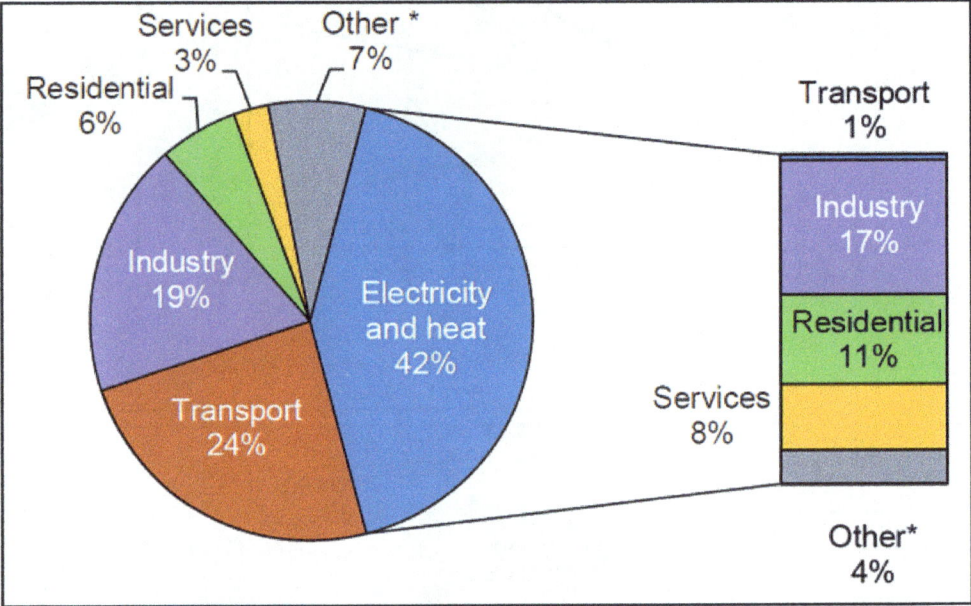

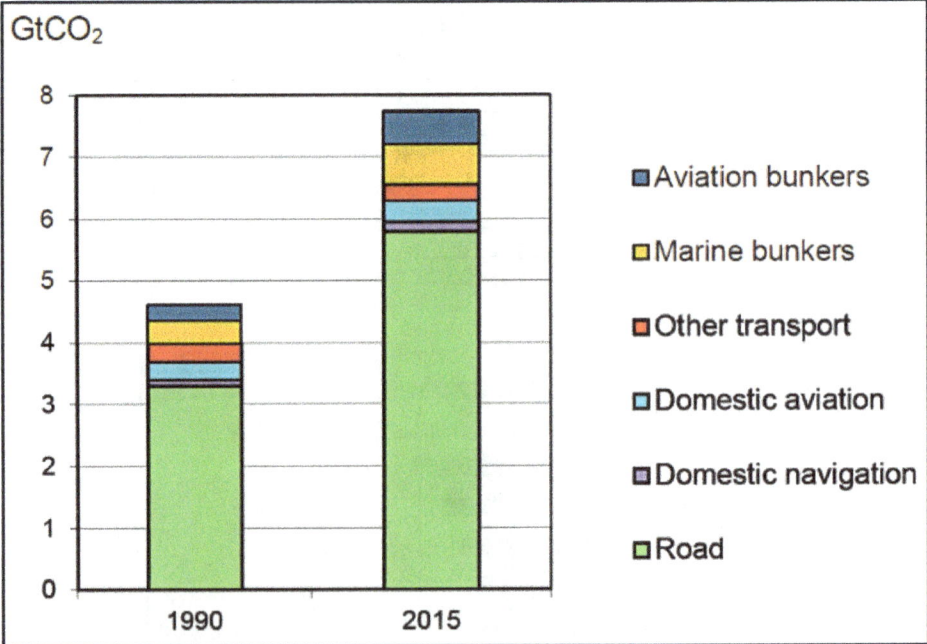

Figure 4.14 Energy-related CO$_2$eq emissions by end use *(IEA, 2017e)*.

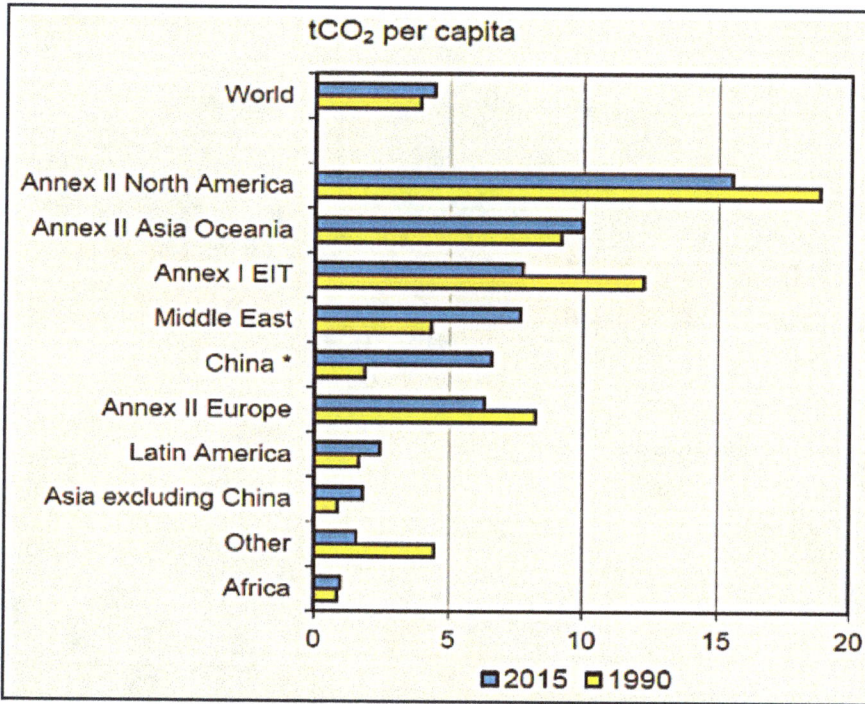

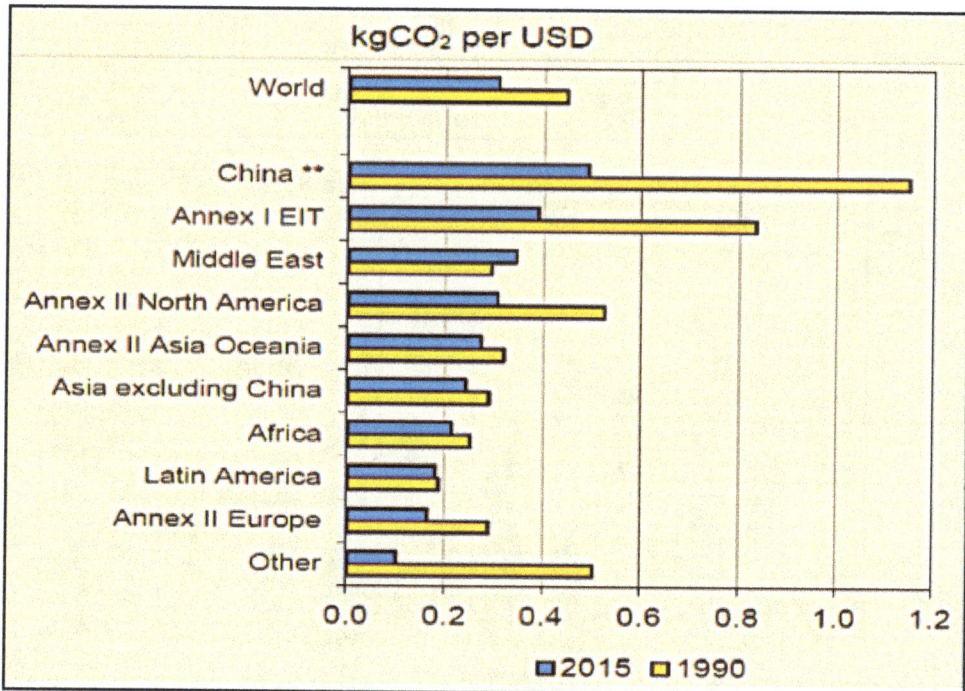

Figure 4.15 CO_2 emissions per capita and per GDP by major world regions *(IEA, 2017e)*.

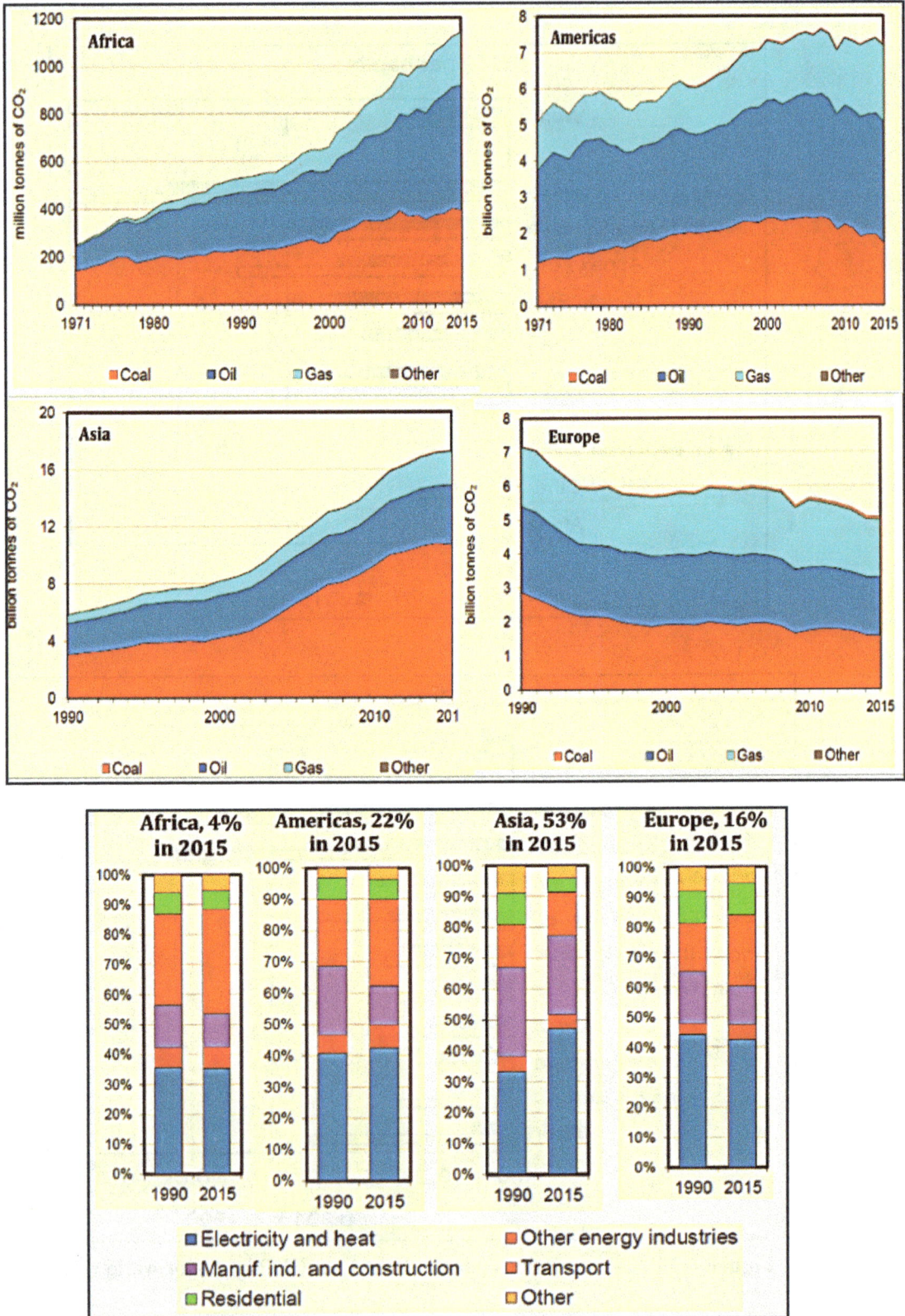

Figure 4.16 CO_2 emissions by fuel, region and sector *(IEA, 2017e)*.

4.7.1 Emissions from mining and production

Although oil and gas produce significantly less pollution at the end-use point, emissions during production and transportation can be very high. Natural gas (mainly methane) is always associated with oil and is routinely flared (burnt-off) before oil recovery begins. In some locations the gas is simply vented into the atmosphere. It is estimated that 150-170 billion cubic meters of gas is flared worldwide every year, contaminating the environment with about 400 million metric tons of carbon dioxide (Andersen et al., 2012). Just ten countries accounted for three quarters of the amount of gas flared in 2018 (World Bank, 2019). (Flaring also releases fine and ultrafine carbon particles known as black carbon (BC) into the environment. Black carbon emission from gas flaring may account for up to 40% of the total anthropogenic BC emissions in some countries, for example, Russia. Other sources of BC include power generation, residential, transportation and industry. Significant amounts of natural gas escape into the environment (fugitive losses) during transportation by pipelines or liquefaction. Compared with natural gas, life-cycle GHG emissions attributed to liquefied natural gas supply chain is significantly higher, from liquefaction, tanker transport and regasification processes. Coal deposits are always associated with methane gas which is routinely vented into the atmosphere before and during mining for safety reasons. It is estimated that, on the average, around 100 cubic meters of methane is vented for every metric ton of coal mined. Modern coal mining techniques and post-mining transportation and preparation activities generate considerable dust. A recent LCA study of fossil fuel production for power generation showed that around 33% of life-cycle GHG emissions from an LNG-fired power plant come from extraction, processing and transportation stages, compared to 32% for natural gas and only about 2% for coal (Figure 4.17).

4.7.2 Emissions from power generation

There are many different power generating methods with different merits and demerits with respect to capital and operational costs, available primary energy source, environmental impact, strategic policies, etc. All generation methods produce GHGs in varying quantities and at different phases of their life cycles. Coal-fired plants release around 98% of total LCA emissions at the power generation facility, natural gas plant releases a significant proportion during production and transportation, while others - wind power, solar power, hydropower and nuclear power release the majority of emissions during component manufacture, construction and decommissioning. About 42% of the total global anthropogenic emissions come from heat power generation (Figure 4.14). However, there are significant regional differences: emissions from electricity and heat generation in emerging and developing countries have doubled in the last two decades, compared with a significant fall in industrialized countries. China accounted for around two-thirds of this increase, coming mainly from large increases in coal-fired power generation and energy-intensive primary materials and chemicals production. Fossil fuels (mostly coal and natural gas) supplied 66% of global power generation in 2016, and the most dramatic growth has been in China, (Figures 4.18 a&b). There was a slight drop to 64% in 2018. Electricity is the world's fastest-growing form of end-use energy consumption and demand is expected to grow by about 70% over the next two decades, the strongest growth being in developing and emerging nations. Most of these nations rely on internally-sourced coal, which is the highest pollution emitter of the fossil fuels (EIA, 2016). Emissions from power generating plants depend on the fuel and are unique to the individual facility because efficiency counts. Furthermore, published life-cycle analyses studies differed significantly in terms of assumptions and the technical specifications (technology adopted, efficiencies etc.) of the focal generating plant. This explains the wide range of emissions quoted for each fuel in published literature (Figure 4.18c). There is little doubt however, that coal-fired power plants are the greatest sources of pollution from power generation, accounting for about 30% of the global energy-related anthropogenic emissions in 2018, and this level of pollution is estimated to be responsible for over 0.3°C of the 1°C increase in global average annual surface temperatures above pre-industrial levels. This makes coal the single largest source of global temperature increase (IEA, 2019i).

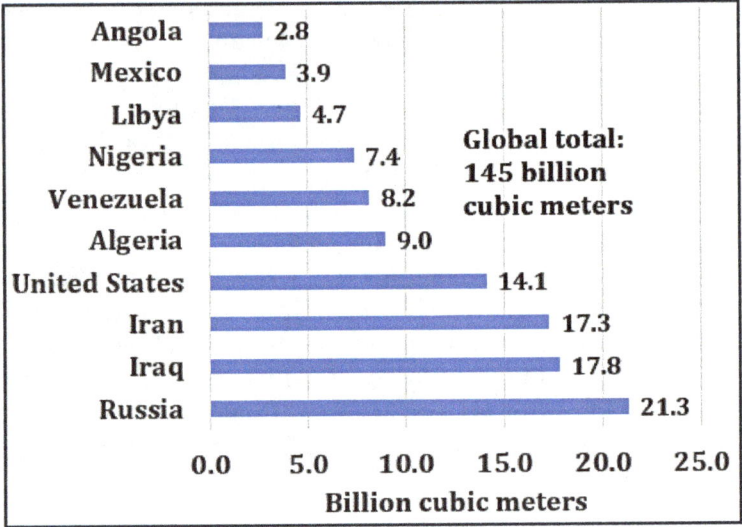

Figure 4.17a Top ten gas flaring countries in 2018 *(World Bank, 2019).*

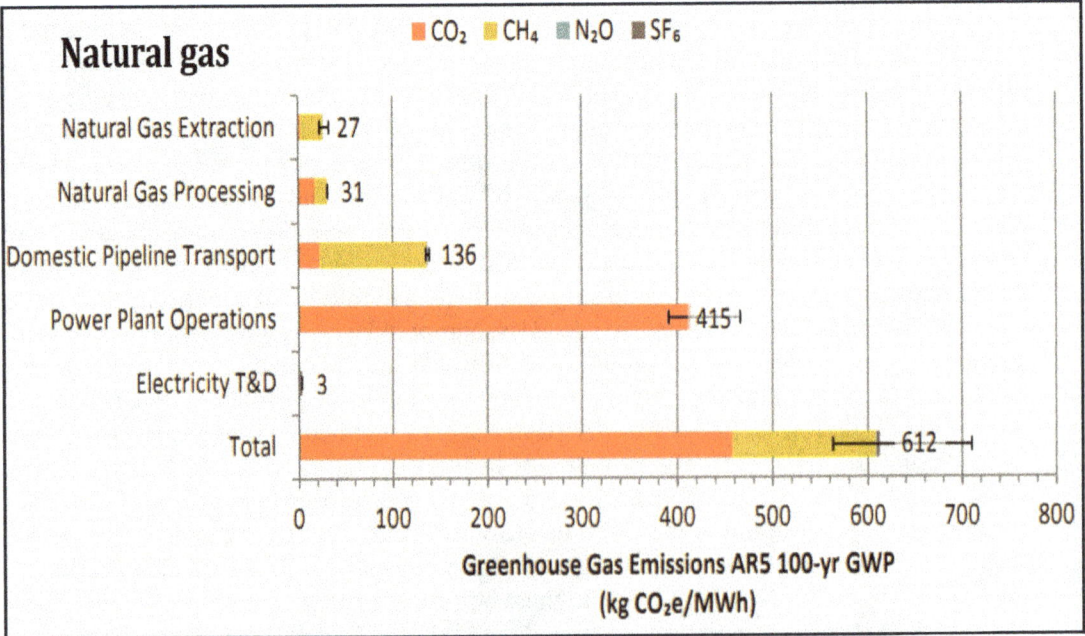

Figure 4.17b Life Cycle greenhouse gas emissions of natural gas power plant *(DOE-NETL, 2014).*

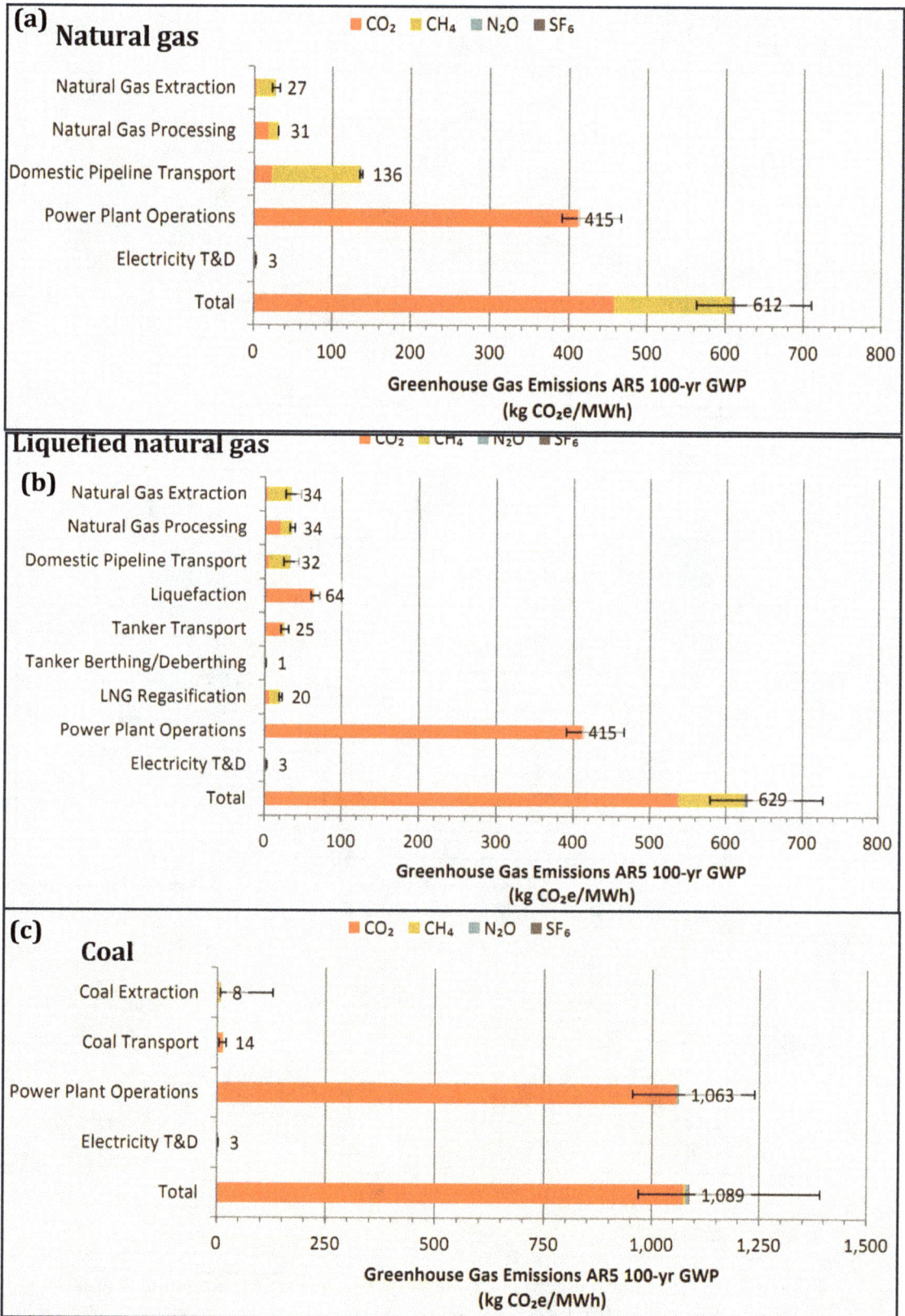

Figure 4.17c Life Cycle greenhouse gas emissions (a) Natural gas power plant (b) Liquefied natural gas powerplant   (c) Coal power plant *(DOE-NETL, 2014).*

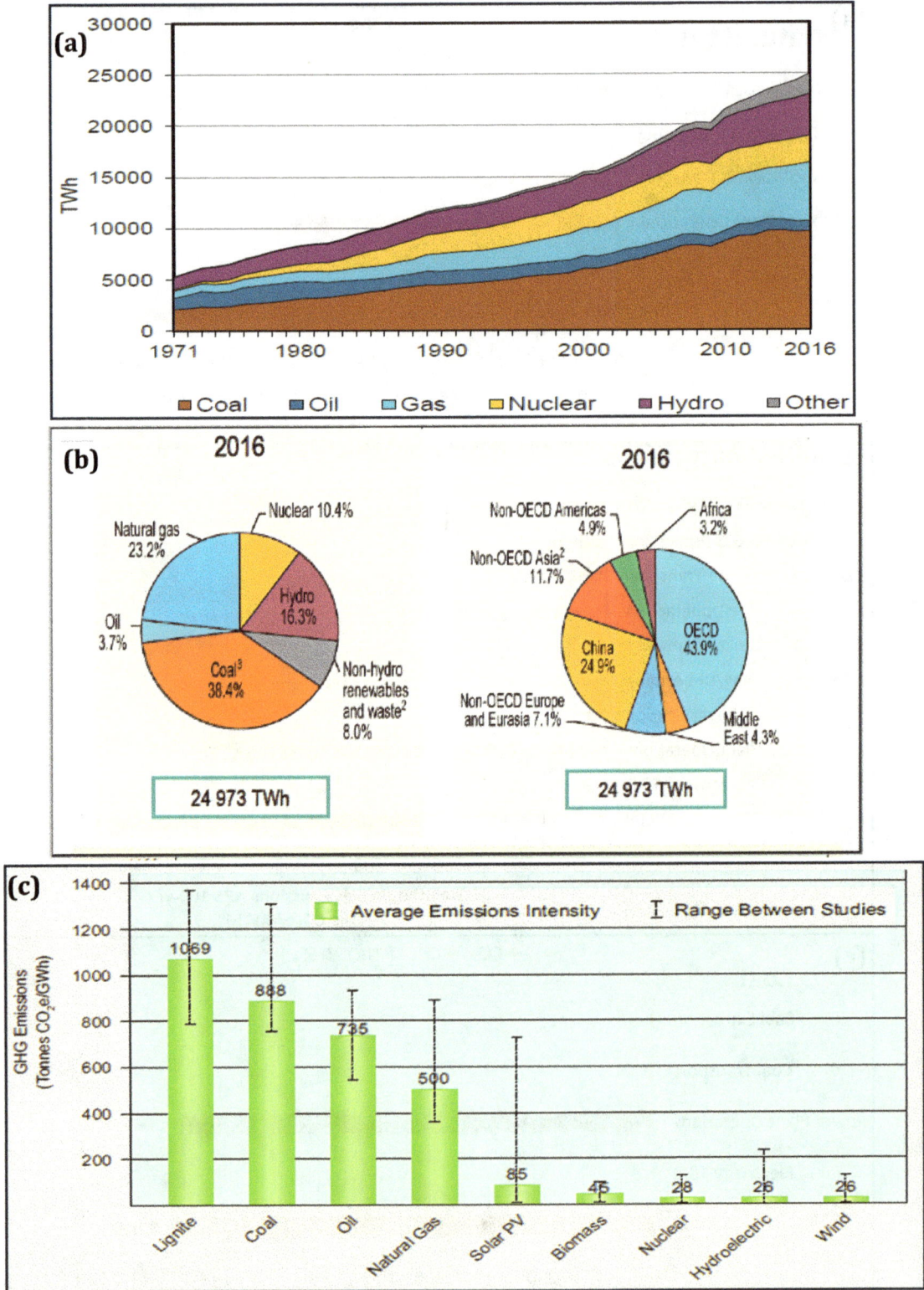

Figure 4.18 (a) Global electricity generation by source (b) Regional shares of electricity generation (c) Lifecycle greenhouse gas emissions intensity of electricity generation methods *(EIA, 2017a; WNA, 2011).*

It is clear however that fossil fuels (coal, oil and natural gas) are by far the highest pollutants. Lignite, the youngest coal produces nearly fifty times the emissions produced by nuclear, hydroelectric or wind power per unit power output. The power sector will account for two-thirds of the global primary energy increase in the next two decades, and developing and emerging nations will account for most of the growth. These countries depend on local deposits of low-grade coals and all projections predict that they will also account for most of the growth in GHG emissions (see Figure 4.12).

4.7.3 Emissions from industry

Energy and industry have been closely interdependent since the Industrial Revolution, and their future remain intertwined. Energy powers all kinds of industries - manufacturing, non-manufacturing/services, construction, agriculture. On the other hand, rising consumer demand for the wide range of varied products of industries provides the impetus for massive investments in energy infrastructure, and continuous development of new energy technologies. Energy cost is always a major component of virtually any industrial production process, and this has provided a strong incentive for a persistent pursuit of higher efficiencies across the energy production and end-use spectra. Industrial energy demand growth would be at par with economic growth but for the extensive gains in manufacturing and other energy end-use efficiencies over the last two decades or so.

The industrial sector energy demand is higher than any other end-use sector, and accounts for 50-55% of total global final energy use. The mix and carbon intensity of fuels consumed in the industrial sector vary across regions and countries, depending on the level and mix of economic activities, and on technological development. Energy-intensive manufacturing (iron and steel, cement, paper, food, glass, chemicals, primary metals) accounted for nearly 70% of the energy use by the sector in 2016. Manufacturing energy intensity has been declining over the last decade or two especially in the industrialized countries as a result of gains in production efficiencies. However, growth in demand in non-OECD countries has been relatively high (0.8%/year compared with 0.2%/year in OECD countries and a world average of 0.7%/year). Around 90% of the rise in industrial sector energy use over the next two decades will occur in non-OECD nations, (EIA, 2017).

In spite of the fact that industry accounts for more than half of the total global end-use energy demand, the sector is the least polluting (per unit of energy use), accounting for only 19% from fuel combustion in 2015, compared with 42% by the power sector, 24% by the transport sector, and 6-9% by the buildings sector (See Figure 4.14). Fuel mix in the sector is diverse across regions: industrial use of coal is declining in OECD countries, but rising in non-OECD regions, with the region consuming around seven times as much coal as OECD regions. For industrial processes that use coal (and most are energy-intensive processes located in emerging nations), fuel switching opportunities are rare. Also, projections show that the current gradual migration of energy-intensive, most polluting industries from mature regions (OECD North America, Europe, Asia Pacific) to emerging markets - Asia, Africa, the Middle East, and Latin America - will intensify over the next two decades. The highly-polluting chemicals industry is the industrial sub-sector with the highest rate of growth, driven by the rapidly rising global demand for plastics, fertilizers, and other petrochemical products, and most of the growth is in the emerging nations, but much of the products end up in the developed world. The economies of the mature regions have been transitioning from energy-intensive manufacturing to higher value manufacturing and services. In effect, the bulk of the growth in GHG emissions from industry will come from emerging nations over the next two decades or so.

4.7.4 Emissions from transportation

The transportation sector accounts for about 20% of the total global final energy use, filled mainly by oil-derived gasoline, diesel and jet fuel. Around 24% of energy-related emissions come from the sector. Nearly 95% of the current transportation energy needs are met by oil, but natural gas is becoming increasingly popular as fuel for trucks, buses, as well as ships, accounting for around 2% of the total end-use energy demand by the sector in 2016. Natural gas is expected to play a more prominent role in transportation and could account for up to 11% by 2040. Gas is projected to supply around 50% of bus energy consumption, 17% of freight rail, 7% of light-duty vehicles, and 6% of domestic marine vessels by this date (EIA, 2016). Projections also indicate that global energy demand for transportation will rise by only about 30% over the next two decades in spite of the fact that global vehicle population is expected to double, increasing gains in fuel economy being the major mitigating factor (ExxonMobil, 2018, 2019).

OECD-32 countries account for half of global transportation demand currently and demand will fall by about 10% over the next two decades despite a significant growth in transportation, due to major advances in technology which continue to create more efficient transportation and significant fuel savings. In effect, all the growth in demand over the next two decades will come from emerging and developing countries, up by 50-80%, due largely to the projected rapid growth of the economies, industrialization and population of automobiles, but also because of non-existent or weak policies on emissions and fuel efficiency. Commercial transportation (trucks, air, rail, ships) is growing rapidly all over the world and will rise by about 55% over the next two decades as the global economy doubles. Two recent studies concluded that gasoline and diesel will continue to dominate the transportation fuel mix in all regions of the world and the growth of electric transportation is not expected to make any significant impact. Currently, there are about 2 million electric vehicles in the global fleet, representing about 0.2 percent of the total. Various projections indicate that the population will grow to between 100 and 350 million by 2040, assuming that most of the problems can be resolved. This will still be insignificant compared with the projected population of 1.9 billion vehicles by 2040. In any case, about 60-65% of global electricity for charging EV batteries will still come from fossil fuels, hence the carbon footprint of electric vehicles will still be substantial. Diesel oil is the most polluting of all transportation fuels, and the main source of nitrogen oxides, which are the highest climate forcers of all the major anthropogenic greenhouse gases. While many developed countries have strict pollution emission control policies, most in the emerging nations do not. Modern diesel vehicles now feature advanced catalytic converter units which are capable of reducing emissions to levels comparable with or even lower than petrol engines. However, most of the vehicles that fail to meet environmental requirements in the developed countries end up in the emerging nations where demand for diesel and indeed all transportation fuels will also be highest over the next two decades. In effect, transportation pollution problem cannot be resolved by individual country actions since the negative impact on climate will be global, therefore international support for the developing countries will be crucial.

4.7.5 Emissions from buildings

The building sector comprising residential and commercial users accounts for about 20% of the total delivered energy consumed worldwide, but only 6-9% of the total global energy-related emissions come directly from the sector, although the emissions level more than doubles when emissions associated with electricity used by the building sector are factored in. The projected world average growth rate in energy demand by the sector over the next two decades is 1.5% but there will be significant variations by country and region, not only in growth rates but in type and amount of energy consumed. For example, growth rate in the United States will be about 0.1%. OECD, 0.6%, Mexico, 1.9%. The lower growth rate in the industrialized countries is a result of the expected relatively slow growth rate in the economies and population, as well as improvements in building insulation shells and efficiency of appliances. On the other hand, the economies, population and urbanization will grow in non-OECD countries where over 80% of the world's population resides. The region's projected population as well as economic growth rates are more than twice that of OECD.

Delivered energy consumption in the residential sector is expected to grow by 48% and will account for around 13% of total world delivered energy consumption in 2040 but the bulk of the growth (around 80%) will be in non-OECD countries. Growth rate in non-OECD countries is around 2.1%/year compared with 0.6%/year in OECD countries. Considering the expected faster growth rate in the building sector, particularly in the emerging world, growth in energy demand should be much higher than the projected values. However, buildings and appliances are becoming increasingly energy-efficient and new technologies coming on stream will improve efficiencies at even faster rates.

Natural gas is the dominant residential and commercial source of energy in OECD countries but electricity is more important in the emerging economies of the non-OECD countries. Even in OECD countries, demand for electricity is growing at a fast rate of 1.0%/year compared with 0.6%/year for natural gas, and is expected to surpass natural gas as the leading source of residential delivered energy within the next ten years. Electricity share of world residential energy consumption is around 40%, increasing to about 45% in 2040. In spite of the vibrant current and projected growth in global housing demand over the next two decades, energy efficiency is playing a big role in driving down end-use energy demand by the building sector as modern appliances and advanced building technologies and materials proliferate across all regions of the world.

The building sector is the least polluting of all the end-use sectors, and accounts for less than 10% of GHG emissions from energy end-use. It could be significantly higher when considered on a lifecycle basis since the sector is a major consumer of electricity, much of which comes from fossil fuels. The dynamic growth of infusion of energy-efficient technologies into all aspects of the sector, - from energy-efficient building to fittings, appliances and smart remote control systems - should help to further drive down emissions from the sector. However this will depend on the extent of decarbonization of power generation, since around 90% of the energy demand growth in the sector will be met by electricity. Current policies in many developed countries promoting self-power generation with modular solar systems should also help in reducing pollution from the sector to a very small proportion of the total end-use emissions. Some international organizations and private enterprises are also helping to spread self-power solar power systems across the developing world in order to fast track access to electricity, from small personal units that charge phones and power lamps, to rooftop PV systems that can power homes and small communities.

4.8   ENERGY-RELATED POLLUTION BY TYPE

Two main types of pollution emanate from fossil energy use: gaseous chemical compounds and particulates, most of which are released at ground level into the lower atmosphere (troposphere), the main exception being airplanes that cruise in the stratosphere to minimize turbulence and air resistance, and save fuel. While most particulate pollutants remain in the troposphere and drop off within days, gaseous pollutants (particularly the greenhouse gases) can rise to great heights, ending up in the middle atmosphere (stratosphere) where they can reside for very long periods. Anthropogenic ozone and particulates mostly stay in the troposphere close to the Earth's surface, largely in the vicinity of the sources, but very fine particulates can travel over long distances depending on the prevailing wind dynamics. Gases that reach the stratosphere are relatively well mixed with the natural GHGs and are largely independent of the sources. The impact of the two levels of pollution on the environment differs significantly. While stratospheric pollution affects climate mainly by warming the Earth, tropospheric (ambient) pollution largely affects the weather and impacts negatively on city environment, public health, and other lives in the ecosystem. Furthermore climate change is a very gradual process while the effect of ambient pollution could be instant.

4.9 STRATOSPHERIC POLLUTION

The natural greenhouse gases in the stratosphere 20-50 kilometers above the Earth control the global average temperature while the natural ozone layer prevents most of the Sun's harmful rays from reaching the Earth. Most of the anthropogenic greenhouse gases emitted into the lower atmosphere eventually end up in the stratosphere, increasing the natural concentration of GHGs, thereby enhancing their ability to absorb mainly infra-red radiation reflected from the Earth, and reflect it back to Earth, causing global warming. They are long-lived, and some can last for centuries or longer in the atmosphere. Both natural and anthropogenic GHGs are very well mixed and the concentration in any atmospheric location is largely independent of the sources. However, the impact on climate can vary with latitude. The average global temperature has been rising steadily since the start of the Industrial Revolution some three hundred years ago, and its effect on climate is becoming evident and multi-dimensional. The oceans are warming, thereby energizing winds that cause increasingly severe and extensive damage when they hit land; rising temperature could intensify extreme weather such as heat waves and interfere with natural processes of evaporation, transpiration, and rainfall particularly in arid regions; ocean levels are rising due to rising global temperature and melting Arctic ice, causing coastal flooding; warmer oceans acidify by absorbing carbon dioxide, (the main energy-related pollutant) causing damage to aquatic life. Warmer climate could also promote breeding of disease vectors such as mosquitoes, with consequential health risks. While many of these events can be caused by natural processes such as changes in the Earth's rotation orbit, there is ample scientific evidence that anthropogenic pollution is playing a key role.

Natural ozone is formed in the lower part of the stratosphere by a continuous breakdown of oxygen molecules into unstable atoms which recombine with oxygen molecules to form ozone. The ozone formed is unstable and reacts with oxygen atoms to form two oxygen molecules. This continuous process of ionization and recombination is powered by the dangerous components of the Sun's ultraviolet radiation (UV-A and UV-B) which would have caused major health issues if they reached the Earth. In effect, stratospheric ozone, in spite of the low concentrations acts as an effective shield, protecting the Earth from radiation damage. However, human activities in particular, refrigeration and air conditioning have been releasing some chemicals, notably fluorocarbons (including chlorofluorocarbons) into the atmosphere. These compounds end up in the stratosphere and react with ozone, thereby depleting the concentration and creating gaps in the protective shield through which harmful radiation can pass and reach the Earth. Ozone depletion has been declining, largely as a result of global action in the 1990s, banning the use of fluorocarbons. However, the chemicals are still very much in use in developing countries which are the main recipients of used equipment from the developed world.

4.10   LOWER ATMOSPHERIC (AMBIENT) POLLUTION

Some pollutants emanating from human activities are also present in the troposphere, the layer of the atmosphere that is in direct contact with life on Earth. Most of these pollutants last only a few days to a few weeks in the atmosphere, hence the name: short-lived climate pollutants (SLCP). However, they contribute a substantial proportion of the additional positive radiative forcing that is believed to be causing anthropogenic climate change. Also, the effect of ambient pollution on the immediate human environment (up to 10-15 kilometers above the Earth's surface) can be serious. Apart from causing a wide range of human health issues, the smog formed over towns and cities interferes with the natural cloud systems and local weather, obscures vision and affects agricultural crop yields. Tropospheric pollution, often referred to as ambient air pollution or atmospheric aerosols, is the contamination of the indoor or outdoor environment by any chemical, physical or biological agent that modifies the natural characteristics of the atmosphere. It originates from both natural and anthropogenic sources. Natural sources include dust storms, volcanoes and forest and grassland fires, but the contribution is small compared to human activities. Heat and power generation, motor vehicles, industrial facilities (manufacturing, mining, oil production and refining, etc.), municipal and agricultural waste incineration, landfills, and household biofuel combustion for cooking and lighting, are common sources of anthropogenic ambient pollution.

Pollutants of major public health concern include particulate matter, carbon monoxide, ozone, nitrogen dioxides and sulfur dioxide. Tropospheric pollution alters the natural balance of the environment in many ways and also impacts negatively on human health. Outdoor and indoor air pollution cause respiratory and other diseases, which can be fatal (WHO, 2018). Although most emissions of ambient air pollution are from local or regional sources, under certain atmospheric conditions, ambient air pollution can travel long distances across national borders over time scales of 4-6 days, thereby affecting people far away from its original source. For example, windblown dust from desert regions of Africa, Mongolia, Central Asia and China can carry large concentrations of particulate matter, fungal spores and bacteria that impact health and air quality over long distances, across country boundaries, settling on snow in the temperate regions. Also, radioactive emissions from the Chernobyl nuclear accident in 1986 covered many countries in Europe and beyond.

Atmospheric aerosols are a complex colloidal mixture of minute solid particulate matter suspended in liquid droplets, air or another gas. Aerosols reside in the lower atmosphere (up to around 15 km) and may be visible or otherwise depending on the density and composition. There are many sources of aerosols, some are natural (background aerosols), while others are anthropogenic (ambient aerosols). Atmospheric aerosols, whether natural or anthropogenic, originate from two different pathways: emissions of primary particulate matter, and formation of secondary particulate matter from gaseous precursors. The main constituents of the atmospheric aerosol are inorganic species (such as sulphate, nitrate, ammonium, sea salt); organic species (also termed organic aerosol or OA); black carbon (BC), a distinct type of carbonaceous material formed from the incomplete combustion of fossil and biomass-based fuels under certain conditions; mineral species (mostly desert dust); and primary biological aerosol particles (PBAPs). Aerosols are always present in the atmosphere but in extremely variable concentrations, due to the very large heterogeneity in aerosol sources and their relatively short residence in the atmosphere (a few hours to a few weeks). The vast majority of aerosols are not visible to the naked eye because of their microscopic size, but they become visible as haze or smog when the concentration is sufficiently high (Boucher, 2015). The major component of atmospheric aerosol (sulphate aerosols) emanates from the burning of fossil fuels, (in particular, coal and oil), biofuels, and biomass.

Mineral dust, sea salt, BC and PBAPs are introduced into the atmosphere as primary particles, whereas non-sea-salt sulphate, nitrate and ammonium are predominantly from secondary aerosol formation processes. Organic aerosols have both primary and secondary sources, influenced by both natural and anthropogenic events. The majority of BC, sulphate, nitrate and ammonium come from anthropogenic sources, whereas sea salt, most mineral dust and PBAPs are predominantly of natural origin. Atmospheric aerosols are largely generated locally, but can be transported a long way from their source if the winds are strong. Once they are airborne, particles can change in size and composition as a result of condensation, evaporation, chemical reaction or coagulation with other particles. Also, the effect on the environment is largely local and the most noticeable impact is the formation of haze/smog which scatters and absorbs sunlight, and obscures visibility. The health impacts are even more serious, accounting for millions of deaths annually and interfering with the weather and natural balance of the ecosystem.

Particulate matter (PM) refers to inhalable particles, composed of sulphates, nitrates, ammonia, sodium chloride, organic carbon, black carbon, mineral dust or water. The health risks associated with particulate matter are substantial and well documented. Particulate matter is also the most widely used indicator to assess the health effects from exposure to ambient air pollution. Fine particulate carbon also known as soot originates from incomplete combustion of carbonaceous fuels - fossil fuels, wood, biomass etc. The finest component, (2.5 microns and below), known as $PM_{2.5}$ is the major component of soot and the most hazardous, and the World Health Organization (WHO) has set a safe limit of 10 µg/m3 for $PM_{2.5}$ and 10 µg/m3 for PM_{10} in ambient environment. Black carbon (BC) is the most strongly light-absorbing component of $PM_{2.5}$ and indeed the most light-absorbing airborne particle in the atmosphere. It is also the most potent health hazard because it is fine enough to be inhaled. BC represents about 10% of PM mass globally but concentrations can vary widely depending on the sources, location, time of day, and atmospheric transport and loss mechanisms. For example, concentrations in urban areas could be over a hundred times of values for remote, rural areas and emissions from prolific sources such as diesel engines could be up to 80% BC (Bond et al., 2007; 2013). Other components of outdoor pollution are brown carbon, volatile organic carbon, sulphur dioxide, nitrous oxide, methane, ozone and carbon dioxide.

Aerosols are products of natural phenomena as well as human activities, in particular, fossil fuel use. Those that are the results of direct emission (black carbon, organic carbon, sea salt, dust) are often grouped as primary aerosols. However, aerosols also get into the atmosphere as products of chemical reactions (secondary aerosols). Secondary aerosols are in two main groups: secondary inorganic aerosols (SIAs) - sulphate, nitrate, ammonium - and secondary organic aerosols (SOAs). Secondary inorganic aerosols are products of reactions involving sulphur dioxide, ammonic and nitric oxide emissions while SOAs are the results of chemical reactions of non-methane hydrocarbons and their products with the hydroxyl radical (OH), ozone (O_3), or nitrate (NO_3). Although many hydrocarbons in the atmosphere are of biogenic origin, anthropogenic pollutants are believed to act as catalysts that promote their conversion into SOAs. Aerosols containing black carbon, organic carbon and sulfates as well as methane (CH_4) and ozone (O_3) do not last long in the atmosphere, often only a few days, hence they are commonly identified as short-lived climate forcers (SLCF) or short-lived climate pollutants (SLCP). However, in spite of the short atmospheric lifetime, black carbon is one of the largest contributors to global warming, surpassed only by carbon dioxide. Aerosols are often classified into two categories depending on the source - natural (volcanic, desert dust), and man-made aerosols. Human activities enhance the greenhouse effect directly by emitting GHGs such as CO_2, CH_4, N_2O and chlorofluorocarbons (CFCs). In addition, pollutants such as carbon monoxide (CO), volatile organic compounds (VOC), nitrogen oxides (NO_x) and sulphur dioxide (SO_2), which themselves are negligible GHGs, have an indirect effect on the greenhouse effect by altering, through atmospheric chemical reactions, the abundance of important gases and the amount of reactive gases such as CH_4 and ozone (O_3), and/or by acting as precursors of secondary aerosols. Clouds affect the climate system in a variety of ways: they produce precipitation (rain and snow) that is necessary for most life on land; they warm the atmosphere as water vapor condenses. Clouds strongly affect the flows of both sunlight (warming the planet) and infrared light (cooling the planet as it is radiated to space) through the atmosphere. Clouds also contain powerful updraughts that can rapidly carry air from near the surface to great heights. The updraughts carry energy, moisture, momentum, trace gases, and aerosol particles. Ambient pollution can alter the natural cloud system significantly, thereby contributing to weather variability, and, eventually to climate change.

4.10.1   Natural aerosols

There are many natural sources of atmospheric aerosols but the most important are volcanic eruptions and desert dust (Figure 4.19). Volcanoes are a result of internal pressures in the molten core of the Earths crust, often strong enough to force the flow of molten lava, volcanic ash and hot gases to the surface through gaps or cracks in the Earth's seventeen major tectonic plates that seal off the hot, molten core, and have always been a natural feature of global natural phenomena. Desert dust is a normal feature of desert areas of the world, although the intensity and mobility depend on wind activities. When winds are strong, desert dust can travel over thousands of kilometers.

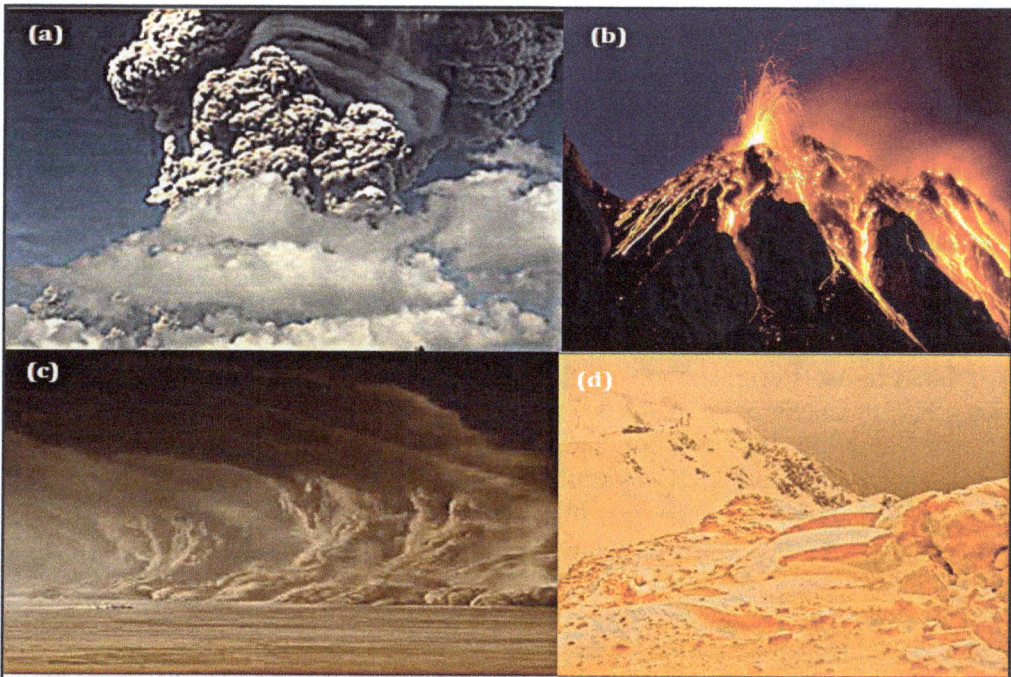

Figure 4.19 (a) Volcanic eruption in Karo, Sumatra Island, 2018; (b) Hawai eruption, 2018, (c) Desert dust (d) Dust from Sahara desert turns snow in Eastern Europe orange *(watchers.news; outerplaces.com; scientificamerican.com; bbc.com).*

4.10.1.1 *Volcanic dust*

Volcanoes are common, and particulate dust, hot lava, gases and chemical compounds are ejected into the atmosphere (troposphere) after a major eruption. While most of the coarse particles fall back on the Earth's surface within a few days, strong winds can carry fine particles into the stratosphere (20-50 km above the Earth's surface), and these compounds may remain there for up to two years. Sulphur dioxide gas in volcanic aerosols reacts with moisture to form clouds of sulphuric acid droplets in the stratosphere within a few days to several months after the eruption. Winds in the stratosphere spread the aerosols all over the globe in form of clouds which reflect sunlight back to space, preventing its energy from reaching the Earth's surface, thus cooling it. These middle atmospheric sulphuric acid clouds also locally absorb energy from the Sun, the Earth and the lower atmosphere, thus heating up the stratosphere and impacting on the weather. Although volcanic eruptions inject both mineral particles and sulphate aerosol precursors into the atmosphere, it is the latter, because of their small size and long lifetimes, that are responsible for RF that impacts on climate.

4.10.1.2 *Desert dust aerosol*

Desert dust plumes comprise minute grains of dirt and dust suspended in strong winds which originate mostly from the deserts of North Africa. Desert dust particles remain mainly in the troposphere and normally fall out of the atmosphere after a short flight. However, intense, strong dust storms can blow them to high altitudes (5,000 meters or higher) and over thousands of kilometers (Figure 4.19 c&d). In such situations, the dust particles which contain minerals, absorb sunlight as well as scatter it. When they absorb sunlight, the layer of the atmosphere where they reside becomes warmer and interferes with the formation of storm clouds which control rainfall, with consequent desertification (NASA, 2018).

4.10.2 Anthropogenic aerosols

There are two main sources of man-made aerosols: ambient and household aerosols. Both comprise smoke, (a product of incomplete combustion of carbon), and many other potentially harmful chemical compounds such as black carbon, methane and carbon monoxide. Technically, household pollution is part of ambient pollution, but the intensity is higher because it occurs in confined space and the impact on human health is more serious.

4.10.2.1 *Ambient aerosols*

Ambient aerosols originate mostly from many human activities, in form of chemicals-laden smoke. Ambient air pollution occurs mainly as a result of incomplete combustion of fuels and main sources are inefficient automobiles, heat and power generation, bush fires, industrial facilities (manufacturing, mining, oil and gas production), burning oil spills, and household fuel use (Figure 4.20). Complete combustion of carbon, the main element in fossil fuels should yield only carbon dioxide, but combustion is never complete and, apart from CO_2eq, other potentially hazardous compounds are formed - carbon monoxide, (CO), volatile organic compounds (VOC), organic carbon (OC) particles, and black carbon (BC). While a sizeable proportion comes from burning vegetation, the major component comes in the form of sulphate aerosols created by the burning of fossil fuels. Sulphate aerosols survive in the atmosphere for only a few days, black carbon a few weeks, and methane over ten years. The sulfate aerosols absorb no sunlight but they reflect it, thereby reducing the amount of sunlight reaching the Earth's surface. Black carbon is a major component of soot and the most potent form of carbon pollution. There are many sources of black carbon but the most prolific are all forms of fossil and biofuel combustion in mining, building and construction, industrial processes, and household heating, cooking and lighting (Figure 4.21).

Figure 4.20 (a) Gulf burning oil spill, 2010 (b) Coal-fired power plant (c) Traffic pollution (d) (e) A brickmaking factory in Mexico (f) Wild forest fire *(Sources: cnn.com; timesofindia.indiatimes.com; azocleantec.com; cnn.com; livebunkers.com).*

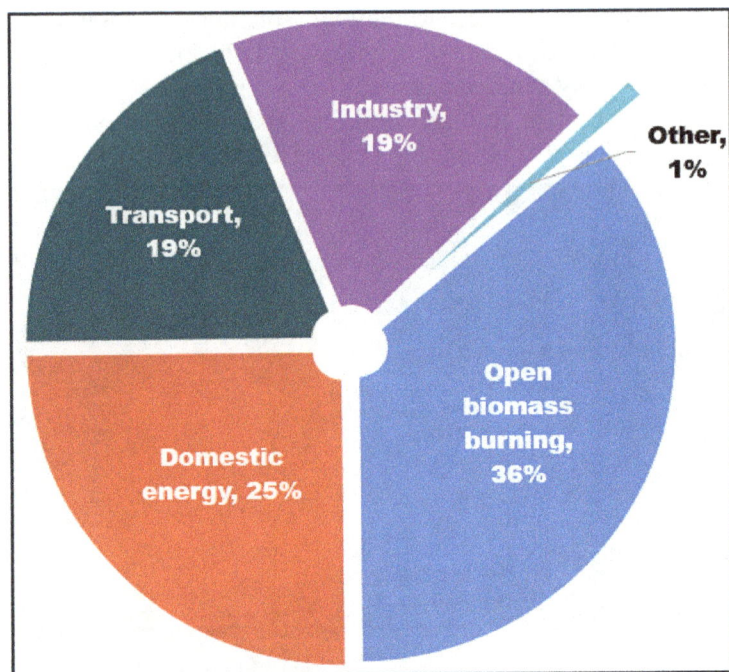

Figure 4.21 Main sources of global black carbon *(UNEP, 2015, Lamarque et al., 2010).*

Black carbon is often co-emitted with long-lived GHGs (carbon dioxide, methane), as well as other short-lived pollutants, including organic carbon (OC), brown carbon (BrC), nitrogen oxides (NOx), non-methane volatile organic compounds (NMVOC), carbon monoxide (CO) and sulphur dioxide (SO_2) (Anenberg, et al., 2012). It is estimated that three regions contribute around 75% of global BC emissions - Asia (40%), Africa (23%), and Latin America (12%) (UNEP, 2015). Major sources of BC vary by source, region and country. The main sources in developing countries are open biomass burning, residential solid fuel combustion, traditional charcoal and brick kilns, whereas transport dominates in Europe and North America, although solid-fuel household heating is also a significant source because many households in North America and Europe still use open wood and coal-burning stoves for home heating (WHO, 2015). Around 8% of British homes still use wood-burning stoves, accounting for nearly 40% of the country's particulate matter pollution. In East Asia, most emissions stem from industrial use of coal for power generations and primary materials production, while domestic energy use is the main source in Africa. Black carbon is mainly a local/regional pollutant, very heterogeneous and variable in composition and spatial distribution. Also, its behavior in the atmosphere is subject to many variables, including wind pattern, season of emission, the moderating effects of oceans and physical obstructions, etc. The net impact of any individual emission source on weather and climate depends on its composition, primarily the relative amounts of co-emitted warming and cooling pollutants.

Black/brown carbon warms the environment (positive Radiative Forcing, RF) while co-emitted organic pollutants (organic carbon and sulphur dioxide, etc.) have a cooling effect (negative RF). The net effect on climate depends on the balance between these to opposing forces. The heavier particles will settle on plants and land surfaces while the fine particles remain suspended in the atmosphere, mixing with clouds and often forming dense haze or smog. The most health-damaging particles are those with a diameter of 10 microns or less, ($\leq$ PM_{10}), also known as aerosols.

There is ample evidence that the use of manures and fertilizers and animal husbandry are also significant sources of $PM_{2.5}$ and emissions can be very high particularly in agricultural countries of Europe and North America. Ammonia (NH_3) released from agricultural processes reacts with other compounds in the atmosphere, in particular, sulphuric and nitric acid to form fine particles of ammonium sulphate and nitrate salts. It is estimated that a reduction of agriculture-related ammonia by 50% would reduce $PM_{2.5}$ particulates by 20-30% and deaths attributable to air pollution would drop by around 8%, equivalent to about 250,000 deaths per year.

4.10.2.2   *Household aerosols*

On the average, people spend around 90% of their day indoors, hence, the air they breathe is vital for their health and well being, yet, according to the World Health Organization, over 40% of the global population spend their time in polluted indoor environment. Household pollution (also known as household aerosols) are in fact part of ambient aerosols but much more localized because they are mostly generated in confined an poorly-ventilated spaces. Around 3 billion people in across all regions of the world cook and heat their homes using polluting fuels - wood, coal, charcoal, agro-waste, dung, kerosene, mostly in crude, inefficient, smoky stoves, open hearths, lamps and space heaters which emit smoke laden with fine particulate matter that penetrate deep into the lungs, and health damaging chemical compounds. In many developing countries it is not uncommon for small portable electric generators to be operated in confined, poorly ventilated space, causing many deaths annually. The most dangerous constituents are tar, carbon monoxide, a wide range of other chemical compounds, and fine particulate matter. The home is usually poorly ventilated and both adults and children are exposed to highly toxic environment. Other indoor air pollutants include mold, building materials, home products, volatile organic compounds (VOCs) and naturally occurring gases like radon. Most household pollutants pose serious health risks, exacerbated by poor ventilation.

In poorly ventilated dwellings, PM in indoor smoke can be 100 times higher than acceptable levels set by the World Health Organization (WHO, 2018), and women and children are the most exposed because they spend the most time indoors or around the inefficient burners (Figure 4.22). Black carbon is perhaps the most deleterious of household emissions and, while annual average concentrations range between <0.1 $\mu m/m^3$ in remote locations to approximately 0.1 $\mu m/m^3$ in urban areas, household indoor levels could be as high as 30 $\mu m/m^3$. Household pollution has been associated with many non-communicable diseases, including stroke, ischaemic heart disease, chronic obstructive pulmonary disease (COPD), and lung cancer. The World Health Organization (2018) estimates that close to 4 million people die prematurely every year from illnesses attributable to household air pollution from inefficient cooking and home heating practices, in addition to the 3 million or so who die as a result of ambient aerosols. Also, close to half of deaths due to pneumonia among children under 5 years of age are caused by particulate matter (soot) inhaled from household air pollution.

4.10.2.3   *Ambient ozone*

Natural ozone resides mainly in the stratosphere but also occurs in small concentrations in the lower atmosphere (troposphere), originating from the stratosphere. This is often referred to as ground-level, tropospheric, or good ozone. However human activities that release nitrogen oxides, hydrocarbons and volatile organics into the atmosphere cause complex photochemical reactions, powered by sunlight, to produce ozone, thereby increasing the concentration of the gas in the lower atmospheric layer.

Figure 4.22 Household pollution *(Sources: (a) kopernik.info (b)   google images (c)   who.int (d) mvccolumbia.co).*

Ozone is produced when carbon monoxide (CO), methane, or other volatile organic compounds (VOCs) are oxidized in the presence of nitrogen oxides (NOx) and sunlight. This type of ozone is often referred to as secondary or synthetic ozone because it is not emitted directly but produced as a result of chemical reactions. Ozone is also produced from oxygen for many applications including removal of anthropogenic nitrogen oxides from fossil fuel combustion systems, water purification, and many other industrial and agricultural applications, and is also a byproduct of fossil fuel combustion. Much of this synthetic ozone also known as anthropogenic or bad/dirty ozone is released into the atmosphere and is a potent component of photochemical smog, a key health risk. The distribution of tropospheric ozone is highly variable between regions, with the highest concentration in the northern hemisphere, believed to have increased threefold in the last hundred years. Tropospheric ozone is short-lived, with a lifetime of only a few weeks, hence its distribution is highly variable by season, altitude and location, but even in its short lifetime, the gas exhibits an estimated radiative forcing of 0.40

± 0.2 Wm/m^2. Tropospheric ozone is a major component of urban smog and a potent health hazard. When inhaled, ozone can damage human lungs and relatively low doses cause lung and eye irritation, chest pain, coughing, and shortness of breath. It also causes significant damage to crops and vegetation. Several chemical compounds in tropospheric pollution act as ozone precursors: CO, VOCs and NOx, although they are also potent anthropogenic greenhouse gases and injurious to human health.

4.10.2.4 *Ambient NOx, SOx and CO aerosols*

Nitrogen dioxide emanates mainly from power generation, internal combustion engines, particularly diesel vehicles, and industrial processes. It is a very important constituent of particulate carbon and there is considerable medical evidence that, independently, it can cause severe respiratory problems. Sulphur dioxide comes mainly from the burning of coal and oil, but also from the smelting of mineral ores that contain sulphur. The gas reacts with water to form corrosive acid, the main component of acid rain, and can affect human eyes and respiratory system. Carbon monoxide is produced when oxygen supply is inadequate for the complete combustion of carbon, and this occurs in virtually all fossil and biofuel combustion processes. The gas is poisonous at both low and high concentrations, it impairs oxygen transport in the blood stream and can be fatal if the concentration is high. Even at low concentrations, the gas has been associated with a wide range of health problems.

4.10.2.5 *Solid waste aerosols*

The global economy is rising and access to modern materials and electronic gadgets is growing exponentially. As a result of rising prosperity and rapid technological advances that are driving innovation and rapid obsolescence, people can afford to upgrade frequently and this is leading to the dumping of end-of-life electronic waste (e-waste). Electronic waste is produced in staggering quantities, estimated at about 45 million metric tons globally in 2016, and projected to increase to around 52 million metric tons in 2021 (ITU, 2017). Asia accounted for the largest share (18.2 mmt), followed by Europe (12.3 mmt), the Americas (11.3 mmt), Africa (2.2 mmt), and Oceania (0.7 mmt). Electronic waste contains many valuable metals such as gold, copper, silicon, but also contains potentially hazardous chemicals and materials - lead, cadmium, chromium, plastic, brominated flame retardants, polychlorinated biphenyls (PCBs), etc. Few countries have formal collection and recycling policies in place and as little as around 20% of e-waste was formally recycled in controlled environment in 2016, with Europe and the Americas accounting for 35% and 15% respectively, followed by Asia (15%). However, the majority of e-waste collected in the developed countries is shipped to developing countries for dumping or recycling by an unregulated informal sector using primitive recycling techniques such as burning cables in poorly ventilated homes to recover copper. This results in a significant risk of release of toxic aerosols into the atmosphere, and exposure of recyclers who are often women and children to hazardous environment (WHO, 2018; Grant *et al*. 2013; Noel-Brune, 2013). Also, e-waste used as landfill may eventually contaminate soil, agro-products and water.

Polymers are products of petrochemicals and, since invention in the 1950s, have taken over virtually every aspect of human life, from packaging to clothing. One of the most attractive properties is that they are unreactive, which makes them suitable for food and chemicals storage. Unfortunately, this property also makes it difficult to dispose of polymer waste. Most polymers are not biodegradable and can remain intact for several hundred years in landfills. When incinerated, polymers release a lot of energy, black smoke, carbon dioxide, and black carbon aerosols into the atmosphere (Figure 4.23). Furthermore, the bulk of plastic waste from the developed countries is being exported to emerging nations where they are sorted manually and recycled or incinerated, usually in unregulated environment, thus releasing potentially dangerous aerosols into the atmosphere and contributing to ambient pollution. Fossil fuels (oil, natural gas, coal) are precursors to polymers and enormous energy (mostly supplied by fossil fuels) is required for production.

Figure 4.23 (a) Burning polymer waste (b) used plastic dump sites around the world *(dailymail.co.uk; theguardian.com).*

Recycling plastic waste would save a lot of raw materials and energy use thereby reducing associated pollution. Unfortunately, different polymer wastes have to be separated from each other, and this can be difficult and expensive. Also, most recycled polymers cannot be processed into the same products. Some recycling processes freeze plastic waste and process into granules for use in playgrounds, or reprocessing into useful products like floor mats. Also, some polymer biodegradable products such as carrier bags and refuse bags now contain chemicals that cause polymer to break down within a few weeks. The fact that plastic waste needs sorting and washing before recycling makes the business unattractive in most developed nations, hence they are exported to emerging nations for processing, encouraged by local policies which favor waste exporters and do little to promote local recycling. While these nations may succeed in minimizing local pollution, stratospheric greenhouse gas pollution which is mostly responsible for global warming and climate change is largely independent of source of pollution. Many polymers become brittle on exposure to the Sun's ultraviolet rays over time, and tend to crumble and pieces less than five millimeters in size are classified as microplastics. Some microplastics are also manufactured, such as microbeads added to health and beauty products.

Polymer bottles are now a first choice for bottled water and soft drinks, and the bottled water/soft drinks market is now the fastest growing drinks market in the world, fueled by the current 'on the go' lifestyle which has spread across all regions of the world. More than 480 billion of plastic bottled drinks were sold worldwide in 2016, and projections show an increase of nearly 40% in 2021. China alone accounted for a quarter of the total global consumption.

Most plastic bottles for bottled water and soft drinks are produced from virgin petroleum feedstock (Polyethylene terephthalate, PET), and it is estimated that 162 grams of oil and seven liters of water are required to manufacture a one-liter disposable PET bottle, with a release of 100 grams of carbon dioxide into the environment (NUETC, 2010). This carbon footprint is comparable to that of driving a car for one kilometer. Fewer than half of the bottles sold are recycled and only about 7% is processed into new bottles, the balance ending up in landfill or in the oceans. It is estimated that between 5 and 13 million metric tons of plastic waste end up in the oceans every year, mostly from plastic bottles dropped on beaches or plastic waste dumped in rivers and oceans. A recent research by World Economic Forum and Ellen Macarthur Foundation (2016) projects that the oceans will contain more plastic by weight than fish by 2050.

Evidence abounds that micropolymers are being ingested by sea birds, fish and other sea animals, and some of it is already finding its way into the human food chain. The results of a recent study reported the presence of microplastic in fish caught on the coast of the United Kingdom, including cod, haddock, mackerel, whale, sharks, shellfish (Thompson *et al.* 2009), and studies in other parts of the world have reported similar results. Many plastics contain toxic chemicals and the danger to humans who consume seafood is real. Whales and sharks are particularly vulnerable since much of the microplastics floats on ocean surfaces from where these sea creatures feed mostly on small preys by straining them out of ocean water. In the process, they swallow hundreds to thousands of cubic meters of surface water daily, thereby exposing them to ingestion of substantial amounts potentially toxic chemicals present in microplastics such as heavy metals and phthalates. A whale found dead recently off the coast of Thailand had ingested around 85 plastic bags weighing over 8 kilograms, and there have been similar occurrences in other parts of the world.

The war against plastics pollution is intensifying globally. Many countries, notably in Western Europe are introducing *refuse-reduce-reuse-recycle* (R⁴) policies which have been shown to be very effective within months. These include taxes on shopping bags, deposits on drinks bottles, and assistance with sorting and recycling of domestic waste. The European Parliament is discussing measures that could require member states to cut plastic bag use by 80% by 2019. The United States consumes around 100 billion plastic bags a year (around one per person per day), and many cities, counties and states have either banned or imposed taxes on plastic bags. One major flaw in global efforts to reduce plastics pollution is the fact that policies lack incentives for local recycling. The bulk of plastics collected in the developed world is either incinerated, used as landfill, or shipped to third world countries for processing. All these options have significant global warming and ambient pollution impacts, especially in the emerging countries where processing is often carried out by the informal sector, often in household environments. The main raw materials for plastics production are oil and natural gas and 8-10% of global annual consumption is accounted for by the petrochemicals sector which produces plastics, fertilizers, chemicals, etc. Also, demand for oil and gas feedstocks by the sub-sector will account for most of the rise in demand for primary energy by industry over the next two decades or so. It is not surprising therefore that the petroleum/petrochemicals industry is hostile to policies on plastic reduction, and some have even gone to the extent of challenging plastics reduction ordinances in court. It is also not surprising that projections published recently by oil companies indicate that recycling will have an insignificant impact on global oil demand. However, the reality is that a significant reduction in plastic use will have multi-dimensional mitigation impact on public health, the environment, and climate change.

Chapter 5

Outlook on energy use
and related emissions

5.1 INTRODUCTION

Many independent and partisan organizations publish outlooks over two to three decades on global energy demand and use, based on historical antecedents, and various assumptions on the future pathways of the main determinants of energy demand: global economy, population growth and demographics, global energy market dynamics, emerging technological innovations, urbanization, emerging policies, etc., all of which are subject to significant uncertainties. The energy industry is in transition, shaped by growing global economy and population, rising prosperity and urbanization, new technologies, government policies, social preferences, challenges of producing more energy with less carbon emissions, and the net effect on future energy demand and consumption is impossible to predict today. This is evident from significant changes which are often made in projections and forecasts from year to year by the same organizations, and the evolving strategies of exploring different potential scenarios. Nevertheless, even though there may be differences in specifics, there is general consensus on the likely growth pathways of global energy demand and energy-related emissions. This chapter presents a summary of in-depth evaluation of projections and related analyses published by many independent and partisan organizations in the last five years, with a view to identifying the most likely scenario of the global energy demand, utilization and growth of associated emissions over the next two to three decades.

5.2 THE GLOBAL PRIMARY ENERGY DEMAND OUTLOOK

Many variables none of which can be predicted accurately will shape the global energy demand outlook: economic and population growth dynamics, prices of options, potential investments, new technologies, national priorities and policies, changing dynamics of fuel mix, changing consumer preferences etc., hence any energy projection should be regarded as 'intelligent speculation based on currently available facts' and, indeed, some projections have had to be revised within a year. Nevertheless, it is useful to identify the likely future direction in the context of trends over the last two decades, and various known and anticipated technological and policy innovations that could impact on global trends in energy supply and demand. There are many annual projections by respected independent organizations as well as energy companies on primary and end-use energy consumption and associated emissions. Most of these publications in the last five years or so have been evaluated critically and the projected trends are very similar. The main sources are listed in Box 5.1 and a composite outlook on energy supply, consumption and pollution emissions is discussed in the following sections of this chapter.

5.2.1 Fossil fuels

Global primary energy demand will grow about 25-30% by 2040, driven largely by the growing economy which is projected to double over the same period. Shift in world demographics, another major driver of energy use, over the period will also be significant, with a projected 25% growth to around 9.1 billion by 2040. Rising prosperity and urbanization should push up energy demand at roughly the same rate as the economic growth, but emerging, more efficient technologies, particularly those that impact on energy production and use have played a major role in the gradual de-coupling of economic and energy demand growth over the last decade and the mitigation is expected to continue at an even faster rate. Without this development, energy demand would probably more than double by 2040. Projected improvements in efficiency across the energy chain from production to utilization should fully de-couple economic growth and energy demand over the next two or three decades.

BOX 5.1   SOURCES OF ENERGY OUTLOOK DATA

- Energy Information Administration (EIA). *International Energy Outlook, 2017.*
- Energy Information Administration (EIA). *International Energy Outlook, 2018.*
- International Atomic Energy Agency (IAEA). *International Status and Prospects of Nuclear Power 2017.* International Atomic Energy Agency.
- International Energy Agency (IEA). 2013. *Resources to Reserves.*
- International Energy Agency (IEA). *The Power of Transformation 2014.*
- International Energy Agency (IEA). *Energy and Air Pollution 2016.*
- International Energy Agency (IEA*). CO$_2$ Emissions from Fuel Combustion 2017, 2018.*
- International Energy Agency (IEA*). Global EV Outlook 2017, 2018.*
- International Energy Agency (IEA). *World Energy Statistics 2017, 2018.*
- International Energy Agency (IEA). *World Energy Balances 2017.*
- International Energy Agency (IEA). Key World Statistics 2017, 2018.
- International Energy Agency (IEA). *World Energy Outlook 2017, 2018.*
- International Energy Agency (IEA). *Energy Access 2017.*
- International Energy Agency Energy (IEA). *Energy Technology Perspectives 2017.*
- International Energy Agency Energy (IEA) Energy Technology Roadmap: Bioenergy 2017.
- International Energy Agency Energy (IEA). *Energy Efficiency 2017.*
- International Energy Agency (IEA, 2017). *Future Scenarios for Energy Efficiency.*
- International Energy Agency Energy (IEA). *Emissions from fuel combustion highlights 2017*
- International Energy Agency Energy (IEA). *Renewables 2017, 2018.*
- International Energy Agency(IEA). *Global Energy and CO$_2$ status 2017, 2018.*
- International Energy Agency (2017). Sustainable Development Goal 7: Ensure access to affordable, reliable, sustainable and modern energy for all.
- International Energy Agency (IEA*). Electricity Information 2018.*
- International Energy Agency (2018) World Energy Outlook 2018
- International Energy Agency (2019). Global EV Outlook
- International Hydropower Agency (2017). *Briefing: 2016 Key Trends in hydropower.*
- World Bank, (2017). *State of Electricity Access Report 2017.* worldbank.org.
- World Energy Organization (WEO). *Energy World, Energy Outlook 2016.*
- World Energy Organization (WEO). *World-Energy-Scenarios 2016.*
- World Energy Organization (WEO). *World Energy Focus 2017.*
- International Renewable Energy Agency (IRENA). *Rethinking Energy 2017.*
- International Renewable Energy Agency (IRENA). "Renewable power generation costs in 2018." .
- ExxonMobil. *Energy Outlook 2018, 2019.*
- BP. *Statistical Review of World Energy 2017, 2018.*
- BP. *Energy Outlook 2018, 2019.*
- UN (2019) World Population Prospects 2019. Population Division.

Emerging nations (non-OECD) will account for much of the growth in both the economy and population, and for about 70% of the global total primary demand, with China alone accounting for about 20%. Increasing urbanization is projected to continue worldwide, with around 2 billion additional people likely to live in urban centers by 2040. Much of this urbanization will occur in Africa where the urban population is projected to grow by around 600 million, about a third of the global increase. However, the impact on the region's energy consumption and intensity will depend on the extent to which this urbanization facilitates increased levels of industrialization and prosperity. The global trend towards rising income and population growth greatly stimulates energy demand rise, but that growth is moderated significantly by increasing energy use efficiency as indicated by energy intensity (energy consumed per dollar GDP) (Figure 5.1). The expanding middle class in Asia will also make a major impact on energy demand and consumption over the next two decades or so.

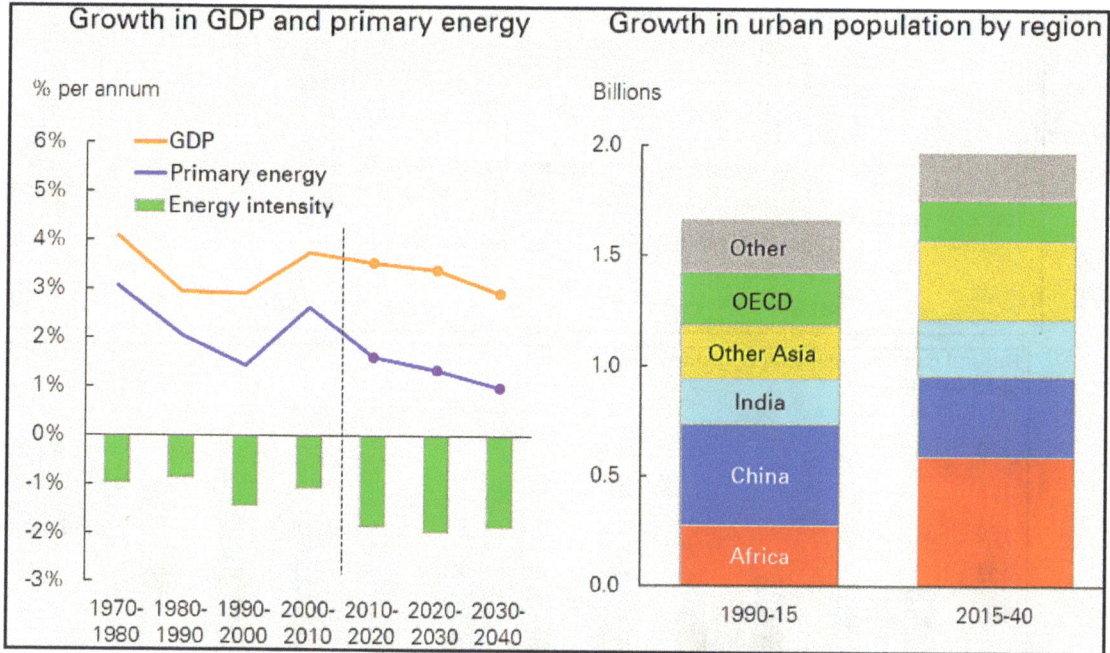

Figure 5.1   Projected growth in GDP, population and primary energy *(BP, 2018).*

Recent technology developments particularly in production of unconventional primary energy (shale, bitumen, coalbed methane) and liquefied natural gas production are changing radically the pattern of global energy supply and diversifying trade in fossil fuels. All recent projections show that fossil fuels will continue to provide the bulk of primary energy demand in 2040, accounting for around 78% compared with about 82% currently (Figure 5.2). However, there will be a significant shift in mix, with natural gas and renewable energy (in particular, hydro, solar and wind) playing a more prominent role in power generation at the expense of coal.

5.2 .1.1 *Oil*

Oil has been and will remain the world's leading primary energy resource, and global demand will rise steadily by about 20% in 2040 compared with today's consumption level. Around two-thirds of the oil is used by the transportation sector (road, 49.3%; aviation, 7.8%; navigation, 6.7%; rail, 1.7% in 2016) but non-energy use, in particular, the petrochemicals industry that converts oil and gas to high-premium chemicals, fertilizers, and polymers, is rising (16.6% in 2016). The transport sector consumes most of the world's liquid fuels and the share will remain flat over the next two decades or so. Transport accounts for two thirds of the projected growth in liquid fuel demand, shared almost equally by all its sub-sectors – cars, trucks, ships, trains, planes. Emerging economies will account for most of the growth as a result of the projected strong economic, population, and transportation growth in the region.

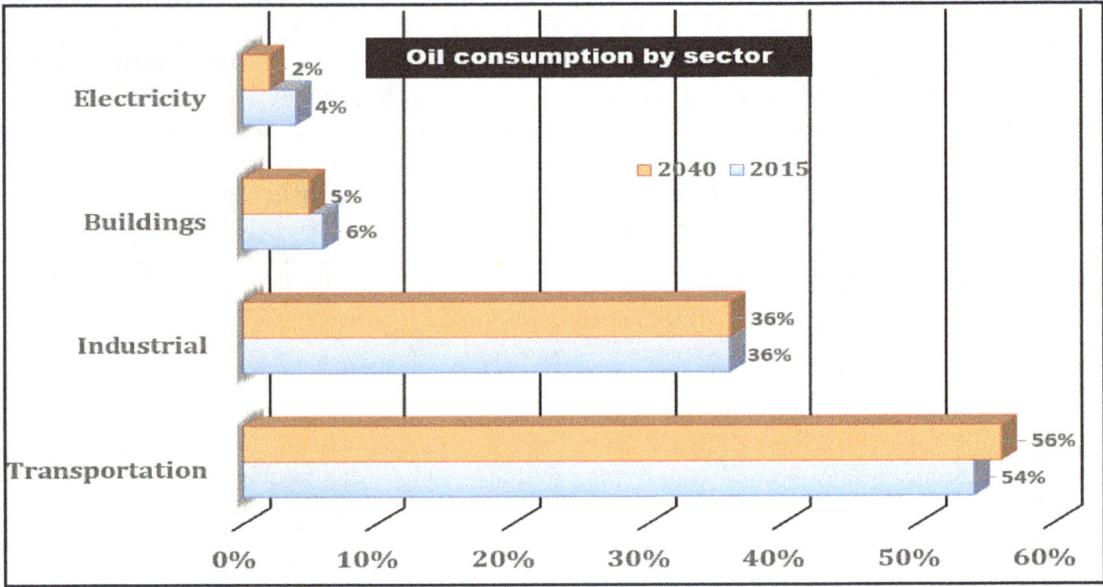

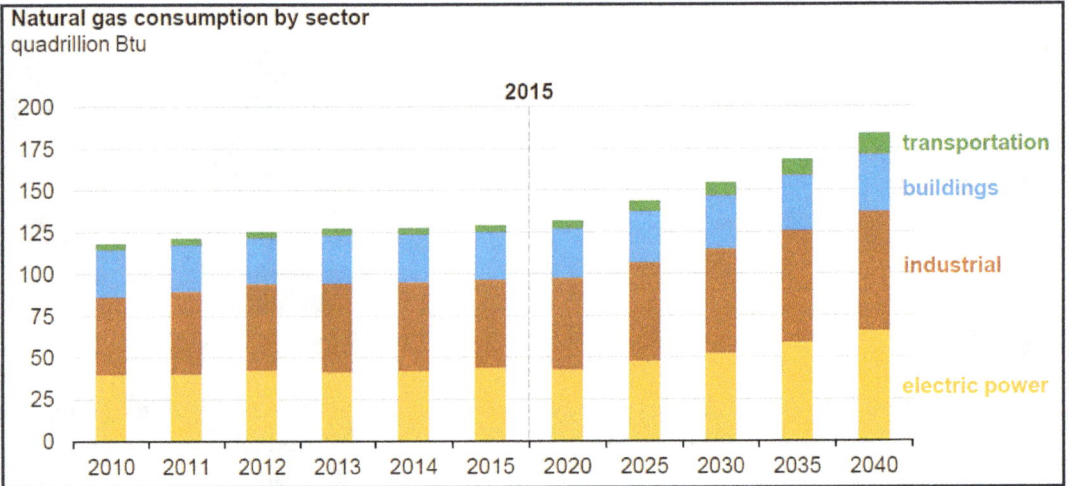

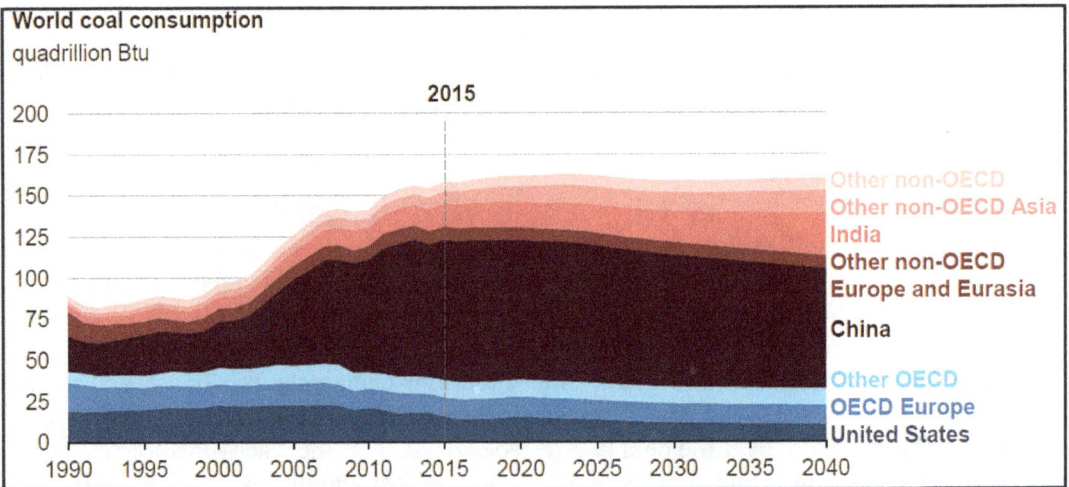

Figure 5.2 Outlook on fossil fuel use (oil, natural gas and coal) *(EIA, 2017).*

While overall demand is projected to rise by nearly 40% in the region, overall consumption in OECD countries will actually fall by around 3% due to increasing transportation energy use efficiency. Much of the growth in the global transportation energy demand will be accounted for by Asia (China, 36%, India, 142% growth from 2015 level). In spite of the anticipated growth in global electric vehicle population, oil will continue to dominate transportation in the foreseeable future, simply because there is no viable option. Electricity will account for less than 2% of global transportation energy in 2040, and natural gas (used mostly by heavy trucks and buses) will also account for only 2-4%.

5.2.1.2 *Natural gas*

Natural gas is gaining increasing prominence in the global primary energy scene across all regions of the world, primarily because of its lower carbon foot print compared with coal, but also due to relative ease of production and transportation by pipeline or shipment in liquefied form. All reviewed projections show that gas consumption worldwide will grow by 43-45% over the next twenty years or so, both in OECD and non-OECD countries, but more than twice as fast in the latter, because of the expanding industrial sectors and electricity demand. The region will account for around 60% of the total global demand on natural gas in 2040. All regions are finding natural gas increasingly attractive for the power and industrial sectors, both of which will account for about 75% of the projected increase in total consumption by 2040. This has greatly stimulated extensive investments in the development of international gas pipeline networks and liquefied natural gas production and transportation.

Natural gas when available at competitive prices is the first choice for many new power plants all over the world because of its low capital costs, favorable heat and efficiency rate, and relatively low fuel cost. Also, demand is growing in the industrial sector, in particular, chemicals, refining and primary metals production. Natural gas use by heavy trucks, buses and ships has been rising in recent years and demand is expected to increase by around 43% from 2015 to 2040, be it from a very low starting base. Developments in shale oil production (hydraulic fracturing) technologies have opened up access to previously unrecoverable shale gas, led by the United States and China which will account for 20% and 18% of the projected growth respectively. Around 50% of the natural gas output by the United States already comes from shale and tight gas deposits, moving the country to the world's largest producer of oil and gas, and projections show that these sources will account for around 70% in 2040. China is also aggressively developing shale resources which could account for around 50% of the country's natural gas output in 2040. Many other countries also have shale gas deposits which are under development, in particular, Canada which is expected to source most of its future gas output from the rich deposits from several regions in British Columbia and Alberta. Production of conventional natural gas will also grow due to increased output by Russia and the expansion of its international pipeline networks. Demand for liquefied natural gas (LNG) is growing in every region particularly in countries that have no deposits and are not near any natural gas pipeline networks. World LNG trade has increased by about 10% per year in the last two years and stood at 350 million metric tons in 2018, projected at around 600 million metric tons in 2040. Qatar led five other countries (Australia, Malaysia, United States, Nigeria, and Russia), accounting for around 75% of the total global exports in 2018, but most of the fourteen remaining exporters are expanding their capacities, and many other countries are developing their resources. However, projections show that existing and planned new capacities will not be sufficient to meet the global demand for LNG in 2040 (IGU, 2019).

5.2.1.3   *Coal*

Worldwide coal use will not change very much over the next two decades or so, because projected decrease in China and the United States (the two leading global consumers of coal) will be offset by growth in India and other developing nations (Africa, the Middle East, and other non-OECD Asia). Although coal use in China has been declining over the last few years due largely to increasing use of solar and nuclear energy for power generation, but also because of transition to less energy-intensive manufacturing, overall demand is expected to continue to rise, and the country will still remain the largest consumer of coal in 2040. India's coal use has continued to grow very strongly and the country surpassed the United States as the second largest global consumer of coal in 2016 (EIA, 2017). Coal demand in India is expected to nearly double by 2040 as the country continues to install new coal-fired electricity generating plants, as well as rising demand by industry, in particular, primary metals and cement production. Despite the significant increase in coal consumption, coal's share in overall energy consumption in India is projected to decrease from 49% in 2015 to 43% in 2040, due in part to policies promoting renewable and nuclear-based generation.

Coal demand in other non-OECD Asia, the Middle East and Africa will nearly double by 2040 because many countries in the region are taking advantage of the relatively low cost of coal, sourced locally or by importation, in efforts to meet the rapidly growing demand for electricity. The use of coal for home and industrial heating is expected to decline in most regions of the world due to increased use of natural gas and biogas. Coal use has been declining in OECD countries, and the trend is expected to continue because of increasing competition from natural gas and renewables, and also because increasing energy efficiency is driving down electricity demand. A slight drop in coal demand is also expected in China due to increasing use of renewables and natural gas for power generation. However, coal use will continue to be significant in the future because of the projected rise in demand in the rest of Asia. Most of coal use worldwide is for power generation, industrial processes (iron and steel production, cement production, conversion to premium chemicals and petrochemicals), and home heating. Over the last two decades or so, coal energy-intensive production has been moving to emerging countries and the trend is expected to continue.

5.2.2   **Renewable energy**

Renewable energy is an important and growing part of the ongoing global energy transformation, and is now recognized by many countries as the first-choice for providing quick access to electricity, expanding, upgrading and modernizing power systems. Most countries in the world have established renewable energy policies and targets, and potential investors from the private sector are playing increasingly prominent roles. Solar and wind are particularly popular, accounting for most of the investments in recent years, driven largely by falling costs: the cost of wind turbines has fallen by nearly a third in just six years, and solar photovoltaic (PV) modules now cost 20% of the prevailing prices six years ago. Renewables accounted for around 14% of the total global final energy consumption evenly split between traditional biomass and modern renewables, and 24% of global power generation in 2017. Projections based on current active policies and investments around the world, show a growth to around 20 - 35% contribution to power generation by 2030. Just six countries led by China account for nearly 80% of current renewable energy use, but many other countries have active renewable energy development programs and policies. The International Renewable Energy Agency has developed an aggressive yet feasible roadmap for global energy transition that positions the world on the path to a sustainable environment by keeping the rise in average global temperatures well below 2°C, ideally to 1.5°C in the present century. This would require aggressive decarbonization of energy through major improvements in efficiency, electrification of energy services, and deployment of renewable energy which together can deliver more than 90% of needed reductions to energy-related CO_2 emissions (Figure 5.3) (IRENA, 2019a). However, this would require that the share of electricity in final energy use increases from 19% in 2016 to around 50%, around 70% of all cars, buses, two- and three-wheelers, and trucks would be powered by low-carbon electricity.

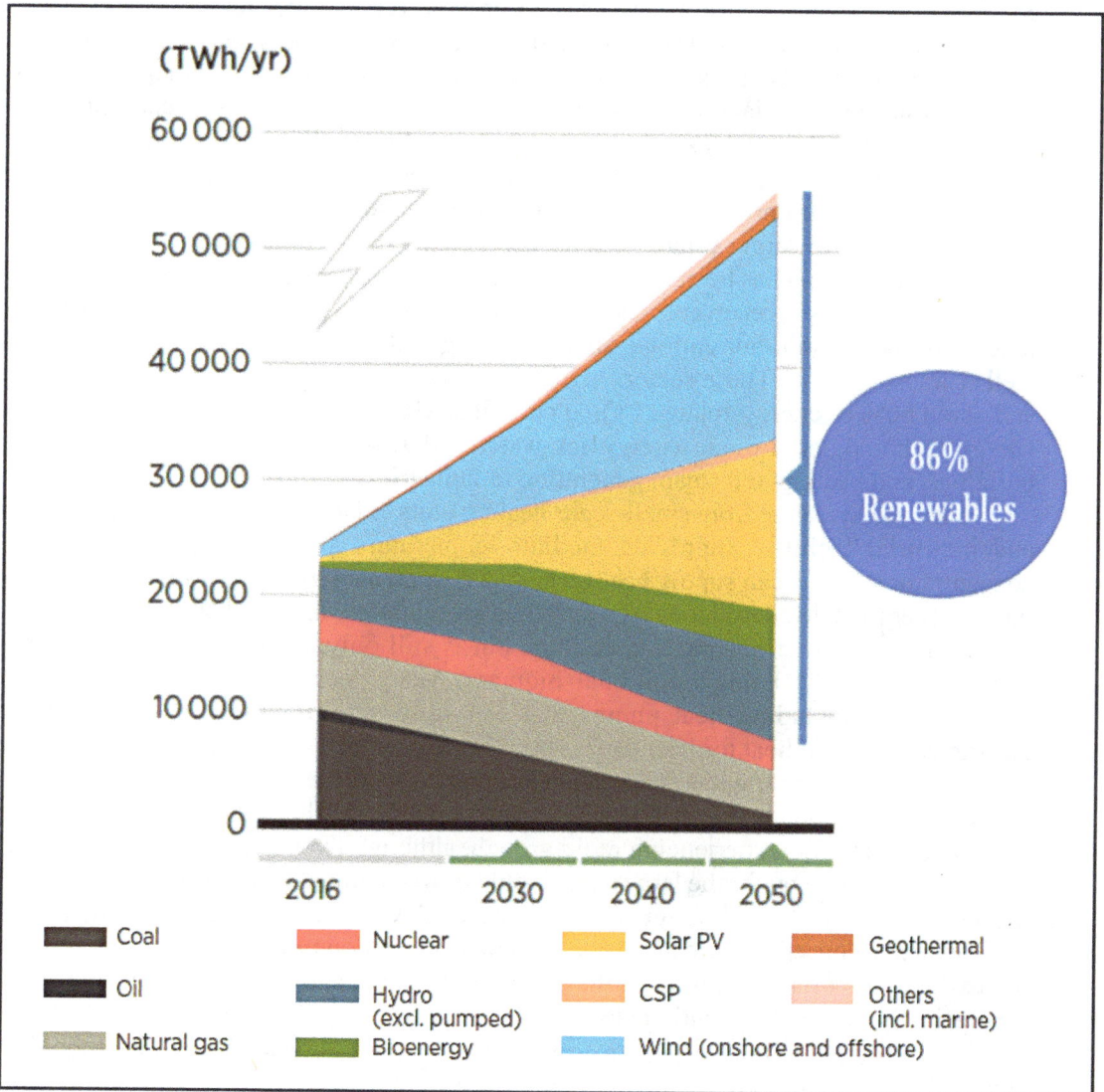

Figure 5.3 Renewable Energy Roadmap (Remap) for environmentally sustainable power generation energy mix by 2050 . *(IRENA, 2019a).*

Much of the projected growth in renewable energy for power generation will be in emerging nations, many under pressure to fast-track 'electricity for all by 2030'. Most of the expected growth in hydro-energy will be in sub-Saharan Africa, while China, United States, and Europe are leading the world in the deployment of solar and wind energy, often referred to as variable renewable energy (VRE). Solar energy is particularly suitable for deployment as utility units serving mini-grids, individual households, or small communities, making them particularly attractive in sub-Saharan Africa which will account for most of the people without access to electricity globally in 2030. One major advantage of PVs is modularity which allows for very small start-up and easy scale-up, from mini-units that fill personal needs to large-scale power plants that serve communities and vital establishments, and feed grids.

The United States leads the world in terms of VRE mini-grid capacity deployment, serving sensitive installations: hospitals, universities, industrial plants, military facilities, and small communities, driven largely by the need to provide reliable and resilient energy service to critical units in the face of extreme weather and frequent grid outages. Many states of the US and other countries also have strong policy incentives that promote installation of small units for homes and commercial establishments. The State of California recently enacted a law which requires all new homes to feature solar roof panels. Most utility projects range in size from one megawatt to several hundred megawatts, and deployment lead times vary from days to months depending on project size. Small units have battery backup to resolve the problem of intermittency, while larger establishments use standby diesel generators.

Power generation has been the primary target of most renewable energy policies, and most of the developments in renewable energy use have been in the power sector. However, extensive potentials abound in other end-use areas - heating and cooling in buildings and industry, as well as transportation. These sectors together account for a large portion of global energy use, and about 60% of energy-related CO_2eq emissions (IEA 2017a). Already, modern renewable energy technologies provide energy for water and space heating, cooling and cooking in buildings, and heating and steam generation in industries.

Solar systems range from small-scale heating units to large-scale district heating networks, and industrial facilities supplying medium-temperature (100-250°C) process heat. Solar concentrator systems can supply heat up to 400°C, high enough to raise superheated steam for power generation. Modern bioenergy accounts for by far the largest share of renewable energy use for direct modern heating and cooling, but is still only about 10% of the total biomass energy, the balance being traditional biomass. Many technologies are already available commercially, including heat pumps that can store surplus renewable energy-generated electricity in form of heat for later use in space heating and a wide range of industrial processes. The major challenge for wider deployment, in particular, those that are bioenergy-based, is the low price of alternative fossil fuels in recent years. Solar thermal and geothermal heating technologies are also experiencing rapid growth, although from a relatively low starting base. Hydroenergy is still by far the largest renewable power source globally and will remain so over the next two decades, although deployment in developed countries will reduce due to environmental concerns, but will grow in the emerging nations. Geothermal energy is becoming increasingly important in countries that have the resource and use for heat raising and power generation is expected to continue to rise.

Penetration of renewable energy in the transport sector has been relatively slow, but potential opportunities are substantial. Renewable biofuels are already in use commercially: bioethanol is being blended with conventional gasoline in some developed countries, helped by blending policy mandates; biodiesel is playing an increasing role in transportation because of its lower carbon intensity compared with oil-derived diesel; biokerosene is being blended with jet fuel for aircraft. Extensive opportunities exist for a greater use of renewable fuels in all forms of heavy transportation - aviation, marine and road freight - which together account for around 35% of transport energy demand. However, in spite of the clear potential in decarbonizing transport, investment in new biofuel production capacity has been declining, due largely to the prevalent low prices of oil in the global market. Other transportation technologies are approaching commercialization, including biomethane for heavy-duty vehicles and hydrogen produced with renewable electricity, which can be used to power fuel cells for transportation.

Linking electricity with other sectors, in particular, heating, cooling, transport (sector coupling) will help to promote more extensive use, solving the problem of variability in solar and wind energy, and integrating a rising share of variable renewable energy (VRE) into existing power systems. It will provide continuous energy for various applications, while also expanding the use of renewables in other end-use sectors. For example, a VRE system can supply grid power when availability coincides with demand, but can also use surplus electricity to power heat pumps that feed thermal grids for district heating and cooling, or individual building thermal systems. Effective control of the production and distribution systems that couple variable renewable electricity with thermal applications will require 'smart' electrical and thermal energy control systems, many of which are already available and in commercial use. Availability of affordable power storage systems will be crucial for the development and wide deployment of variable renewable energy (VRE) technologies, and there are many different energy storage technologies for different applications. These include pumped/hydro storage, compressed air energy storage, thermal storage, fuel cells, supercapacitors, batteries, etc. Pumped hydropower is the most mature which is in use worldwide and accounts for the vast majority of global electricity storage capacity estimated at 145 GW in 2015 (IHA, 2016), and projections show that capacity will more than double by 2030. In a typical installation, surplus electricity generated from any energy source during night time or idle energy from a VRE plant can be used to pump water to a high level for generation of hydropower during high-demand periods, or to cover periods when solar and wind energies are not available, and can provide 24 hours or more of backup. However, hydro-storage is only suitable for large installations.

Electrical battery storage is the most mature for small-scale applications and the demand is increasing rapidly, driven primarily by the growing market for electric vehicles (IRENA, 2016). Batteries are being deployed to store unused electricity from VREs, for later use when direct solar electricity is not available. These include household and solar PV systems, island wind systems, and off-grid VRE systems for rural electrification. Total global electric battery capacity is low currently, around 0.8 GW, but is projected to increase to around 250 GW by 2030 (IRENA 2015). There are many types of electric batteries but lithium-ion batteries are set to dominate the market because of their high energy density, efficiency, and relatively long life, followed by relatively low-cost advanced deep cycle lead-acid batteries, sodium sulphur and advanced flow batteries. Lithium-ion batteries are used widely in consumer electronics as well as plug-in electric vehicles, airplanes, ships and stationary solar PV systems. Cost of battery storage systems is expected to continue to decline rapidly due to increasing demand driven largely by the auto industry, and could become widely deployed as residential utility-scale storage in a few years. Already, in countries where grid power supply is erratic, many homes and commercial enterprises rely on deep-cycle lead-acid battery-inverter backup systems to store power when available.

5.2.3 Nuclear energy

The future of nuclear electricity has been uncertain for the last decade due to a combination of factors: declining price of natural gas; the impact of policy-driven rapid expansion of renewable resources on electricity prices; and national policies in several countries following the Fukushima Daiichi nuclear power plant accident in 2011 are all affecting the growth prospects of nuclear electricity. Japan has the world's third largest installed capacity of nuclear power (11.2% in 2013). Prior to the accident, nuclear power provided about 26% of the country's primary energy supply (IEA 2015). However, 52 of its 54 reactors were shut down after the accident, and share of nuclear in the domestic electricity generation had dropped to less than 2% in 2013. A few reactors have been reactivated and many more are being evaluated, but it would take many years to reach the pre-Fukushima level and in any case Japan is actively diversifying its power generation fuel mix to include more natural gas and renewables. Also, many other countries that have historically accounted for the majority of nuclear power development are de-commissioning old nuclear power plants and slowing down on new construction. The future prospects of nuclear power generation in Western Europe are unclear: some countries in the sub-region have either established firm policy not to build new plants or to de-commission existing plants - Belgium, Spain, Switzerland, Germany, while others, in particular France and the United Kingdom have not made clear policy statements on the future of nuclear electricity. Most of the United State's 99 operable nuclear power reactors are nearing their lifespans and four were decommissioned recently. Only one new reactor has been added in the last 20 years, although the licenses of some reactors have been extended. The United States Energy Information Administration (EIA, 2018) projects a continuing decline in nuclear power generation over the next two decades, largely because of the increasing competitiveness of natural gas.

On the contrary, the future of nuclear electricity in Asian and East European countries is bright. Starting around 2000 and driven by aggressive economic development in Asia, nuclear energy has become an important part of the energy mix in the region. Another motivating factor has been the global pressure on the region to reduce dependence on fossil fuels (mostly coal). Nuclear power offers an attractive low-carbon alternative for rapidly adding large baseload power generation required to power the aggressively increasing economies in the region with minimal greenhouse gas emissions, and most of the inputs can be sourced locally. Around 450 nuclear power plants are operating in different countries of the world, about 60 are under construction, and over 500 are planned or proposed (WNA, 2018). China is accelerating the deployment of nuclear electricity plants and nuclear capacity is growing at around 11% per annum. According to the International Atomic Energy Agency (IAEA, 2016), 40 of the 68 nuclear reactors which were under construction in 30 countries in 2015 were in China, India, Korea and Russia, and China alone accounted for 28 plants. Clearly, the cooling effect of the series of nuclear accidents on new nuclear power projects is receding.

Demand for electricity in the industrialized nations is stagnating or even declining due to increasing efficiency across the energy supply and use spectrum, and growth has been shifting from OECD to non-OECD countries. Demand is projected to surpass that of OECD within the next few years, rising to around 60% of the total global demand by 2040. This rapid growth will no doubt stimulate the development of locally available and appropriate electricity generating options, including nuclear power. It is not surprising that most of the nuclear electricity plants under construction or planned, are in emerging nations. Nuclear power is a reliable and secure energy source that can be deployed in many countries, considering the small amount of fuel required to generate enormous energy. Furthermore, the industry is largely insensitive to changing dynamics of the global oil market. Nuclear power generation is considered environmentally benign and with zero carbon emissions but this is not true when associated carbon emissions in the production of infrastructure inputs and decommissioning are taken into account.

The energy intensity of nuclear fusion, that is, the amount of energy produced per unit weight of fuel is very large, around 10,000 to 20,000 times larger than in fossil fuel combustion. This is one of the major attractions because countries which have no uranium resources can easily import small quantities to generate very large amounts of energy. Furthermore, fuel accounts for only about 5% of the total generating costs, which makes it the lowest-cost baseload electricity supply option in many countries. A typical, large 1000 MWe nuclear power plant requires about 75 metric tons of low-enriched ore in its core at start-up, with annual top up of only 25-27 metric tons of fresh enriched fuel each year. It means that nuclear power plants can maintain continuous operation for decades without worrying too much about turbulence in the global primary energy market. The recent shift away from nuclear power generation as a result of accidents appears to be in reverse and investment is picking up again, with nuclear energy accounting for 2% of the growth of total primary energy demand in 2017. There are sufficient uranium reserves to power the current level of capacity for over a hundred years and resources are believed to be at least ten times the known reserves. Moreover, the design of reactors in terms of safety and efficiency has improved tremendously in recent years and commercialization of fast breeder reactors which have been under development for many years could sustain nuclear power production indefinitely.

Around thirty (newcomer) states are considering, planning or starting nuclear power programs, while twenty more have expressed interest, and, with the support of the International Atomic Energy Agency (IAEA), are working with technical partners in planning infrastructure development. However, it is unclear whether these expected new additions will be sufficient to stall or reverse the downward trend in the global nuclear electricity generation. More than half of the around 450 reactors currently in operation are over 30 years old and, while there may be new constructions (probably not in the same regions), IAEA expects a continued decline over the next decade or two, in spite of the fact that capacity in Asia is expected to increase by around 43% over the same period. There has been a major shift of global energy investments from fossil fuels to low carbon technologies in the last few years, but the shares of energy efficiency and renewables were ten times and fifteen times the share of nuclear power generation. On the other hand, the continued growth in energy demand in the developing world could stimulate the re-emergence of nuclear electricity as a prominent component of the global low-carbon electric power generation mix.

Many nuclear power countries in the world are committed to nuclear non-proliferation and are concerned about the possibility that newcomers could move from peaceful use to nuclear weaponry, but some, in particular, Russia and China are involved in virtually all the planned new capacities in the developing world. Another major issue is the capability of these newcomers for coping with safety and waste management issues, considering the seriousness of the three accidents in the last three decades, all of which occurred in the developed world. Used nuclear fuel is intensely radioactive and although the radioactivity diminishes quite quickly, it remains at levels which could be dangerous to life for thousands of years. Used fuel requires shielding, can only be handled remotely by experts and requires special measures for its final disposal. It also continues to generate a significant heat load, both inside the reactor and out, for several years and requires active cooling during this time. If fuel gets too hot it will melt, and the radioactive materials contained will be released. In a very serious fuel melt event the radioactive materials may escape the reactor and containment structures, posing high risk to people and the environment, as happened at both the Fukushima Daiichi and Chernobyl plants.

It is noteworthy that the United States is still debating the potential final safe resting place for the nuclear waste accumulated in the atomic bomb project of the early 1940s, still potent and currently stored on temporary sites. The same is true of waste stored on sites of nuclear plants that are currently being de-commissioned. The Chernobyl nuclear accident in 1986 is the world's most serious and, after nearly forty years, decontamination is still in progress, zones remain excluded and large areas are desolate. Most of the radionuclides (short-lived iodine-131 and long-lived caesium-137) were deposited as dust in the vicinity of the plant and over much of Europe but fine material was carried far and wide, and high into the stratosphere which made possible long distance migration detected in many countries all over the world. Recent studies have shown a significant rise of thyroid cancer in people who were children in the area at the time of the accident, and settlements as far away as 200 kilometers from the nuclear plant are still radioactively contaminated. Milk produced has been found recently to contain five times the acceptable level of radiation for adults, and twelve times for children, and this level of radiation is expected to persist until at least 2040. About 200 tonnes of highly radioactive material is still in the reactor core, encased in concrete. A New Safe Confinement (NSF) structure with a lifetime of 100 years is currently under construction, funded largely through international donation. This will facilitate a safe decommissioning of the plant and management of the radioactive waste. There is little doubt that a lot of lessons have been learnt from the three nuclear accidents in the last three decades or so and current global efforts and cooperation will likely make nuclear energy safer but the fact that the most advanced nuclear nations of the world still have no clear plans about safe, permanent disposal of nuclear waste should be of major concern. Also, it is unclear whether the around twenty emerging nations/newcomers can safely and effectively manage nuclear safety and waste disposal, or respond appropriately to potential emergencies.

Another major cause for concern on nuclear electricity proliferation is the potential effect of radioactive nuclear waste on the environment, in particular, the oceans. Over the past half-century, radioactive material has been dumped or buried in ocean beds all over the world, including de-commissioned reactors, and large amounts of liquid and solid nuclear waste. Nuclear waste is currently encased in copper and concrete containers and stored on site, or buried underground and under the sea. Since the waste remains potent and dangerous for hundreds and may be thousands of years, experience (around fifty years) on the potential failure of containers and contamination of soil and aquatic life is very limited. How this continuing contamination will affect marine life or humans, is still unclear, but there is considerable concern in the scientific world that both short-lived radioactive elements, such as iodine-131, and longer-lived elements such as cesium-137 and plutonium can be absorbed by marine bacteria and other marine life, and transmitted up the food chain, to fish, marine mammals and humans (Grossman, 2011).

5.3   OUTLOOK ON POWER GENERATION

Electric power demand is growing twice as fast in emerging countries compared with OECD countries, with a total projected growth of around 70% by 2040, and around 60% of the rise in demand will come from Asia Pacific. Most of the projected growth will be accounted for by natural gas, nuclear energy and renewables, mostly solar PV and wind. The future of nuclear energy in power generation is mixed. Overall, global nuclear capacity is projected to grow by around 25% (75% in non-OECD countries) from current level by 2040 and most of the growth will come from China which accounts for around 50% of the growth. However, its share of total global power generation will decline because other sources, notably renewables are growing much faster. Furthermore, Europe is shutting down ageing plants and not replacing them, hence nuclear generation's share is expected to decline from around 25% in 2015 to less than 15% in 2040. China is rapidly expanding its nuclear power capacity, with around 40 plants under construction, and by 2040, nuclear share of power generation in the country would have increased from 3% in 2015 to 11%. India is also expanding its nuclear power plant capacity, expected to be around 8 times the current capacity by 2040. Japan is reactivating some of its shut-down plants and contribution of nuclear energy to power generation is expected to grow from the current 3% to around 11% by 2040. Growth in the developed world is less predictable, largely because most of the existing plants are old, many close to the end of their lifespans and there are no clear decisions on whether to scrap them or replace them with new plants. Another potential problem arises from the fact that many of the emerging countries that are planning nuclear power generation capacities do not appear to have the required infrastructure, in particular, grid systems that can handle power output of nuclear power plants.

Utilization of wind and solar is also projected to grow significantly but the intermittency which limits capacity utilization to 30% and 20% respectively will be a major constraint. This problem is being overcome gradually by combining VRE and fossil fuel generating plants (usually natural gas) in a hybrid configuration. Also, considerable progress is being made in the development of efficient backup power storage systems. Hydropower's share of renewable power generation, around 71% in 2015, will fall to about 53% in 2040, due largely to dwindling potentially suitable sites and increasing environmental concerns in OECD countries. However, new maxi and mini hydro-dams are under construction or planned in many countries, particularly in sub-Saharan Africa. Perhaps the most important feature of the projected growth in power generation over the next two decades is the increasing diversity in fuel use by mix and region. There will be a significant shift in the U.S. and Europe from coal to wind, solar and natural gas, nuclear power. Natural gas, nuclear, solar and wind will dominate the growth in Asia, while coal, hydropower and solar PV will feature most prominently in the projected growth of power generation in sub-Saharan Africa.

5.4   OUTLOOK ON END-USE ENERGY

End-use energy is the total delivered energy consumption by industry (including mining, energy conversion, manufacturing, agriculture and construction), transportation and building sectors of the economy. The industrial sector has accounted for the largest share of end-use energy for decades and the dominance is projected to continue, with the sector accounting for more than half of the total consumption by 2040. However, the growth rate over the period will be lower than the other sectors, around 18% compared with 26% for transportation and 28% for buildings. Around 90% of the industrial sector energy use increase will occur in non-OECD countries, with a growth rate four times the rate for OECD nations. Industrial activities largely determine economic growth, and also account for around 55% of total end-use energy consumption, split approximately evenly between manufacturing and services. Although economic output from the sector will more than double by 2040, energy demand by the sector is not expected to rise by more than 30-35% due to increasing moderation by energy efficiency across the sector, from energy-intensive manufacturing (metals, chemicals, cement, paper, etc.) to non-energy-intensive manufacturing (agriculture, construction, extraction), and non-manufacturing services.

World transportation has been growing because of worldwide rising prosperity. As incomes rise, people seek access to personal mobility, and demand for car travel is expected to double over the next twenty years or so. Growth in the global economy stimulates commercial transportation and aviation; the global car fleet is projected to double from 0.9 billion cars in 2015 to 1.8-2 billion by 2035 (BP, 2017), and almost all of this growth is in emerging markets as incomes rise and road infrastructure improves. The non-OECD fleet is projected to triple from 0.4 billion cars to 1.2 billion. Motorcycles and tricycles facilitate a lower cost entry point to personal transportation, and global population of these machines is projected to rise, particularly in Asia and sub-Saharan Africa where they are also used extensively for commercial transportation.

The worldwide transportation sector accounts for 25% of total end-use sector fuels consumption in 2040, about the same as its share in 2015. Rising global population of vehicles should require a commensurate rise in end-use energy demand, but significant efficiency gains will help limit growth in fuel demand to only 25%. The average car today returns around 10-12 kilometers per liter of gasoline compared with a projected 15 kilometers or higher by 2040. The impact of fuel economy will be more significant in OECD countries and will outpace projected increases in vehicle miles traveled, resulting in a decline of around 2% in transportation energy by 2040. On the contrary, demand in non-OECD countries will continue to rise and the region will account for around 60% of the world's transportation related energy use by 2040. Furthermore, most of the new and pre-owned fuel-inefficient vehicles that can no longer meet the increasingly stringent OECD efficiency and emissions regulations end up in the emerging economies where there is little effective regulation.

Refined petroleum and other liquid fuels currently account for around 95% of the total fuel use by the transportation sector due to widespread availability, economic advantages, and high energy density, and projections indicate a drop of a only few percent to 88% by 2040, but there will be significant changes in fuel mix. While gasoline demand growth is expected to be neutralized by fuel economy improvement, demand for diesel will grow by around 30% to meet the expected significant growth in trucking and marine transport, and demand for aviation fuel is projected to grow by around 50%. Most of the growth in transportation fuel will occur in emerging economies and China and India will account for more than 70% of the increase as rising demand for personal transportation from growing middle classes largely outpaces vehicle efficiency gains.

Freight transportation has been growing worldwide in response to the growing economy and globalization, and the trend is expected to continue. However while fuel efficiency impact will flatten transportation energy demand in OECD countries, growth in non-OECD countries will be substantial. More than half of the increase in the world's freight travel energy is attributable to marine transportation due to increasing inter-regional supply chains for raw materials and finished goods. LNG-powered heavy trucks and buses are becoming increasingly competitive and the population is projected to grow substantially by 2040 on a percentage basis but will account for only a few percent of the total transportation energy demand. Air transportation (passenger and freight) has been growing steadily and accounts for an increasingly large portion of transportation energy consumption. Projections show that the share will double in OECD countries and more than double in non-OECD region by 2040. Although aviation fuel is being mixed increasingly with biofuel, the proportion is still very low, accounting for less than 0.1% in 2018.

All recent projections reviewed show that conventional internal combustion engine (ICE) vehicles will continue to dominate the global auto market for decades, powered mainly by fossil oil derivatives. ICEs have been around for over a hundred years, with a solid research and development base that keeps coming up with a wide range of technology innovations, many in direct competition with the unique features of electric vehicles. The latest models of both gasoline and diesel vehicles return around 50% more kilometers per liter compared with a decade ago and CO_2eq emissions have also been reduced substantially in order to comply with limits that are being reviewed downwards regularly in many developed countries. Diesel vehicles have the reputation for higher pollution but the latest models have catalytic converter units which limit emissions to similar levels (per kilometer) as for petrol cars, although nitrogen oxide emissions are still higher. Also, increasing availability of biodiesel (which is much less polluting than fossil-diesel) should help reduce the current stigma on diesel cars. It appears that the only potentially effective short-term environmental mitigation options with respect to conventional vehicles are continuing efficiency improvement, and more stringent emission control regulations.

It is clear from the analysis presented above, that demand for fossil fuel-sourced transportation fuels will continue to increase in the foreseeable future, although its dominance will start to diminish with the expected increases in the market shares of electric vehicles and natural gas-powered heavy transportation. However, the progress will be slow considering the very low starting base as well as the highly competitive improvements in internal combustion technologies. Electric vehicles have a huge potential for cutting the emissions that come from the transport sector, currently around 25% of the total energy-related emissions, but the net gains will depend on the extent to which electricity is decarbonized. Tropospheric (ambient/lower atmospheric) pollution is significantly more hazardous than stratospheric (middle atmospheric) pollution because it is in direct contact with people and the ecosystem and, although residence in the atmosphere is only a few days to weeks, the impact can be very severe. The main source of pollution is transportation, particularly in city traffic because vehicle pollutions are highest in slow motion or when engines are idling at traffic stops. Other sources of urban pollution are power generation, bush burning and household use of traditional biomass. The most obvious consequence of ambient pollution is the smog that envelopes most cities worldwide, but the more serious impacts are on health, believed to result in over 7 million deaths a year worldwide. Battery/hydrogen vehicles operating mainly in cities could provide the ultimate solution because, although the electricity may still come from fossil fuels, emissions can be contained at power plants, and the only city tailpipe emissions would be environment-friendly heat and water.

The potential for reducing lower atmospheric (tropospheric) pollution which casts toxic smog over cities round the world is particularly high, and many features of EVs make them particularly suitable for intra-city transportation. There has been growing worldwide optimism that electric vehicles (EV) will virtually take over the current role gasoline-powered light passenger transportation by 2040, and for good reason: there is no pollution at point of use, although they could still have a substantial carbon footprint depending on the source of electricity. In fact, some European countries have recently announced plans to ban the sale of conventional gasoline/diesel cars by 2040. However, none of the projections reviewed supports this optimism, based on the following facts (see also Appendices 1 & 2):

- A fully charged lead-acid battery delivers less than 500 kilometers and the average recharge time is around 6-8 hours depending on the size, and facilities are mostly private and are slow chargers. Lack of adequate public battery recharging/swapping infrastructure will remain a major constraint to EV proliferation, especially for people who live in apartments or park in the streets. However, battery technologies are on a fast track and new lithium-ion batteries that can be fully charged within an hour are coming on the market, but they require special charging bays. Policies for providing fast public charging facilities or proliferating commercial battery swapping facilities are currently inadequate in most countries of the world.

- Conventional gasoline cars are significantly cheaper than equivalent EV cars, fuel-efficiency is rising, and attractive features of EVs, like start-stop engines are beginning show up in conventional cars, creating doubt whether consumer preference, (currently heavily in favor of conventional vehicles and hybrids which combine ICE and EV features) will change very much over the projection period.

- Most recent projections show that refined petroleum and other liquid fuels will remain dominant, supplying 90-95% of global transportation fuels in 2040, and conventional liquid fuel-powered vehicles will remain the most popular, especially with low global oil prices, rising transportation efficiencies, and rapidly growing commercial, marine and air transportation.

- The most optimistic projections expect the population of electric vehicles to grow substantially, making up 15-35% of the projected 1.8-2 billion vehicles in 2040, but this would depend critically on upgrading current policy supports. EVs will penetrate mainly the light-duty vehicle sub-sector which accounts for less than half of transportation energy, and most of the growth will be in OECD countries while emerging countries will account for most of the rise in transportation energy demand.

The building sector (residential and commercial) energy consumption currently accounts for about 21% of the total end-use energy consumption, and, in spite of the expected exponential growth in housing globally, projections show an increase of a modest 32% by 2040, reflecting the impact of modern energy-efficient buildings and appliances. However, the sector's share of the total end-use energy consumption will remain unchanged. Most of the increase will occur in large non-OECD countries where there is strong economic growth and rural-urban migration. Non-OECD Asia alone will account for more than 50% of the total increase of the global end-use energy demand by the sector. China accounts for 46% of the increase in residential energy demand and 30% of the increase in commercial consumption in non-OECD countries by 2040.

The building sector is dominated currently by electricity and natural gas, accounting for around 75% of total energy use. Residential electricity use will grow by around 70-75%, accounting for nearly all the growth in total energy demand from 2016 to 2040, while natural gas consumption rises by 20%, but there will be significant regional/country differences in terms of fuel mix. Growth in non-OECD countries will be around 150%, and as high as 250% in India and Africa, filled mostly by fossil fuel-based power generation. China and India will account for a quarter of the world's total building sector electricity demand in 2040. Most developing nations have relatively warm climate, hence the need for space heating is small. Instead, the majority of demand is driven by the need for space cooling/refrigeration, and electricity is the most efficient source of energy for meeting these demands.

Household energy is being modernized all over the world, with coal being replaced gradually by natural gas for space heating, promoted largely by the increasing global availability of liquefied natural gas, which is being delivered by truck and marine transportation from source to all parts of the world. This resolves the perennial problem of the need for proximity to a natural gas supply pipeline. Progress is slower in the developing world, although China and India have strong policies for promoting the use of natural gas and liquefied petroleum gas in household heating and cooking. Other regions of Asia and sub-Saharan Africa will still depend largely on traditional energy for household needs over the next two decades.

5.5 OUTLOOK ON ENERGY-RELATED EMISSIONS

In spite of their poor environmental credentials, fossil fuels continue to meet more than 80% of total primary energy demand and over 90% of energy-related emissions are CO_2eq from combustion. Carbon dioxide comes mainly from power generation and transportation, methane originates mainly from oil and gas extraction, transformation and distribution, and accounts for about 10% of energy sector emissions. Much of the balance is nitrous oxide from energy transformation, industry, transport and buildings. Although methane and nitrous oxide gases account for very small proportions of the total anthropogenic greenhouse gases, their global warming potentials (GWPs) are much higher. The level of polluting emissions from energy use is closely related to the amount and mix of both primary and end-use energy consumption, and the main focus of global mitigation effort is to decarbonize the entire energy value chain.

5.5.1 Energy demand

Energy supply, mix and use largely determine the nature and extent of energy-related emissions. All recently published outlook reports on future global primary energy demand predict a rise of around 35% over the next two decades or so, in response to the growth in the world economy which is expected to double over the same period, and growing prosperity across all regions (see Chapter 3). Fossil fuels which currently account for over 80% of the total primary energy demand will remain dominant throughout the period, still accounting for nearly 80% of the demand. Much of the projected growth in primary energy demand is attributed to non-OECD nations, many of which continue to rely heavily on fossil fuels to meet the fast-paced growth of energy demand.

5.5.2 Energy-related emissions

World energy-related emissions currently account for around 65% of the total global emissions, and will continue to grow over the next two decades, be it at half the pace of the previous two decades. A growth of only about 10% is projected by 2040, despite the fact that population will have risen by about 25% and global GDP will have more than doubled. However, growth in emerging countries will be significantly faster compared with the OECD countries. Emissions are projected to peak in OECD countries by 2030, and then start to decline, falling by around 15% by 2040 compared with 2016, but the contribution of non-OECD countries (excluding China) is expected to rise by around one-third, accounting for about 50% of the total global emissions by 2040. China alone contributed about 60% of the growth in emissions from 2000 to 2016, and will account for about 35% of the total global energy-related emissions in 2030, after which energy decarbonization policies announced recently by the country should cause a gradual decline to about 25% in 2040, still more than double that of the United States.

Energy-related emissions growth will be mitigated by increasing energy efficiency and gradual shift from coal to lower carbon energy sources, in particular, natural gas and renewables for power generation. By 2040, the carbon-intensity of the global economy is likely to fall by half, with substantial contributions across all regions. Energy efficiency gains are expected to be a major contributor to this achievement because they translate to lower energy input and associated emissions for a unit of economic output. The greatest impact of energy decarbonization will be in the power generation sector where the use of coal is expected to decline in favor of natural gas and renewables. However, most of this shift in fuel mix will be in the developed nations.

Energy-intensive manufacturing currently accounts for around 27% of total global delivered energy to the industrial sector and the proportion is unlikely to increase over the next two decades despite an overall increase in the demand by the industrial sector. This is because of the ongoing shift largely by the developed countries and China to less energy-intensive manufacturing, a trend which is expected to intensify. However, these processes are shifting to the emerging nations which rely heavily on coal, and often cannot afford to deploy more efficient manufacturing technologies, hence the achievement of the developed nations in lowering emissions from industry will be largely offset by significant increases in the emerging nations. Emissions from liquid fuels, mostly from transportation will flatten in the OECD countries, again due mainly to the positive effects of improved energy efficiency, but will grow substantially in the developing nations, in particular, India and China due to the expected rapid growth in both personal and commercial transportation. The building sector contributes less than 10% to global energy-related emissions and further reduction is projected.

The fossil fuel share of total global primary energy supply (TPES) was 81% in 2016, roughly the same level for nearly thirty years (EIA, 2018). Over the period, coal and oil jointly represented about 60% of TPES and almost 80% of CO_2 emissions. The contribution of gas has also been stable at around 20% while the balance of 19% accounted for by non-emitting sources. Coal has the highest carbon intensity and accounts for about 44% of global energy-related emissions. Details of emissions in terms of trends over the last three decades, fossil fuel source, economic sector and region are presented in Figure 5.4. While the use of coal has declined significantly in Europe and North America over the last few years, it is still the predominant fuel in Asia, accounting for around 60% of the emissions from that region in 2016. The region accounted for about a quarter of the total global energy-related emissions in the same year.

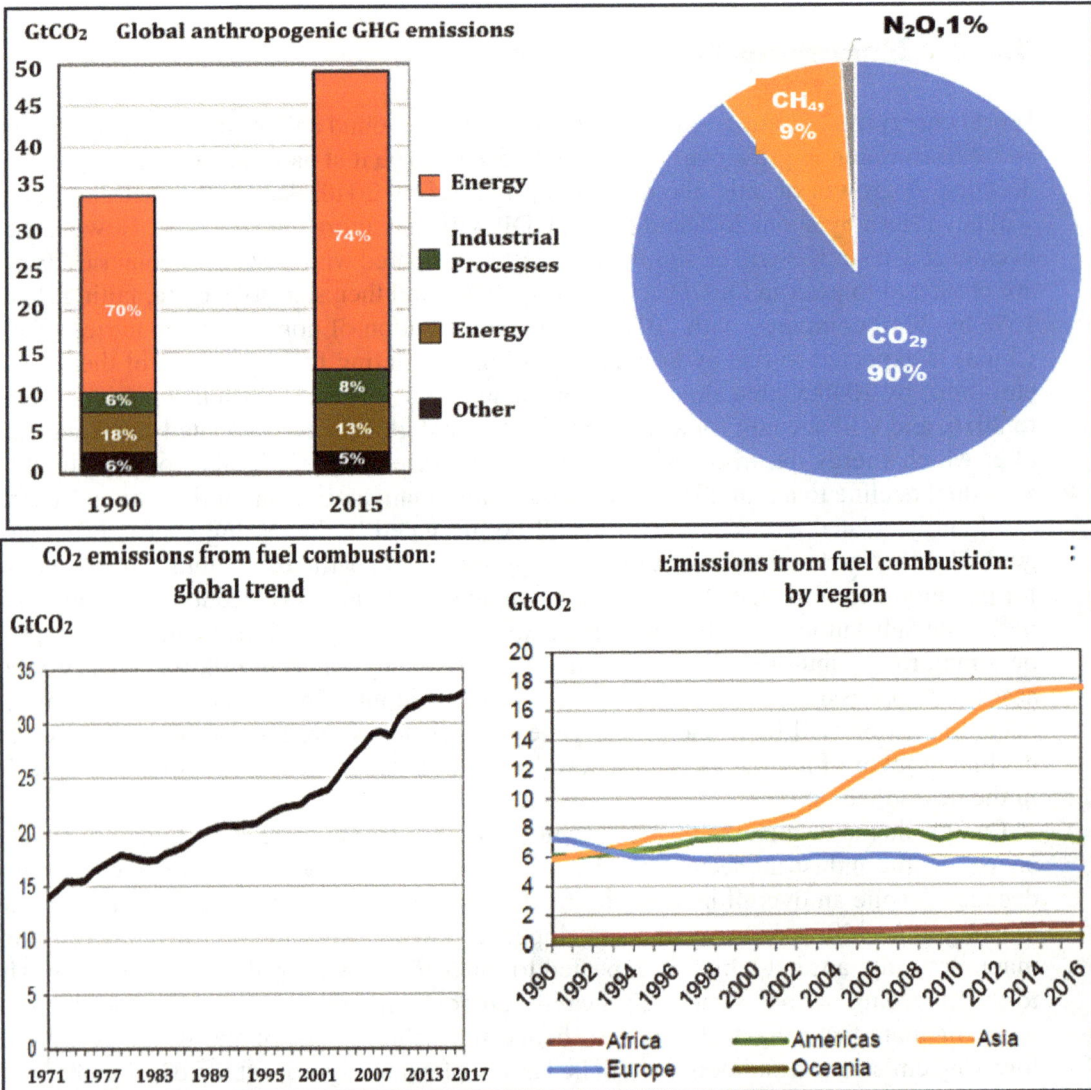

Figure 5.4a Global anthropogenic emissions by source; global trend of energy-related emissions and by region in 2016 *(IEA, 2018a).*

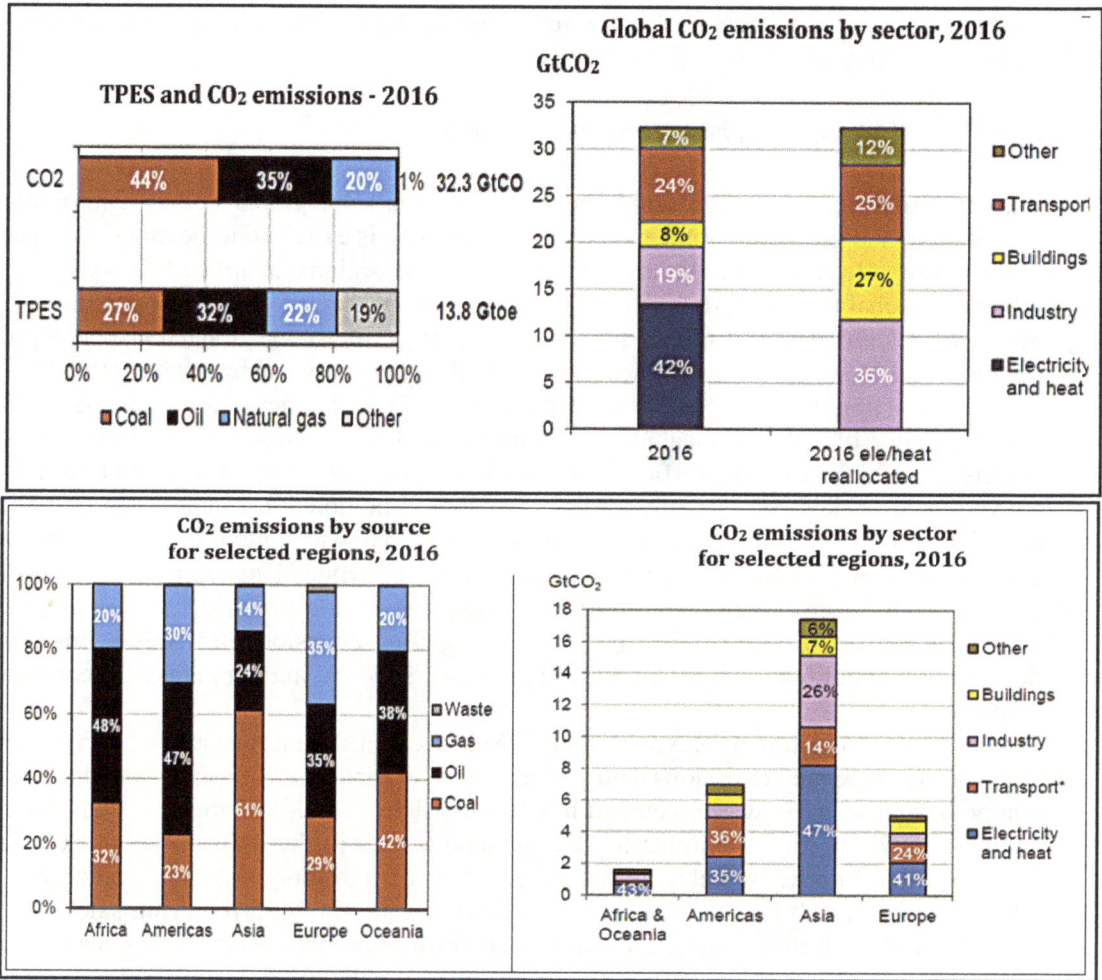

Figure 5.4b Energy-related emissions by fuel, economic sector and region in 2016 *(IEA, 2018a).*

5.5.2.1   *Emissions from power generation*

Electricity generation accounts for about 42% of energy-related CO_2eq emissions, but the expected shifts in energy mix to less carbon-intensive sources (wind, solar, nuclear, natural gas) will help reduce primary energy use per unit of power output and the CO_2eq intensity of delivered electricity by more than 30%. Natural gas which has the lowest carbon footprint of all fossil fuels, will likely play a critical role in helping to decarbonize the power sector in future. The fuel is reliable and efficient for power generation when available at competitive prices. Also the ease of handling, transportation, and flexibility make it well suited to meet peak demand and backup intermittent renewables. However, the extent of its role depends on several uncertainties: stronger public sentiments against nuclear, hydro and coal power could enhance the role of natural gas as a fuel of choice, while a strong increase in the deployment of renewables could reduce its growing role in power generation. Also, the role of natural gas in the electricity generation mix will vary across countries and regions, prominent in natural gas-rich countries and regions, but relatively low in countries that have limited access to inexpensive natural gas. Renewable energy, in particular, solar and wind will play an increasingly significant role in decarbonizing power generation, with growth in deployment across all regions, although at different rates.

5.5.2.2 *Emissions from transportation*

Transportation currently represents about 25% of the total global CO_2eq emissions from combustion, and growth over the next two decades or so is expected to be modest, responding to increasing efficiency of conventional vehicles across regions, in spite of the expected rapid growth in global transportation. The expected growth of electric vehicle population over the period, representing 6–35% (depending on level of optimism) of the global transport population by 2040 will likely have the greatest impact in the light-duty vehicle sub-sector. However, growth in EV population is one of the greatest uncertainties in energy use projection analyses, largely because of lack of critical infrastructure, policy uncertainties, and entrenched consumer preference which may be difficult to switch in view of increasing competitiveness of conventional vehicles on virtually all fronts. It should be noted also that the largest share of electricity for EVs will still be sourced from fossil-fueled power generation, unless renewables' use can grow at a much faster pace than currently projected. Uncertainties in government policies and the pace of market penetration of various technologies could also have a significant impact on the future role of electric vehicles in CO_2eq emissions mitigation. Furthermore, declining oil prices in the global market tend to depress the EV market in favor of conventional vehicles.

In a recent projection by ExxonMobil (2018b) several different scenarios of energy demand, end-use and associated emissions were tested, based on various assumptions and uncertainties, a hypothetical scenario was developed in which all light vehicles on the roads are electricity-powered by 2040, totally eliminating the demand for oil in the light-duty vehicle sector of transportation. In order to achieve this, global sales of light-duty vehicles would need to be 100% all electric (50 times from existing levels), starting in 2025 requiring sales of around 110-140 million electric vehicles annually henceforth (more than a hundred times the number of electric vehicles sold in 2016). Even with this unrealistic scenario, total energy-related emissions would be reduced by no more than about 5% and, although emissions from light-duty vehicles would reduce to zero, emissions from power generation would rise with the increase in electricity demand, and CO_2eq emissions from the power sector could increase by about 15%, with coal accounting for 60% of the increase. Even though this analysis may seem biased, coming from a petroleum marketing company, several independent organizations have come to similar conclusions, notably the International Energy Agency (IEA, 2017c), and Energy Information Administration (EIA, 2017) as exemplified by the following statement from IEA:

> "*Electric vehicles (EVs) are in the fast lane as a result of government support and declining battery costs but it is far too early to write the obituary of oil, as growth for trucks, petrochemicals, shipping and aviation keep pushing demand higher…… rising oil demand slows down but is not reversed before 2040 even as electric car sales rise steeply.*"

Perhaps the most optimistic scenario on electric vehicle market penetration was presented in a recent IEA publication (2019b), based on the EV30@30 initiative launched at the Clean Energy Ministerial meeting in 2017 which aims to promote market penetration of EVs to a 30% share of the global new vehicle stock in 2030. This translates to about 250 million EVs available for sale in 2030, with passenger light duty vehicles (PLDV) and battery vehicles

(BEVs) accounting for around 50%. The outlook on EVs is discussed in some depth in Appendix 1 of this chapter.

In the recent past several countries in Europe have announced policies to stop the sale of diesel cars by 2040. This was clearly based on the assumption that diesel engines are more polluting than petrol equivalents. This is in fact not true of modern diesel vehicles which feature advanced catalytic converters that reduce emissions. Furthermore, diesel vehicles deliver more kilometers per liter of fuel, hence pollution rate could be actually lower. In order to check this, manufacturer's specifications of five 2018 models of a popular car were analyzed, all with 1.6 liter engines, four petrol and one diesel. Emissions for the four petrol models range from 110 to 140g/km CO_2eq, compared with 93g/km CO_2eq for the diesel model. The petrol engines deliver around 20 km/liter compared with around 30 km/liter for the diesel car. It should be noted however that the quoted figures are from auto manufacturers, and independent verification is needed. Also, even if the announced policies of banning sales of ICE cars by 2040 is unrealistic, any potential reduction in the nitrogen oxides emitted by diesel engines would be enormously beneficial since most of the effect is local and the negative impact on human health can be substantial. Furthermore, it has accentuated consumer awareness, interest and reactions. Sales of diesel cars have plummeted, but there has been no major rise in the sale of EV, largely because people are confused and prefer to delay investments in new vehicles.

Interest in hydrogen-powered fuel cell electric vehicles (HEV/FCEV) has been strong for some time but the major issue was the development of a capable hydrogen fuel cell. This problem has now been largely resolved, and the market for HEVs is emerging. Several big auto manufacturers are now offering a variety of models. The difference between BEVs and HEVs is that BEVs run on batteries that have to be recharged, while HEVs have fuel cells on board which draw hydrogen from the tank and produce electricity on board. This means that they can stop at a hydrogen refueling station to buy liquid hydrogen, just like a conventional vehicle, and a full tank can deliver 400-450 kilometers, more than most electric cars currently on the market. Liquid nitrogen is currently being produced mainly from fossil fuels, but can also be produced from biomass, and water, and, like EVs, the only emissions are heat and water. Electric and hydrogen vehicles hold very high promise as a very effective mitigation for transport pollution, particularly in cities currently plagued by incessant and toxic smog. One potential problem is the fact that production of hydrogen by any method requires a lot of electric power most of which will likely come from fossil-fueled power stations. However, most power stations are located on the outskirts of, or well away from cities, and emissions can be captured and processed on site. There is little doubt that the electric vehicle market will continue to grow and moderate emissions from the transportation sector. However, most of the projected growth will occur in OECD countries, while emerging economies will account for most of the future growth in vehicle population. Furthermore, intra-country policies that are promoting the replacement of ICEs with electric vehicles are also stimulating (be it inadvertently) the export of inefficient used vehicles to the emerging countries, most of which have no strong emissions' policy control policies. Since there are no stratospheric country boundaries, pollution from any source anywhere in the world will likely have the same effect on global climate change.

5.5.2.3   *Emissions from industry*

Economic growth is the primary driver of energy use and associated CO_2eq emissions. The industrial sector is the highest polluting in the energy end-use sector, contributing around 25% of CO_2eq emissions in 2016 and not much had changed in 2018. Projections show that, in spite of the fact that GDP is expected to double by 2040, growth in emissions will be modest due to efficiency gains and growing use of less carbon intensive energy and manufacturing technologies which will help reduce emissions from the sector relative to GDP by about 50%, and contribute to a nearly 45% decline in the carbon intensity of global GDP. However, there will be significant regional variations: most of the emerging countries that will largely host future energy-intensive industries rely heavily on coal, the most carbon-intensive fossil fuel for power generation as well as the production of primary metals, cement, glass, chemicals, petrochemicals, all of which are indispensable globally for economic growth. It is interesting to note that high proportions of these products are exported to the developed economies.

5.5.2.4   *Emissions from buildings (residential and commercial)*

The building sector is the least polluting of all energy end-use sectors, accounting for only about 6-9% of the total global emissions despite the fact that the sector currently accounts for around 20% of the total global delivered energy consumption. However, considering that buildings rely strongly on electricity, this contribution rises to around 27% when emissions from electricity are allocated to the consuming sectors. Rapid population growth, rising urbanization and incomes are all fueling demand for modern homes, which in turn stimulates growth of commercial services. Residential energy demand is projected to grow by about 20%, consistent with the projected population growth over the next two decades or so, with non-OECD nations accounting for around 40% of the total growth, and China and Africa alone will account for about 30% of the growth in demand (ExxonMobil, 2018). Around 90% of the demand growth will be met by electricity, hence, the average worldwide household electricity will rise by about 30% across all regions over the period, moderated by rising efficiency in the sector - energy efficient building construction, fittings, appliances, and consumer products, and increasing awareness among consumers of the multiple benefits of optimized energy use. Residential electricity use is projected to rise by about 75% by 2040, driven largely by around 150% increase in non-OECD countries (about 250% rise in India and Africa), and coal will likely be the main power generation fuel.

5.5.2.5   *Emissions and global access to modern energy*

Energy is an indispensable propellant of human development and, with the sector accounting for around three-quarters of global pollution emissions, decarbonizing energy will be critical in achieving global efforts to slow down the negative impact on the environment. Traditionally, the industrial (Annex I) countries have emitted the large majority of anthropogenic emissions but there has been a rapid shift to the emerging countries which will account for around 70% of energy-related emissions in 2040. Since there is no physical atmospheric boundary and the effect of any emissions in one area of the world will resonate across the world, any mitigation efforts to decarbonize energy must target all countries to be effective. The International Energy Agency (IEA) currently supports countries through the provision of energy emissions statistics and training country officials in policy, modeling and energy statistics. IEA also has a Clean Energy Transition Program which targets emerging countries. As laudable as these programs are, the effect will be minimal unless the countries have financial and technological support to improve efficiencies across the energy supply and consumption chain. Electricity which accounts for around 40% of energy-related emissions is increasingly becoming the fuel of choice across the world, filling for around 20% of global final energy consumption, and is projected to rise even faster in future, with the emerging world accounting for the bulk of the demand. Decarbonizing the sector particularly in the developing world will be critical to the achievement of global effort to mitigate the impact of anthropogenic emissions on the climate. Around a third of current global investments in energy is in electricity generation in developing economies but low, regulated consumer prices and the need for cost recovery are disincentives for the much-needed strong investments, and compromising the adoption of high-efficiency technologies.

Projections on the the feasibility of the United Nations' target of access to clean domestic energy all by 2030 are not very optimistic. Currently, around 3 billion people - more than 40% of the global population and about 50% of the population in the developing countries - lack access to clean domestic energy. Most of them rely on traditional wood, charcoal, and biomass for cooking and heating, often in poorly-ventilated environment. This has been linked by the World Health Organization (WHO, 2018) to around 3 million premature deaths in 2016 in addition to the 4.2 million who died as a result of outdoor (ambient) pollution. More than 90% of air pollution-related deaths occurred in low- and middle-income countries, mainly in Asia and Africa. While progress on access to clean domestic energy has been gathering momentum in parts of Asia, backed by targeted policies focused mainly on proliferation of the use of liquefied petroleum gas (LPG), sub-Saharan Africa is far from being on track. Projections show that the number of people globally without access to clean household energy will decline by only 18% by 2030, with around 820 million of them (56% of the population) in sub-Saharan Africa. In effect, increased access to electricity, mostly in emerging countries will likely continue to depend heavily on fossil fuel-powered generation, although renewables will also play a significant role, and the continued dominance of fossil fuel and traditional biomass use in emerging nations accounts for the projected rise in anthropogenic emissions across the regions. Also, any dramatic improvement in household ambient pollution over the period, particularly in the sub-Saharan region is unlikely.

5.6   OUTLOOK ON ENERGY DEMAND PATHWAY TO PARIS-2015 TARGET

The United Nations has played a key role in bringing environmental degradation to the forefront of global discourse in the last three decades or so. The Montreal Protocol of 1987 which sought to restrict the use of ozone depleting substances (halocarbons) was a milestone. This action has been very effective and use is now restricted mainly to developing countries. The substitutes for CFCs [hydro-fluoro-carbons (HFC's), per-fluoro-carbons (PFCs), and sulphur hexafluoride (SF_6)] do not harm or breakdown the ozone molecule and currently have a relatively small aggregate radiative forcing impact, although their contribution to the greenhouse effect is still significant. The United Nations Framework Convention on Climate Change (UNFCCC), established in 1992 has led international effort to understand and address the risks of climate change. The Paris-2015 conference, also convened by UNFCCC ended with a Protocol on the environment which set a target of limiting the global average temperature to well below 2°C above pre-industrial levels and pursue efforts to limit the rise to 1.5°C. Nearly every nation in the world attended the conference and committed to action on climate change by developing nationally determined contributions (NDCs), and reporting on related progress every five years.

The Paris Agreement came into force in 2016 and around 200 have committed to the agreement (UNFCCC, 2018). However, the UNFCCC's report on NDCs in 2016 concluded that the estimated aggregate greenhouse gas emission levels in 2025 and 2030 resulting from the intended nationally determined contributions would not achieve the 2°C target. In view of the complexity and scale of the world's energy system and its interaction with societal aspirations, the pathway to the UNFCCC goal would depend on multiple variables, including population growth, economic growth, and the infusion of new technologies which drive down energy intensity, enhance efficiency of energy production and use, and mitigate CO_2eq emissions. Furthermore, there are wide variations in country needs, available resources, constraints, and pathways to energy security. Unless there is a dramatic improvement in national and well coordinated international policies that help developing economies overcome obstacles to decarbonization of energy, achievement of a global emissions level which would make the achievement of this goal feasible by 2040, is unlikely.

Given a wide range of uncertainties, no single pathway to the achievement of the Paris-2015 Agreement can be reasonably predicted. As a result, many international organizations, university experts, governments have defined different potential pathways to achieving this goal. One key uncertainty is unpredictability of advances in technology that may impact assumptions one way or the other, future trends in the global economy or emergence of much stronger policy instruments, hence many of the studies tested a wide range of technology options and scenarios. All the studies recognized the inevitable fact that global economy and population will continue to grow, and so will global demand for energy and associated emissions, in spite of all anticipated improvements in energy efficiency and intensity. Therefore, all scenarios targeted a reasonable range of pathways focused two main emission mitigation options:

- More aggressive reduction in global energy demand through improved energy intensities and efficiencies, and energy conservation (for example, through reduced use of products by adopting the refuse-reduce-reuse-recycle model.

- Major decarbonization of energy through strong infusion of lower carbon energy.

A recent analysis by the International Energy Agency (IEA, 2018) evaluated two scenarios: New Policies Scenario (NPS) which projects energy-related emissions over the next twenty years or so, based on existing policies that may become effective before 2040 on decarbonization of energy, and the 450S scenario which assumes that more aggressive policies will emerge before 2040. The study concluded that, although most countries are on track to meet national pledges made as part of the Paris Agreement, these are insufficient to meet an early peak in global emissions anytime soon. The NPS scenario projects emissions level of about 36 billion metric tons in 2040, but the 450S scenario shows that the pathway to Paris-2015 target requires more effective international mitigation actions that reduce the projection by around 50% to around 18.3 billion metric tons (Figure 5.5). In order to achieve this target, further decarbonization of total primary energy supply and reduction of total primary consumption by about 15% compared with currently projected 2040 levels would be required. For example, processing and transportation of oil and gas to consumers accounts for around 15% of energy sector greenhouse gas emissions, mainly carbon dioxide and methane, and there is substantial scope for reduction by 25-30%, for example reducing pipeline leakages, eliminating flaring, capture and use of carbon dioxide to enhance oil recovery, use of low-carbon electricity to support operations, and on-site processing of hydrocarbons with carbon capture to produce hydrogen. Also, so far, there has been low emphasis on what the population can do to lower personal carbon footprint, and the potential is substantial: judicious acquisition and disposal of consumer goods and support for reuse and recycling, optimization of personal transportation, running energy efficient homes, etc.

Total final energy consumption by sector and scenario in 2040

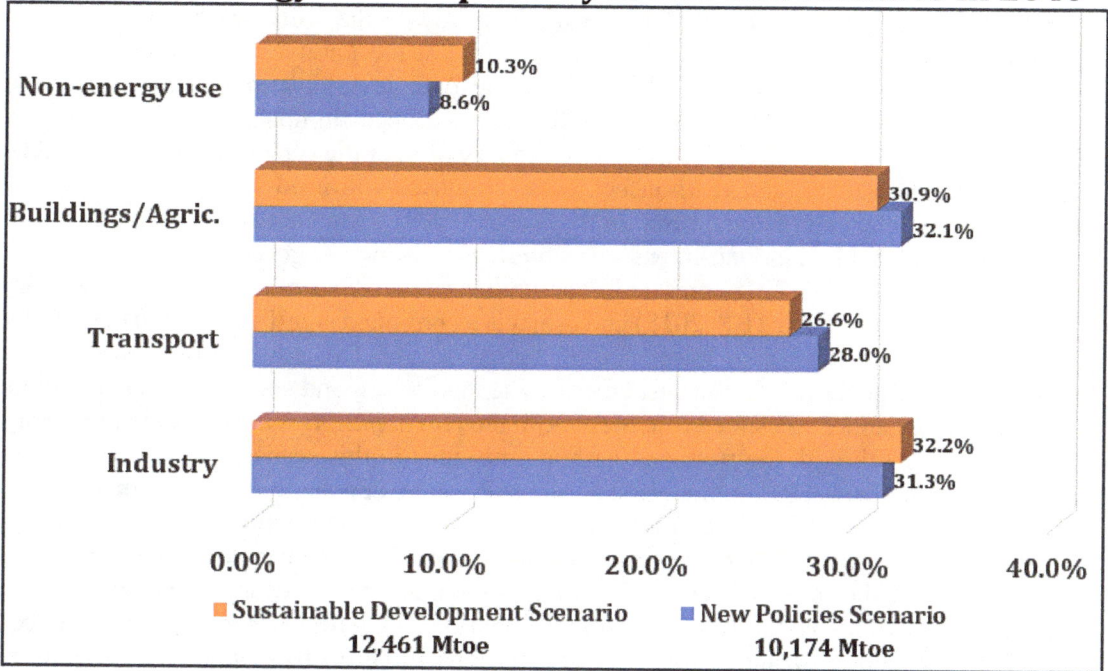

Total CO_2 emissions by region and scenario in 2040

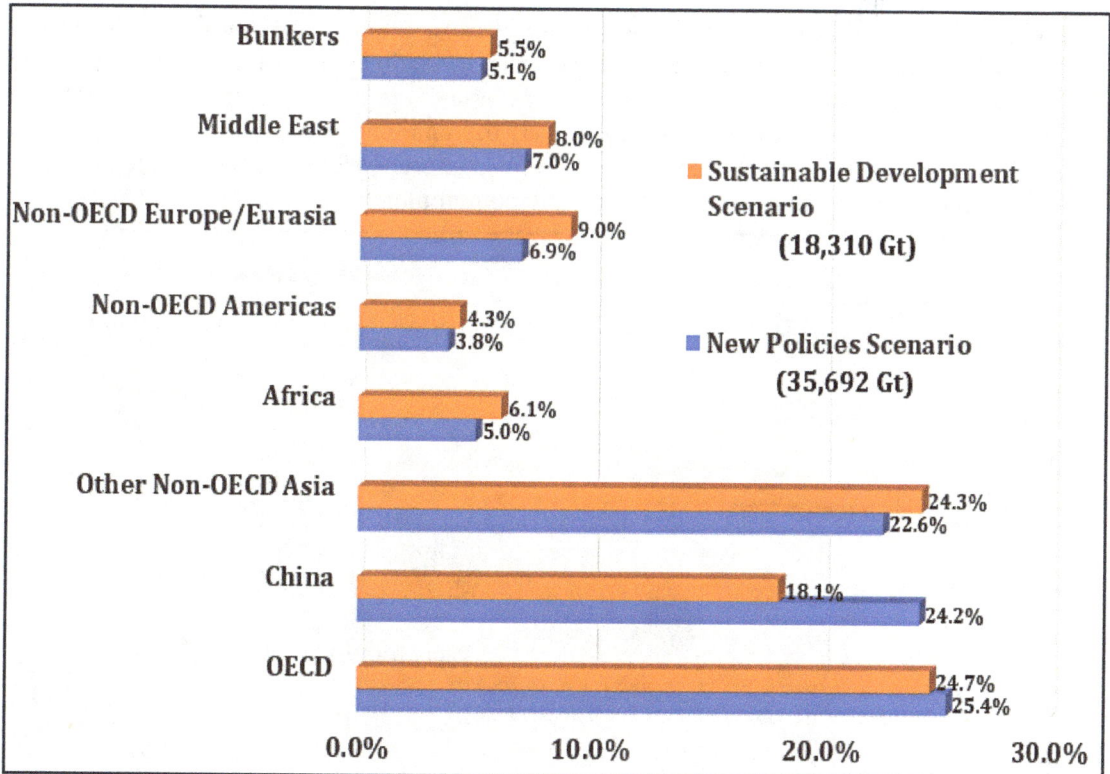

Figure 5.5 Two scenarios on global total final energy consumption and CO_2 emissions outlook to 2040 *(Data from IEA, 2017a).*

Many other studies, notably by BP (2018) and ExxonMobil (2018), IRENA, (2018) have developed different pathways based on energy mix (Figures 5.6 - 5.10). One uncertainty that may be important in assessing the feasibility of any of the scenarios is the extent to which emerging nations can achieve set goals presented at the Paris-2015 conference. All recent projections show that most of the expected growth in total primary energy demand and associated emissions over the next two decades will occur in non-OECD countries, many of which do not have much choice in terms of energy mix, or the capability to bear the added cost of environmental mitigation. In effect, these countries will need technological and financial support from the international community to enable them adopt more efficient energy utilization technologies and less carbon-intensive technologies for power generation.

The study by BP (2018) presents a comprehensive analysis of possible scenarios for future global energy transition (BP, 2018) and projects a continued significant rise in energy demand, but with a shift towards a lower carbon fuel mix. The Evolving Transition Scenario (ET) assumes that government policies, emerging technologies, and social preferences continue to evolve in a manner and speed consistent with the recent past. But the outcome hardly signifies a decisive break with the past, and carbon emissions in this scenario are not consistent with achieving the Paris-2015 climate goals. The study considers various other scenarios involving different developments that could fast-track CO_2 emission mitigation, notably, more effective policy push to decarbonized electric power generation, renewable technology, and more dynamic global electric vehicle market such as banning the sale of internal combustion engine cars (Figure 5.7). The Evolving Transition Scenario (ET) indicates that energy use will continue to grow through 2040, and energy-related emissions will grow by around 10%, whereas a sharp decline is believed to be necessary to be consistent with achieving the Paris-2015 climate goals which, according to the International Energy Agency's 'Sustainable Development Scenario,' requires a reduction in emissions by around 50% by 2040. The Even Faster Scenario (EFT) results in a decline in carbon emissions over the projection period. It is clear from the different scenarios that, given the wide range of uncertainties, there is no predictable single pathway to global sustainable environment. However, the unanimous message is that there is a very urgent need for implementation of much more comprehensive emission abatement strategies for energy production and use than currently contemplated.

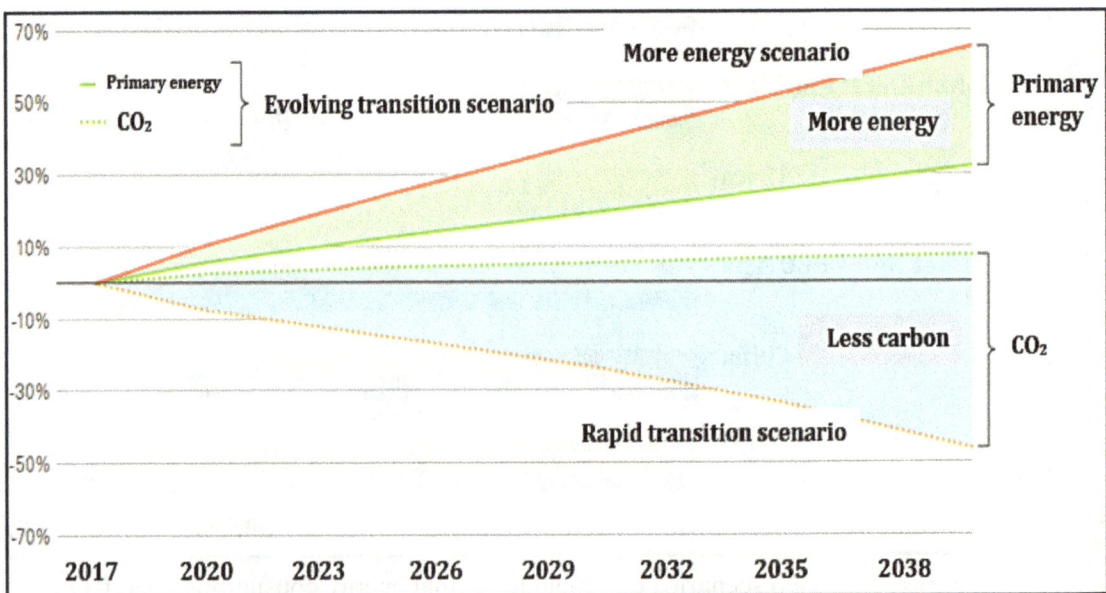

Figure 5.6a Global dual energy challenge: need for more energy and less carbon
(BP, 2019)

(a)

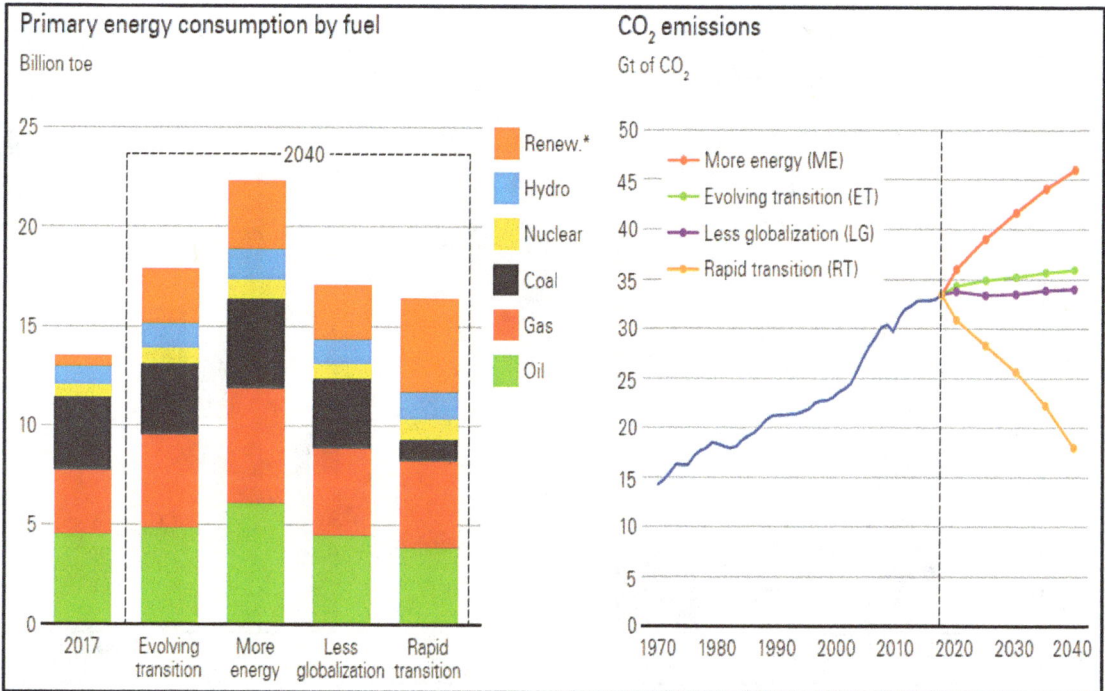

(b)

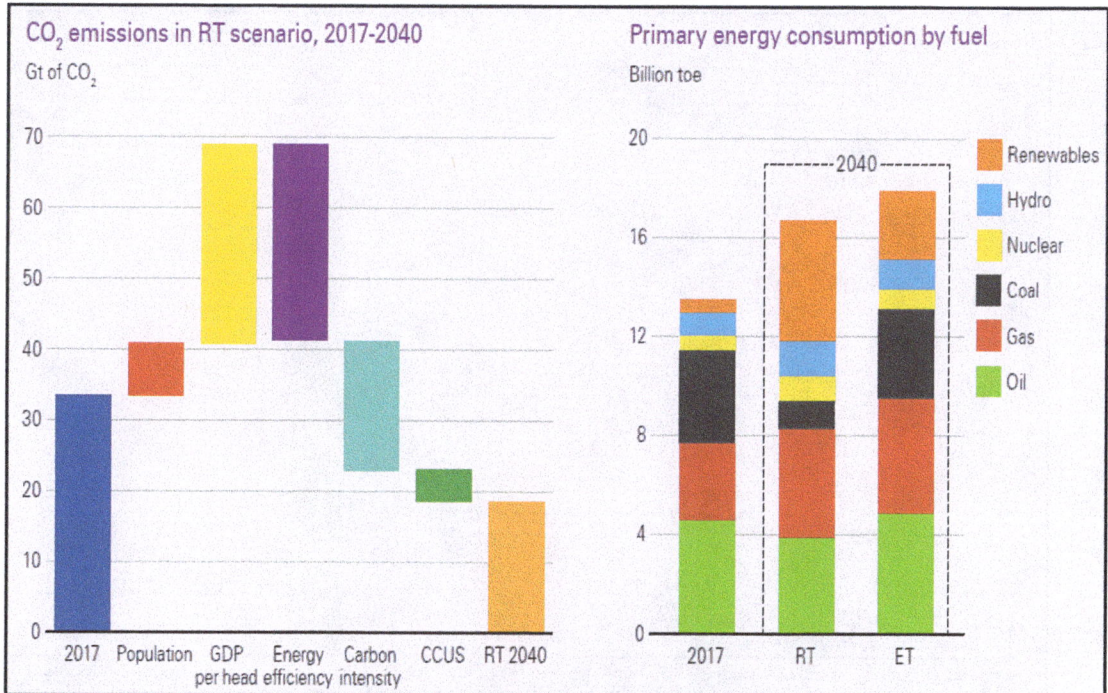

Figure 5.6b Pathways to the Paris-2015 Climate Change Goal: (a) Different scenarios of primary energy transition by fuel and CO_2 emissions by fuel mix; (b) Effect of population and economic variables, efficiency gains and carbon capture and use on CO_2 emissions *(BP, 2019).*

(a)

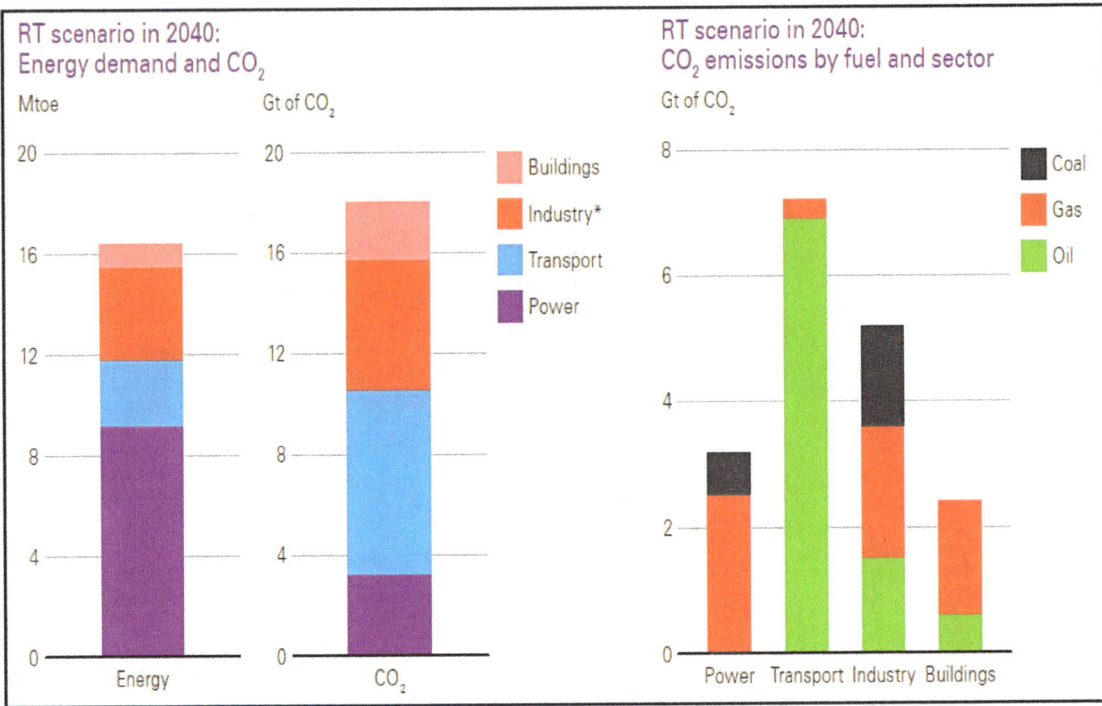

(b)

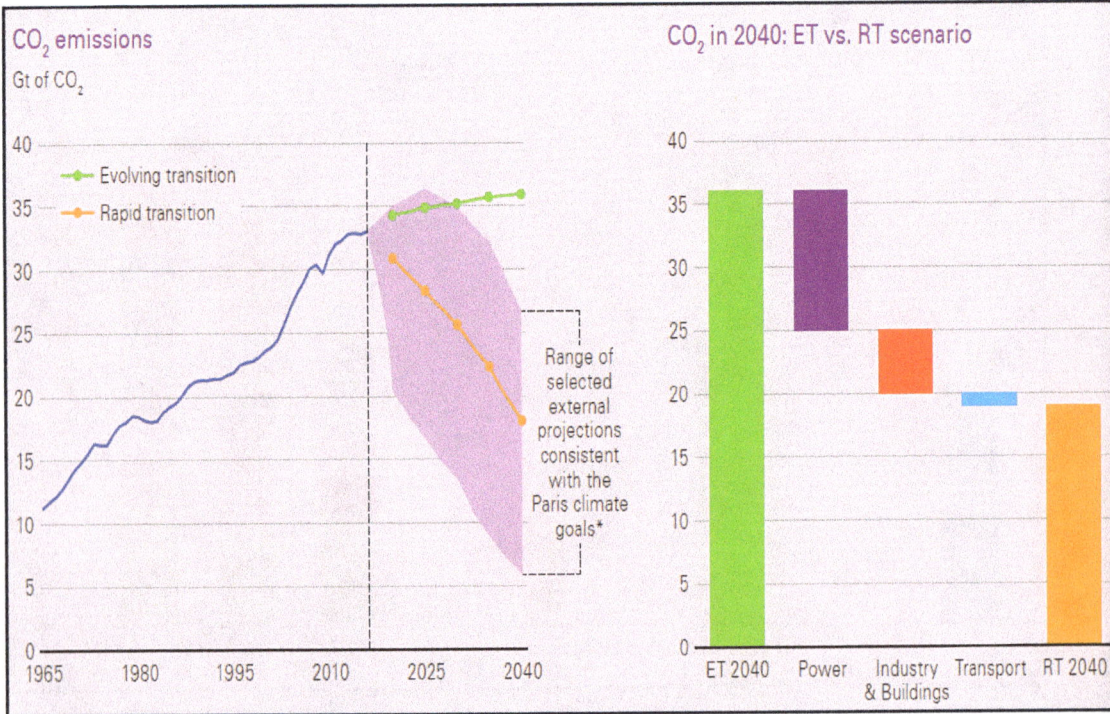

Figure 5.7   (a) Sustainable CO_2 emissions by fuel and sector and primary fuel mix in the Rapid Transition Scenario (b) Emerging Transition versus Rapid Transition Scenario to a lower carbon energy system   *(BP, 2019).*

2040 global demand by model by energy type in the assessed 2°C scenarios and the IEA SDS

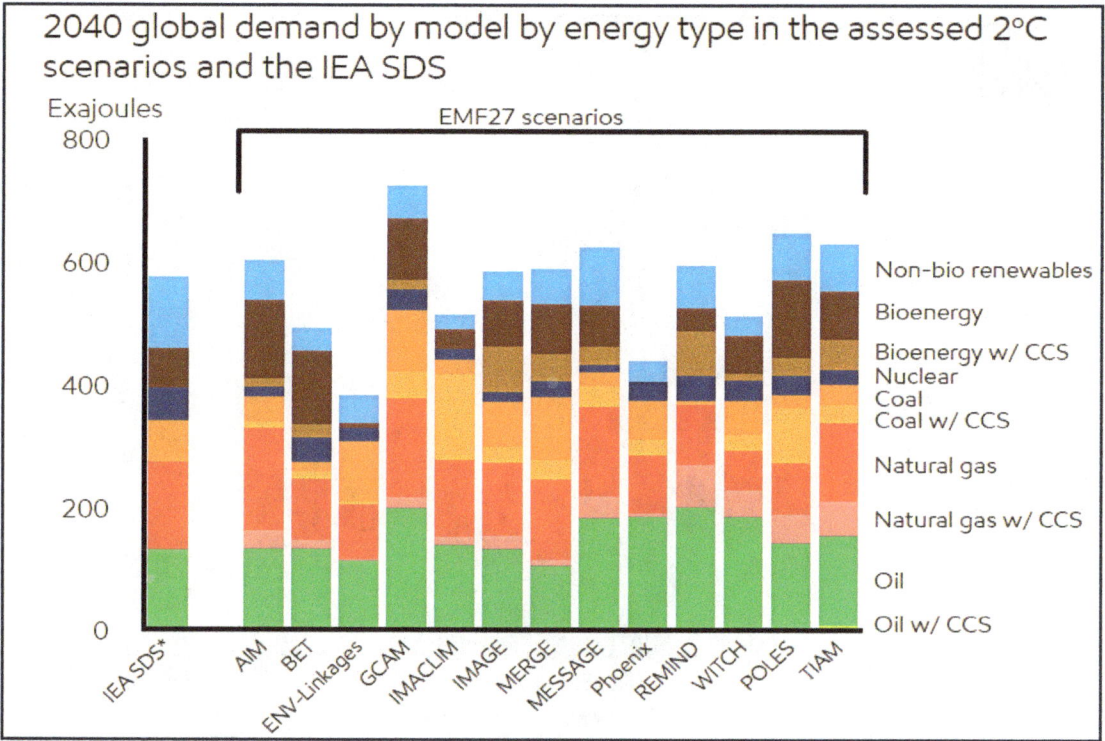

Global energy-related CO₂ emissions

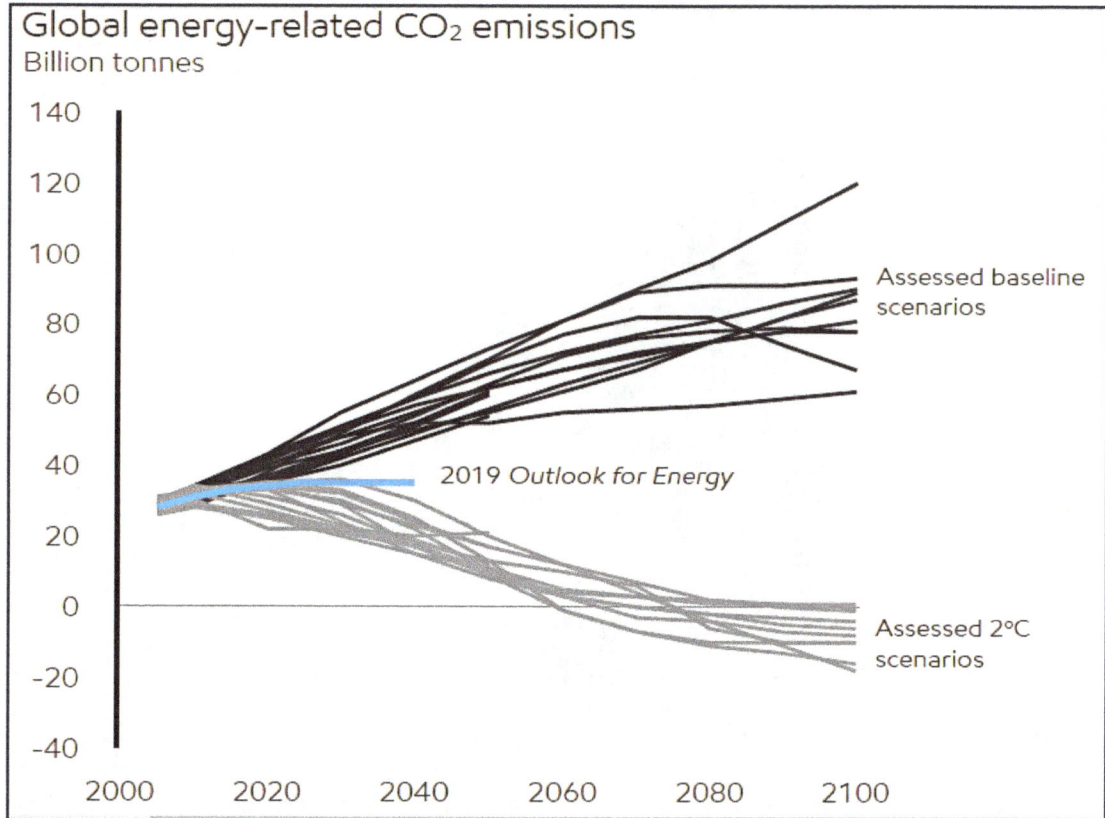

Figure 5.8 Alternative potential pathways to IEA Sustainable Development Scenario developed by the Energy Modeling Forum 27 (EMF27) *(ExxonMobil, 2019).*

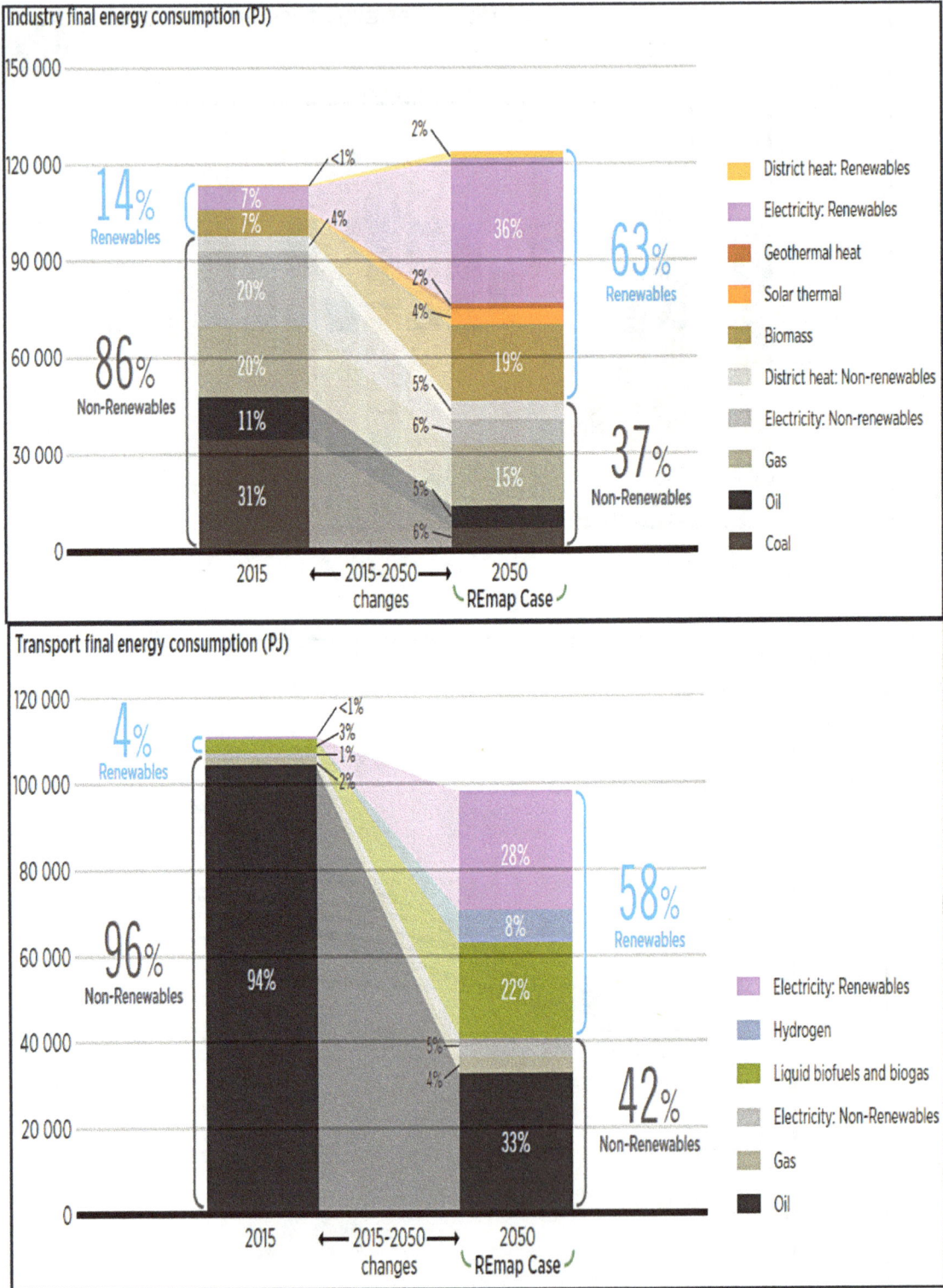

Figure 5.9a Pathways to decarbonizing industry and transport energy through increased renewable energy (Remap), compatible with IEA Sustainable Development Scenario *(IRENA, 2018).*

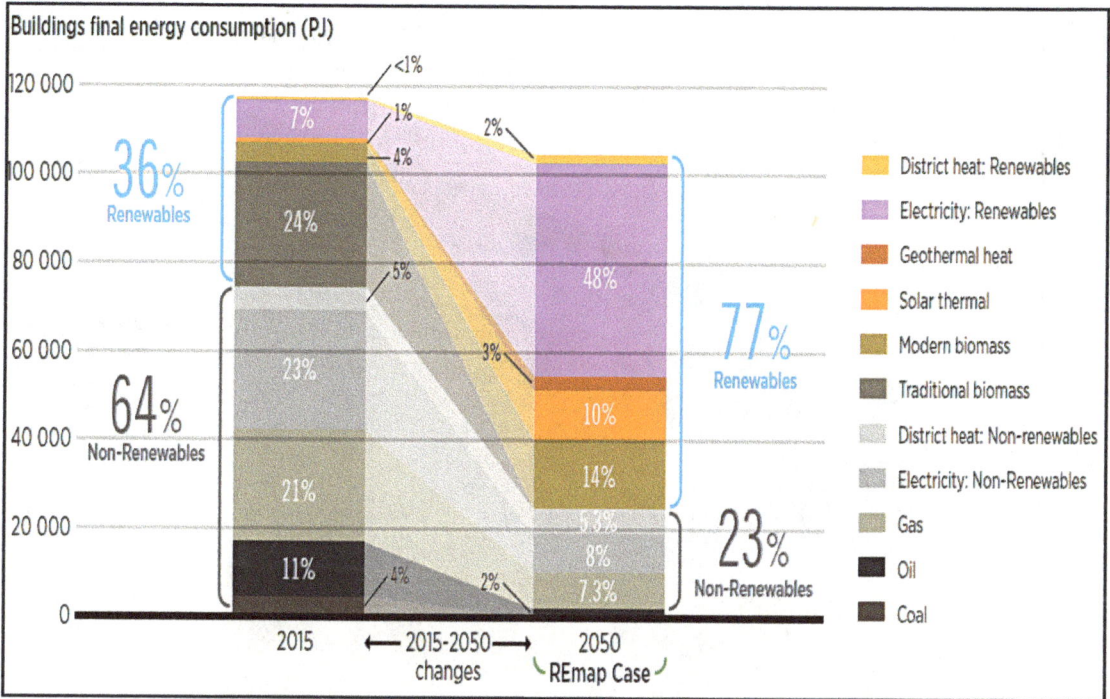

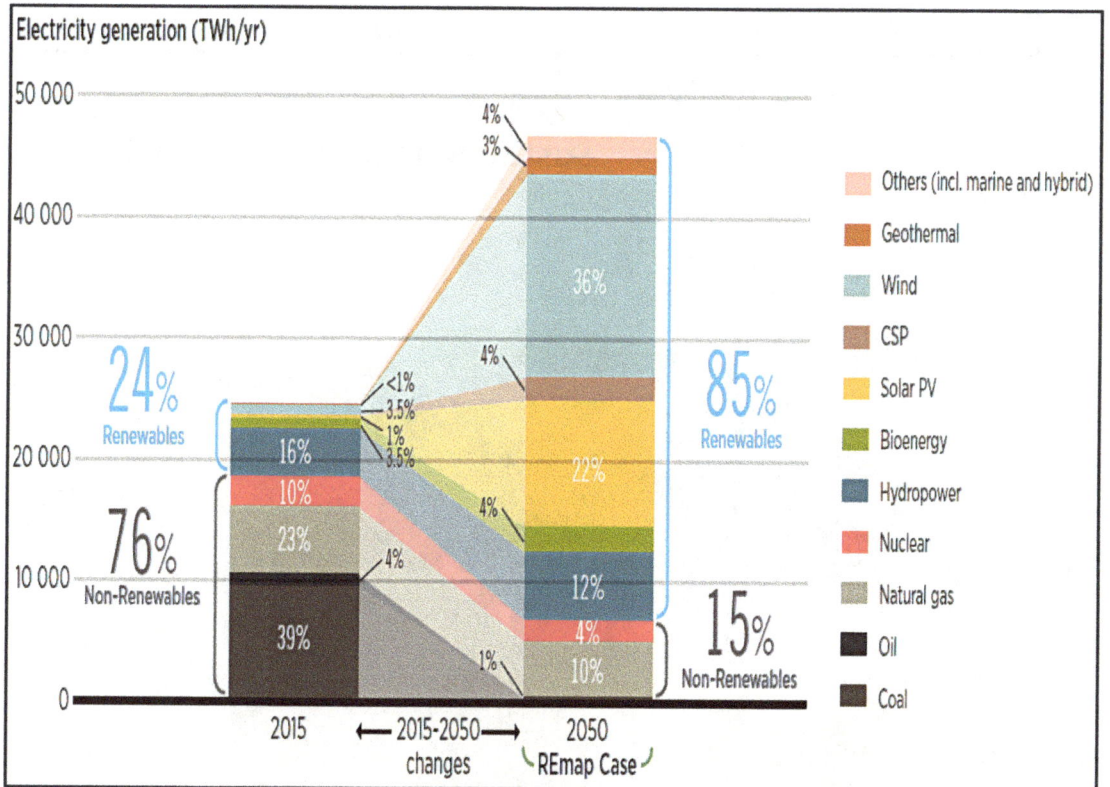

Figure 5.9b Pathways to decarbonizing building and power generation energy through increased renewable energy (Remap), compatible with IEA Sustainable Development Scenario *(IRENA, 2018).*

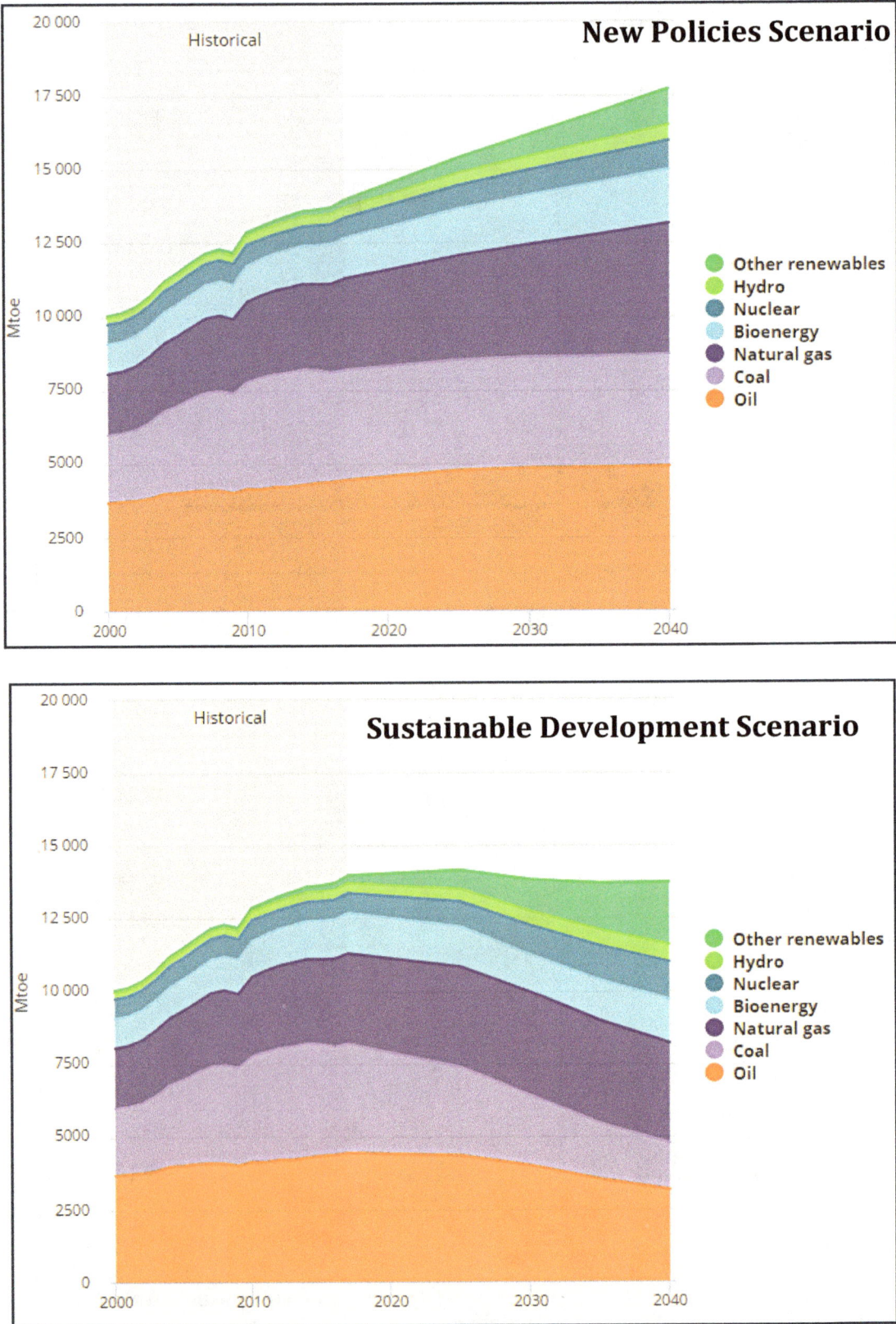

Figure 5.10a International Energy Agency's total global primary energy energy mix for two scenarios *(IEA, 2019g).*

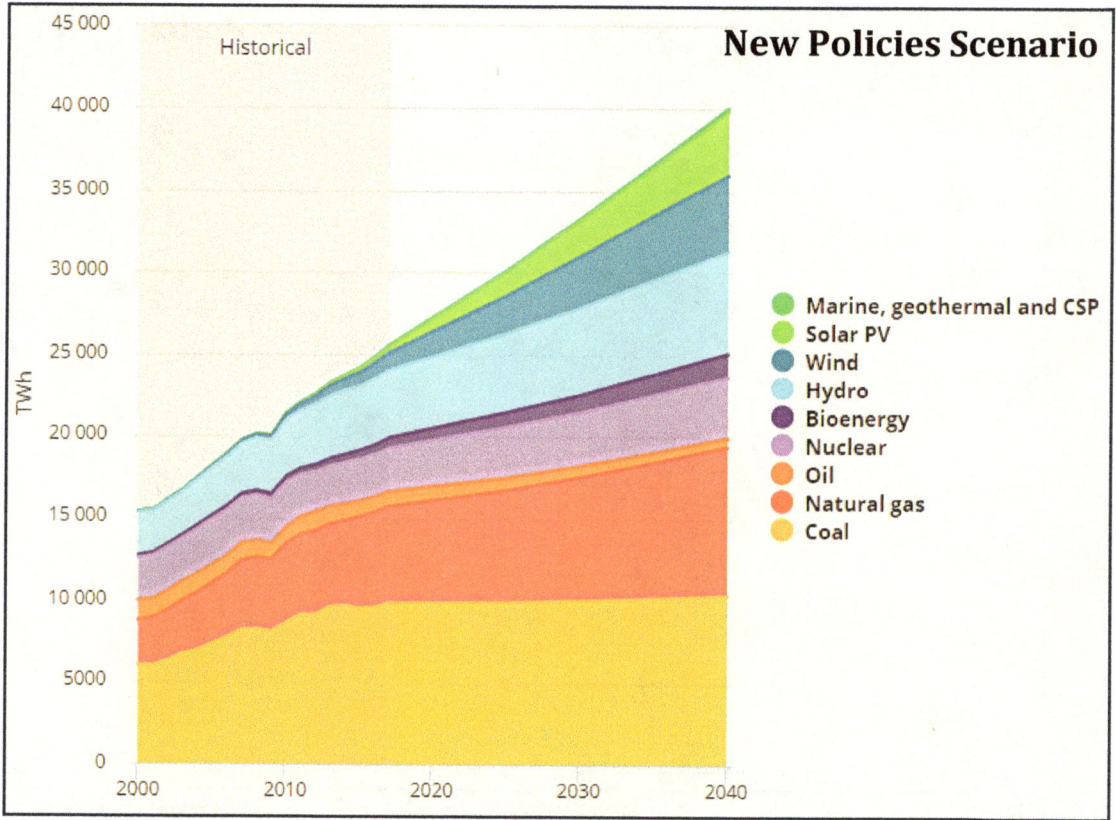

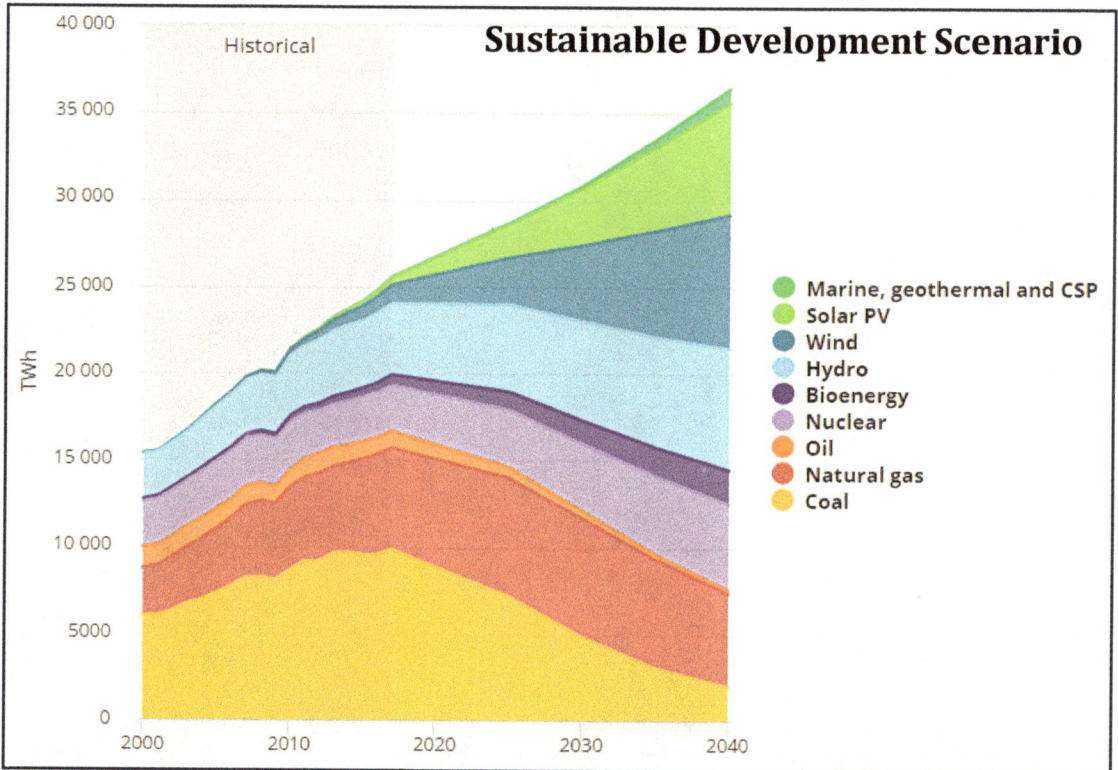

Figure 5.10b   International Energy Agency's global electricity fuel mix for two scenarios *(IEA,2019g).*

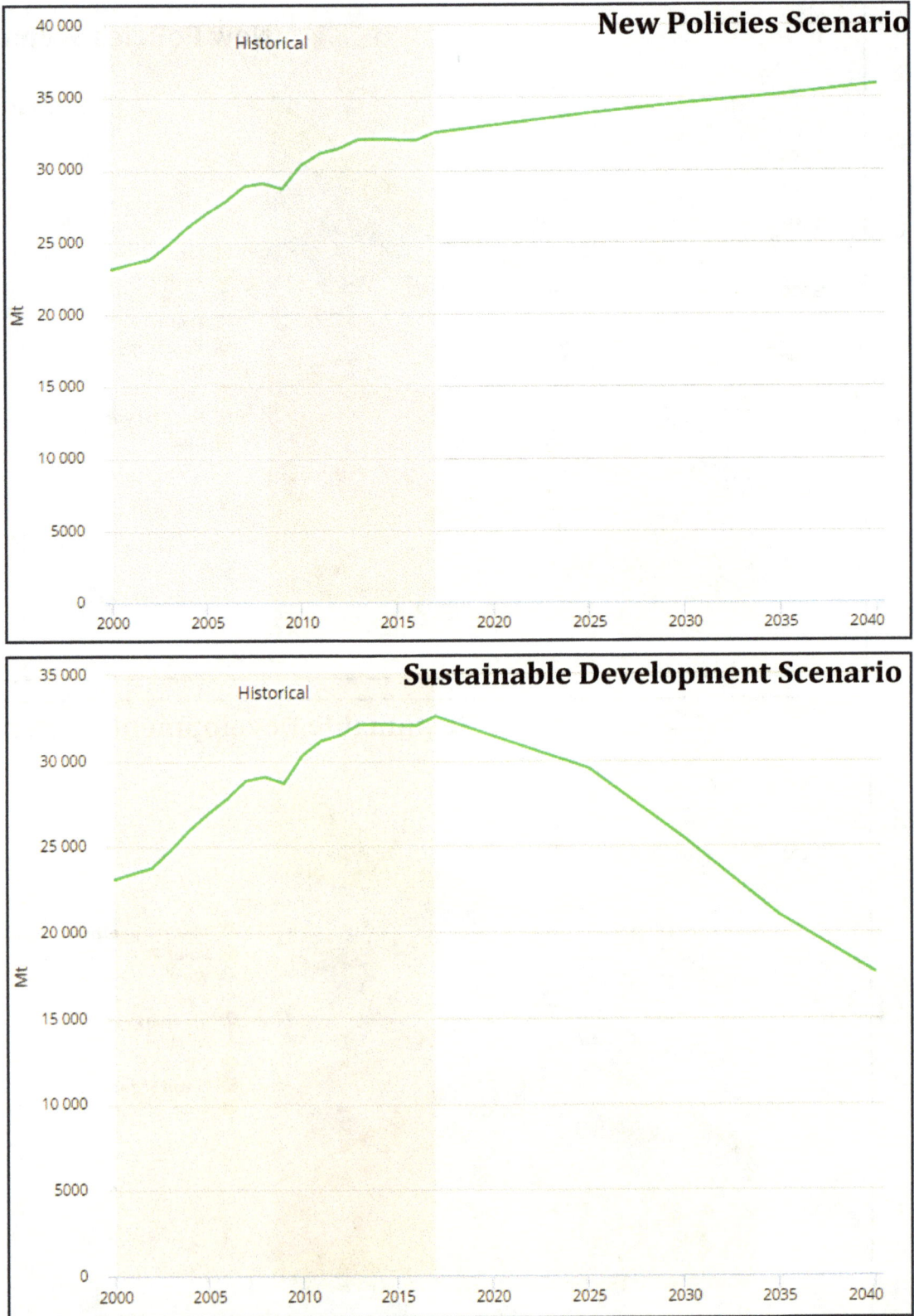

Figure 5.10c   International Energy Agency's energy-related emissions for two scenarios *(IEA,2019g).*

Chapter 5:
Appendices

Appendix 5A
Outlook on electric vehicles

There are two main types of electric vehicles and many variants (see Box 5A-1). Battery electric vehicles differ from normal internal combustion engine vehicles in that they are propelled by a large electric motor which is powered by a rechargeable onboard battery system (Figure 5A-1). One main advantage of the EV is that there are no tailpipe emissions which have been the main sources of urban pollution, and total lifecycle emissions of EVs can be reduced significantly if the electricity for manufacture and operation is generated by low-carbon sources such as renewables and nuclear energy. Deteriorating urban air quality is a serious problem for human health, and is a prominent feature of most cities around the world. Hybrids are variants of EVs which combine internal combustion engines and electric motors, and have some tailpipe emissions depending on type and driving habits. The transportation sector of the economy is the second highest source of energy-related pollution, after power generation, and accounts for about a quarter of the global total. Hence the sector has been a major focus of emissions abatement policies. Pollutions from the sector also have a deleterious impact on human health and the ecosystem because most of the emissions remain in the lower atmosphere. Electromobility (electric vehicles, e-bikes, electric scooters, trains and trams) has a very high potential for decarbonizing the sector because the tailpipe anthropogenic emission of a battery electric vehicle (BEV) is zero. Electric vehicles are seen by governments as an important part of cutting emissions and reducing global warming, but an intensive discussion is emerging in the scientific world on the environmental credentials of electric vehicles largely because variables such as source of local electricity, tariffs and access to cheap off-peak electricity could have a significant impact on the relative advantage of EVs. For example, when most of the electricity comes from coal, as in China, EVs could be far more polluting than conventional cars on a lifetime basis. On the other hand, in a country like Norway which obtains most of its electricity from hydropower, pollution credentials of EVs are superior to those of conventional cars.

The United Nations Paris-2015 world conference on the environment set the objective of limiting the increase in the global average temperature to well below 2°C above pre-industrial levels, and pursuing efforts to further reduce this limit to 1.5°C (UNFCCC, 2015b). Global response has been strong, and many organizations have proposed potential pathways to achieving this goal. The International Energy Agency (IEA) has defined two emissions reduction pathways that could be compatible with these goals, providing a 50% chance of meeting the respective temperature rise limit targets: the Two Degree Scenario (2DS) which targets the 2°C limit, and the Beyond Two Degree Scenario (B2DS) that focuses on the 1.5°C limit (IEA, 2017h). In both scenarios, energy-related GHG emissions will need to reach net-zero in the second half of this century; close to 2060 for the B2DS and close to 2090 for the 2DS (Figures 5A-2 & 5A.3). The International Energy Agency (IEA) is coordinating the Electric Vehicles Initiative (EVI) comprising ten member governments (nine OECD countries and China). Collectively, these countries account for 95% of the global electric vehicle market registrations, and the group is dedicated to promoting electromobility as a major climate change mitigation option, and accelerating the deployment of EVs worldwide (CEM, 2017). The EVI has established partnerships with many independent international organizations and automakers that have set similar goals. In 2017, EVI launched the *EV30@30* campaign that set the collective aspirational goal of a 30% market share for electric vehicles in the total population of passenger cars, light commercial vehicles, buses and trucks in all member countries by 2030. The body also identified several implementing actions aimed at helping achieve this goal in accordance with the priorities and programs developed in each EVI country (IEA, 2017l).

BOX 5A-1
TYPES OF ELECTRIC VEHICLES

Electric vehicles are light-duty vehicles (LDVs) which comprise cars, sports utility vehicles and light trucks that run fully or partly on electricity. There are numerous classifications, definitions and terminologies which can be confusing. An attempt is made here to present a simple classification that eleminates as much of the technical jargons as possible, with emphasis on relative advantages and problems. Basically, there are two main classifications of electric vehicles: all-electric vehicles (AEVs), and hybrids (PHEVs/ EREVs). There are several variants of each type, designed to accommodate different needs. The main difference between the two types is that AEVs are powered entirely by electricity from on-board batteries, while hybrids combine internal combustion engines (ICEs) with battery power.

All-electric vehicless can be sub-classified into two main types: battery electric vehicles (BEVs) and fuel cell electric vehicles (FCEVs), the main difference being in the source of electricity. BEVs are powered by on-board batteries that must be charged from the electrical grid, while FCEVs draw power from an on-board fuel cell that generates electricity on board from liquid hydrogen stored in the vehicle tank. Both types also derive some electric power from some of the energy normally lost when braking (regenerative braking). Depending on driving habits such as use of heating/air conditioning, a battery-electric vehicle has a typical range of up to 150 kilometers between charges, while some luxury models have ranges up to 400 kilometers after which the vehicle must be plugged into a power grid. Charge time ranges from 30 minutes with fast charging, to nearly a full day depending on the type of charger and battery. Fuel cell vehicles draw electricity from an on-board fuel cell that runs on liquid hydrogen drawn from a tank which is replenished in much the same way as for ICEs. The range of an FCEV depends on the size of the liquid nitrogen tank and is typically around 450 kilometers. An FCEV costs around three times the cost of a comparable BEV, and also has the relative disadvantage of few liquefied hydrogen refill stations.

In broad terms, a hybrid has a conventional internal combustion engine as well as one or more electric motors which draw electric energy from a battery pack. Energy for charging the battery may come from conversion of momentum energy that would have been lost in braking into electricity, from external power grid, or from an on-board generator fueled by petrol or diesel, and there are two common classifications: full hybrid and extended-range (EREV). Full hybrids are not considered as electric vehicles in some classifications because they cannot run purely on electricity and only use electrical energy recovered from the braking system to supplement internal combustion engines, for example, by providing more power when needed and reducing fuel consumption, typically by around 10%. Plug-in hybrids have both conventional ICE, one or more electric motors, and a battery that provides electrical energy to the motors. This configuration provides an option of operating the vehicle solely on energy provided by the battery, or on the ICE drawing petrol or diesel fuel from the tank. The main advantage is the option of driving on electricity as often as possible, for example in urban areas where EVs are most efficient, and on ICE in inter-city travel. PHEVs have a driving range of 20 - 60 kilometers of battery-powered driving and a combined range of around 300-450 kilometres depending on the model. Most plug-in hybrids have on-board computers that help select the best option for different driving conditions. The fact that the exhausted battery must be recharged from an external grid has led to the development of variants known as extended-range electric vehicles (EREVs) which feature on-board continuous battery charging facilities that eliminate the need for external grid plug-in. This extends the combined range of travel to the range of the ICE.

The main advantages of electric cars are the relatively lower energy consumption (around 30% energy economy compared with gasoline powered vehicles). For example, hybrids feature the stop-start technology which shuts the engine off while the vehicle stops in traffic, and the electric motor automatically restarts when the driver releases the brake, thereby conserving fuel and eliminating emissions from an idling engine which is the highest of all driving modes. Electric vehicles have relatively low carbon footprint because, when the vehicle is being used in electric power mode, there are no tailpipe emissions. However emissions levels are standard when a hybrid is being driven in traditional fuel mode. Also, EVs are particularly well-suited to intra-city driving (one of the main sources of ambient pollution and city smug), where the start-stop system helps reduce energy consumption when the vehicle stops, thereby reducing energy consumption. Apart from being environmental-friendly, plug-in cars offer a number of other benefits compared with conventional vehicles. Depending on the relative costs of gasoline and electricity, specific energy cost (energy cost per kilometer) for BEVs could be only a quarter of the cost of a similar internal combustion engine (ICE) vehicle, especially when owners have access to low off-peak electricity tariffs; there are fewer mechanical components in an electric vehicle when compared with conventional vehicles, which often results in lower servicing and maintenance costs. However, there are also several limitations: emissions during manufacture are higher than for comparative conventional vehicles; a BEV costs around 30% more than a comparable ICE vehicle; a PHEV costs around 50% more, while an FCEV) costs around three times. However, the cost gap is expected to close as battery costs decrease over the next few years. Proliferation of battery charging and liquid hydrogen refill facilities is still severely limited. Most existing support policies (such as tax-free purchase) target BEVs only because of their zero tailpipe emissions, and also promote the provision of home and work charging facilities.

Figure 5A-1 Chassis of Tesla Model S, displaying Batteries and Motors.
(Source: large.stanford.edu).

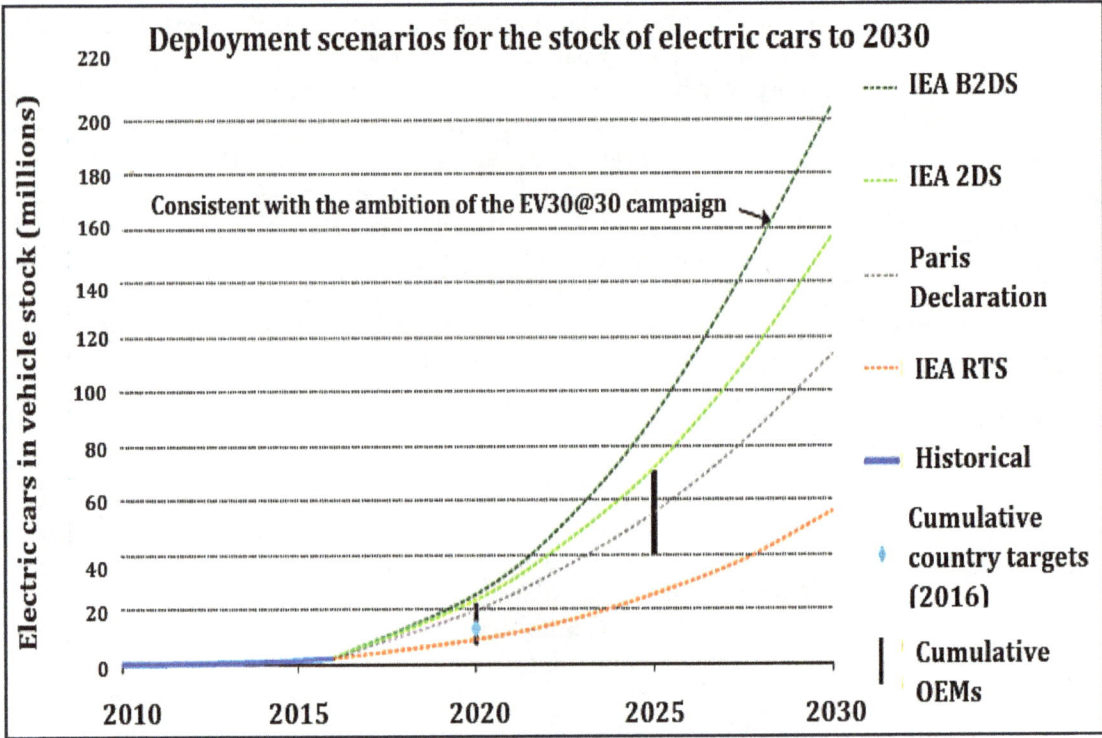

Figure 5A-2   Projections of the electric car market compatible with various emission mitigation goals *(IEA, 2017l).*

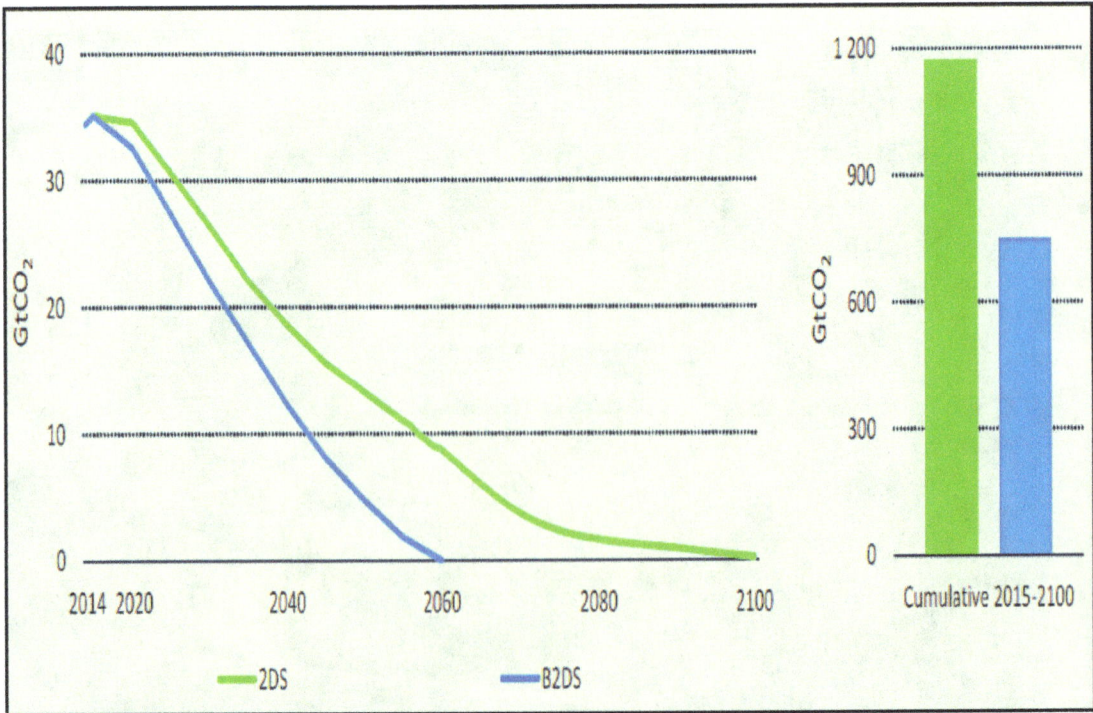

Figure 5A-3   GHG emission budgets and emission trajectories to 2100 for the energy sector, 2DS and 2BDS *(IEA, 2017h).*

There are clear signs that emerging technologies in EV battery technologies will continue to narrow the cost competitiveness gap between EVs and ICEs, and many countries particularly in the OECD region are introducing strong support policies to stimulate EV market growth and proliferation of charging bays, while some have even announced total ban on sales of internal combustion engine cars by 2040. Many auto manufacturers (OEMs) have also announced significant electric car production capacity scale-up plans over the next five to ten years, a development which is expected to increase the electric car stock up to tenfold the current level by 2020. The global stock of electric passenger vehicles has increased consistently by about 60% annually over the last three years reaching about 5 million in 2018, with BEVs accounting for around 65%. Projections of the electric car market compatible with various emission mitigation goals are shown in Figure 5A-3, and the actual market growth rates are presented in Figures 5A-4 to 5A-8. The Reference Technology Scenario (RTS) takes account of the past market growth pathway, emerging technologies that are likely to fast-track improving energy efficiency, and EV support policies that have been announced or are under consideration, to arrive at a projection of nearly 60 million EVs in the vehicle stock by 2030, while a projection based on the announced plans of the Organized Equipment Manufacturers (OEMs) comprising mainly auto manufacturers shows a rise to between 9 million and 20 million by 2020. Both fall short of IEA's 2DS projection of around 70 million by 2020 and over 160 million by 2030, which is consistent with 50% probability of achieving the 2°C target. An even faster growth rate would be required to achieve an EV population of around 200 million by 2030, which is consistent with 50% probability of achieving the 2°C target. Another projection by IEA (2019) indicates that total EV stock could reach 250 million by 2030 under the EV30@30 initiative.

It is difficult to see a clear, feasible pathway to the achievement of most of these scenarios, considering the fact that the total global population of the three variants of electric vehicles is currently around 5 million, accounting for about 0.2% of the nearly one billion vehicles on the world's roads (CNEV, 2017). However, achievement and sustenance of this level of growth rate could lead to 40-70 million cars on the road by 2025. There are many other projections on the outlook of the global EV market over the next two decades, based on the Reference Technology (RTS) Scenario, and the most optimistic forecast a share of 6-15% of the global vehicle market, projected to be around 2 billion by 2040, and the share of electricity in the total global transportation energy will still be no more than about 5%.

Recent studies identified the high cost of electric vehicles is the main disincentive to consumers: a standard medium size EV costs around 40% more than a similar size internal combustion engine (ICE) vehicle. However, when compared on a total life basis which includes purchasing, operating and maintenance costs the gap closes significantly and may be similar depending on local fuel and electric power prices. In effect, the economic advantage of an EV over an ICE may be insignificant. The main problem is the battery cost which could account for up to half the price of the vehicle. Another disincentive is the inadequate battery charging infrastructure. Although there were as many charging points as there were EVs in 2018, most were private, slow systems and a typical battery would require 6-8 hours to charge fully (Figure 5A-8). People who live in apartments or park in the streets would hesitate to buy an EV and even those who have home charging points could be stranded if they travel too far. Furthermore, the low price of fuel and improving specifications of ICEs have depressed sales of EVs over the last year or two. However, battery cost has dropped by about 25% in the last twelve months or so and this could improve the economic benefits of EVs. Even then, strong policy support, in particular, proliferation of fast public charging facilities would be required to motivate potential consumers. The potential for commercial battery swapping bays is also high but batteries would need to be standardized.

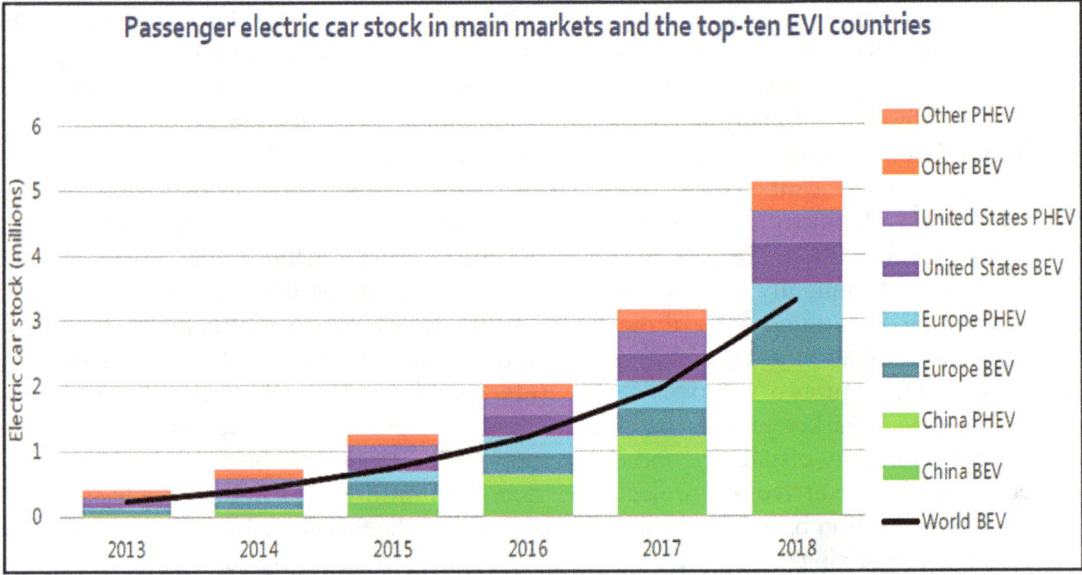

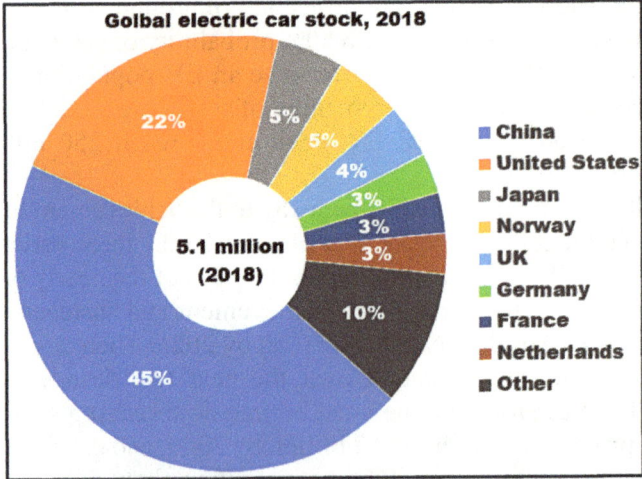

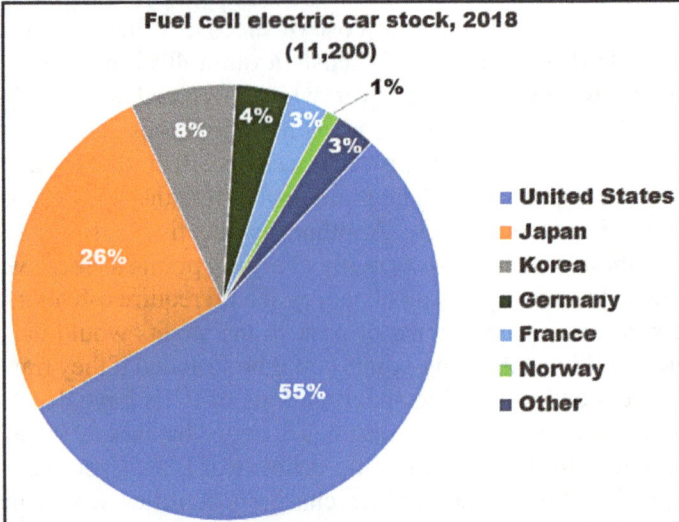

Figure 5A-4   Passenger electric car and fuel cell electric car stocks in main EVI markets in 2018 *(Data from IEA, 2019).*

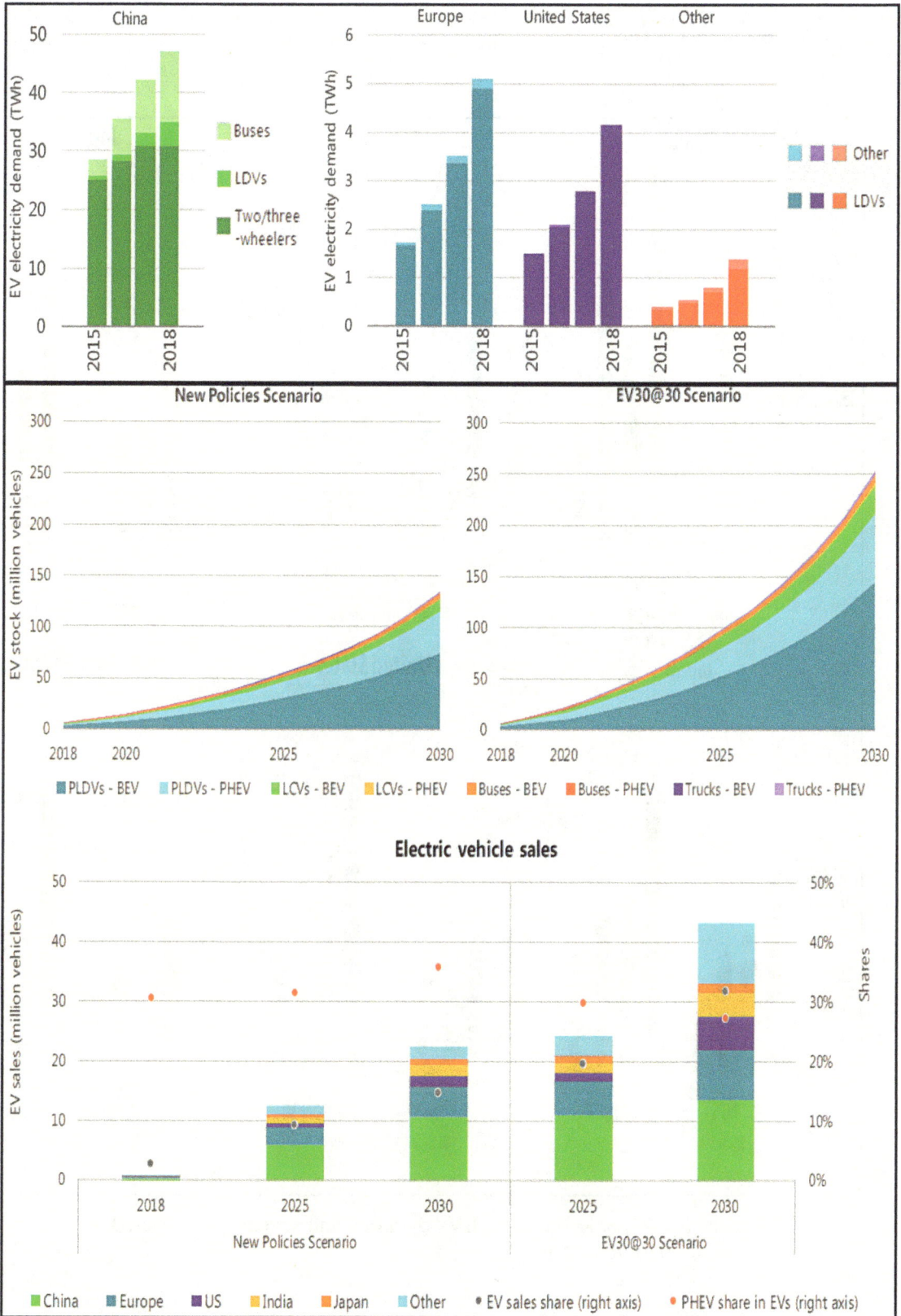

Figure 5A-5   Global EV stock and sales by scenario, 2018-30 *(IEA, 2019).*

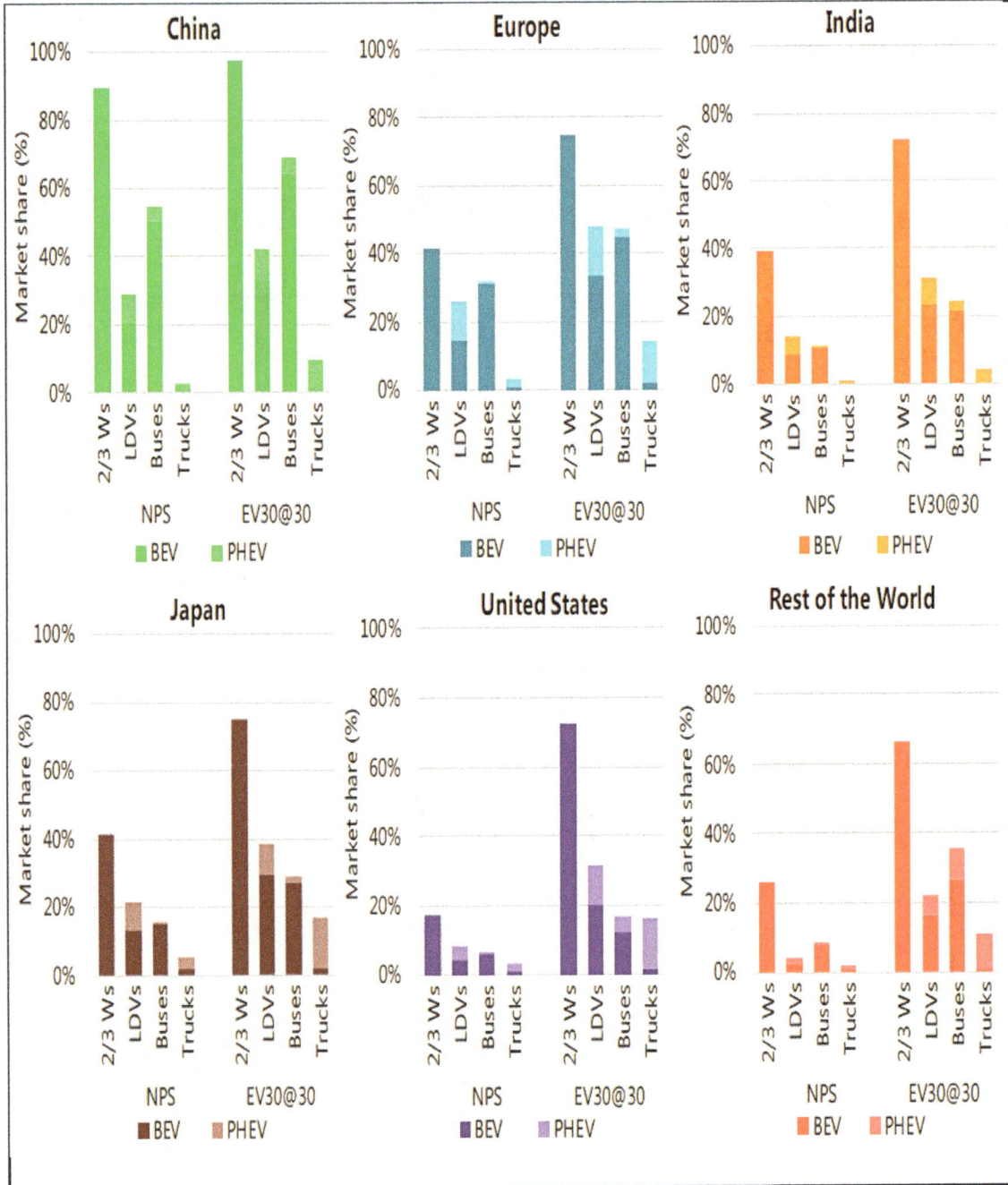

Figure 5A-6 Sale shares of EVs by mode and scenario in selected regions, 2030 *(IEA, 2019)*.

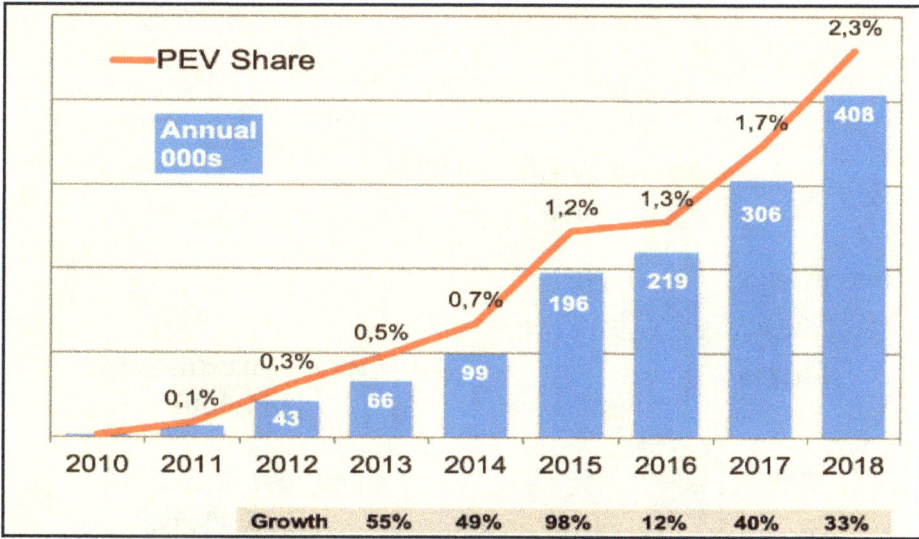

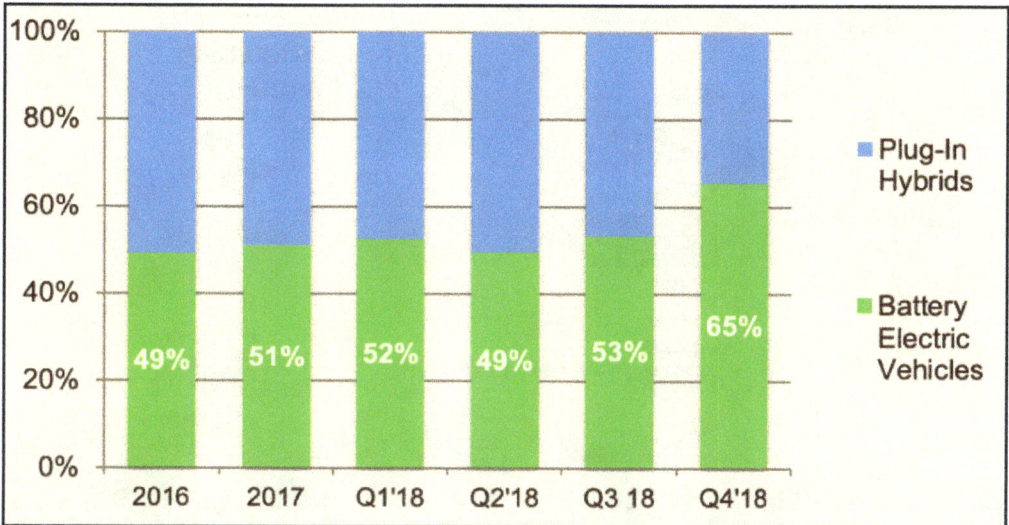

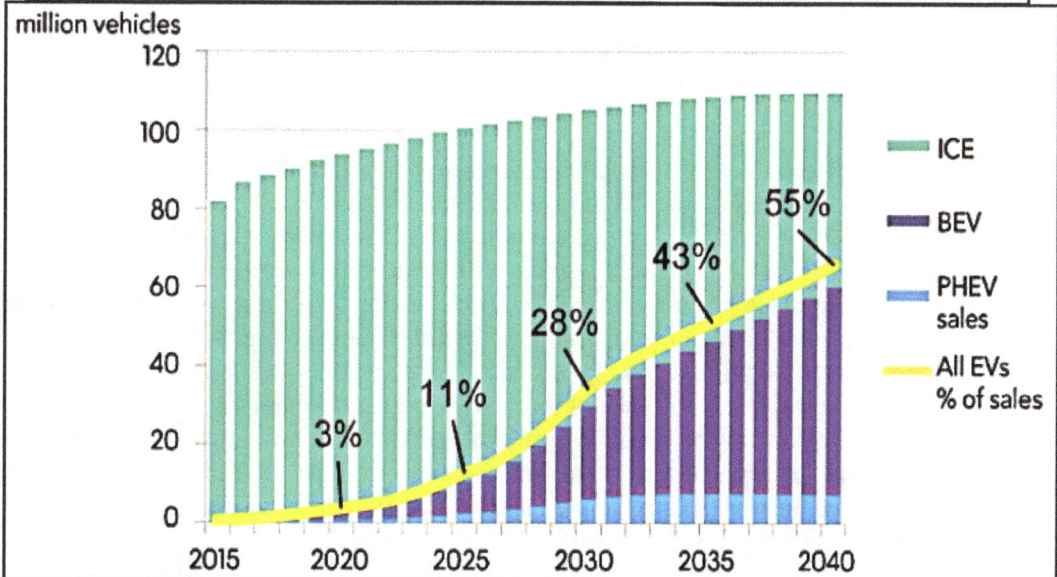

Figure 5A7 Projected EV market to 2040 *(BNEF, 2018).*

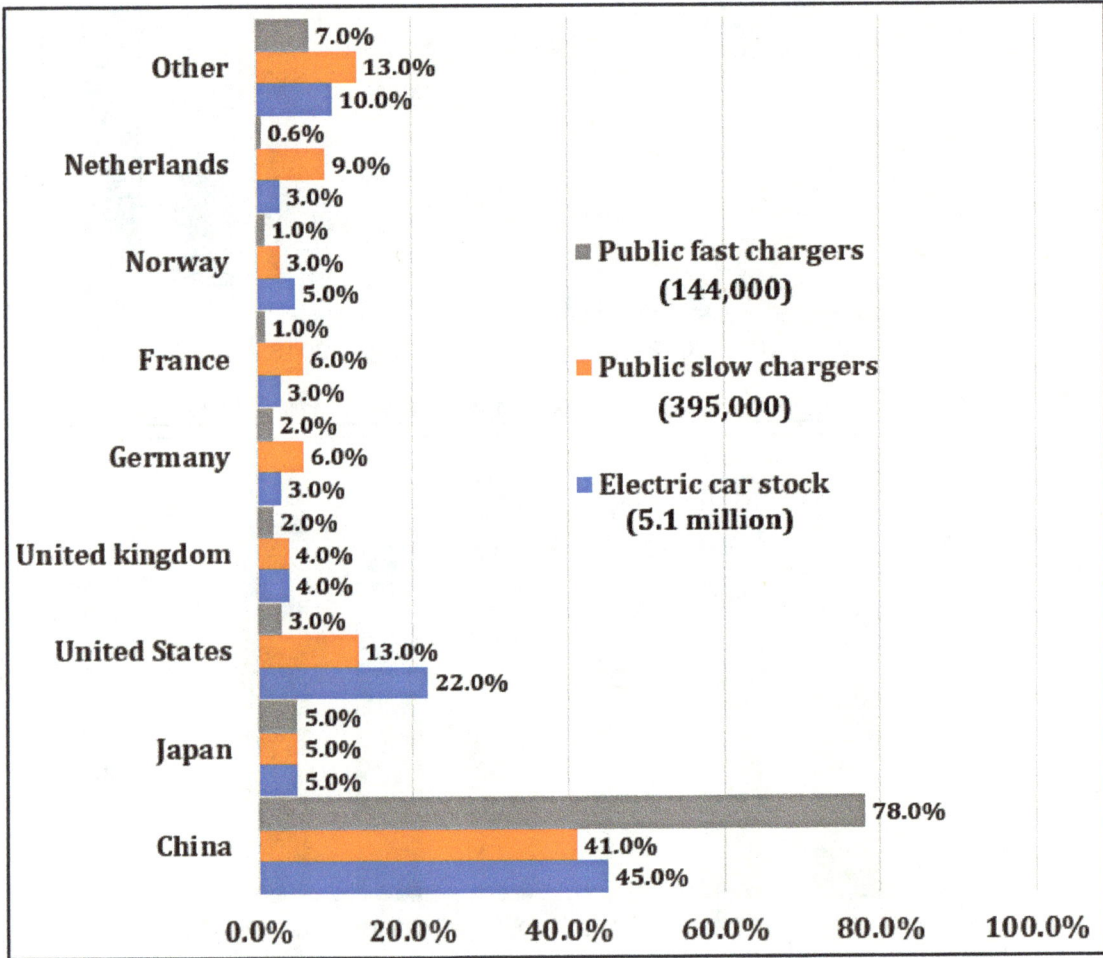

Figure 5A-8 Electric car stock and publicly accessible chargers by country, 2018 *(Data from IEA, 2019).*

So far sales of electric vehicles have been restricted mainly to Western Europe, Japan and the United States, although China also has a large population of mostly 2/3 wheel vehicles. Most of the projected growth in the EV market will be in the emerged countries whereas the bulk of the growth in the global vehicle market will be in the emerging countries most of which have no strong policies on mitigation of environmental pollution. Many have low access to electricity for primary use, and provision of facilities for recharging EV batteries are of low priority. Also, by most projections, coal which is the most polluting fossil fuel will still be the primary electric power generation fuel in most emerging countries in 2040, in which case the impact of electric vehicles on emissions mitigation would be minimal. Furthermore, commercial, marine and air transportation sub-sectors will account for most of the growth in the transportation sector over the next two decades, largely in the developing world, and there are no viable alternatives to diesel and jet fuel as the main fuels.

A recent study (BEN, 2018) presents a more optimistic projection on the global electric market over the next twenty years or so. Sales of electric vehicles are projected to increase from 1.1 million in 2017 to 11 million in 2025 when manufacturing costs are expected to drop to the level for making comparable petrol/diesel cars. Henceforth, sales will rise at an even faster rate, reaching around 30 million in 2030, when EVs should become competitive even without subsidies. China will lead this transition, accounting for about 50% of the global EV market in 2025, and will continue to lead the global market through 2040. By 2040, 55% of all new car sales and 33% of the global fleet will be electric. Electrification of buses is projected to grow at an even faster rate than cars. China currently has around 300,000 E-buses (99% of the world total in 2017)and the global market is expected to grow by about 84% above current level by 2040. Achievement of the projected growth of the EV market is expected to raise global electricity demand by around 6%, displacing about 7.3 million barrels a day of transport oil. The projections were based largely on manufacturer's plans for model roll-outs, new regulations on urban pollution, and the assumption that consumer interest can be sufficiently stimulated. The study also identified several formidable hurdles that need to be surmounted to achieve these optimistic projections, notably shortage of vital alloys (cobalt, lithium) required for battery manufacture, and the challenge of charging infrastructure. The study also ignored the important factor of persistent decline in global oil price which has always impacted negatively on EV sales.

The net environmental impact of the EV depends on the fuel source for electricity: an electric vehicle that runs on fossil-fueled electric power still has a carbon footprint of around 80% of that of a comparable internal combustion engine vehicle in addition to the significantly higher emissions in the manufacturing stage. China is projected to account for around half of the global EV sales in 2040, but the country's carbon intensity of electricity is one of the highest in the world. Furthermore, there has been a sharp rise in the demand for lithium, cobalt and nickel, the most important constituents of lithium-ion batteries which power the latest generation of EVs. Apart from the fact that the production of these primary alloys is energy-intensive, most lithium ion batteries which are emerging as the battery of choice, are sourced from Asian countries which depend on fossil energy from mining to finished products. The battery is the most important component of the EV and, currently, there are major issues with size, cost and charging cycles. However, emerging technologies show that these issues could be resolved in the next few years (see Appendix 5B). Another issue with batteries is that, while nearly a hundred percent of lead-acid batteries which power ICE vehicles are being recycled, only around five percent of lithium-ion batteries are currently recycled, and the anticipated number of discarded batteries could create new environmental issues unless policies that promote reuse or recycling are put in place expeditiously.

In spite of the low prospects of EVs emerging as a significant component of the global vehicle mix in the next two decades, they still have a high potential to clean-up the polluted ambient environment of most cities fouled by emissions from urban traffic which constitute a very serious threat to public health. While transportation pollution may have some long-term effect on climate change which is controlled largely by events in the middle atmosphere, most of the pollution is retained in the lower atmosphere and, although most of the pollutants only last a few days to a few weeks in the atmosphere, they have the potential to cause severe public health problems which when combined with household pollution, account directly or complimentarily for around 7 million (one in eight) premature deaths a year globally, mainly from heart disease, stroke, chronic obstructive pulmonary disease, lung cancer and acute respiratory infections, particularly in children (WHO, 2018). Many cities and municipalities have been at the forefront of stimulating EV deployment by instituting policy and financial incentives, support for the proliferation of charging infrastructure, and leading by example by electrifying public vehicles.

In recent times, the green credentials of electric vehicles when considered on a lifecycle basis have come under intense scrutiny. Multiple studies have found that electric cars are more efficient, and therefore responsible for less greenhouse gas and other emissions than cars powered solely by internal combustion engines. Also, a recent EU study found that an electric car using electricity generated solely by an oil-fired power station would use only two-thirds of the energy of a petrol car traveling the same distance. Although an electric car powered in this way is still ultimately burning the same fuel as the petrol car it replaces, it is burning much less of it. Although around three times energy is required to generate electricity that an EV requires to cover a given distance, compared with energy required to get an equivalent amount of petrol into the tank of an ICE, for the same distance, an ICE expends nearly four times as much energy as an EV in covering the same distance, which means around four times higher pollution. Furthermore, although greenhouse gas emissions are similarly harmful wherever they occur, some other emissions which are harmful to human health are less dangerous when they happen in a high stack at a power plant outside the city than in densely populated cities. However, many other studies reviewed reached conflicting conclusions. For example, two recent studies by MIT (2017) and T&E (2017) considered a wide range of variables and showed that, in some cases, EVs could be as polluting or even more polluting than ICEs of comparable specifications, but, on the average, electric vehicles emit significantly less CO_2eq than petrol cars. On the other hand, some recent lifecycle analyses of the global warming impact of production and operation of EVs compared with the production and operation of conventional cars indicate that electric cars can have a greater impact (in some cases, almost double) on global warming than conventional cars, depending on the sources of electricity for the energy-intensive manufacture of EVs, especially the batteries, as well as for recharging the batteries, largely because mining and production of the main components of lithium-ion batteries are energy-intensive and are largely concentrated in the emerging world (see Hawkins *et al.*, 2012).

Another study by Ricardo Consulting (2011) estimated that production of an average petrol car involves emissions amounting to the equivalent of 5.6 metric tons of CO_2eq, while for an average electric car, the figure is 8.8 metric tons. Of that, nearly half is incurred in producing the battery. Despite this, the same report estimated that over its whole lifecycle, the electric car would still be responsible for 80% of the emissions of the petrol car. Reports of many more studies have been published in recent times and it is interesting to note that these reports form the basis for arguments of proponents and opponents of e-mobility. All that is needed is to take an appropriate statement in a report out of context. Most of the reports highlight issues that could raise the lifecycle pollution emissions by EVs, such as higher manufacturing energy intensity compared with ICEs, and the sources of electricity for both manufacture and eventual operation, but nearly all conclude that electric cars are more efficient, and therefore responsible for less greenhouse gas and other emissions than cars powered solely by internal combustion engines, even when the electricity for manufacture and operation is sourced from fossil fuels. One major advantage of EVs is the fact that nearly half of the lifecycle emissions of EVs occur before the vehicles have come on the roads, thereby significantly reducing impact on ambient pollution, in particular, urban traffic pollution and travel along major transport axes, and offering important potential options of reducing electricity-related emissions at source.

There is no doubt that the manufacture of electric vehicles emits significantly more pollution than comparable ICE car manufacture, estimated to be around 60% more. However, over the lifetime of each vehicle, many studies have shown that EVs produce around a third less CO_2eq during their life cycle (a mid-size EV produces around 18 metric tons of CO_2eq compared with 24 metric tons for ICEs). However, the comparative cradle-to-grave emissions depend on the sources of electricity for manufacture and subsequent operation of electric vehicles. In effect, a lifecycle emissions analysis of EVs and ICEs of similar specifications is necessary in order to capture and compare the full environmental impact of the two vehicle types. A complete life cycle analysis of a vehicle takes into account emissions associated with four major life stages: the fuel supply chain [Well-to-Tank (WTT)]; energy conversion in the vehicle [Tank-to-Wheel (TTW)]; manufacturing, maintenance and recycling of the vehicle (Glider); and manufacturing the power train of EVs (motor, battery, and electronics (Figure 5A-9 & 10).

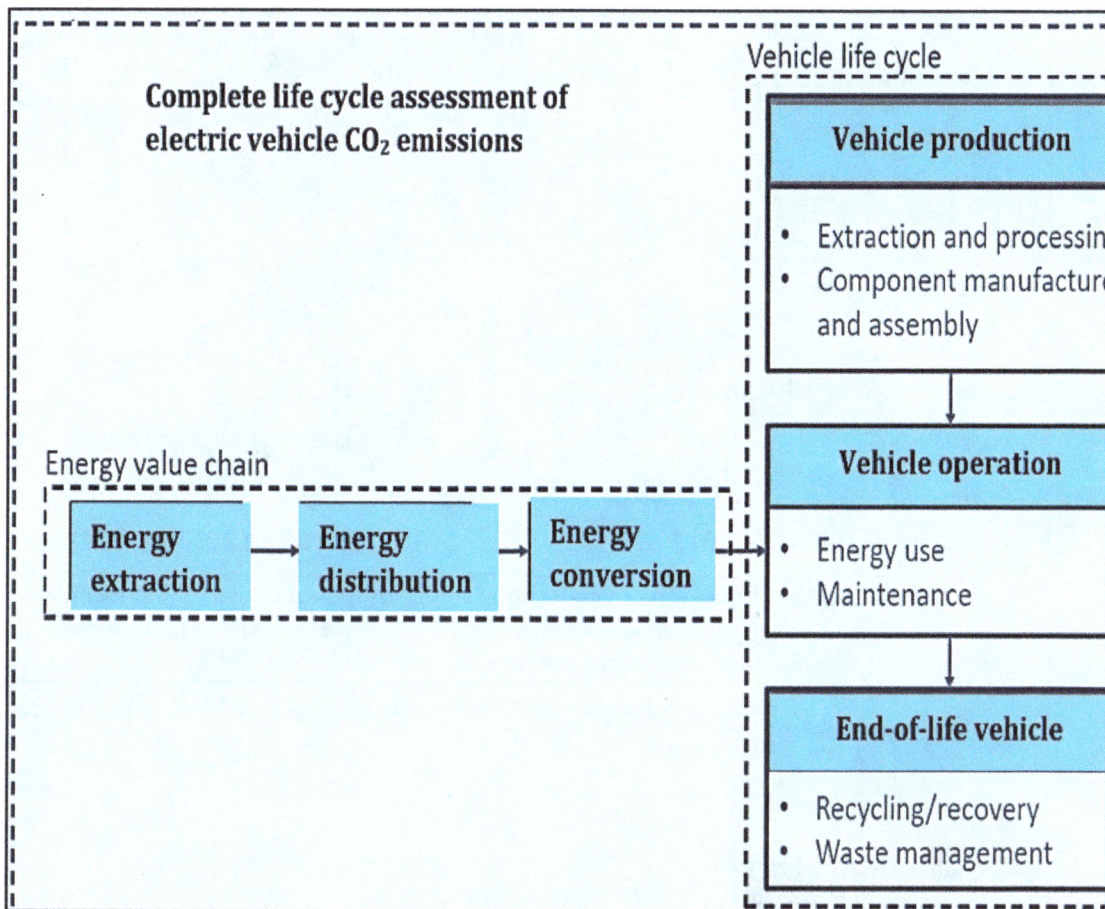

Figure 5A-9 Complete lifecycle analysis of vehicles. *(Ellingsen (2017).*

Many studies have been published in the last five years on EV outlook and environmental credentials and most have been reviewed critically in this study. The important conclusions are presented in Figures 5A-11 to 5A-16 and Table 5A-1. Nearly 70% of emissions associated with EVs come from the sources of electricity for the manufacture and recharging of batteries while around 75% emissions of a diesel equivalent comes from burning diesel over the lifetime of the vehicle (Figure 5A-12). In effect, environmental impact of electric vehicles depends largely on the carbon intensity of the electricity for manufacturing and operation. This is also evident in the results of a study of the impact of electric vehicles operating in different European countries with different power generation carbon intensities (Figures 5A-11 & 5A-12). The life cycle emissions of EVs are lowest when they operate in countries which have low carbon intensity electricity mix due to the inclusion of renewables and nuclear sources, such as Sweden and France, while those that operate in Poland which has the highest carbon intensity due to use of coal in electricity generation, have the highest emissions (Figure 5A-12). The overall conclusions of most of the studies support EV proliferation as the best option for mitigating transport pollution. The most important conclusions from the literature reviewed are summarized in Box A-2. It should be noted however that the summary simplifies the analytical findings, which in many cases have more scenarios, technical details, and results than are presented in the summary.

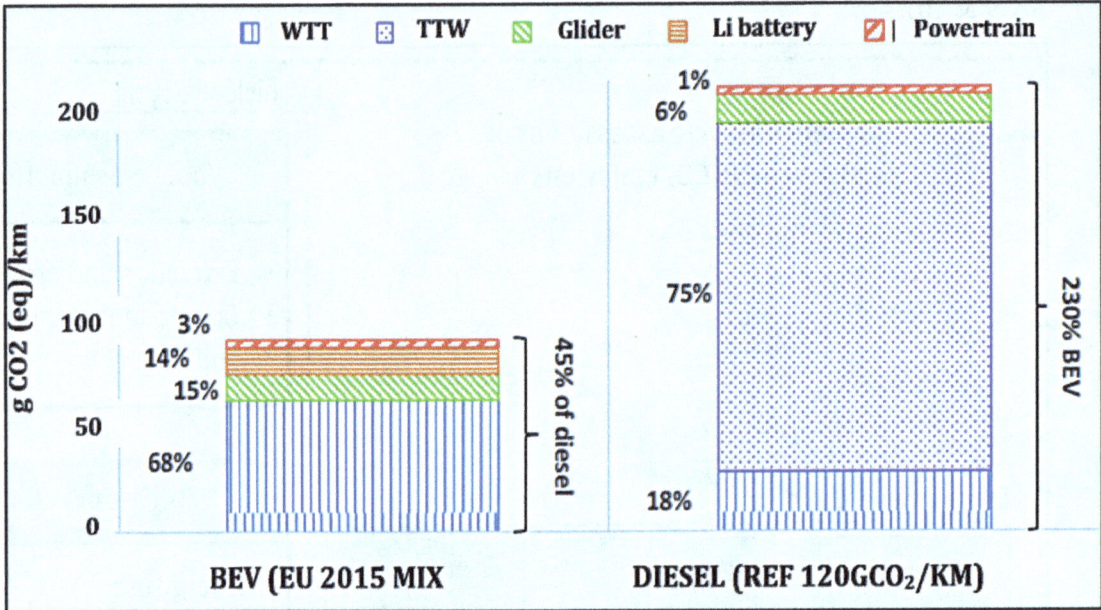

Figure 5A-10   Emissions from various stages of manufacture of and electric vehicle and a comparable diesel vehicle *WTT = Well-to-Tank; TTW = Tank-to-Wheel; Glider = manufacturing, maintenance and recycling of vehicle   (T&E, 2017).*

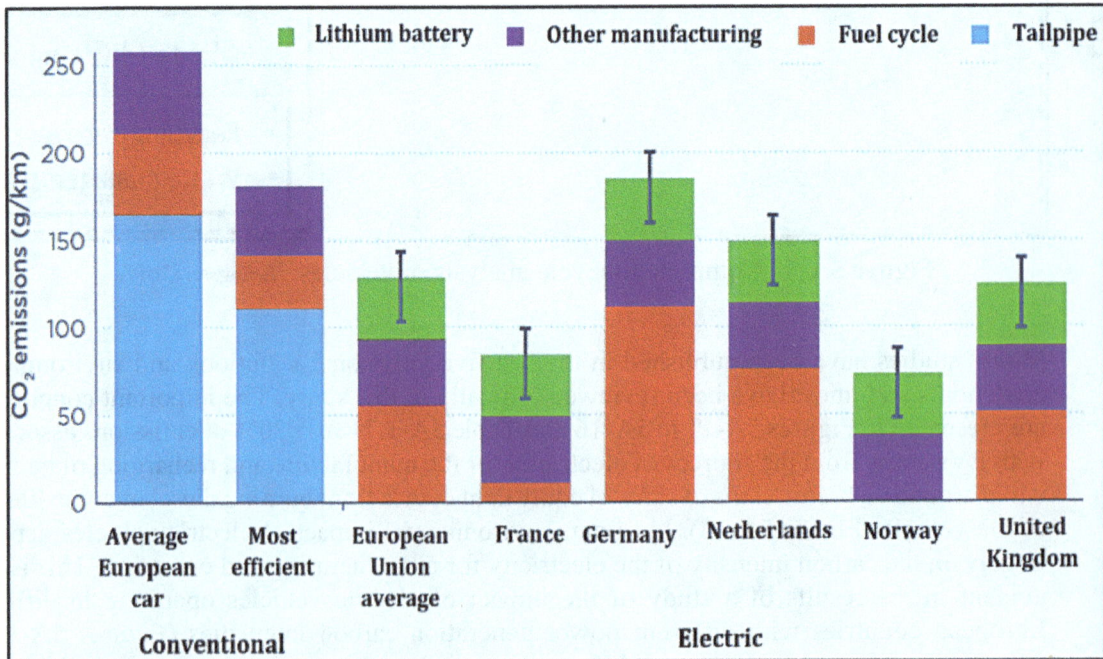

Figure 5A-11   Life-cycle emissions (over 150,000 km) of electric and conventional vehicles in Europe in 2015 *(ICCT, 2018).*

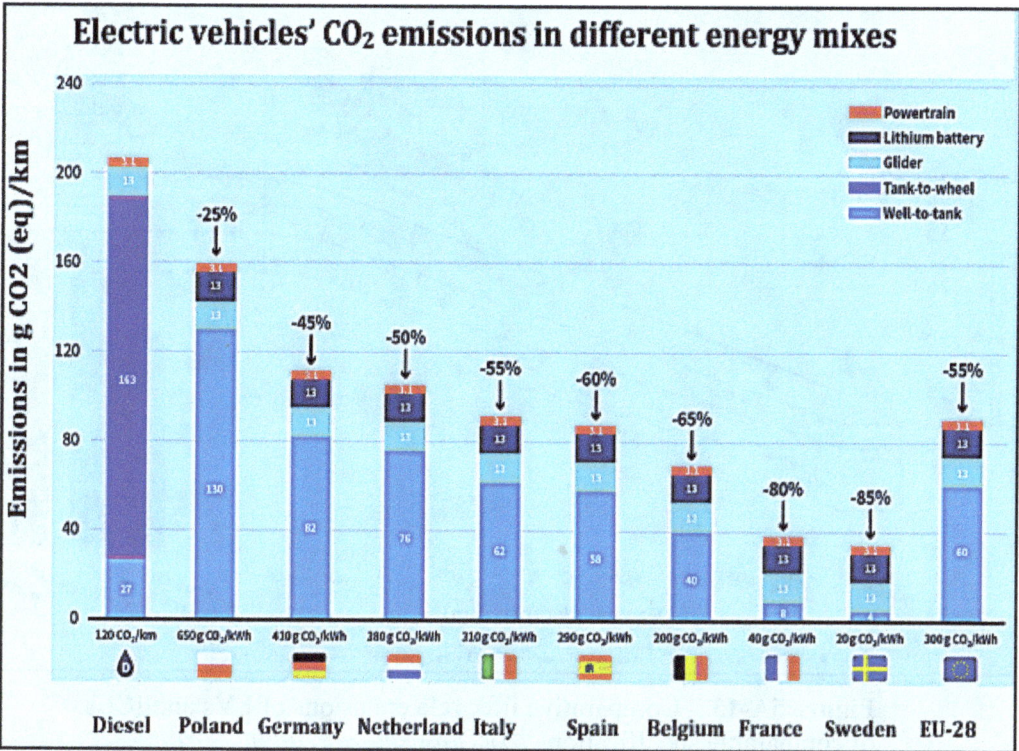

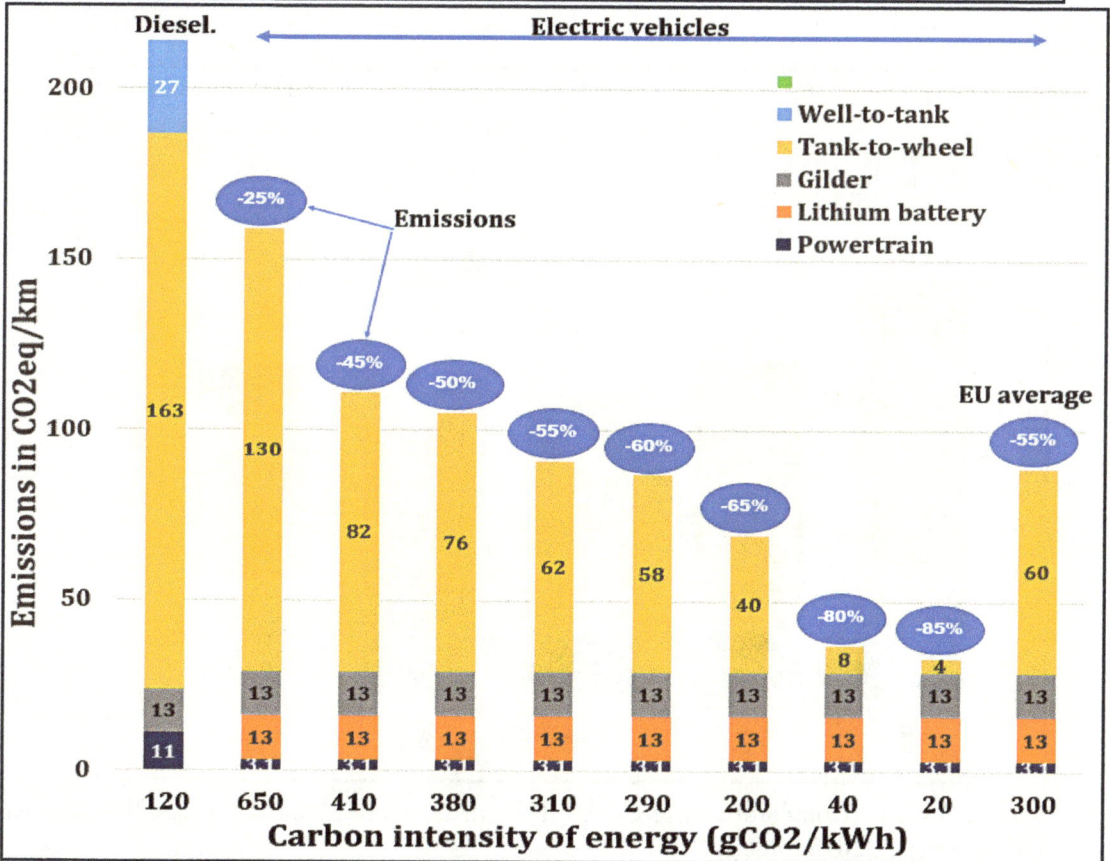

Figure 5A-12 Life Cycle Analysis of the climate impact of electric vehicles by carbon intensity of the electric source *(T & E, 2017).*

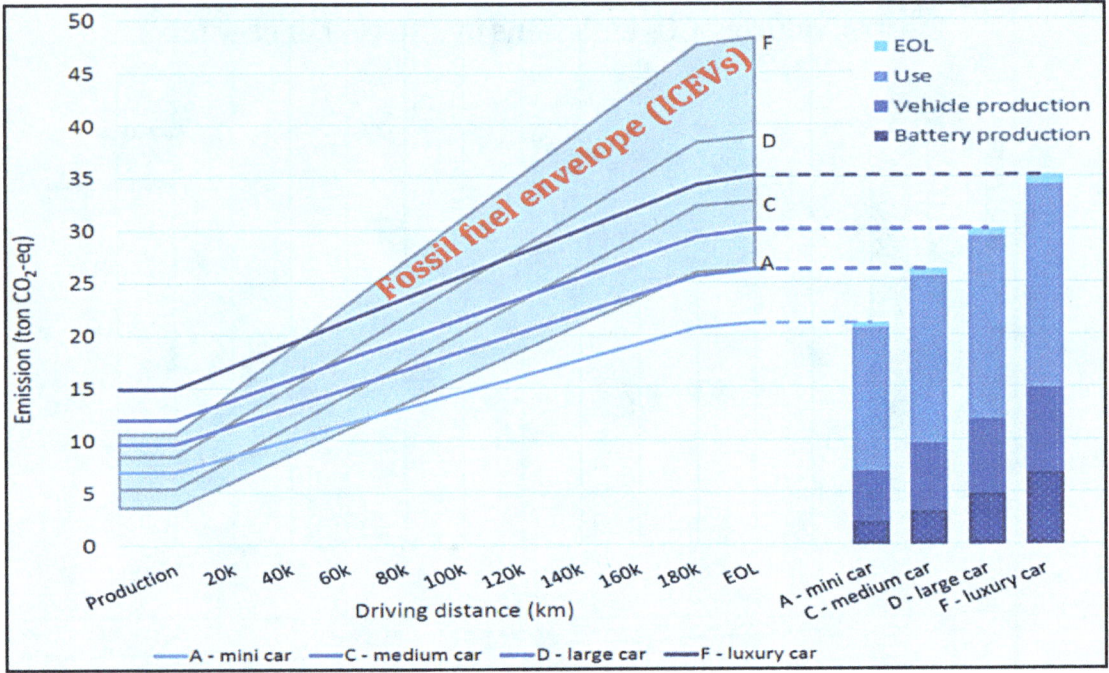

Figure 5A-13 Comparative lifecycle emissions of EVs and ICEs of comparable specifications *(Data from Ellingsen (2017).*

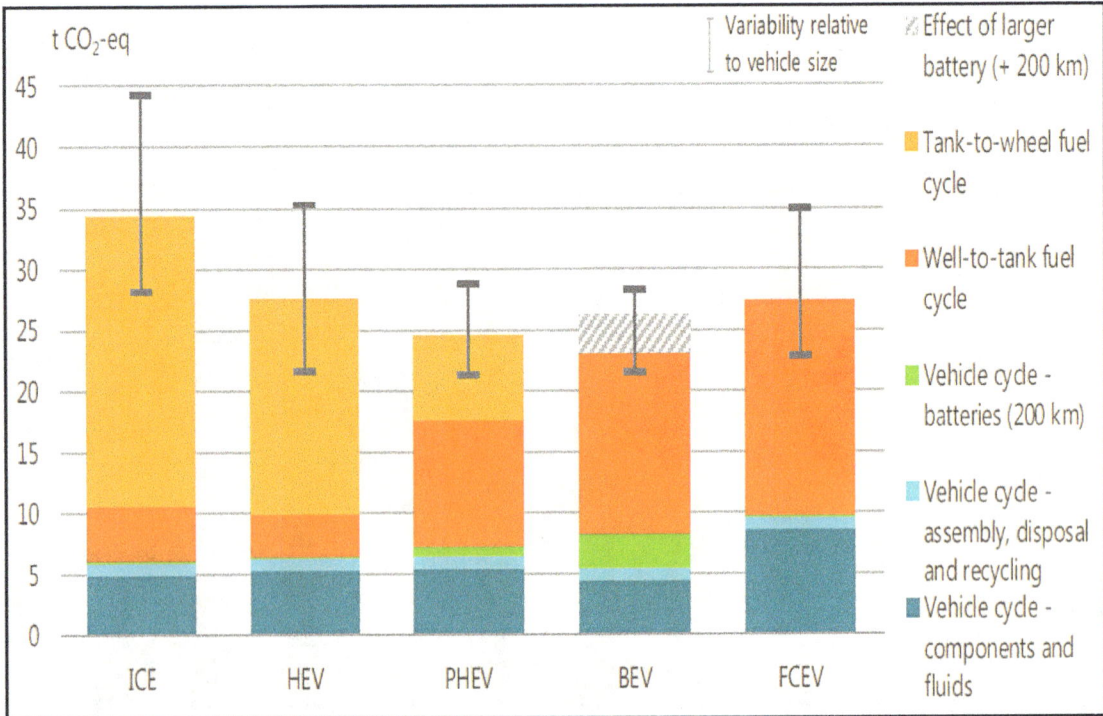

Figure 5A-14 Comparative life-cycle GHG emissions of a global average mid-size car by powertrain, 2018 *(IEA, 2019).*
ICE = Internal Combustion Engine; BEV = Battery Electric Vehicle; HEV = Hybrid Electric Vehicle; PHEV = Plug-in Electric Vehicle; FCEV = Fuel Cell Electric Vehicle

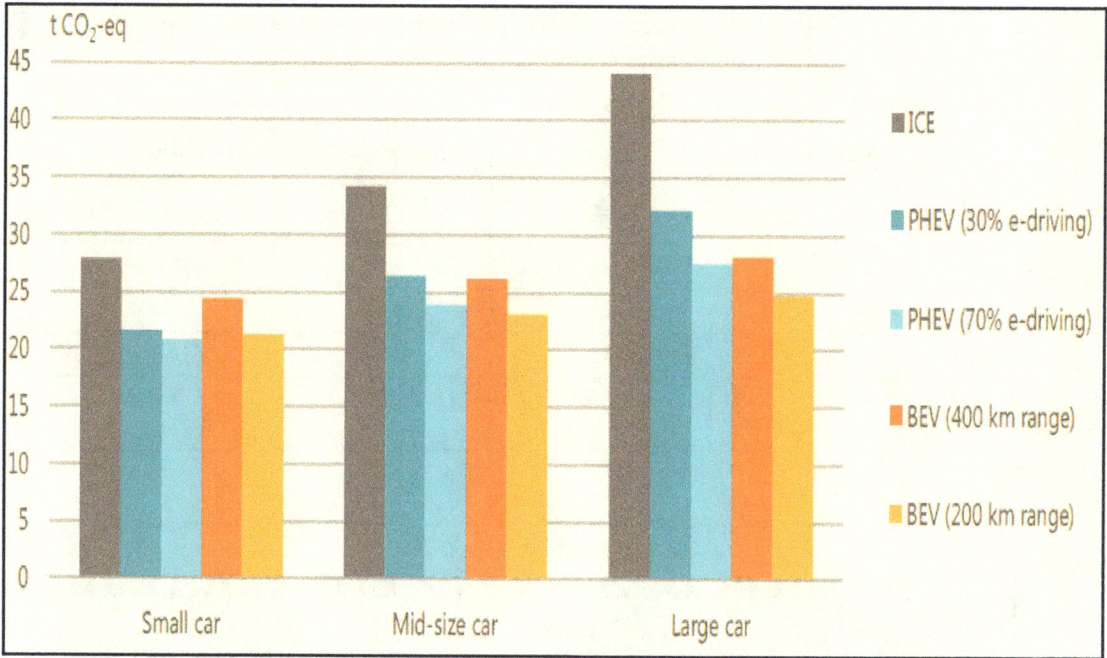

Figure 5A-15 Life-cycle GHG emissions, PHEVs and ICEs by market segment, 2018 *(IEA, 2019)*

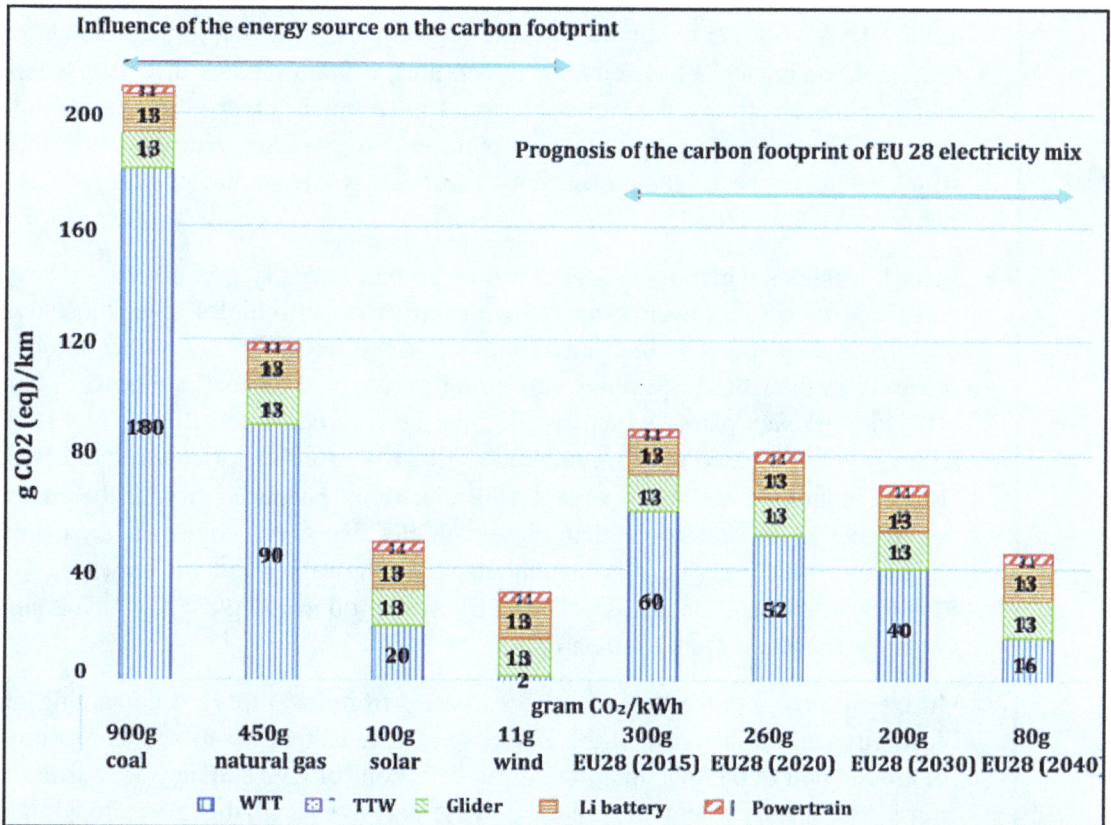

Figure 5A-16 GHG emissions of electric vehicles depending on the energy sources and the prognosis of the reduction in carbon intensity *(T&E, 2017).*

Table 5A-1 Lifecycle emission of internal combustion engine and electric vehicles (*lowcvp.org.uk*).

Vehicle type	Estimated lifecycle emissions (tonnes CO_2e)	Proportion of emissions in production	Estimated emissions in production (tonnes CO_2e)
Standard petrol	24	23%	5.6
Hybrid	21	31%	6.5
Plug-in hybrid	19	35%	6.7
Battery electric	19	46%	8.8

BOX 5A-2
SUMMARY OF OUTLOOKS ON ELECTRIC VEHICLES

- The most important variable in assessing the lifetime emissions of an Electric vehicle is the carbon intensity of the electricity used both for manufacture and charging of batteries once it gets on the roads. Carbon intensity varies from around 1,000 CO_2/kWh for coal-sourced electricity, compared with 20-40g CO_2/kWh for renewable or nuclear electricity. In effect, an EV that operates in China where much of the electricity is from coal would have much lower green credentials compared with one that operates in countries like Sweden, Norway or France which source most of their electricity from renewable or nuclear energy (see Figures 5A-12 & 5A-13).

- Electric vehicles which have access to low carbon intensity grid electricity have emissions up to 85% lower compared with similar diesel vehicles. Even those that operate on coal-sourced electricity could still have around 25% lower emissions below ICEs over their life times depending on the type of coal and power plant efficiency. Power plants which use lignite have carbon intensities over 1100g CO_2eq/kWh compared with around 900 CO_2eq/kWh for those that burn hard coal. An EV manufactured and operated with electricity produced from either plant could have higher emissions than diesel vehicles. However, countries like China and Poland which rely heavily on coal also use renewables and nuclear energy to lower the overall carbon intensity of electricity to around 700 g/kWh, which is still very high by international standards.

- On the average, relative benefit of EVs varies from 28% to 72% depending on electricity carbon intensity of the electricity. Since battery manufacture accounts for around half of the total manufacturing emissions of EVs and they are currently sourced from Asia where most of the electricity is from coal, the green credentials of EVs should be greatly improved when current plans to manufacture batteries in American and European countries with lower carbon intensity electricity mature.

- The battery which accounts for nearly half of the associated emissions during manufacture has a life of 150,000 to 200,000 km, and end-of-life (EOL) treatment could have a significant impact on the lifecycle emissions of EVs. A typical EOL battery still has around 80% electric storage capacity which could serve as storage for surplus electricity from variable solar power generation (for example in solar-powered buildings) for many more years. Around 5% of EOL batteries are currently recycled to recover valuable lithium, nickel, cobalt and graphite.

- China is the world's leading market for automobiles, both in terms of supply and demand, and accounted for nearly half of the total global sales in EVs in 2017, adding around 600,000 EVs in just 2016 and 2017, more than the EV population on all of Europe's roads. China had 215 million vehicles on the road in 2017, compared with the United State's 255 million. However, that amounted to only one vehicle per seven people, compared with nearly one-to-one in the United States, and a dynamic growth in vehicle ownership is expected in the coming years.

- China also leads the world in electrifying two-wheel (motorcycle) and commercial transportation with over 200 million e-motorcycles and more than 300,000 e-buses on the roads in 2017. China is projected to lead the world EV market, accounting for around 50% of global sales in 2025, but the share is expected to drop to around 40% by 2040 because of the expected growth of other markets. Clearly, the extent to which the country succeeds in decarbonizing electricity will largely determine the overall mitigation effect on global transport pollution.

- EU-28 is the world's second largest EV market after China, but sale in 2017 was only 307,000, accounting for only 1.74% of the total light vehicle sales.

- Manufacture of EV batteries currently accounts for nearly half of the emissions from manufacturing operations, and Asia where most electricity is coal-sourced, is also the main global source of EV batteries. Many auto manufacturers in the US and Europe have reached various stages in domesticating EV battery manufacture, using lower carbon electricity.

- Transportation accounts for around a quarter of the total global energy-related emissions, higher than other end-use sectors - industry and buildings - and electromobility offers unequalled opportunities for decarbonizing the sector. Nearly all major global auto manufacturers already have inventories of a wide range of EV options and models, and many more are at various stages of planning. Many of the manufacturers and petroleum companies are also supporting charging infrastructure development.

- Many publications and policy statements have predicted the demise of ICE mobility within the next twenty years or so, fully replaced by EVs, and some countries have even announced ban on sale of ICE diesel cars by 2040. However, an objective, in-depth analysis of the outlook does not support this level of optimism. There is no doubt that EV market penetration will continue to grow, but the rate is unpredictable (growth rate fell in the last two years due largely to low global oil prices). The most optimistic projections expect that EVs could account for 25-30% of total global auto fleet in 2040-2050. Also, although sales of diesel cars have plummeted particularly in Western Europe where severe tax penalties on diesel cars have been introduced and some countries have announced future ban on diesel car sales, there has been no significant spike in the sale of EVs, largely because most people are unsure of the outlook on EVs and ICEs, considering the conflicting views, and are simply deferring investment in autos.

- Analysts should not lose site of the fact that ICEs have been around for over a hundred years and immense resources are committed to continuous development. Major improvements in fuel efficiency, infusion of many features that are the selling points of EVs such as low emissions, stop-and-start capability, and the dynamics of the global oil market will continue to keep the ICE vehicle market in the lead in the foreseeable future. Furthermore, e-mobility has been penetrating mostly the light vehicle market whereas most of the future growth in auto transportation will be in both the light and heavy commercial sub-sectors (mostly in emerging nations) where there is relatively little emissions regulation or e-mobility penetration. Another unpredictable variable is the impact of emerging technologies in the ICE sector. Many new technology innovations such as catalytic converters fitted to the latest diesel cars, and a recent invention announced by Bosch are bringing diesel vehicle emissions to comparable or even lower levels than petrol engines.

- Outlook of consumer preferences on the types of electric vehicles is unclear and no pattern is identifiable so far. Globally, around 66% of electric vehicles sold in 2017 were pure battery electric vehicles while the balance was a mix a different types of hybrids. In 2016, the share was around 50:50. China, France, Japan, Denmark, Korea all have BEV oriented markets: around 60% to nearly 90% of their 2016 electric car sales were BEVs, In contrast, in the Netherlands, Sweden and the United Kingdom, the majority of cars registered in 2016 were PHEVs, while sales of the two EV types were split roughly equally on the average in most other countries. Most government policy incentives target pure battery electric vehicles which are also cheaper than hybrids. However, relative local costs of electricity and petrol/gasoline also play an important role. For example, hybrids are more popular in the United States because petrol is much cheaper than in Europe where sales of BEVs are higher.

- Many projections on EVs over the last five years or so have tended to be over-optimistic. For example, most projections published five to six years ago predicted a market share of around 11-12 million EVs by now, but the share at the end of 2017 was only 2 million vehicles, accounting for just 0.2% of the nearly one billion vehicles on the world's roads. The outlook for EVs has been aptly summarized by the International Energy Agency (IEA,2017l) which coordinates Global Initiative on EV Development in the following statement: "Declining year-on-year increments are consistent with a crowing electric car market and stock size, but the scale achieved so far is still small: the global electric car stock currently corresponds to just 0.2% of the total number of passenger light-duty vehicles (PLDVs) in circulations. Electric vehicles still have a long way to go before reaching deployment scales capable of making a significant dent in the development of global oil demand and greenhouse gas (GHG) emissions."

- In spite of starting from a very low base compared with ICE vehicles, the exceptional growth rate of over 60% per year in the last few years will likely be sustained, driven by very strong in-place and emerging policy instruments and increasingly positive response by the private sector, especially auto manufacturers and associated industries. The International Energy Agency (2019) estimates the global population of EVs in 2030 under the New Policies Scenario would emit about 230 Mt CO_2-eq compared with about 450 Mt CO_2-eq GHG emissions if the same fleet was powered by fossil-powered ICE power trains. In the EV30@30 Scenario, EV emmissions would be 230 Mt CO_2-eq compared with 770 Mt CO_2-eq compared with an equivalent ICE fleet. However, this projection assumes that the rapid decarbonization of power generation envisioned in the scenario and consistent with IEAs Sustainable development scenario can be achieved otherwise EV emissions could be as high as 340 Mt CO_2-eq. The impact of either development on the urban environment and human health would be very significant.

Appendix 5B
Outlook on electric vehicle battery technologies

The battery is the most important unit in an EV and it currently presents a formidable obstacle that needs to be surmounted in order to fast-track market penetration of EV cars. A battery electric car (BEV) is currently around 30% more expensive than an equivalent internal combustion engine vehicle, and a hybrid (PHEV) is even more expensive. A typical EV will return less than 200 kilometers (much less with use of air-conditioning) between charges. While this may be adequate for intra-city commuting, inter-city travel would be problematic if there is no easily accessible commercial battery charging infrastructure. Many EV owners do not have home-based charging access because they either live in apartments or park in the streets. However, the greatest constraint is the battery charging time which averages 5-8 hours depending on battery size (the more expensive models now feature Li-ion batteries which can be fully recharged within an hour). While this may be no problem for home-based charging, it would be expensive time wasting for those who do not have home charging facility. 'Range anxiety' - the fear of running out of charge in the middle of nowhere - has become a common syndrome among users of fully electric vehicles. This explains why the hybrid electric vehicle (which features an electric drive and an alternative gasoline/diesel engine drive, or an alternator that continuously recharges the battery like in a conventional car) is competing evenly with or even outselling BEVs in some countries.

The earliest models of electric vehicles featured lead-acid batteries but they are heavy and a full charge takes 6-8 hours. Nickel metal hydride (NiMH) batteries are also mature technologies and have featured in the early electric vehicles. They have double the specific energy of lead acid batteries (68Wh/kg), making them significantly lighter than lead acid batteries of the same power rating. However, NiMh batteries also have major drawbacks - low charging efficiencies and high self-discharge (up to 12.5%/day under normal room temperature conditions) (Mok, 2017). Although they still feature in some inexpensive EVs models, both lead acid and NiMH batteries are now considered obsolete in BEVs. Most modern EVs now feature lithium-ion/lithium-polymer batteries which have been in common use in laptops and cell phones for over a decade. They have high specific energy (140 Wh/kg) and 80 to 90% charge/discharge efficiency, but they lose the ability to hold charge, with time and cycle lives of only a few hundred to a few thousand charge cycles, especially if they are fully discharged frequently.

Lithium-ion (Li-ion) batteries are now considered to be the standard for modern battery EV vehicles. They are well-established in the electronics market; they are lighter, more efficient than other alternatives, and can be charged to 80% full capacity in less than an hour. Lithium-ion batteries combine a wide range of materials as anode and cathode in batteries, and there is a wide range of combinations, each with distinct advantages and disadvantages in terms of safety, performance, cost, and other parameters (BCG, 2018). The most prominent technologies for automotive applications are lithium-nickel-cobalt-aluminium (NCA), lithium-nickel-manganese-cobalt (NMC), lithium-manganese-spinel (LMO), lithium titanate (LTO), and lithium-iron phosphate (LFP) as anode and graphite as cathode, while lithium-cobalt oxide (LCO) is the most common in consumer applications.

Compared with other mature battery technologies, Li-ion offers many benefits: for example, it has excellent specific energy (140 Wh/kg) and energy density, making it ideal for battery electric vehicles. Li-ion batteries are also excellent in retaining energy, with a self-discharge rate (5% per month), an order of magnitude lower than NiMH batteries. However, apart from the fact that lithium-ion batteries are much more expensive than other alternatives, there are several technical issues: lithium-ion batteries have memory issues - they lose the capacity to hold charge over time, and have to be replaced. Furthermore, Li-ion batteries have issues with overcharging and overheating, with thermal runaways that have triggered fire incidents in Boeing 787 airplanes, cell phones, and BEVs in recent times. Global research on lithium-ion batteries is intensive and extensive and the overheating and fire issues have now been largely resolved. Other issues such as size and cost of Lithium-ion EV batteries will be resolved and the batteries will become lighter and more affordable in the near future. Research in lithium iron battery technology is very prolific and intensive: the 2019 Nobel prize for chemistry was won by three researchers who pioneered the development of lightweight batteries for use in many applications: portable electronics, mobile phones, pacemakers, etc., and new innovations are coming on stream that increase the energy intensity and lower costs.

The production of Li-ion EV batteries is the emission-intensive aspect of the EV manufacture, accounting for nearly half of the associated emissions, and the main sources are Asian countries most of which have carbon-intensive electricity, sourced mainly from coal. In effect, irrespective of global location, most EVs currently have significant carbon footprint which, in some situations could be as high or even higher than comparable ICE vehicles. However, many auto manufacturers have plans for local production of Li-ion batteries and, although they will still be sourcing most of the critical raw materials from Asian and other emerging countries, they can take full advantage of local low-carbon electricity to reduce the lifecycle emissions of their products. For example, the carbon intensity of electricity in some emerging countries could be as high as 1000 gCO_2/kWh, especially when they are fueled by low-grade coals, compared with 20-30 gCO_2/kWh in some European countries that generate electricity mainly from renewables or nuclear energy.

Fuel cells are becoming increasingly viable alternatives to batteries in electric vehicles. The main difference is that fuel cells are onboard electric power generators fueled mainly by liquid hydrogen which has zero tailpipe emission, although most hydrogen production processes have a significant carbon footprint. However, there is considerable controversy especially among auto manufacturers on whether battery electric technology or the hydrogen fuel cell (FCEV) provides the best long-term option. Some manufacturers have large FCEV development programs, some focus on battery electric technology, while others are committed to both technologies. FCEVs hold great promise because of certain key advantages over batter electric: they have much higher range, often twice that of a typical EV; and refueling is much faster, often in just a few minutes. However, there are also significant challenges, including sustainable, low-carbon sources of mass-produced hydrogen, development of fueling stations (much more expensive than charging stations), and safe onboard storage. Furthermore, fuel cells are much less energy-efficient that electric batteries. Some municipal establishments such as police forces are already using FCEVs but future projections are difficult largely because they are much more expensive and will be competing in an already established and growing BEV market whose range and cost limitations are fast diminishing.

The specific energy of batteries (their capacity of storing energy per kilogram of weight) is still only 1% of the specific energy of gasoline/petrol (140-170 watt-hours/kg compared with 13,000 watt-hours/kg for gasoline), hence the severe limitation to distance covered per full charge, which is still only 250 to 300 kilometers. Unless there is a technology break-through, batteries will continue to limit the driving range of electric vehicles (BCG, 2018). However battery energy density has been improving by around 5%-8% per year in the last few years and could rise by around 50% within the next decade. Another major issue is the charging time of batteries. Depending on the specifications of an electric outlet and type of battery, A 15-kWh battery plugged into a 120-volt outlet could take around 10 hours to charge fully, while a 40-amp, 240-volt outlet can take just 2 hours. A three-phase charging socket will take as little as 20 minutes. All these options have weight and cost implications. Several other companies are developing battery swapping station models that could replace a discharged battery in just a few minutes. For this to succeed, auto-manufacturers would have to standardize battery sizes and specifications.

Battery manufacturing is undergoing significant transitions, with substantial investments in new production facilities as well as commercial charging points. Advanced battery technology development is attracting huge capital investments which will push both Li-ion and new battery technologies across competitive thresholds for the electric vehicle industry and other applications, in particular, storage systems for variable renewable energy (VRE - solar, wind). These massive investments will drive battery performance higher while pushing costs to less than half of the current level within the next five years, making VRE power plants competitive with natural gas power generation (Bloch *et al.*, 20119). It is noteworthy that major energy companies that traditionally focus on fossil fuels are now diversifying investments to include renewable energy EV battery manufacture and public battery charging facilities. Another emerging issue is what to do with batteries when they reach end-of-life, which is when they have lost around 20% of total energy retention capacity (after around 150,000 to 200,000 kilometers or about 7 years in operation. While nearly all lead-acid batteries used in conventional ICEs are recycled, only about 5% of Li-ion batteries are currently recycled, yet they contain potentially environmentally hazardous metals and chemical compounds. Furthermore, most of the critical raw materials - lithium, nickel, cobalt graphite, and rare earth elements (REE) - have high carbon footprints, coming from emerging nations, produced from raw materials sourced from small mines, extracted with high energy and carbon intensive technologies. It is unclear whether these sources can cope with the rapid expansion of demand for batteries which will inevitably result from the growing EV market, or whether supply sources of other critical input materials for EV manufacture can cope with the expected exponential demand expansion. For example, the average EV uses around four times of copper in the electric motor and wiring compared with a conventional ICE, and cobalt and nickel are prime alloying elements in the steel industry.

When an EV battery loses a significant part of its electricity storage capacity, typically 20%, it can no longer power a vehicle effectively and has to be changed. However, the battery still has high storage capacity that can be reused in stationary storage applications, for example, in storing surplus power in utility-scale intermittent renewable energy installations such as in buildings, thereby increasing the lifetime use of the battery by 72% and therefore reducing the battery greenhouse emissions attributable to the vehicle on a per-kilometer basis by about 42% (ICCT, 2018). This *second-life* use could extend the total life of the battery by around ten years, provide valuable electricity storage capacity, and significantly reduce the lifecycle emissions associated with the battery. Furthermore, li-ion batteries contain very expensive and carbon-intensive materials such as lithium, cobalt, nickel, copper, graphite which could be recovered by recycling batteries after second-life use. There are of course potential challenges, notably collection strategies, reconditioning, and redeployment, and the economics of re-use. Furthermore, there are currently not enough batteries to sustain the expensive investment that would be required for setting up recycling plants. However, there is little doubt that these issues will gain more attention in the near future, and new, low-cost recycling technologies will emerge as sufficient volumes of discarded batteries become available.

Charging infrastructure, (electric vehicle supply equipment, EVSE) whether at home or at public locations, is indispensable and a significant constraint to the fast market penetration of electric vehicles. Various studies have shown that the availability of chargers is one of the key factors that will determine the future market penetration of electric vehicles. One issue is the need to match public investments in in EVSE with growth rate in EV population. At the same time, access to EVSE is a significant consideration in the decision of a consumer to invest in an electric vehicle, considering that the larger population live in apartments or park in the streets. Data presented in Figure 5B-1 shows that Electric cars worldwide outnumber public charging stations worldwide by more than six to one and facilities in most countries with the exception of China are slow systems which take 6 - 8 hours, indicating that current owners rely primarily on private or workplace charging facilities. Public charging outlets need to grow by 10 to 25 times the current rate in order to meet the requirements of the projected EV population in 2030 under different scenarios.

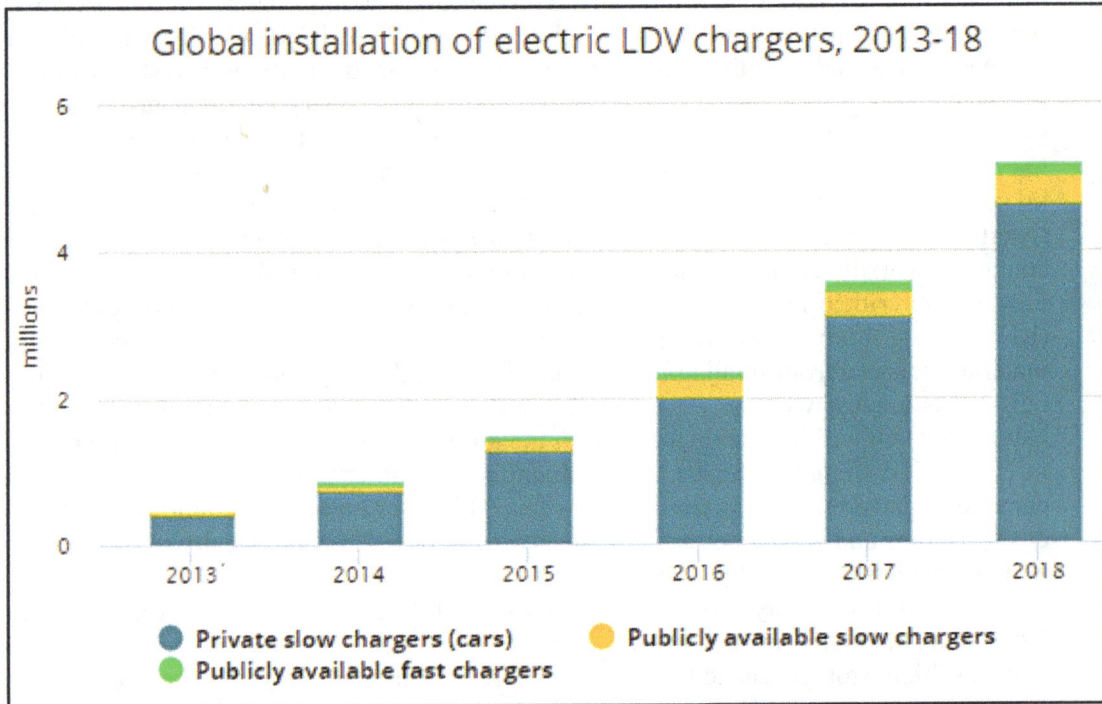

Figure 5B-1 Global installation of Light-duty Vehicle chargers, 2013-18 *(IEA, 2019m)*.

Chapter 6

Consequences of environmental pollution

6.1 INTRODUCTION

Environmental pollution is of two types: stratospheric (middle atmospheric) pollution and tropospheric (atmospheric/ambient) pollution. The stratosphere (15-50 km above the Earth's surface) houses the greenhouse gases which regulate the global average temperature. They do this by trapping part of the Sun's reflected radiation from the Earth's surface in day time, and reflecting it back at night time when temperatures fall. Without this natural regulation process, the Earth would be too cold to sustain life. Natural ozone which is found in the lower part of the stratosphere also plays a critical role in screening off damaging ultraviolet rays of the Sun, thereby protecting life on Earth from the potential dangers. Human activities, especially fossil fuel combustion and agriculture produce many of the greenhouse gases which can significantly enhance the concentrations of greenhouse gases, thereby increasing their ability to trap part of the Sun's energy that is reflected from the Earth's surface, and reflect it back, causing an increase in the world's average temperature (see Chapter 4). Global warming is the most important indicator of potential climate change, and the consequences are multi-dimensional.

Some gases collectively known as F-gases, also released into the stratosphere by human activities are capable of reacting with the natural ozone, thereby depleting its concentration and reducing its ability to protect the Earth from the Sun's dangerous rays (ozone depletion or ozone hole phenomenon). F-gases are also greenhouse gases and can contribute to global warming. Human activities can also cause an accumulation of a cocktail of gases and particulate matter in the troposphere (lower atmosphere), 10-20 km above the Earth's surface. This is known as atmospheric or ambient pollution. Although there are natural phenomena that can cause global warming and ambient pollution, emissions from human activities (anthropogenic emissions) are believed to be the main causes. The anthropogenic and natural gases in the stratosphere become well-mixed and pollution emissions can remain in the atmosphere from ten years to thousands of years, hence impact on global climate is largely independent of the source of emission. Most of tropospheric/Ambient pollutions remain in the vicinity of the source and large particulate constituents drop back to the Earth within days or weeks, but fine or gaseous components can be spread far and wide by strong winds. When the particles are large enough, they are visible, suspended in air as smog, a common feature of many cities all over the world. Depending on the physico/chemical nature, constituents of atmospheric pollution can cause global warming or cooling.

The effect of environmental pollution on human health and welfare is very prominent in current global discourse, and is in many dimensions. Global warming, climate change and severe weather impact negatively on human safety and welfare, and agriculture/aquatic life on which mankind depends for survival. Also, environmental pollution has been identified as the cause or enhancer of many human health issues. The effect of particulate matter (PM) concentrations on human health has been the focus of extensive research but the effect on soils, plants and crops, vegetation, wild life and aquatic life can also be quite devastating. There is also substantial evidence concerning the negative effect on the ecosystem of other pollutants, in particular, ozone, nitrogen dioxide, sulphur dioxide, carbon monoxide, fluorides, organic and synthetic chemicals.

Anthropogenic interference with the natural equilibrium of stratospheric greenhouse gases and the ozone layer, and creation of tropospheric smog impact negatively on life on Earth, and the consequences are still subjects of intensive research. Climate change through global warming is one of the major negative effects of atmospheric pollution, others are solar radiation damage, acid rain and public health issues. All types of pollution (air, water and soil) can cause serious damage to human and animal health, marine life, and other living organisms, plants, vegetation, etc. The impact of anthropogenic tropospheric pollution on the environment can be viewed in two dimensions: the effect on the weather and climate; and the impact on humans and the ecosystem. Unlike stratospheric natural and anthropogenic gases which are well-mixed, long-lived, and exert essentially the same effect on climate irrespective of the emission location or season of release, concentrations of ambient pollution are highly variable across regions, and so are their impacts on local populations and ecosystems. This variability is due in part to the spatially heterogeneous particulate matter loadings which are usually highest near the emission sources, but also because local atmospheric transport dynamics determine the extent and speed of spreading pollution. The effect of stratospheric pollution, (quantified as carbon dioxide equivalent, CO_2eq) becomes significant when quantified over many decades but the impact of ambient pollution can be felt in days, particularly in its obstruction of visibility, and effect on human health. Ambient pollution has a short atmospheric lifetime of a few days to weeks, and concentrations are typically highest in urban areas where sources are densely located. However, in spite of the short lifetime, constituents can travel long distances in the atmosphere and the effect can be felt far from the emission source (Shindell *et al.*, 2008; Anenberg *et al.*, 2014).

6.2 IMPACT OF ENVIRONMENTAL POLLUTION ON WEATHER AND CLIMATE

Climate science has developed very rapidly in the last five decades or so and a lot is now known about the variables which determine global and regional climate, weather patterns, etc. This has made it possible to forecast weather patterns with considerable degree of accuracy. However, there are still many unknowns and uncertainties, and this explains why forecasts are often dead wrong. For this reason, predictions and projections about the climate and weather are always qualified with levels of confidence, from very low to very high (from exceptionally unlikely to virtually certain). There are many natural phenomena that can cause climate change, for example, changes in solar irradiance, atmospheric trace gases and natural aerosol concentrations. Paleoclimatic reconstructions have generated extensive data on probable climate changes over the previous thousands of years, prior to the instrumental period. It is therefore possible to place currently observed changes in the perspective of natural climate variability.

Evidence abounds showing that the Earth system has always responded and will continue to respond to natural forcings - solar, volcanic, orbital, natural variations in atmospheric composition. For example there are known variations in the Earth's orbital parameters as well as changes in its axial tilt, both of which affect the seasonal and latitudinal distribution and magnitude of solar energy received at the top of the atmosphere, which in turn determines the Sun-Earth energy balance and intensities and durations of local seasons. Solar irradiance (TSI) is the nearly periodic 11-year cycle of changes in solar radiation which shifts from maximum solar conditions (with many sunspots) to solar minimum (with, basically, none). Over the period, the radiation ejected from the Sun varies in quantum, spatial distribution, and composition.

During the solar maximum conditions, there is a slight increase in the total solar irradiance but the ultraviolet (UV) content is increased significantly, leading to generation of stratospheric ozone in the mid-to-upper stratosphere, which ultimately results in greater ozone concentration in the tropical lower stratosphere. The combined effect of increased UV and long-wave radiation helps warm that region. How this impacts on climate is still a subject of intensive research but there is ample evidence that solar cycles influence tropospheric rainfall patterns in different ways across the globe (Rind *et al.*, 2008). Also, total solar irradiance, though small in magnitude, does appear to affect sea surface temperatures, especially at latitudes (such as Northern Hemisphere sub-tropics during summer) where cloud cover is small and irradiance is abundant. However, satellite observations of total irradiance (TSI) changes from 1978 to 2011 show a small positive radiative forcing (RF) (0.05 W/m^2) compared with values for anthropogenic GHGs (3.0 W/m^2) and will become increasingly insignificant in the coming years. The recognition of the fact that natural phenomena can cause significant climate forcing has facilitated current conclusions about climate change which are determined based on comparison of current phenomena with paleoclimatic database to determine whether or not they are unusual.

Perhaps the most convincing evidence of climate change over the last 200 years or so is the result of in-depth analysis of observational records of the global atmosphere, land, ocean and cryosphere systems over the period. There is incontrovertible evidence that the atmospheric concentrations of GHG gases, in particular, CO_2, CH_4 and N_2O have increased substantially; land and sea surface temperatures have increased over the last 100 years; observations from satellites and *in situ* measurements indicate reductions in glaciers, Arctic sea ice and ice sheets; data on radiative budget and ocean heat content suggest a small imbalance; and satellite data on atmospheric temperatures show increases in tropospheric temperatures and decreases in stratospheric temperatures. The fact that the stratosphere and troposphere exhibit opposing temperature changes is a strong indication that the Sun is not the main driver (otherwise the response would both either positive or negative). The opposing response is consistent with known GHG effects.

6.2.1   Weather and Climate

There is a clear distinction between weather and climate. Weather describes the conditions of the atmosphere at a certain place and time with reference to temperature, pressure, humidity, wind, and other key parameters, collectively known as meteorological elements. Weather is also described by the presence and movement of clouds, precipitation, and the occurrence of spatial phenomena, such as thunderstorms, dust storms, tornados, etc. Climate takes into account all the above variables over a period of time ranging from months to years, projected in either direction to thousand and millions of years in order to arrive at a global climate pattern. The World Meteorological Organization (WMO) defines climate as the average of variables (mainly temperature, precipitation and wind) over 30 years, but more detailed analysis also considers frequency, magnitude, persistence, trends, etc. in the determination of global climate. Although weather and climate are interdependent, the clear distinction is the fact that weather defines the state of the atmosphere at a given time and place, and can change from hour to hour, day to day, month to month, or year to year. Climate on the other hand, generally refers to an aggregate of weather statistics over a decade or more. It is therefore easier to predict climate than weather. Short-term changes in climate are known as *weather* changes but climate change refers to changes in the statistical properties of the weather pattern over of long period of time, covering several decades. In effect, climate is the long-term pattern of weather in a particular area.

6.2.2   Climate change

Climate is determined by a complex mix of natural processes, such as the rotation of the Earth around the Sun, changes in the equilibrium of energy transfer between the Sun and the Earth, the pattern of distribution of surface moisture by winds, ocean currents and other mechanisms, and activities in the five different 'spheres' that make up the Earth's system: the *atmosphere, lithosphere, hydrosphere, cryosphere, biosphere*. The atmosphere surrounds the Earth and is filled with gases. This is the most variable part of the climate system where the composition and movement of gases surrounding the Earth can change radically within short periods of time. The lithosphere (land) contains all the world land surface (and the semi-solid and liquid land deep beneath the surface). The lithosphere is very uneven, made up of mountain ranges, hills, deep valleys, plains. All these surfaces have different solar absorption and reflection characteristics (albedo) hence contribution to global warming is variable. Furthermore, uneven terrain acts as wind barriers or speed enhancers, and therefore can significantly affect wind movements in terms of intensity, variability and direction.

Changes in the hydrosphere (a collection of oceans, rivers and lakes), include variations in temperature, salinity, acidity, but the pace is very slow compared with the atmosphere. The cryosphere (really a sub-division of the hydrosphere), comprising frozen water, ice sheets and glaciers is also relatively consistent. The main climate-related activity in this sub-system is the continuous process of reflection or absorption of the Sun's rays (albedo) which determines the melting rate of the ice. Any changes in the natural balance are slow and become evident after long periods of time. The biosphere which comprises all life on Earth profoundly influences climate through the complex control of natural activities, especially the natural balance of CO_2eq in the atmosphere. Living humans and animals inhale oxygen and exhale carbon dioxide which plants need for photosynthesis; forests and oceans absorb substantial proportions, serving as 'carbon sinks'.

Climate varies across regions because of variations in the determinant interactive factors: variations in the intensity and distribution of the Sun's energy that reaches the Earth and the amount that is reflected; elevation and topography, land use pattern, the latitudinal location which largely determines how much of the Sun's energy reaches the region and for what duration during the Earth's annual rotation cycle around the Sun. For example, regions located near the Equator (zero latitude) receive sunlight year-round, whereas those near the poles hardly receive any and temperatures rarely rise above freezing. Other variables are vegetation on the Earth's surface, proximity to oceans, and, not the least, the biosphere which is the sum total of all life on Earth. The most familiar features of weather are the day-to-day and day-to-night variations in temperature, humidity, precipitation, rainfall, windiness, atmospheric pressure, and cloud structure. Weather can change from day-to-day, even within the day but climate change happens slowly over decades, hundreds, maybe even thousands of years. and for many reasons. Global climate is taken as the average across the world, and could be in terms of average temperature or precipitation patterns.

The Earth's climate system is powered by solar radiation, around 56% of which reaches the Earth's surface in the visible part of the electromagnetic spectrum. Global temperature is determined by the balance between incoming solar energy and the radiation reflected back into space by the Earth's surface, and has been relatively constant over many centuries (Figure 6.1). The radiative balance between incoming solar shortwave radiation (SWR) and outgoing longwave radiation (OLR) is influenced by several global climate variables. Natural fluctuations in solar energy output due to the Earth's rotation, and consequent fluctuations in the amount of incoming SWR can cause changes in the energy balance, so can human activities that release emissions of gases and aerosols into the atmosphere which modify natural concentrations of greenhouse and ozone gases. It is estimated that about half of the incoming solar shortwave radiation is absorbed by the Earth's surface. The fraction of SWR reflected back to space by gases and aerosols, clouds, and by the Earth's surface (albedo) is about 30%, and about 20% is absorbed in the atmosphere. The Sun provides its energy to the Earth primarily in the tropics and the subtropics, from where it is partially redistributed to the middle and high latitudes by atmospheric and oceanic transport processes. The global climate system has been monitored over the last hundred years or so, especially since the 1950s when sophisticated instrumentation and satellite technology became available, and the changes observed have led to the conclusion that global average temperature, the main indicator of climate change, is rising, with potentially serious consequences for life on Earth. Apart from global warming, there are many other negative effects of anthropogenic pollution, notably acidification of the oceans, acid rains, tropospheric smog, etc. Activities of greenhouse gases in the stratosphere regulate the global average temperatures which have been around 14°C for over a century up to 1980 or so (Figure 6.2). Without this natural control, the Earth would be too cold to sustain life. However, there are wide temperature variations across the globe, from around zero in the Antarctica to around 50-60°C in the Libyan desert. There is ample scientific evidence that the average global decadal temperature has increased by nearly around 0.6°C in the last four decades. This may seem insignificant but small changes in the Earth's average temperature can have big impacts on climate. Records of global anthropogenic emissions from 1850 show an exponential rate of increase from the 1950s (Figure 6.2). Recent reports of some of the world's leading authorities on climate change (Box 6.1) have been evaluated and the important conclusions are summarized in Box 6.2.

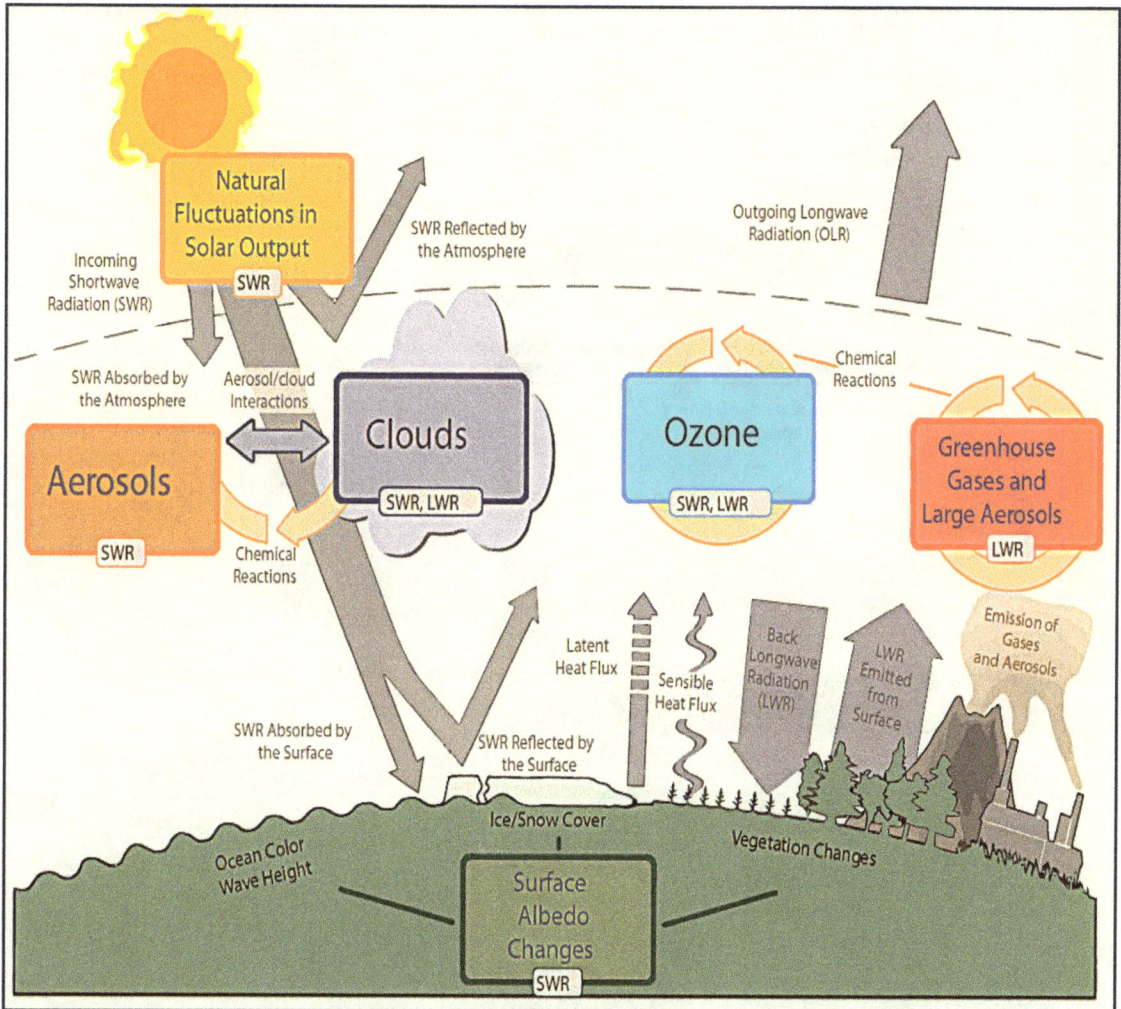

Figure 6.1 The radiative balance between incoming solar short-wave radiation and reflected long-wave radiation by the Earth's surface *(IPCC-WG1, 2013).*

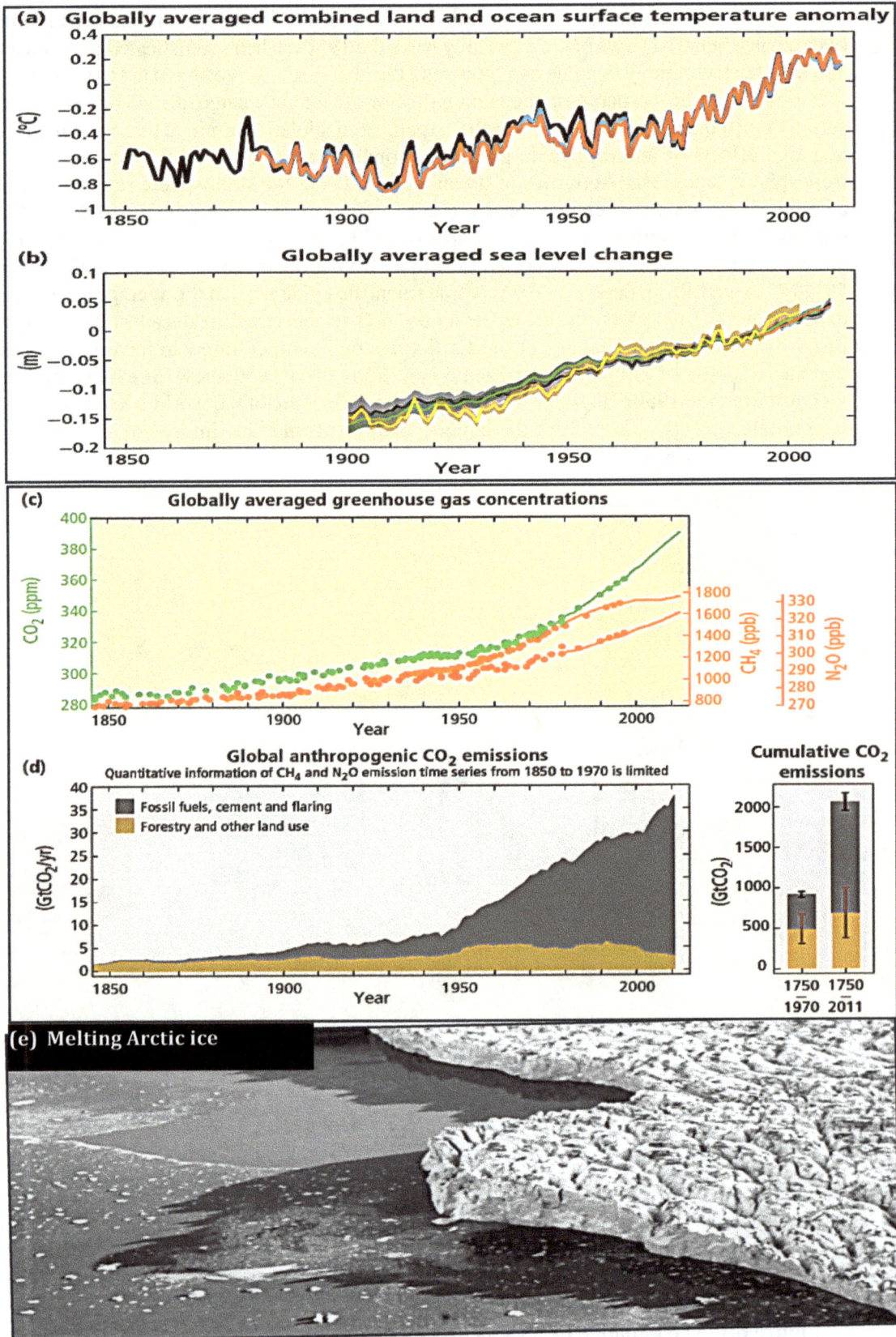

Figure 6.2   Globally averaged warming indicators *(IPCC Synthesis Report, 2014).*

BOX 6.1
MAJOR SOURCES OF DATA ON FOSSIL ENERGY USE AND CLIMATE CHANGE

- ExxonMobil Energy and Carbon Summary 2018.

- Intergovernmental Panel on Climate Change (IPCC)-2011 Renewable energy sources and climate change mitigation.

- Intergovernmental Panel on Climate Change (IPCC)-2013 ClimateChange-2013 The physical Science Basis.

- Intergovernmental Panel on Climate Change (IPCC)-2014 ClimateChange-2014 :Impacts. Adaptation and Vulnerability.

- Intergovernmental Panel on Climate Change (IPCC)-2014 ClimateChange-2014

- Intergovernmental Panel on Climate Change (IPCC) ClimateChange-2014: Synthesis Report.

- Intergovernmental Panel on Climate Change (IPCC) ClimateChange-2018 Summary for Policymakers.

- International Energy Agency (IEA),2015. Energy and climate change.

- International Energy Agency (IEA), 2016. Energy and Air Pollution. World
- Energy Outlook Special Report.
-
- International Energy Agency (IEA). World Energy Outlook 2017.

- International Energy Agency (IEA)-2017. CO_2 emissions from fuel combustion: Highlights.

- Max Planck Institute Special Report, 2017. Global warming doesn't stop when emissions stop.

- National Academy of Sciences. 2014. *Climate Change: Evidence and Causes.*

- National Aeronautics and Space Administration (NASA) Do variations in the Solar Cycle affect our climate system?- David Rind.-2009.

- National Aeronautics and Space Administration (NASA, 2018) Climate Change and Global Warming.

- National Aeronautics and Space Administration (NASA, 2018) Climate Change: Vital signs of the Planet: Evidence, Causes, Effects, solutions.

BOX 6.2
SUMMARY OF EVIDENCES OF GLOBAL WARMING AND CLIMATE CHANGE

- It is virtually certain that human influence has warmed the global climate system, resulting in changes in temperatures near the surface of the Earth, in the atmosphere and in the oceans, as well as changes in the cryosphere, the water cycle, and some extremes. While some of these observations may also have been caused by natural phenomena, there is ample evidence that the contribution is insignificant.

- A rise in global average surface temperatures is the best-known indicator of climate change, and there has been a steady rise over the last century or so.

- More than half of the observed increase in global mean surface temperature (GMST) from 1951 to 2010 is very likely due to the observed anthropogenic increase in greenhouse gas (GHG) concentrations.

- The atmosphere and oceans have warmed, the amounts of snow and ice have diminished, sea levels have risen, and the concentrations of greenhouse gases have increased. Each of the last three decades has been successively warmer at the Earth's surface than any preceding decade since 1850.

- Ocean warming dominates the increase in energy stored in the climate system, accounting for more than 90% of the energy accumulated over the last fifty years or so. The frozen parts of the planet, known collectively as the cryosphere, affect and are affected by local changes in temperature. Over the last two decades, the amount of ice contained in glaciers globally has been declining every year for more than 20 years, the Greenland and Antarctic have been losing mass, glaciers have continued to shrink almost worldwide, and Arctic sea ice and Northern Hemisphere spring snow cover have continued to decrease in extent.

- The rate of sea level rise since the mid-19th century has been larger than the mean rate during the previous two millennia. Over the period 1901 to 2010, global mean sea level has risen by about 0.2 meter.

- The atmospheric concentrations of carbon dioxide, methane, and nitrous oxide have increased to levels unprecedented, carbon dioxide concentrations have increased by 40% since pre-industrial times, primarily from fossil fuel emissions, but also from net land use emissions. The oceans have absorbed around 30% of the emitted anthropogenic carbon dioxide, causing ocean acidification. This is the strongest evidence of human influence on the climate system.

A large majority of global land areas have experienced decreased frequency of cold extremes and increased frequency of warm extremes since the middle of the 20th century, consistent with warming of the climate. Warm days and nights have become more frequent while frequency of cold days and nights has decreased. There is an increasing trend   in the frequency and intensity of heat waves across most regions. Increases in heavy precipitation have probably also occurred over this time, but vary by region.

- While there is no sufficient evidence that the annual numbers of tropical storms, hurricanes and major hurricanes count in the North Atlantic basin have increased over the past 100 years, there has been a substantial increase in storm activity, the frequency and intensity of the strongest tropical cyclones in that region since the 1970s consistent with warming oceans.

- The total radiative forcing (RF) is positive, and has led to uptake of energy by the climate system, the largest contribution caused by the increase in the atmospheric concentration of CO_2eq since 1750.

- Climate change, whether driven by natural or human forcing, can lead to changes in the likelihood of the occurrence or strength (or both) of extreme weather and climate events.

- It is very likely that anthropogenic forcings, dominated by GHGs have contributed to the warming of the troposphere, while anthropogenic ozone-depleting substances, have contributed to the cooling of the lower stratosphere.

- Continued emission of greenhouse gases will cause further warming and long-lasting changes in all components of the climate system, increasing the likelihood of severe, pervasive and irreversible impacts for people and ecosystems.

- Future climate will depend on committed warming by past as well as future anthropogenic emissions, and natural climate variability. It is estimated that human activities have caused between 0.8°C and 1.2°C global warming above pre-industrial levels, likely to reach 1.5°C between 2030 and 2052 if it continues to increase at the current rate. It is very likely that heat waves will occur more often and last longer, and that extreme precipitation events will become more intense and frequent in many regions. The oceans will continue to warm and acidify, and global mean sea level will continue to rise.

- Climate change will amplify existing risks and create new risks for natural and human systems. Risks are unevenly distributed and are generally greater for the disadvantaged people and communities in countries at all levels of development.

- Even if emissions were stopped today the negative impact on some components of the climate system could persist for centuries, due to the long lifetime of some of the gases and the inertia of the main sinks, in particular, the oceans. Aerosols have a lifetime of weeks, methane (CH_4) about 10 years, nitrous oxide about 100 years, and hexafluoroethane (C_2F_6) about 10,000 years. Carbon dioxide is more complicated as it is removed from the atmosphere through multiple physical and bio-geo-chemical processes in the ocean and on land, all operating at different time scales. It is estimated that about 10-40% of CO_2eq emission remains in the atmosphere for as long as 1000 years.

6.2.2.1 *Global warming*

Observations in the climate system are based on direct physical and biochemical measurements, and remote sensing from ground stations and satellites, which provide a comprehensive view of the variability and long-term changes in the atmosphere, the oceans, the cryosphere and at the land surface, and records date back to the mid-19th century. The database is extended back hundreds to millions of years, based on paleoclimate reconstruction and future directions are determined by modeling. It is relatively easy to predict climate response to anthropogenic emissions over the next two decades or so (near-term/decadal climate prediction) because there is a strong body of knowledge and baseline data on most of the important variables that will determine climate change. There is considerable inertia in the response of the environment to climate forcing, in particular, the oceans. In effect, current data on anthropogenic emissions will largely determine the direction of climate change over the next two decades or so.

There is extensive data dating back to the 19th century on multiple independent climate indicators, from high up in the atmosphere to the depths of the oceans. They include changes in surface, atmospheric and oceanic temperatures, glaciers, snow cover, sea ice, sea level and atmospheric water vapor. Many independent research groups have analyzed these extensive database and all have come to the same conclusion that the world has warmed since the 19th century. Although each year and even decade is not always warmer than the last, global surface temperatures have warmed substantially since 1900. More than 90% of the excess energy absorbed by the climate system since at least the 1970s has been stored in the oceans. As the oceans warm, the water expands, accounting for much of the observed rise in sea levels over the past century. Melting of glaciers and ice sheets due to rising temperatures also contribute to the rising levels of the oceans. A warmer world is also a moister one, because warmer air can hold more water vapor, and global analyses show that specific humidity, which measures the amount of water vapor in the atmosphere, has increased over both land and the oceans.

Regional climates - monsoon systems, tropical phenomena, cyclones - are the complex result of processes that vary strongly with location and so respond differently to changes in global-scale influences. Monsoons are the most important mode of seasonal climate variation in the tropics, and are responsible for a large fraction of the annual rainfall in many regions. Their strength and timing are related to atmospheric moisture content, land-sea temperature contrast, land cover and use, atmospheric aerosol loadings, and other factors. Increasing global temperature will intensify monsoons in the future, affecting larger areas, because atmospheric moisture content increases with temperature.

Anthropogenic greenhouse gases that reach the stratosphere enhance the natural concentration of the gases and their ability to absorb reflected heat from the Earth and reflect it back, thereby warming the Earth. Halocarbons deplete ozone and reduce its concentration in the lower part of the stratosphere, causing more of the Sun's harmful rays to reach the Earth's surface. Because aerosols are distributed unevenly in the atmosphere, they can heat and cool the climate system in patterns that can drive changes in weather, in particular, precipitation. Aerosols from large volcanic eruptions that enter the stratosphere sediment out of the stratosphere within a year or two but, in that short period, can cause stratospheric cooling.

There are uncertainties about the net effect of natural and anthropogenic aerosols on climate. Aerosols can affect climate in many ways. First, they scatter and absorb sunlight, which modifies the Earth's radiative balance. Aerosol scattering generally makes the planet more reflective, and tends to cool the climate, while aerosol absorption has the opposite effect, and tends to warm the climate system. The balance between cooling and warming depends on aerosol physico-chemical properties and environmental conditions. The presence of sulphate aerosols in clouds is believed to modify the morphology of the cloud droplets, making them more reflective. However, most recent studies agree that the overall radiative effect from anthropogenic aerosols is to cool the planet.

Black carbon (BC) is the most potent radiative forcer in atmospheric pollutants, and its relative proportion and composition will largely determine the overall effect on the climate. Black carbon absorbs ultraviolet radiation from the Sun and therefore disturbs the planetary radiation balance. The radiation is converted to heat energy which has the potential for global warming. Because atmospheric residence time of BC is usually very short, contribution to global warming is substantial. The global warming potential over a 20-year time frame (GWP20) is 2,421, that is, one metric ton of BC would have the same integrated radiative effect over 20 years as 2,421 metric tons of carbon dioxide (Sims *et. al.*, 2015). On the other hand, black carbon reduces sunlight that reaches the surface and the amount that is reflected back to space, causing some cooling. Many studies have also suggested that atmospheric black carbon (BC), the most prominent component of energy-related particulate matter (PM) emissions can alter the atmospheric temperature above, within and below cloud, and consequently alter cloud distribution.

Black carbon can reduce cloud albedo and change the density of small particles in clouds which act as seed for water vapor condensation. This could change the lifetime, reflectivity and stability of clouds and alter changes in the hydrological pattern of rainfall (U.S.E.P.A, 2012; Bond *et al.*, 2013; IPCC, 2014). Although the interaction between atmospheric pollution clouds and the effect of atmospheric pollution on weather is largely localized and variable because of its short atmospheric lifetime, the cumulative effect over long periods could result in global warming and climate change. Black carbon, as well as brown carbon, (a relatively absorbing and hence warming portion of organic carbon), as well as others, especially variable organic compounds (VOC) and sulphur dioxide (SO_2) (which forms sulphates) have forcing negative values, indicating cooling effect. Therefore, variability of the concentration and physico-chemical properties of emissions must be taken into account to understand the net climate and health impacts of any emission source or mitigation (UNEP/WMO, 2011).

Black carbon settles on snow and ice, thereby reducing the surface reflectivity (because it is black) and the amount of diffusive reflection of incoming solar radiation (land surface albedo) is reduced. In effect, the surface absorbs more heat, accelerating its melting. This explains the increasing reduction and disappearance of ice in the Arctic, the Himalayas and other glacial and snow-covered regions of the world (Figure 5.2). The Arctic region is particularly vulnerable because the atmosphere is drier and temperature inversions that inhibit atmospheric vertical mixing are frequent, and particulate matter tends to remain in the atmosphere longer than in other regions of the world (U.S.E.P.A., 2012). The melting of snow and ice causes warmer oceans and increased ocean levels, both of which impact on storm patterns and strength, resulting in severe weather (snow albedo feedback) (Flanner *et al.*, 2007; Quinn *et al.*, 2008). It has been estimated that the effect of black carbon on snow albedo may be responsible for as much as a quarter of the observed global warming.

Clearly, the effect of black carbon on the climate is complex and multi-dimensional, and the net effect on climate is not yet well understood. However, there is substantial scientific evidence that BC can have direct and indirect effects that can result in global warming or cooling depending on the season and the composition. There is also evidence that the global net mean human-induced RF due to the change in BC emitted from fossil fuel and biomass combustion between 1750 and 2011 was high, estimated to have been 0.64 W/m^2 (Myhre et al., 2014). This makes BC the third largest contributor to anthropogenic RF, after carbon dioxide (1.68 W/m^2) and methane (0.97 W/m^2). Recent studies indicate that BC has contributed much higher globally averaged RF than previously estimated, perhaps as high as 1.1 W/m^2 (Carmichael, 2008; Quinn *et al*, 2011; Bond *et al.*, 2013).

6.2.2.2 *Extreme weather*

Extreme weather is one that is rare at a particular place and/or time of the year. Climate change, whether driven by natural or human forcings, can lead to changes in the frequency and intensity of extreme weather events, such as extreme precipitation events or warm spells, draught, heavy rainfall, flooding, heat waves, etc. At present, single extreme events cannot generally be directly attributed to anthropogenic influence, but it has been established that human activities are contributing to global warming which in turn can exacerbate such extreme events as thunderstorms and flooding. Hurricanes start as strong winds over warm tropical oceans near the equator, with air picking up moisture from the warm waters of the oceans and rising due to fall in density. The depression that is formed between the rising air and the ocean surface is filled by fresh air which picks up moisture and rises. The rising wet, warm air cools off and the water in the air forms clouds. This process is repeated until clouds are formed and the complex system of forces within the cloud starts off a spin.

Storms formed north of the equator spin counterclockwise while those formed in the south spin clockwise in response to the Earth's magnetic field. The spinning storm begins to move under the influence of winds, and warm waters continue to energize them. Storms can travel over thousands of kilometers across oceans, picking up strength and speed from warm oceans and, by the time they make a landfall, speeds could range from around 100-300 kilometers an hour. Such powerful storms also known as hurricanes can cause catastrophic damage to structures, dump heavy rains and cause heavy flooding. Storms and hurricanes are natural phenomena and catastrophic damage caused by extreme weather has occurred many times over centuries, well before fossil fuel use became established. However, the fact that storms need warm waters to form and strengthen means that they will be stronger, more frequent, and more devastating if ocean waters become warmer. Also, the combined effect of expansion of warmer ocean waters and melting ice caps and glaciers could raise ocean levels to the point that many coastal towns and cities become submerged. In summary, anthropogenic pollution is not the cause of extreme weather but the effects such as rising sea levels and warmer oceans can make bad storms worse and much more destructive. Droughts and wildfires will become more intensive and widespread, crop yields will be lower and the impact on human health would be profound. An increase in the average global temperature can influence weather and climate in many ways, in particular, intensity and frequency of extreme weather. Higher temperatures boost evaporation which dries out soil and increases drought and desertification. Evaporation intensifies as temperatures rise and more moisture in the atmosphere will intensify rainfall. Storms form over warm waters, mainly in the tropical oceans. Winds take more moisture from the warm oceans and bring more rainfall and flooding when they hit land. As temperatures rise, oceans expand, land and ocean ice melts, ocean waters rise above their normal level and are pushed inland by storms, causing severe flooding. A warmer atmosphere holds more moisture and when the temperatures are below freezing, snowfall can intensify. It is likely that the number of heavy precipitation events over land has increased in more regions than it has decreased. North America and Europe are seeing increases in either the frequency or intensity of heavy precipitation with some seasonal and regional variations and similar trends have been observed in Central and North America.

6.2.2.3 *Draught and desertification*

Draught is an extended period of deficient rainfall relative to the statistical multi-year average (long-term mean) for a region - a season, a year or several years. Desertification on the other hand is defined as a process of land degradation in arid, semi-arid and dry sub-humid regions, resulting from various factors, including climatic variations and human activities. The underlying cause of most draughts can be related to changing weather patterns manifested through excessive build up of heat on the Earth's surface, meteorological changes which result in reduction of rainfall and reduced cloud cover, all of which result in greater evaporation rates. The resultant effects of draught are exacerbated by human activities such as deforestation which alters the soil's natural albedo, overgrazing and poor cropping methods, all of which reduce water retention of the soil, and improper soil conservation techniques which lead to soil degradation. The Food and Agriculture organization (FAO) estimates that around 40% of global natural forests have been cleared for agriculture, thereby altering the Earth's natural albedo and topography, both of which have a significant impact on wind movement, weather and draught.

Exposed land masses as a result of uncontrolled agriculture and land use, tree harvesting or bush burning affect the environment in many ways: there are fewer plants and trees that can utilize carbon dioxide for photosynthesis (biosequestration of carbon dioxide); the soil natural ability to absorb and reflect the Sun's rays is disturbed: more energy from the Sun is absorbed than reflected back to space causing global warming; there are fewer barriers to slow down potentially damaging winds; and economically valuable trees are lost. Forests are important parts of the natural carbon cycle and act as effective carbon sinks. It is estimated that 4 billion hectares of forest ecosystems (about 30% of the global land area store large reservoirs of carbon, together holding more than double the amount of carbon in the atmosphere (FAO, 2005). Although the climate protection role of forests is in no doubt, it is complex to determine how much of the forest carbon sink and reservoir can be managed to mitigate atmospheric CO_2 buildup, and in what way. It is estimated however, it is estimated that each hectare of intact of the world's tropical forests across Africa, Amazonia and Asia traps around 0.6 metric tons of carbon a year, which adds up to about 5 billion metric tons, equivalent to about a fifth of the total global energy-related emissions in 2018. Many countries municipalities around the world have strong reforestation programs, not only for environmental reasons but also to restore economically valuable trees. Reduction of current global deforestation rates by 50% by 2050 would be a major contribution to a sustainable environment.

Drought has always been a normal recurrent event in arid and semi-arid lands. There have always been climatic fluctuations but with no long-term trends over the past 2000 years. A possible increase of 1-3°C in arid lands over the next fifty years due to a doubling of the CO_2 content of the lower atmosphere to 700 p.p.m. would increase global potential evapotranspiration (PET) by some 75-225 mm/year. The ratio of mean annual precipitation to PET would then decrease by about 4% assuming that no substantial changes in rainfall took place in arid and semi-arid lands. Thermal load due to global warming is uneven, increasing with latitudes above 40° N and S. The increase is only slight or non-existent in sub-tropical and inter-tropical latitudes where most arid lands lie. Rainfall variability and fluctuations are inherent to arid and semi-arid lands, and there are no indications of any long-term trends in rainfall, either from the global view point, or World Arid and Semi-Arid Lands (WASAL). There are however local and medium-term trends.

Dendro-chronological, archeo-climatic, and archeo-historical studies suggest that long droughts have occurred on several occasions during the past 2000 years, and probably since prehistorical and geological times. The millennium draught of Australia lasted from 1997 to 2010; California and Cape town, South Africa are currently experiencing severe draught; the African Sahel which began in the late 1960s is beginning to show signs of recovery; the drop in rainfall over Argentina which started at the beginning of the last century reversed after around fifty years. There are no consistent trends or periodicity overall. Droughts can be caused by a high level of reflected sunlight (high albedo), an unusual above average strength of high pressure systems, winds carrying continental, rather than oceanic air masses (i.e. reduced water vapor), ridges of high pressure areas which prevent or restrict the developing of thunderstorm activity or rainfall in the region. Oceanic and atmospheric weather cycles such as El Nino-Southern Oscillation (ENSO) make drought a regular recurring feature of areas situated in the sub-tropical high pressure belts of Africa. Discussions on climate change often highlight draught and desertification as obvious consequences of anthropogenic environmental pollution. There is little doubt that both have increased in the Mediterranean and West Africa, but observations show that they have decreased in central and North America and north-west Australia since 1950. The complex variables relating to draught are not well understood, and it is not clear to what extent draughts are due to natural or anthropogenic phenomena. Draught clearly promotes desertification and impacts severely on agriculture, but there is no clear link between drought and global warming.

6.3 IMPACT OF ENVIRONMENTAL POLLUTION ON HUMAN HEALTH AND ECOSYSTEMS

Both stratospheric and tropospheric anthropogenic pollutions can have profound effects on life on Earth, the main difference being the time scale. Pollutions which eventually reach the stratosphere are well-mixed and widely dispersed, and the effects (mainly global warming and increased life exposure to radiative damage) have no regional boundaries. However, the impact is slow and the effect may become noticeable after many years of accumulation. On the other hand, the impact of ambient (tropospheric) pollution can be instant, because it is the layer of the atmosphere that is in direct contact with life on Earth. Global warming is the most important consequence of stratospheric pollution, and its effect on climate change has been discussed above. Tropospheric pollution may cause global warming or cooling depending on the constitution. Both types of pollution are believed to be detrimental to human health, what is not is not clear is how many of the observed health issues can be attributed entirely to pollution. While increases have been observed in human ill-health, heat-related mortality, distribution of water-borne illnesses, much of it could have been the result of other stressors some of which are still not well understood. It is accepted however that climate-related hazards can exacerbate the impact of other stressors.

6.3.1 Stratospheric (middle atmospheric) pollution

The average global temperature has risen steadily since the beginning of the industrial revolution around 1750 and there is little doubt that there has been an uptake of energy by the global climate system. In effect, total radiative forcing is positive, and the largest contribution has been the steady increase in the atmospheric concentration of carbon dioxide (Figure 6.3). As discussed in Chapter 4, different types of anthropogenic pollutants have different effects on the environment, quantified in terms of radiative forcing (RF) which is the difference between the heat energy absorbed by the Earth from solar radiation and energy radiated back to space, resulting from a change in concentration of a particular substance, or the properties of the Earth's system. Each compound has a specific RF value associated with it; the more positive the value, the greater the contribution of warming influence, while negative values indicate cooling influence. Stratospheric pollutants contain compounds which have positive RF (carbon dioxide, methane, NOx), and the effect on the natural mechanisms by which greenhouse gases control the Earth's temperature has been discussed in some depth in

Chapter 4. The effect of anthropogenic halocarbons on the natural ozone layer was also discussed. Volcanic eruptions often release aerosols into the atmosphere and, while much of it stays in the troposphere, a substantial proportion of chemical compounds can rise above 10 km or so into the stratosphere and contribute to radiative forcing, although the effect usually fades off within about two years. The total anthropogenic RF for the main gases (CO_2, CH_4, N_2O and Halocarbons) in 2011 relative to 1750 was 3.00 W/m^2, with CO_2 alone causing over half (1.68 W/m^2). The most important halocarbon is dichlorodifluoromethane (CFC-12) which contributed 0.337 W/m^2 to the total. The Montreal Protocol has been very effective in reducing the use of halocarbons (F-gases) in the developed countries, but the replacement gases (hydrofluorocarbons, HFCs) are potent greenhouse gases although they do not destroy natural ozone. Furthermore, as discussed earlier, F-gases are still in common use in the emerging world and the negative impact will resonate beyond their borders.

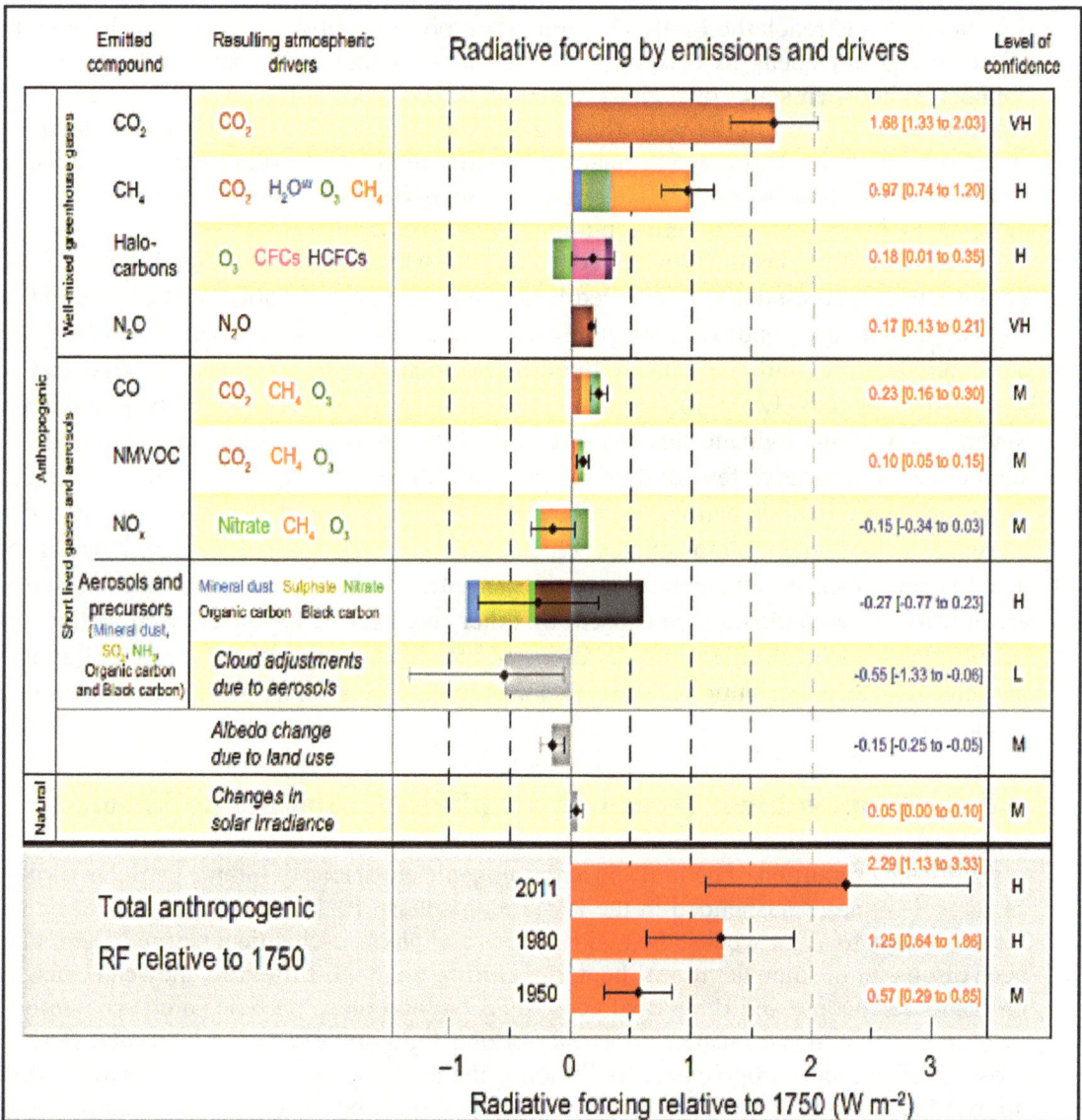

Emitted compound	Resulting atmospheric drivers	Radiative forcing by emissions and drivers	Level of confidence
CO_2	CO_2	1.68 [1.33 to 2.03]	VH
CH_4	CO_2, H_2O^{str}, O_3, CH_4	0.97 [0.74 to 1.20]	H
Halo-carbons	O_3, CFCs, HCFCs	0.18 [0.01 to 0.35]	H
N_2O	N_2O	0.17 [0.13 to 0.21]	VH
CO	CO_2, CH_4, O_3	0.23 [0.16 to 0.30]	M
NMVOC	CO_2, CH_4, O_3	0.10 [0.05 to 0.15]	M
NO_x	Nitrate, CH_4, O_3	-0.15 [-0.34 to 0.03]	M
Aerosols and precursors (Mineral dust, SO_2, NH_3, Organic carbon and Black carbon)	Mineral dust, Sulphate, Nitrate, Organic carbon, Black carbon	-0.27 [-0.77 to 0.23]	H
	Cloud adjustments due to aerosols	-0.55 [-1.33 to -0.06]	L
	Albedo change due to land use	-0.15 [-0.25 to -0.05]	M
	Changes in solar irradiance	0.05 [0.00 to 0.10]	M
Total anthropogenic RF relative to 1750	2011	2.29 [1.13 to 3.33]	H
	1980	1.25 [0.64 to 1.86]	H
	1950	0.57 [0.29 to 0.85]	M

Radiative forcing relative to 1750 (W m⁻²)

Figure 6.3 Global average radiative forcing estimates for the main anthropogenic emissions in 2011 relative to 1750 *(IPCC-WG1, 2013).*

The use of F-gases has been rising in emerging nations where most of the obsolete refrigerators and air conditioners from the developed world end up. F-gases have a very high global warming potential relative to other GHGs in spite of their small concentrations. Their lifetimes in the atmosphere range from one to several hundred years, with an average of 15 years (Blackstock and Allen, 2012). For example, a number of hydrofluorocarbons (HFCs) used primarily in refrigeration and insulating foams, have a warming effect that, on a per tonne-equivalent basis, is hundreds to thousands of times more forceful than CO_2 (Zaelke *et al.*, 2013). According to one estimate, HFCs could account for 9-19% of global GHG emissions by mid-century (Velders *et al.*, 2009).

In summary, anthropogenic contributions enhance the capacity of greenhouse gases mostly in the stratosphere to absorb and reflect back longwave radiation coming from the Earth, causing global warming. Halocarbons react with ozone, thus depleting the natural ozone layer and creating gaps in the shield through which potentially harmful ultraviolet radiation from the Sun can pass and reach the Earth. One important point about stratospheric pollution which is often missed but needs to be emphasized is the fact that there are no hard atmospheric boundaries between states or regions. Gases in the layer are relatively well mixed, hence the group is often referred to as well-mixed greenhouse gases (WMGHG). Their life span is in decades or centuries and they are relatively evenly mixed and evenly spread, and changes in concentrations occur over decades. Global forcing per unit emission and emission metrics for these gases do not depend on the geographic location of the emission.

Global warming leads to several consequences which can impact directly or indirectly on life on Earth. Increased and intensified heat waves can cause death, especially of old people and people with pre-conditions; strengthened tornadoes cause severe damage when they land; ocean temperatures and rise causing flooding in global coastal areas; increasing ocean acidity will impact negatively on global aquatic life; global weather variability could intensity. Several projections indicate that many coastal land areas in several regions could become submerged in the next few decades if the current rise in sea level continues unabated. Increasing temperatures can enhance increased production and migration of some disease-carrying vectors such as mosquitoes which spread malaria and zika virus. It is difficult to quantify the effects of human anthropogenic activities on human health because most diseases are multi-factorial and could be caused by other risk factors. However, there is fairly strong scientific evidence linking respiratory diseases and exposure to anthropogenic pollution. Even in such cases, the contribution of other factors that may cause or aggravate respiratory diseases is unclear.

6.3.2 Tropospheric (lower atmospheric/ambient) pollution

A significant proportion of natural and anthropogenic emissions of minute particles, also known as aerosols remains suspended in the lower atmosphere 10-15 km above the Earth's surface for a few days to a few weeks, the main sources and physico-chemical nature of aerosols have been discussed in some depth in Chapter 4. During their short lifetime, they can cause severe problems for people and the global ecosystem. When these particles are sufficiently large, they are visible, dense, scattering and absorbing light and obscuring visibility. Perhaps the most visible effect of tropospheric pollution is the photochemical smog that hangs over cities all over the world (Figure 6.4). The World Health Organization (WHO) estimates that around 92% of the world's population live in places where air quality levels far exceed WHO limits and, although ambient air pollution affects both developed and developing countries alike, low- and medium-income emerging countries (particularly in Western Pacific and South-East

Asia regions) experience the most severe effects. In a recent study by WHO (2018) covering around 3,000 cities and towns in 103 countries, only four OECD countries (Italy, Poland, Turkey and Mexico) featured in the first 500 most heavily polluted cities, while fourteen Indian cities were among the world's twenty most polluted, with Kanpur topping the list. (Figure 6.5).

Figure 6.4a Ambient/urban pollution (*google images of smog in cities*).

City smog

Household pollution

Figure 6.4b   Ambient/urban/household pollution *(google images of smog in cities and households).*

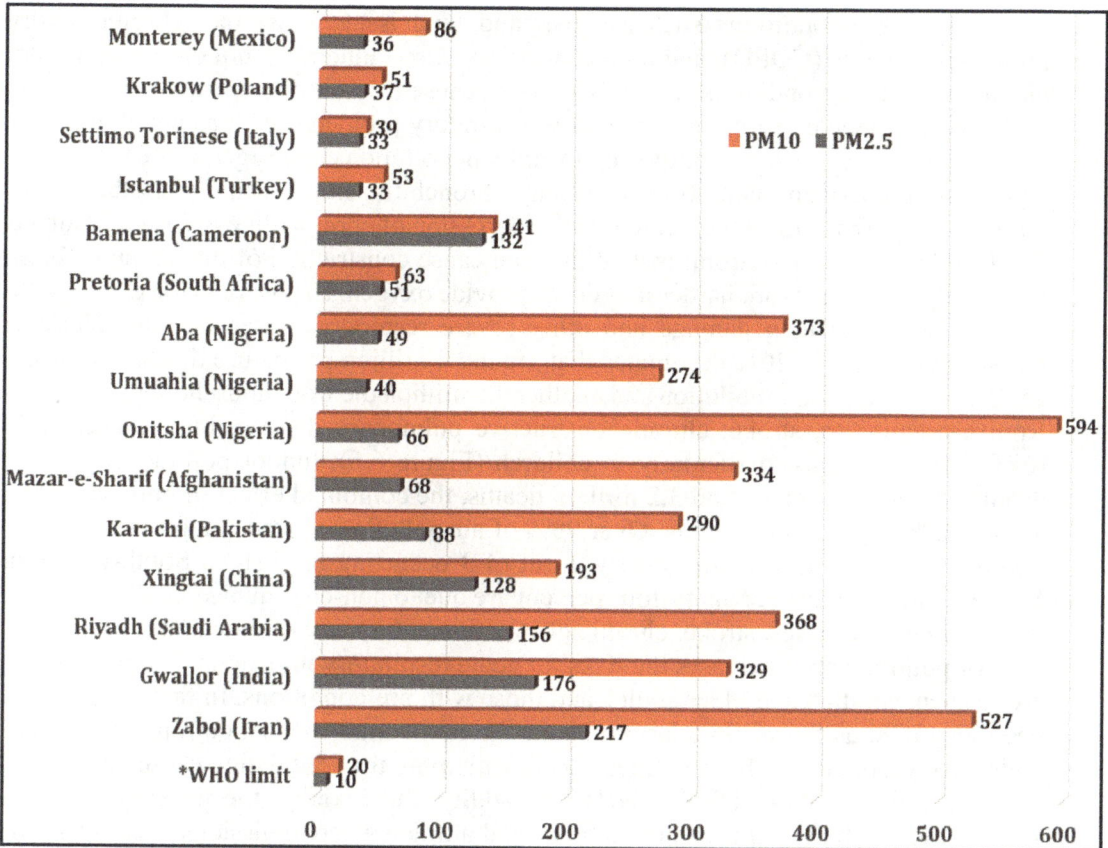

Figure 6.5 World's most polluted cities by particulate matter concentration *(Compiled from WHO Database, 2016).*

The impact of tropospheric ambient) pollution has a much more direct and severe impact on life on Earth because it is the part of the atmosphere that is in direct contact with the Earth's surface. However, because it is mostly locally sourced and there is significant physico-chemical variability in content, the effect on life is much more localized. Mining operations, oil drilling, fossil fuel transportation, conversion, and use, some industrial processes all release toxic gases into the troposphere. Other potential sources include volcanoes, desert dust, power generation, forest fires, transport, oil spills, and household basic energy use. The potential impact of aerosols on health depends on the chemical composition, concentration and proximity, all of which depend on the sources.

6.3.2.1 Impact *of pollutions on human health*

The effect of outdoor/indoor ambient pollution on human health varies, depending on the type of pollution, intensity/concentration and duration of exposure, as well as the age and state of health of the individual. Particulate matter is considered the most dangerous component of atmospheric pollution and the fine particles below 2.5 microns in size ($PM_{2.5}$) pose the greatest problems because they can bypass the body's natural defenses, penetrating deeply into the lungs and bloodstream. Children are particularly sensitive to poor air quality because their lungs are proportionally larger than those of adults in relation to their body weight, hence they breathe more. Furthermore, their immune system is still developing and they are less able to fight health issues that arise from indoor polluted air.

People with pre-conditions such as heart and lung issues - asthma, chronic obstructive pulmonary disease (COPD), and lung cancer - are also vulnerable, not only because polluted air can make their conditions worse, but also because they tend to spend more time indoors. Even healthy people can suffer temporary respiratory problems such as coughing, breathing difficulties and eye irritation. Exposure to ambient pollution can trigger new cases of diseases, exacerbate pre-existing conditions of asthma, bronchitis, emphysema, and other respiratory ailments, or provoke the development and progression of chronic illnesses such as lung cancer and COPD. Ozone is a strong irritant that can cause constriction of the airways, forcing the respiratory system to work harder in order to provide oxygen. Exposure to the gas for extensive periods can cause lung damage and other severe respiratory diseases. The World Health Organization (WHO, 2018) estimates that around 4 million premature deaths a year globally are linked to outdoor air pollution and another 3.5 million die from household pollution, mainly from heart disease, stroke, chronic obstructive pulmonary disease, lung cancer and acute respiratory infections, particularly in children (Figure 6.6). Indoor pollution can be just as deadly, accounting for around 3.2 million deaths, the combined effect of both accounting for around 12% of all global deaths. Over 90% of air-pollution-related deaths occur in low- and middle-income countries, with nearly 2 out of 3 occurring in WHO's South-East Asia and Western Pacific regions. Ninety-four percent are due to non-communicable diseases - notably cardiovascular diseases, stroke, chronic obstructive pulmonary disease and lung cancer.

Air pollution also increases the risks for acute respiratory infections. The most vulnerable are women, children and older adults, and those with pre-conditions. In fact, air pollution had been identified as the world's largest single environmental risk. About one million children under five years die yearly from pneumonia, with more than half being caused by exposure to household air pollution (WHO, 2018). In addition to toxicity, the proximity of emission sources to potentially exposed populations, and altitude where emissions occur can influence the magnitude of their impact on public health. For example, emissions (black carbon, organic carbon) from automobiles, biomass burning, household energy use occur at ground level compared with high power plant smoke stacks. This likely results in greater exposure and therefore larger health impacts compared with nitrogen and Sulphur oxides which come mainly from high smoke stacks, although diesel engines also produce significant nitrogen oxides.

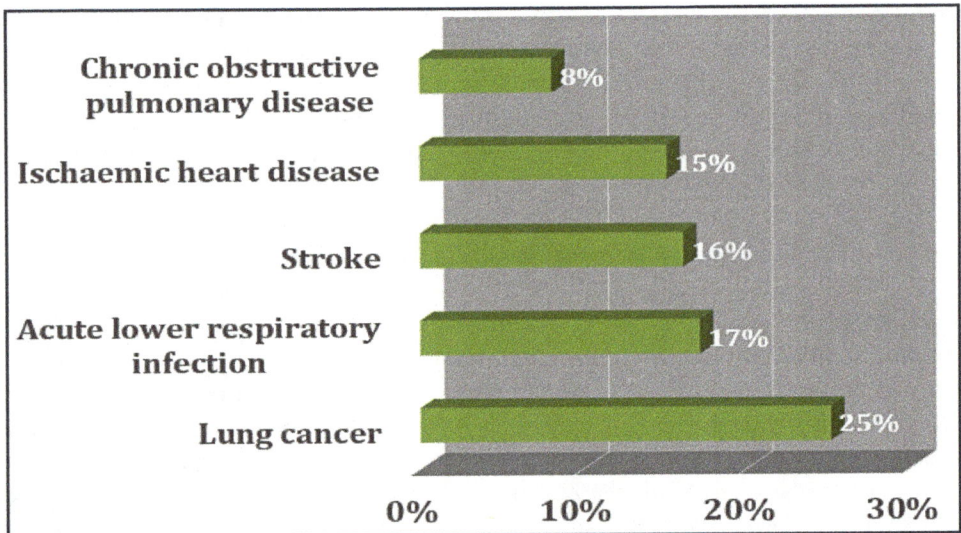

Figure 6.6a   **Annual global deaths from environmental pollution:
Ambient: 4.2 million; Household: 3.8 million** *(WHO, 2018).*

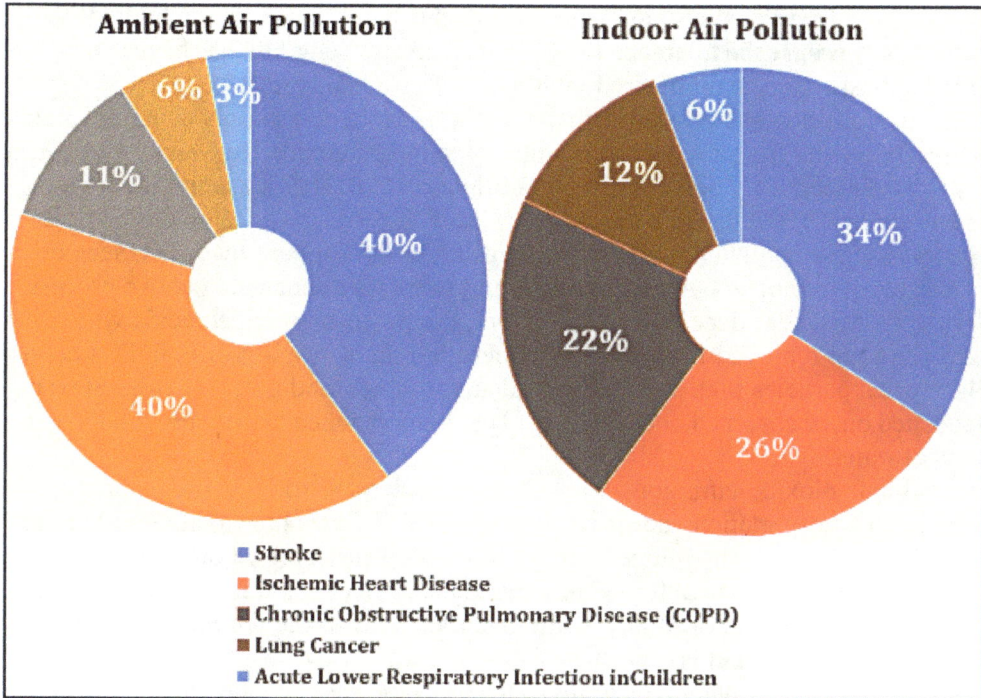

Figure 6.6b Air pollution-Caused Deaths: Breakdown by Disease *(WHO, 2016/2017c)*.

Ozone, nitrogen dioxide and sulphur dioxide are linked to asthma, bronchial disease, reduced lung function and lung disease. Sulphur trioxide causes irritation of the mucous membranes of the respiratory tract, causing coughing and choking. Inhalation of droplets of acid rain formed by the reaction between sulphur trioxide and atmospheric moisture can also harm the respiratory system, and high levels of carbon monoxide are believed to cause problems in the nervous system. Sulphur and nitrogen oxides in gaseous products of coal combustion often react with moisture, soot and fly ash particles to form a dense smog which not only obscures sunlight but can cause eye irritation and breathing problems. Long-term exposure can cause chronic pulmonary diseases such as asthma, bronchitis, lung cancer, and cardiac problems, including arrhythmias and heart attacks (USEPA, 2012).

Depletion of the Earth's protective heat shield formed by the stratospheric ozone layer enables potentially dangerous ultraviolet radiation to reach the Earth's surface, causing skin burns and skin cancer. Incidences of fluorosis, arsenicosis, selenium intoxication, pneumoconiosis and silicosis are high among rural communities that live near coal mines or burn coals with high fluorine, arsenic, selenium, iron and silicon contents in poorly ventilated environment (Zhang *et al.*, 2004; Finkelman, 2017). Global warming and consequent climate change may have wide ranging and mostly negative direct and indirect impacts on human health, including increasing heat waves that can cause heat strokes and deaths. A significant proportion of atmospheric lead comes from traffic emissions, due to the lead content of gasoline, hence lead content of atmospheric pollution can be high in urban areas. This has been reduced significantly in some developed countries in recent years since unleaded gasoline became widely available (though more expensive). However unleaded petrol is not available in most emerging countries and aerial concentrations in urban areas can be very high. Lead accumulates slowly in the blood and when levels reach concentrations of around 800 mg/litre, damage to the central nervous system, kidneys and the brain can occur. Children are particularly vulnerable because the tolerance level is lower and a strong link has been established between high lead exposures and impaired intelligence (Air-Quality, 2017).

Acid deposition is a general term that includes more than simply acid rain. Acid deposition primarily results from the transformation of sulphur dioxide (SO_2) and nitrogen oxides into dry or moist secondary pollutants such as sulphuric acid (H_2SO_4), ammonium nitrate (NH_4NO_3) and nitric acid (HNO_3). The transformation of SO_2 and NOx to acidic particles and vapors occurs as these pollutants are transported in the atmosphere over distances of hundreds to thousands of kilometers. Acidic particles and vapors are deposited via two processes - wet and dry deposition. Wet deposition is *acid rain*, the process by which acids with a pH normally below 5.6 are removed from the atmosphere in rain, snow, slit or hail. Dry deposition takes place when particles such as fly ash, sulphates, nitrates, and gases (such as SO_2 and NOx), are deposited on, or absorbed onto surfaces. The gases can then be converted into acids when they contact water.

Sulphur dioxide emissions come from a variety of sources including coal-fired power generation, transportation, industrial processes, fossil fuel combustion, ore smelting and natural gas processing. The main source of NOx emissions is the combustion of fuels in motor vehicles, residential and commercial furnaces, industrial and electrical utility boilers, engines, and other equipment. Nitrogen oxides also come from air used in fuel combustion chambers of fossil power plants. Acid rain is a problem in many parts of the world because many of the water and soil systems lack natural alkalinity such as a lime base and therefore cannot neutralize acid naturally. Increased acidity of rivers, lakes and streams destroys aquatic life and acidified soils harm crops and forests. Acid rain also causes corrosion of road and building metal fixtures and automotive metal components.

Sulphur dioxide and nitrogen oxides in particulate matter (PM) react with moisture in the atmosphere to form reactive acids which fall on the Earth's surface as acid precipitation, rain or snow (wet precipitation), causing damage to metallic structures, soil, vegetation, etc. On the other hand, the acids may react with other chemical compounds in the atmosphere and form gases and salts which are also deposited on surfaces. These reactions occur naturally since sulphur dioxide is produced from many other sources not related to human activities, for example, volcanic eruptions and decaying vegetation. However, burning of fossil fuels (in particular, power generating plants) has greatly increased the concentration of these gases in the atmosphere over decades and acid rain has become a major hazard to human life and the ecosystem. Once in the air, pollutant gases are carried by the wind, and hence deposition can take place a long distance from the source, and if large quantities of acid are deposited in one place then it may have detrimental consequence for human health, buildings, soils vegetation, crops, freshwater, and aquatic life. Acidic sulphur and nitrogen dioxides are powerful irritants, particularly in the respiratory system, especially when they are mixed with PM (synergistic or cocktail effect). Asthmatics are particularly susceptible because these pollutants can cause serious breathing difficulties.

Most materials used in building and other structural construction are susceptible to acid rain corrosion, although the rate and intensity vary widely. Metal structures are the most vulnerable, and carbon steel which is used extensively in concrete reinforcement and structures leads them all. However, many other materials are also susceptible - limestone; concrete, marble, zinc, nickel, paint and some plastics. Sulphuric acid formed from sulphur dioxide is the most destructive acid rain but the compound has to be oxidized to sulphur trioxide (SO_3) to form acid by reacting with water. The presence of nitrogen dioxide enhances the oxidation process. However, nitrogen dioxide also reacts with water to form nitric acid which is often a component of acid rain. Both exposed structures and those immersed in acidic waters or buried in acidified soils are vulnerable.

6.3.2.2 *Impact of pollutions on agriculture and forestry*

Acid rain can alter soil chemistry significantly, increasing the acidity and destroying many microorganisms which break down organic matter and transform into nutrients for plants. Also, acidified soil may release aluminium metal which is toxic to plant roots. The response of soils to acidification varies depending on the inherent soil chemistry. Soils rich in marble or limestone (carbonates) can effectively neutralize acid rain, while granite-rich soils which are devoid of carbonates have no neutralizing effect. Acidified soils can stunt the growth of plants and trees, and also lower their immunity and resistance to diseases, fungal attack, and invasion of insect pests. Many plants and trees die as a result of soil acidification. Acid rain settles on plants and crops, destroying the protective waxy coating of leaves and allows acids to diffuse into them. The natural plant-atmospheric evaporation and gas exchange system is thereby interrupted, with a negative impact on crop yield and quality. Vegetables are particularly vulnerable to acid rain and many other crops (corn, potatoes, soy beans, etc.) are also susceptible to acid rain damage. Black carbon, in particular $PM_{2.5}$ and associated pollutants such as tropospheric ozone can damage crops and ecosystems, with negative impacts on agriculture, livestock, fish, etc. In addition to affecting precipitation, $PM_{2.5}$ deposited on leaves has been shown to affect crops and ecosystems by reducing the surface albedo, thus obstructing access to adequate sunlight as well as carbon dioxide, both of which are critical to the natural plant biological processes.

6.3.2.3 *Impact of pollutions on aquatic life and wildlife*

Deposits of acid rain, black carbon and other associated aerosols can be toxic to aquatic life. Over time, acid rain increases the acidity of surface water at the catchment or during transport over acidified soil, decreasing the average pH to levels that impact negatively on aquatic life. Oceans are major long-term sinks for carbon dioxide, and may also become acidified. Soil and aquatic organisms have different tolerances to acidity and some die while others have reduced proliferation. Soft-bodied aquatic life - eels, angelfish, snails, shrimps, leeches, crayfish - are usually the early victims of fresh/seawater acidification, but long-term effects have been identified on most aquatic life and insects. All life stages are affected, from egg survival to the survival of young fry and adults. Many studies have demonstrated that surface water acidification can lead to a decline in, and loss of fish populations. Below pH 4.5 no fish are likely to survive. Other aquatic life such as amphibians are also affected: many invertebrate species that contain high concentrations of calcium, such as mollusks and crustaceans, are very sensitive to pH levels and are among the first to disappear during the acidification of wetlands.

It should be noted however that atmospheric pollution is not the only source of freshwater/ocean acidification. Natural imbalance between precipitation and evapotranspiration can cause the leaching of minerals from the soil, in particular, iron, aluminium and mercury (podzolization). Action of atmospheric carbonic acids, use of nitrogen fertilizers, in particular, those that are not sulphur-free, invasion of catchment by livestock can all cause changes in freshwater acidification. In most situations, freshwater acidification is a combination of both natural and anthropogenic factors.

There is also evidence that oxygen concentration is declining in increasing areas of global ocean and coastal waters due to ocean warming which reduces solubility of oxygen in water and the rate of oxygen resupply from the atmosphere to the ocean interior. Increasing discharge of nutrients (nitrogen and phosphorus) and organic matter, primarily from agriculture, sewage and the combustion of fossil fuels into coastal waters is causing increased acidification and accelerated consumption of oxygen by microbial respiration. Deoxygenation of oceans and coastal waters can have a wide range of biological and ecological consequences, including creation of low-oxygen 'dead zones' which cannot support marine life (Breitburg *et al.*, 2018). Wildlife and domesticated animals and birds all feed on plants, vegetation, and fruits, all of which are exposed to the effects of atmospheric pollution, in particular, acid rain and ground-level ozone, hence they are all potentially susceptible to the negative effects of acidification and toxic aerosols. They are affected indirectly because their food chain is disrupted or corrupted. For example, animals and birds that feed on freshwater fish find fewer fish in acidified lakes.

6.4   OUTLOOK ON ENVIRONMENTAL IMPACT OF ENERGY-RELATED POLLUTIONS

There are few signs that energy-related environmental pollution will decrease significantly for decades. Even if stratospheric pollution could be eliminated completely today, the negative effects of past pollution, particularly on climate could persist for decades or even centuries. It appears that the only feasible options are to promote adaptation strategies for the current global population and mitigation strategies and policies that could lower the risks for future generations. These include strategies for coping with the negative impact of climate change and ambient pollution, strong policy support for decarbonization of energy and more aggressive efficiency improvement across the energy value chain, from production to end-use, and promotion of public awareness on how they can reduce their carbon footprint through lifestyle changes. Tropospheric pollution is much easier to deal with because most of the causes are local. Increasingly stringent emission standards for power plants and vehicles would help in reducing ambient outdoor pollution. However, most current policies target the developed world, while the major proportion of future energy-related pollutions will come from the emerging world. Also faster transition to modern domestic energy worldwide would have a significant, positive impact on human health and the ecosystem. However, prospects, particularly in the developing world are not very bright, in spite of numerous and diverse interventions by many organizations to move the regions towards the United Nations' goal of modern energy for all by 2030.

Chapter 7

Mitigation of environmental pollution and climate change

7.1 INTRODUCTION

On the basis of the extensive data presented in the previous chapters, there is little doubt that environmental pollution has increased significantly since the Industrial Revolution, due mainly to intensification of human activities, from agriculture to energy production and use. It is also clear that around two-thirds of the emissions comes from energy production and use, and fossil fuels account for nearly all the energy-related pollution. The potential consequences of environmental pollution impact on the Earth and its environment in many ways, with potentially negative effects on weather and climate, agriculture, and life on Earth. There is little doubt that the Earth's average temperature has been increasing steadily in the last hundred years or so, with potentially severe negative consequences. What is not so clear is the extent to which all the observed changes can be attributed to human activities. However, there is strong scientific evidence that the balances of nature's control systems in terms of regulating the energy balance of the Earth and its environment, are being altered significantly by anthropogenic emissions from various human activities. It is pertinent therefore to seek ways of mitigating this potentially damaging development.

Problems created by energy use go beyond environmental pollution. The energy industry is the second largest consumer of freshwater after agriculture and consumption is high along the energy value chain, both in primary energy production (coal, oil, gas, and biofuels) and in transformation (oil refining, power generation). With the increasing rate of global population and urbanization, greater water stress is predicted in various projections. Also, increasing solid pollution, driven by rising standards of living have repercussions for the environment in many ways: the rate of dumping far outpaces recycling efforts, and discarded plastics and electronic waste are becoming significant sources of hazardous chemicals and compounds in the atmosphere; more energy is required to make new products, and therefore more anthropogenic pollutants are released into the environment.

7.2 WORLD'S FOSSIL ENERGY RESERVE, USE AND POLLUTION OUTLOOK

The outlook on fossil energy use has been discussed extensively in Chapters 3 and 5, and it is clear that the world will continue to rely heavily on these sources for primary energy in the foreseeable future. Most recent projections show that fossil fuel use will drop by only a few percentage points (currently around 82%) to about 79% of total global primary energy by 2040. One important fact that is often overlooked is the increasing use of petroleum (oil and gas) as feedstocks for the production of petrochemicals many of which are precursors to a wide range of consumer products, pharmaceuticals, polymers, fertilizers, high-premium chemicals and materials. Furthermore, many industrial processes are energy-intensive. Around 50% of the world's energy use is accounted for by industry and is projected to grow by around 25% by 2040, led by growth in the chemicals sector, hosted mostly by emerging nations (Figure 7.1). Rising prosperity is driving demand for fertilizers, plastics and other chemical products. Petroleum feedstock is steam-cracked to produce the basic building blocks for plastic and other chemical products which have become an important part of our everyday life. The consumer, pharmaceutical and cosmetic industries depend on derivatives of petroleum for most of their products. Demand for iron and steel and cement used in machine building, building and road construction is rising worldwide, and there is as yet no feasible substitute for coal or natural gas as the primary fuels. In effect, as long as global population, economy and prosperity continue to rise, so will the demand for modern energy and emissions, although moderated significantly by efficiency gains across the whole spectrum of the global economy.

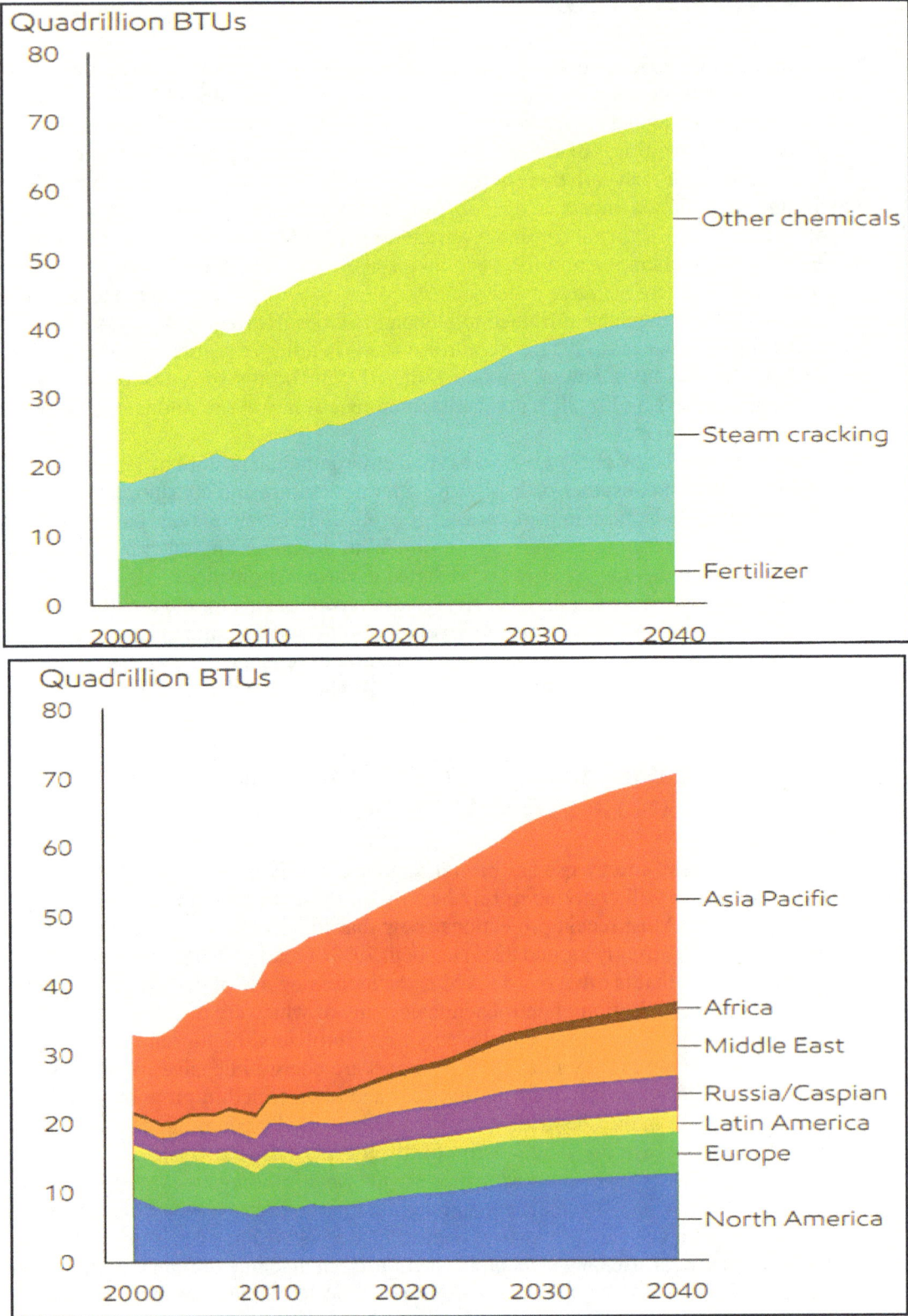

Figure 7.1 Projection of global demand for fossil fuel-sourced chemicals *(ExxonMobil, 2017).*

Silicon and other high-technology materials which are critical to the sustenance of the information technology industry require enormous energy for production. The transportation sector currently relies on oil for nearly all fuel requirements, and this will continue for decades to come because current goals of replacing internal combustion cars with electric cars using low-carbon electricity within the next two decades appear to be over-optimistic (see Chapters 3 and 5). Furthermore there is no potential substitute for diesel which fuels the growing commercial heavy trucking and marine transportation, and aviation fuel for air transportation, and the demand for both fuels could double within the next two decades. The building sector of the global economy relies on mainly on fossil natural gas and mainly fossil electricity. It is not surprising therefore that all recent projections show that fossil energy-related emissions will continue to grow, be it at a slower pace in response to numerous mitigation policies.

Energy-related emissions are projected to rise by around 12% from 33.1 billion metric tons in 2018 to about 37 billion metric tons in 2040, a modest rise compared with a rise of 40% from 2000 to 2016. Coal will account for around 38% (a drop from 45% currently), while oil and gas will account for 36% and 26% respectively (Figure 7.2). Although emissions in all regions are projected to peak before 2040, there will be significant growth disparity between developed and emerging countries. While the projected growth in North America and Europe will decline by about 15%, emission levels in non-OECD countries will grow by about 51%, and the region will account for around 80% of total global energy-related emissions by 2040 (ExxonMobil, 2018).

Just ten countries accounted for 67% of total global energy-related emissions in 2015, and China and United States alone accounted for 45%. On the contrary, the whole of Africa contributed only 5%. Most of the expected growth in energy-related emissions will come from emerging nations whose main preoccupation is to provide better life and access to modern energy for the teaming populations. For many, coal, the most carbon-intensive pollutant is the only locally available or accessible fuel and pollution mitigation is of low priority (Figures 7.3 to 7.6). Also, many of these countries simply do not have the competence or funding to take the actions required to mitigate environmental pollution. In effect, the only feasible options are *adaptation* and *mitigation* which are complementary strategies for reducing and moderating global energy-related environmental pollution and the risks of climate change.

The effects of anthropogenic pollution on climate are very gradual and could take decades or even centuries before the negative impacts become evident. Furthermore, the effects are largely independent of the sources. In effect, current generations are already bearing the consequences and future generations will be the main beneficiaries of current mitigation actions. However, the impact of ambient (indoor/outdoor) pollution is largely localized and instant, and the positive effects of adaptation and mitigation actions can be obvious within short periods. Also, while climate change mitigation requires coordinated global action, state and local governments can introduce and enforce policies that help clean up local environments, thereby reducing risks to public health and the survival of the ecosystem.

7.3 ADAPTATION OPTIONS

It is clear from the foregoing discussion that global energy-related emissions are unlikely to stop any time soon. However, there are many options for slowing down the growth rate in spite of the projected phenomenal growth in, and, eventually reducing total emissions. It is also clear that, even if emissions could be stopped today, the negative impact on the environment could persist for decades, or even centuries to come. In effect, mankind has to develop coping mechanisms (adaptation) to deal with the undesirable consequences that impact on survival and health of people and the ecosystem, while helping to mitigate risks to future generations.

Adaptation options are largely localized and within the jurisdiction of local and state government, hence the impact is more immediate compared with mitigation options which depend on both intra- and international actions. In recent decades, changes in climate have impacted negatively on natural and human systems on all continents and across the oceans - global warming, warming and acidification of the oceans, coastal erosion, extreme weather, toxic environment, variability in precipitation, draught, desertification, etc. (see IPCC, 2014a & b).

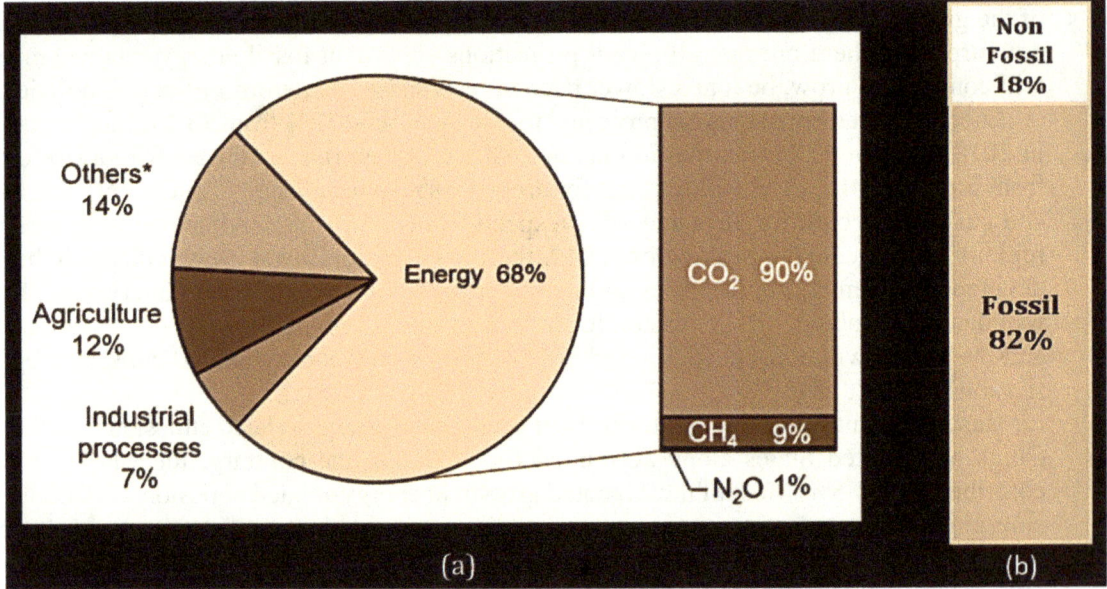

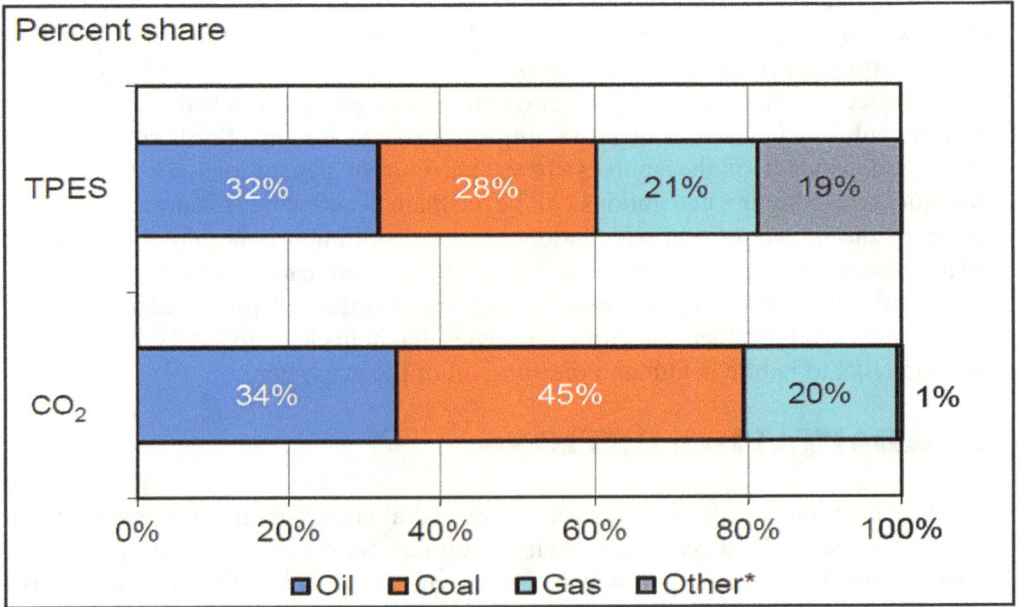

Figure 7.2 Estimated shares of global anthropogenic GHG, 2014 (b) World primary energy supply, 2015 (c) Primary emissions shares by fuel *(IEA 2017k)*.

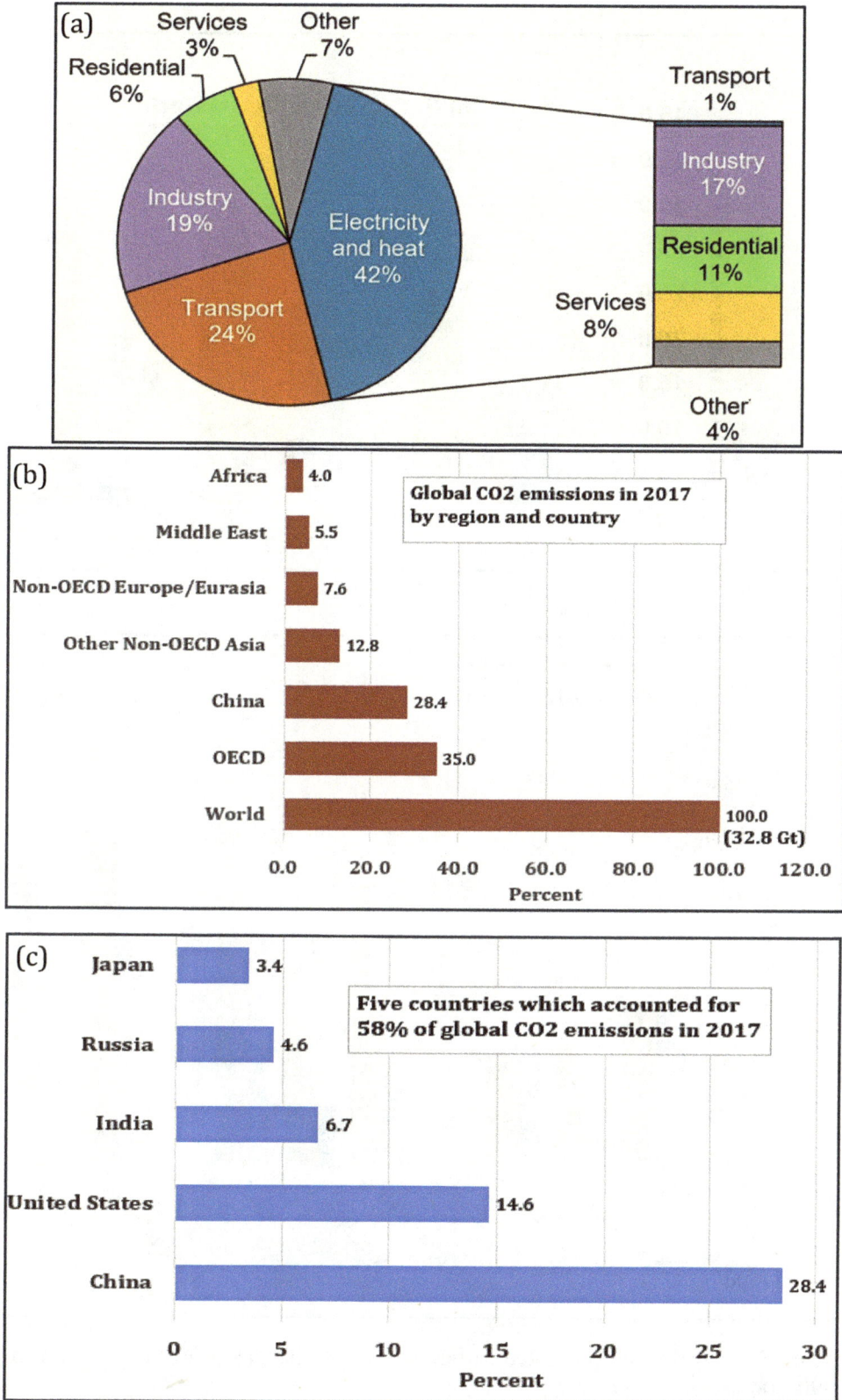

Figure 7.3 (a) World energy-related CO_2 emissions in 2017 (a) by sector, (b) by region and country (c) Top five emitting countries in 2017 *(IEA, 2019).*

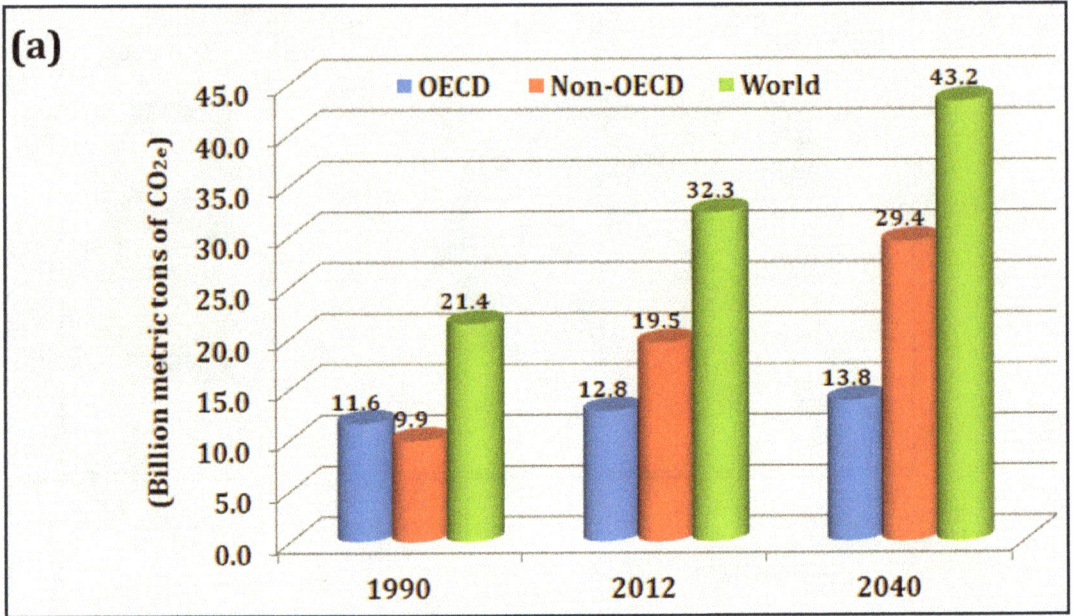

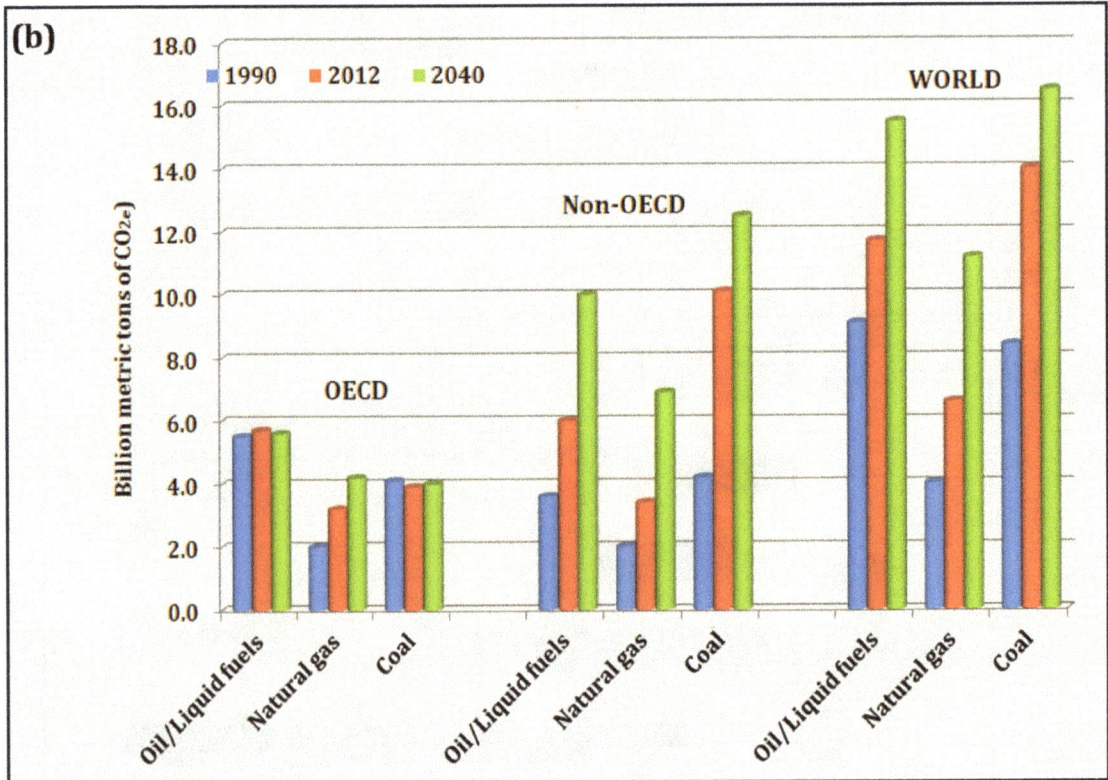

Figure 7.4 World energy-related carbon dioxide emissions by (a) region and (b) fuel, 1990-2040 *(Data from EIA, 2016).*

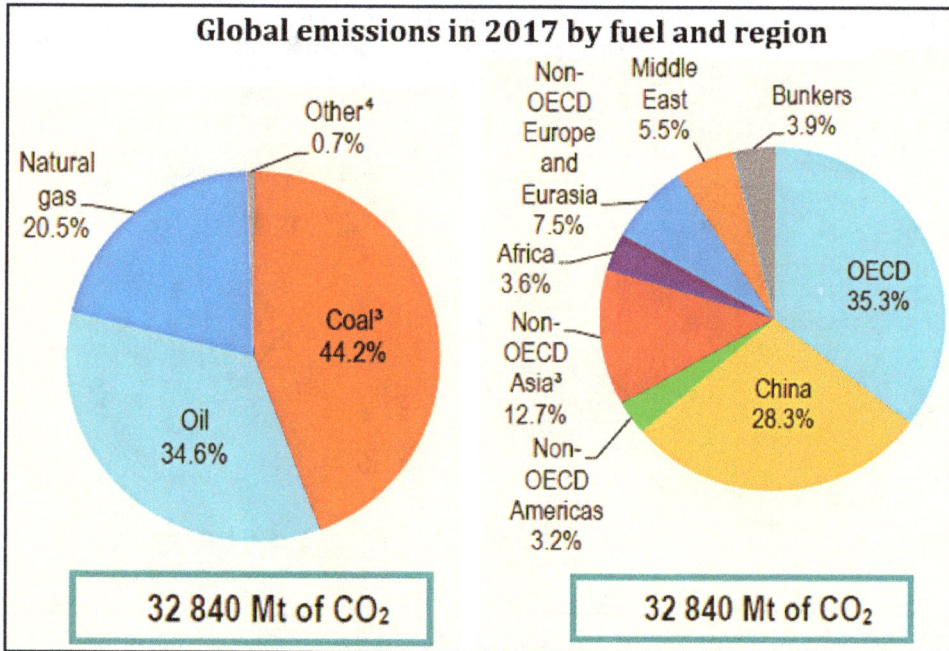

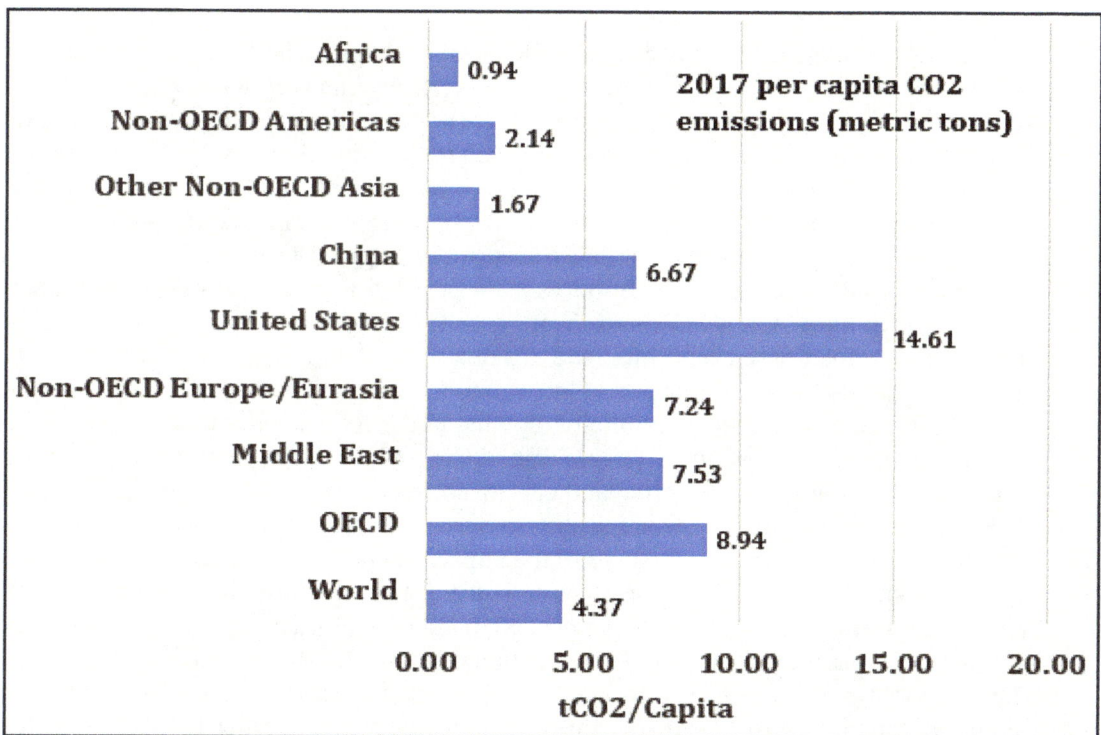

Figure 7.5 World energy-related carbon dioxide emissions by fuel, region and per capita, 2017 *(IEA, 2019n)*.

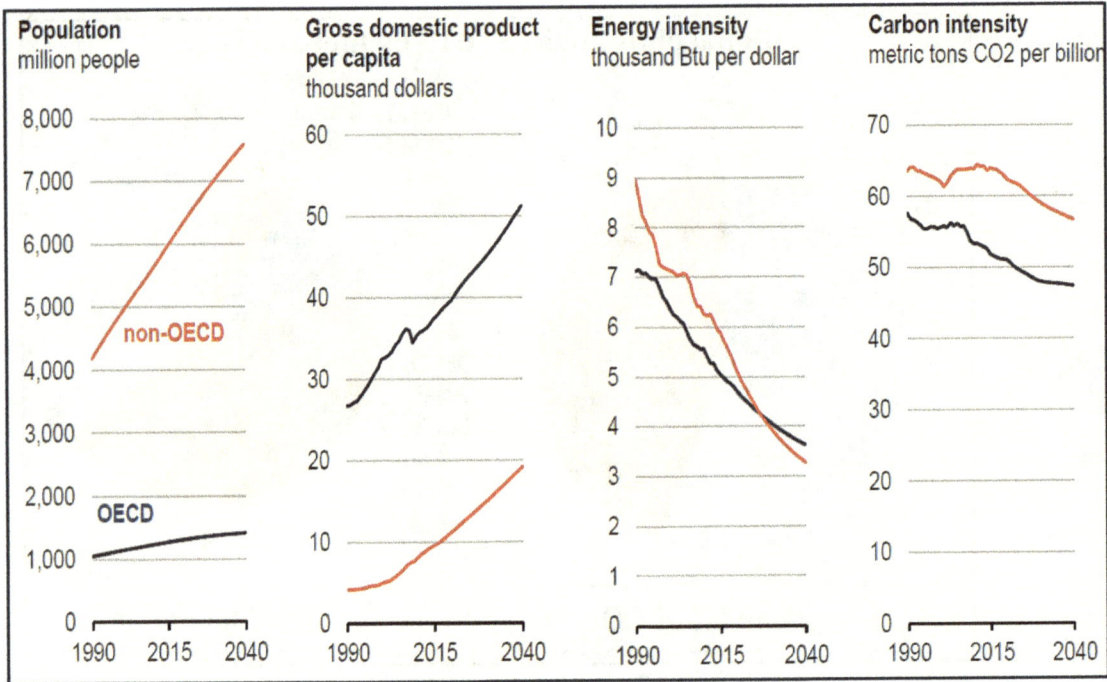

Figure 7.6 Projections for drivers of energy demand, and responses of energy and carbon intensities *(EIA, 2017)*.

It is difficult to quantify the impact of climate change on human and natural systems for many reasons: the nature of hazards, exposure and vulnerability can vary widely across regions and socio-economic stratifications. Furthermore, there are non-climatic stressors that can influence vulnerability, exposure and ability to adapt to changing situations. Adaptation measures are designed to minimize potential consequences of phenomena such as extreme weather and ambient pollution on life. These include provision of early warning systems, storm shelters, first responders, etc. Adaptation policies need to be targeted to be effective. For example, urban areas hold more than half of the world's population and most of its built assets and economic activities. A high proportion of greenhouse gas and aerosol emissions is also generated by urban-based activities and residents. Potential adaptation strategies include improving the design of buildings and structures to withstand extreme weather, multi-level response strategies to disasters, publicity on what to do in the event of extreme weather, and building of awareness on what people can do to reduce their carbon footprint, such as embracing and supporting recycling and energy conservation.

Rural areas have different vulnerability to disasters, for example, obstructions that could slow down tornados are relatively few and the impact can be severe. Furthermore, most disaster response systems are based in urban areas and help can be slow in arriving in case of disasters. Rapid urbanization and migration have peaked in many regions, and poverty/extreme poverty rates in rural areas are falling, with the exception of sub-Saharan Africa where rates are rising. Many in the region live on subsistence agriculture and access to fertile land and adequate rainfall are crucial. Most of these people are subjected to many other non-climatic stressors and any disruption of their livelihood, for example, as a result of draught or excessive rainfall can be devastating. Many rural dwellers devise their local adaptation strategies since there is often very limited institutional support.

The health of human populations is sensitive to shifts in weather patterns and other aspects of climate change. Changes in temperature in terms of increase and variability, precipitation, and the occurrence of heat waves, floods, droughts and fires can impact directly on human health, especially when there are other health pre-conditions. Also, there can be many indirect consequences such as ecological disruptions leading to crop failures, shifting patterns of disease vectors, displacement of populations as a result of extreme weather. Challenges for vulnerability reduction and adaptation actions vary across regions: while many developed countries have good response strategies in place, most in the developing world, many economically, socially, politically, institutionally marginalized, depend largely on interventions by humanitarian organizations for adaptation and mitigation responses.

7.4   MITIGATION OPTIONS

There is little doubt that human activities are influencing the climate system, concentrations of greenhouse gases are the highest they have ever been, and their effects, together with those of other anthropogenic drivers, are evident throughout the whole spectrum of the climate system. Many repercussions of climate change have had widespread negative impacts or human and natural systems (see Chapter 6). Extensive scientific data indicate (with a high degree of confidence) that these emissions have been the dominant cause of the observed global warming since the mid-20th century. The Earth's surface temperature is projected to rise over the 21st century under all possible emission scenarios, and it is very likely that heat waves will occur more often, they will be more intense, and will last longer, flooding will be more frequent and severe, oceans will continue to warm and acidify, storms and tornadoes will be more devastating because they derive their energy from warming oceans, and extreme precipitation events will become more intense and frequent in many regions of the world. The thrust of global mitigation efforts has been to reduce energy-related pollution emissions, but the International Panel on Climate Change (IPCC, 2014a) has concluded (with a high degree of confidence) that current efforts are inadequate, as is evident from its statement quoted below:

> *"Without additional mitigation efforts beyond those in place today, and even with adaptation, warming by the end of the 21st century will lead to high to very high risk of severe, wide-spread and irreversible impacts globally (high confidence). Mitigation involves some level of co-benefits and of risks due to adverse side effects, but these risks do not involve the same possibility of severe, widespread and irreversible impacts as risks from climate change, increasing the benefits from near-term climate change mitigation efforts."*

Considering the inertia of response of the Earth and its ecosystem to climate change, current mitigation efforts may have some near-term benefits particularly in urban areas but the major positive impact on climate change will become evident in the later decades of the 21st century and beyond. The world's primary energy demand is growing, in response to strong economic and population growth, and all recent projections show that the trend will continue. Fossil fuels have been the major sources of primary energy for centuries and the dominance is unlikely to diminish in the foreseeable future. Around two-thirds of the total global atmospheric pollution comes from the use of energy, and fossil fuels account for over 90%. In effect, fossil energy-related pollution will continue to rise as long as energy demand continues to increase, even if all current and planned global mitigation actions are effective. Perhaps the most realistic plan is to continue to reduce the growth rate so that it becomes negative over the next few decades, and there are many options, the four most promising are listed below and the relative potentials shown in Figure 7.7.

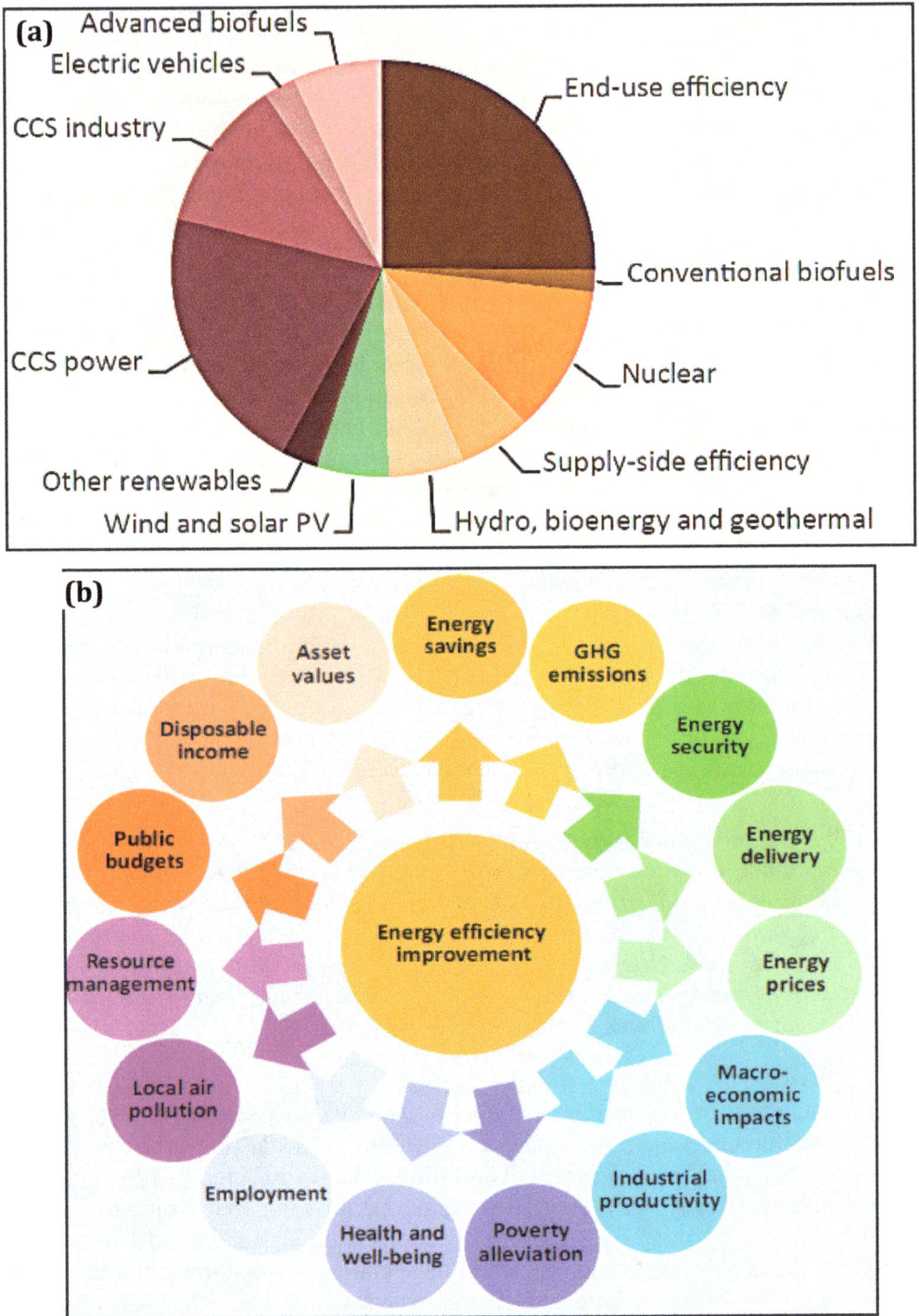

Figure 7.7 (a) Global cumulative CO_2 emissions reductions by mitigation measure, 2015-2040 (113 Gt CO_2 abatement) (b) Multiple benefits of energy efficiency improvements (*IEA, 2015c/2017d*).

7.4.1   Benefits of energy efficiency and intensity

Energy efficiency is simply the ratio of output to input, typically expressed as a percentage. A power plant that converts only 40 kilograms of 100-kilogram feed coal into electric power has only 40% thermal efficiency. Energy intensity is the amount of energy required for a given output and is often expressed as energy/GDP or energy/capita. Efficiency rises either through re-organization of existing methods or as a result of infusion of new technologies and higher values mean lower energy consumption and associated emissions per unit output. Carbon intensity (CO_2 emission/unit energy output or use) is primarily a function of carbon content of the fuel, hence responds to fuel switching. Increasing efficiency over the next two decades or so and switching to lower carbon-intensive fuels should drive down energy intensity and carbon intensity, in spite of strong growth in the main energy demand drivers: population and economic growth (Figure 7.6). Energy efficiency has been evolving as a key resource, with multiple benefits for economic and social development across all economies, and has emerged as a major driver for uncoupling energy consumption from economic development. Apart from the obvious benefits of enhanced energy efficiency - reduced energy demand and lower greenhouse gas emissions - investment in energy efficiency can provide many different benefits to many different stakeholders (Figure 7.7). Whether by directly reducing energy demand and associated cost (which releases funds for other investments), or facilitating the achievement of other objectives (such as more efficient internal combustion engines and household energy appliances that reduce atmospheric aerosols, the enormous multi-faceted potential of energy efficiency has become a prominent topic of global research interest in recent times (IEA, 2017d).

Energy intensity is the primary energy per unit gross domestic product (GDP) and when it declines, it means that the world is able to produce more GDP for each unit of energy consumed. Global energy intensity has declined at an average rate of 2.1% per year for the last ten years or so and the benefits, also known as *energy productivity bonus*, are both enormous and multi-dimensional. It is estimated that, without energy efficiency and intensity improvements, the world would have used 12-15% more energy than it did in 2018, equivalent to the total energy consumption of the European Union and worth over USD 2 trillion. Energy efficiency is regarded as the most effective environmental mitigation effort and the biggest contributor to reduced energy use and emissions. It is estimated that a combination of the decline in energy intensity and the change in energy mix towards lower carbon energy have been responsible for stalling growth and steadying emissions at around 32-33 billion metric tons over the last five years, with energy efficiency and associated fall in energy intensity accounting for the offset of around 77% of the impact on global emissions from GDP growth. Rising energy efficiency and associated decline in energy intensity make big contributions to strengthening energy security since many nations that depend on external sources for primary energy would need to import significantly less, with substantial economic and social benefits. Energy efficiency is considered a major resource that remains largely untapped and believed to hold the key to future sustainable environment. The International Energy Agency (IEA, 2019o) has developed a model that shows the enormous potential of energy efficiency gains in mitigating emissions over the next twenty years or so (Figure 7.8) The model known as *Efficiency World Scenario* (EWS) shows that a strong growth rate in energy efficiency across the global economy could reduce global energy-related emissions by more than 40% of the emissions cuts needed to reach the sustainable climate goals without new technologies. The model also delivers the energy efficiency target one of the United Nations Sustainable Development Goals: energy equity for all. Under the EWS, the amount of global GDP produced for each unit of energy could double over the next two decades, for only a marginal increase in energy demand. However, this would require strong policy actions across all regions, starting from now and focusing on many of the two-thirds of final energy use not currently covered by mandatory energy efficiency policies.

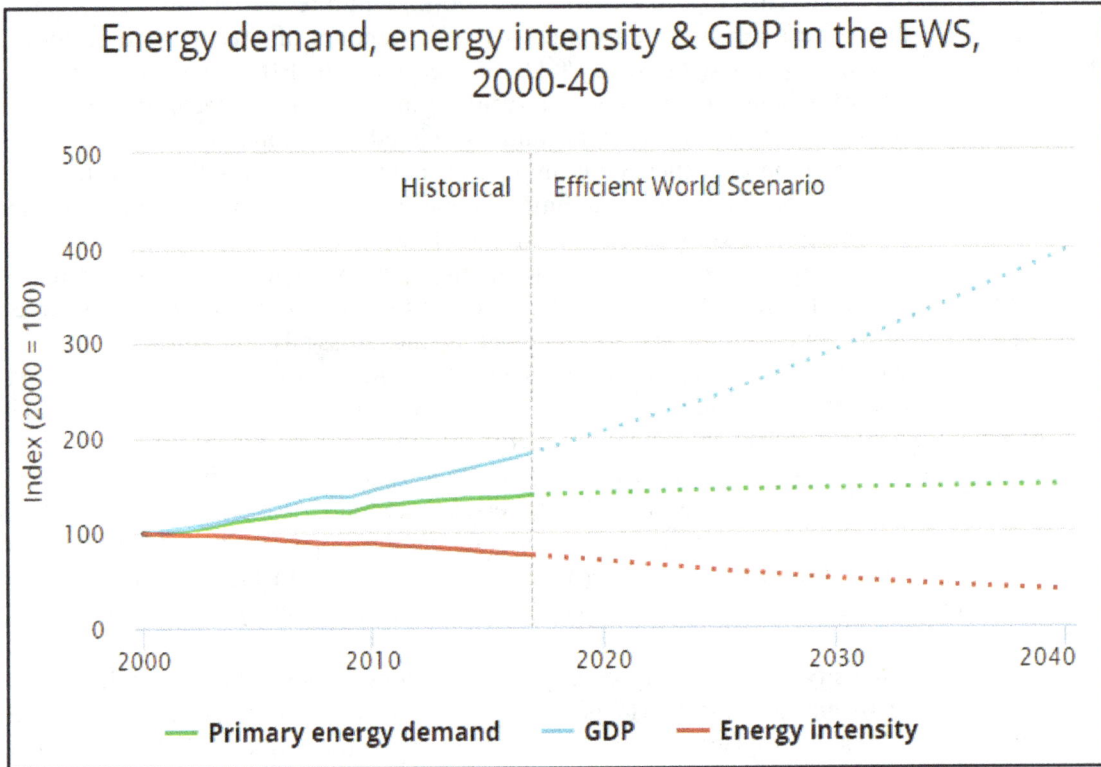

Figure 7.8 The International Energy Agency model of the potential impact of Efficient World Scenario (EWS) on global energy-related emissions over the next twenty years. (IEA, 2019o).

7.4.2   Efficiency improvement in energy production and end-use

The energy industry produces all primary energy - fossil, nuclear, renewable fuels - and converts around 78% of the total primary energy supply (TPES) into other forms namely electricity, heat, refined oil products, coke, but the industry is also the largest energy user, and the least efficient. Energy losses by the sector are around 30% (the end-use energy output is only about 48.7%). Significant losses occur at the production stages, during transportation, conversion, power generation, transmission, and distribution. Many power plants and distribution infrastructure across the world are decades old and, while some are frequently upgraded by retrofitting, many remain grossly inefficient. For example, most coal-fired power plants worldwide operate at below 40% efficiency even though technologies that deliver 45-80% have been available for decades, and power transmission/distribution losses range from 10 - 40%, again in spite of the availability of more efficient distribution technologies.

Higher efficiencies n the energy production and processing sector could resonate in the end-use sectors, and global energy savings multiplication factors could be significant - 4.7

for electric power generation, 2.7 for heat generation, and 1.07 for coal and petroleum products (Bashmakov, 2009). Industry is the largest consumer of end-use energy, accounting for 84% of the final energy use of coal, 26% of petroleum products (for energy and as raw materials), 46% of natural gas, 40% of electricity, and 43% of heat. Transportation consumes 62% of liquid fuels final use, while the building sector accounts for 46% of final natural gas consumption, 76% of combustible renewables and waste, 52% of electricity use, and 51% of heat (IPCC 2014b).

Efficiency is low across the total energy value chain, from production to end-use, and there is a growing awareness that improving efficiency in energy production and end-use is the most promising mitigation option that helps decouple growing energy use and energy-related emissions in the shortest possible time. Emissions growth rate can be reduced significantly through improved efficiencies across all sectors, from production to transportation, conversion and use. Improved energy efficiency over the whole value chain is now regarded as a central pillar of pollution mitigation, an energy resource, and the easiest and least expensive way of conserving primary energy resources. For example, a power plant operating at 40% efficiency is wasting 60% of the primary fuel input, and an LED lamp would deliver the same light (lumens) as an incandescent lamp for 80% less energy consumption is saving much more than 80% energy if losses in the production chain are taken into account. A simple illustration is the production of one unit of electric light (Figure 7.9). About 320 units of thermal energy is required to produce one unit of useful electric light. In effect, energy equivalent to 319 units is lost in generation, transmission, distribution, wiring and fittings. Investment in more efficient power generation, transmission and distribution technologies could reduce fuel input by around 30%, while conversion to more efficient compact fluorescent and LED lighting could save up to 80% of fuel inputs.

Projections indicate that up to 40% of future global primary energy requirements could come from improved efficiencies in production and utilization (WEC, 2016c; IEA, 2017e, 2017f and 2017g). It is estimated also that energy efficiency represents about 40% of the reduction potential of anthropogenic greenhouse gases (McKinsey, 2010). In fact, energy efficiency is now often referred to as 'first fuel' because of its immense potential for reducing process energy and carbon intensities. Energy efficiency is emerging as a key area of innovation, and Improved energy intensity has been the biggest factor responsible for the downward trend of the global anthropogenic emissions over the last two decades or so. Furthermore technology innovation has been dynamic and is playing a key role in creating and actualizing new opportunities.

Energy efficiency has been driving down energy intensity - the amount of primary energy demand needed to produce one unit of gross domestic product (GDP) - which has declined at an average of 2% per year over the last decade. However, there are significant variations between countries and regions, with OECD countries and China accounting for most of the achievement. The most important implication of this development is the fact that the world is generating more value from its energy use thereby making huge savings in primary energy use, and holding back the growth of anthropogenic emissions. Efficiency improvement is impacting positively on the economies across all regions of the world. In emerging economies, energy efficiency gains are limiting the increase in energy use associated with economic and population growth, thereby making more energy available and improving access by their populations. The worldwide improvement in energy efficiency is helping to decouple both energy use and emissions from global economic and population growth. Global total energy use peaked in 2007 and has been declining since then in response to rising efficiency. Improved efficiency has also impacted significantly on the third dimension of the global energy trilemma - energy security (the others are energy access and environmental sustainability).

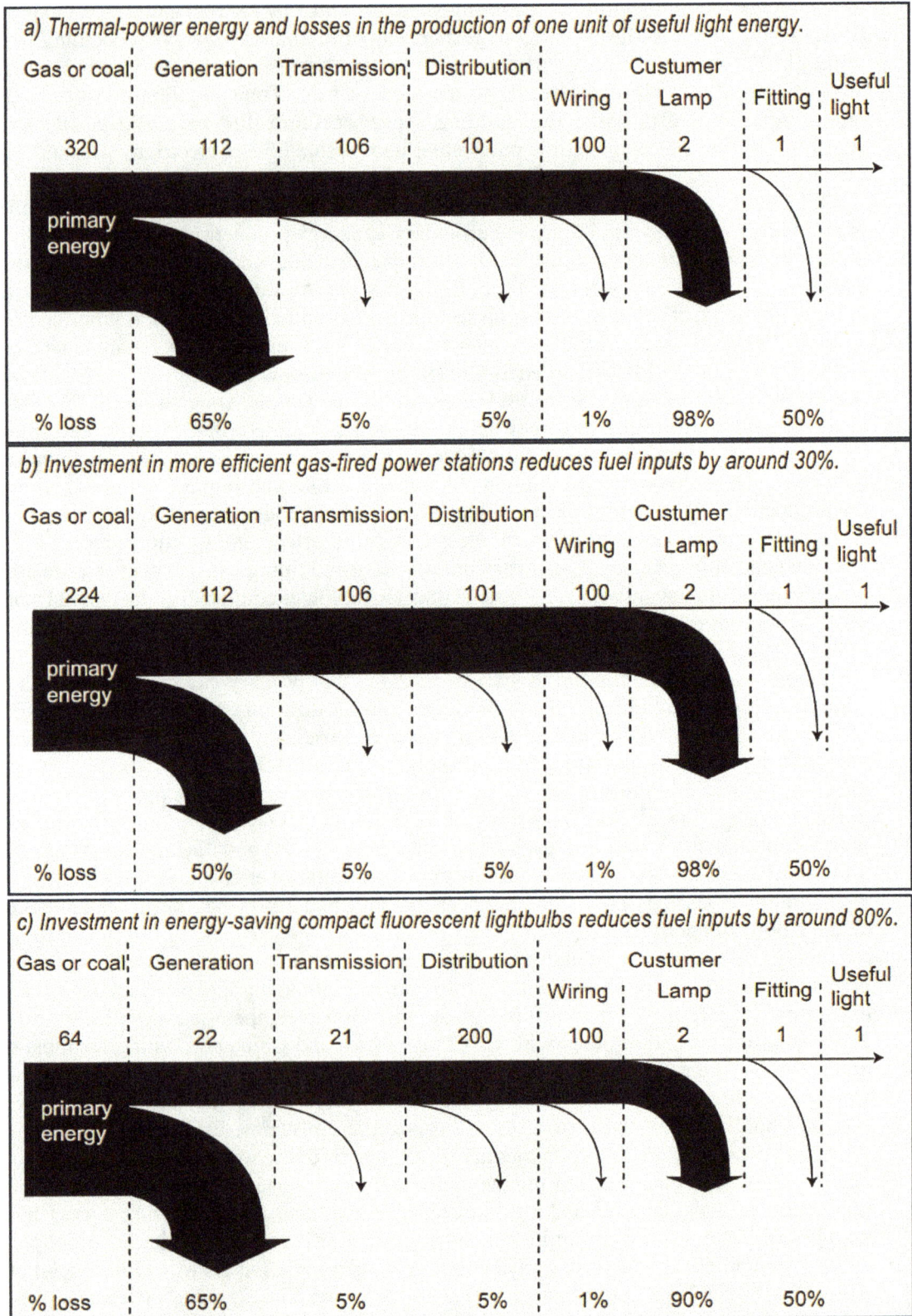

Figure 7.9 Thermal energy and losses in the production of one unit of useful light energy *(Cleland, 2005).*

As a result of declining demand for primary energy, propelled by increased efficiency, many countries were able to cut coal, oil and gas imports by as high as 30% in 2016, thereby improving internal energy security. All sectors of the primary energy value chain have efficiency improvement potential: coal, oil and gas exploration, industrial production and transportation, power generation, energy transmission and distribution, domestic and commercial utilization. For example, electric motors, lighting and home appliances account for approximately 70% of the total electricity consumption in the industrialized nations but efficiency is between 10 and 50%. Estimates vary widely but it is believed that between 30% and 50% of primary energy produced worldwide is wasted through inefficient production processes, transmission and utilization.

7.4.2.1 *Efficiency improvement in fossil energy production*

The recovery efficiencies of coal mines, oil and gas deposits are currently low, sometimes limited by available technologies, but more often by the typical inertia that plagues the adoption of new technologies. At the production stage, oil is spilled and gases are flared. These and many other energy sectors present very high opportunities for efficiency improvement. Oil wells are being pressurized with carbon dioxide to primarily to improve production efficiency, but this also helps sequestrate anthropogenic emissions. Fracking technology has been available for decades, but it is only in the last decade or so that its use became widespread and this has greatly boosted oil and gas production efficiencies in some countries. Also, the large quantities of gas being flared worldwide could be processed for power generation or liquefied petroleum gas to boost overall production efficiencies. Energy conserved through the improvement of efficiencies of production and consumption translates to lower energy demand, lower fuel consumption and lower energy-related harmful emissions. It also means that more fossil fuels are left in the ground. Unfortunately, in spite of the large number of efficiency-improvement technologies already available, deployment has been slow and the enormous potential of improved energy efficiency in reducing primary energy consumption and pollution emissions is still largely untapped.

7.4.2.2 *Efficiency improvement in power generation*

Around 86% of the total global electric power generation is powered by fossil fuels but power plants operating around the world are only 33% to 40% efficient, in particular, coal-fired power plants, the most polluting and most of which utilize sub-critical technology, whereas much more efficient ultra- and super-critical technologies have been available for decades. In effect, 60-70% of the carbon energy locked in the coal is wasted, in addition to transmission and distribution losses which could be 10-40% of the generated power. Efficiency improvement translates to lower coal consumption per kilowatt-hour of generated power and lower emission. One percent increase in efficiency of conventional pulverized coal-fired plants results in 2 to 3% reduction in CO_2eq emissions. Highly efficient modern coal plants emit up to 40% less CO_2eq than the older plants.

Many of the coal power plants operating in the developed world are old and nearing end of life of around 40 years and few new ones are being built. However, most plants in the emerging world are only 10 - 14 years old and are generally more efficient because they adopted more modern technologies. Newer power plants and many retrofitted installations now operate at around 40% efficiency; adoption of pre-gasification of coal could raise efficiency to around

60%; while combined heat and power complexes have even higher efficiencies. Even at the higher levels of efficiency there is still a tremendous amount of energy lost in the generation process which could be reduced through the adoption of more advanced combustion optimization technologies and control systems. The economic and ecological benefits of recovering and utilizing more of the energy in the input coal are also enormous, and many technologies are already available, for example, cogeneration plants, energyplexes and other advanced power generation technologies could return efficiencies as high as 80% (see Afonja 2017).

7.4.2.3 End-use efficiency improvement: industry

Every sector of the economy has significant potentials for exploiting the benefits of enhanced energy efficiency (see Figure 7.10). Industry accounts for over half of the total end-use energy and there is extensive scope for efficiency improvement. About 40% of the global electric power generation and 65% of industrial power are used to run electric motors in a very wide variety of applications ranging from driving industrial equipment to domestic fans microwaves and washing machines, with efficiencies as low as 30%. Most electric motors deployed in industry and incorporated in domestic appliances are fixed speed motors which consume substantially more energy than required at light load, and variable speeds are obtained through gear reduction units. Variable speed motors are much more efficient than fixed speed motors and could reduce the world's demand for electricity substantially. However, the relatively high cost has restricted use even in industry. Virtually every industrial process that utilizes electrical motors has a potential for improved efficiency. These include food and consumer product manufacture, machinery and automotive industries, pulp and paper, metallurgical, chemical and petrochemical industries and service industries. There is considerable technology inertia in the electric motor industry, with little change over the last few decades. Wider use of variable-speed motors will stimulate research and development, and drive down costs.

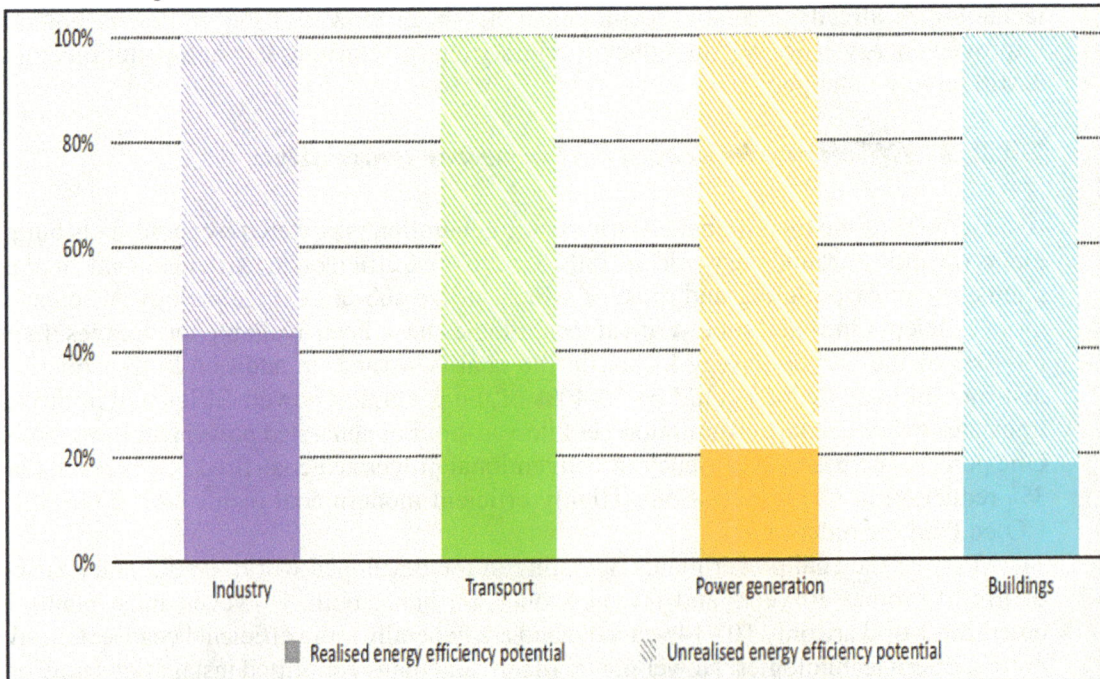

Figure 7.10   Long-term energy efficiency economic potential by sector *(IEA, 2017).*

Industrial energy intensity (amount of energy required per unit product) has been declining in recent years (about 20% since 2000) due to wider adoption of newer, more energy efficient technologies across all sectors and regions. There has also been a notable shift in the regional pattern of industrial energy demand in favor of emerging regions which now use 66% of global industrial energy. Five energy intensive sectors - chemical and petrochemical, iron and steel, aluminium, cement and pulp and paper - account for 67% of the world's total industrial energy consumption, and many are located in emerging countries. Very few innovative energy-saving technologies have emerged in these sectors in recent years but the fact that many of the facilities are new or have been recently expanded/upgraded, provided some opportunity to adopt more efficient technologies. However, opportunities for efficiency gains in the industrial sector are still very high. Application of current best available technologies (BATs) could reduce energy use in the energy-intensive industries by 10% to 25%. Options include process and plant modification, waste and heat energy recycling, more efficient process control, and adoption of proven but under-utilized technologies.

7.4.2.4 *End-use efficiency improvement: transportation*

Improvement in fuel economy of transportation vehicles is considered one of the most cost-effective ways to reduce GHG emissions. The average kilometers per liter of fuel is expected to double by 2040. This should reduce demand for oil and transport-related pollution in 2040 substantially. However, the population of vehicles is growing exponentially and is expected to double by 2040, and the bulk of the growth will be in the emerging world where vehicles are much less fuel-efficient and pollution control is lax. Furthermore, the increasing demand for personal transportation is fueling the growth in highly-polluting single-stroke, 2-3 wheeled motorcycles, and growth in diesel-fueled commercial trucking, aviation, and shipping will also be dynamic. The impact of energy efficiency in the transportation is well below the potential, due largely to the instability in the global oil market. Significant improvements in fuel efficiency of conventional vehicles combined with low global oil prices have led to faster increase in sales, in particular, less efficient large passenger vehicles such as sport utility vehicles and family vans in the last few years, thereby dampening the global rate of improvement in passenger vehicle fuel efficiency, which has been evident over the last two years or so. Another problem is the relative lack of improvement in the truck sub-sector which accounts for around 43% of total oil consumption for road transport, and uses the bulk of diesel, the most polluting of fossil end-use fuels. While many countries have mandatory efficiency standards for passenger vehicles, only a few have fuel economy standards for commercial trucks in place.

It is widely accepted that electric vehicles (EVs) hold great potential in decarbonizing the transport sector (assuming access to lower carbon electricity source), and global sales have been strong, rising from around 2 million total population in 2016 to about 5 million in 2018. However, in spite of the high relative advantages of electric vehicles compared with combustion-powered vehicles (better fuel economy, zero anthropogenic tailpipe emissions and lower carbon footprint), the penetration of electric vehicles into the global auto market has been very slow: the global population of EVs in 2018 was only around 0.5% of the population of nearly one billion vehicles. Although growth in the EV market is expected to be strong over the next two or three decades, EVs will still account for no more than 15% to 35% of the total vehicular population in 2040 (based on several recent projections and different levels of optimism), and even this modest growth will depend on the extent to which battery size and cost can be reduced and charging/swapping bays can be proliferated, and, crucially, consumer interest which may be dampened significantly by increasing fuel economy in conventional vehicles and cheaper fuels. Most of the growth in EVs will be accounted for by Europe, North America, China, and Japan, while nearly all the growth in vehicle population will be in emerging countries. Furthermore, the electric vehicle market is very sensitive to global oil prices which have been low for some years, thereby fueling the growth of internal combustion engine vehicles at the expense of EVs. However, some countries, particularly in Europe have introduced punitive taxes on ICEs and are offering strong incentives to promote the EV market. It should be noted however that every EV vehicle on the road contributes to emissions mitigation especially when low-carbon electricity is available for manufacture and operation. The most obvious and immediate impact is on the highly problematic city pollution and the negative effect on human and ecosystem health.

7.4.2.5   *End-use efficiency improvement: buildings*

Buildings (residential and commercial) accounted for around 21% of final energy consumption globally in 2017 and, although demand is projected to increase by 32% by 2040, the share of end-use energy will still be around 21%. The projected increase will be driven by rising population, urbanization and increasing wealth in emerging countries. Substantial potential exists (up to 25%) for improvement of the energy efficiency of buildings, particularly in heating and cooling which is the largest end-use energy sub-sector, and accounts for one-third of all energy consumed in buildings and 60% for heating in cold climates (IEA, 2017d) . A major area of efficiency improvement is the upgrade of the building thermal envelope through design and use of best available technologies (BATs). Potential areas of innovation include insulated walls, windows, cool roofs, improved natural lighting, and efficient lighting and appliances. The impact of improved efficiency in the building sector is cutting across all regions, and it is estimated that household energy efficiency gains saved consumers 10-30% of their annual energy spending in 2016.

The energy utilization efficiency of buildings also known as Energy star rating (ESR) is becoming a critical variable for both home builders and buyers in many countries. New buildings feature effective wall and roof insulation, double glazed windows, compact fluorescent lighting (CFL), which can provide the same light power for 20% of the power consumption of a conventional resistance bulb, control systems which switch off lights in unoccupied rooms, energy-efficient heating and air conditioning systems, and energy star-rated appliances. Also, consumers are becoming increasingly energy-conscious and look critically at energy efficiency ratings of appliances before they buy. Furthermore, the "smart home" concept also known as "Internet of Things (IoT)" is spreading rapidly and involves connecting homes and appliances to the Internet to facilitate the ability to control home appliances remotely on cell phones, which means that home heating and other appliances can be switched off when every one is out and reactivated shortly before return. Also, temperatures can be monitored and controlled remotely.

Consumer education will also play an important role in mitigating energy use in the building sector, and many home and appliance buyers now list energy star rating as one of the important factors that determine choice from options. Consumers need to be aware of the potential cost savings that could come from energy conservation and should be sensitized on how to access the potential gains. Already, packaging of energy-efficient light bulbs and appliances carry details of potential energy and cost savings. Also, energy marketing companies in some developed nations offer detailed analysis of day-to-day energy use of each customer including use for heating, cooling, lighting, etc., which serves as a useful domestic energy management tool. Many also offer lower unit pricing and bonuses for low energy consumption, and strong incentives for using more energy in low-peak periods such as nights and weekends.

7.4.2.6 *End-use efficiency improvement: the digital revolution*

The information and communications sector (ICT) has literally exploded in the past decade or so: Internet connectivity is growing at an exponential rate. There were around 4 billion Internet users in 2017, over half of the global population, and Internet traffic could double in the next few years (Statistica, 2018; Internet World Stats, 2018). About 5 billion connected devices were in use in households worldwide and may double by 2020. Deployment of network-enabled devices in offices and homes has increased exponentially - broadband connectivity, wireless mobility, cloud computing, e-commerce, social media, video-enabled devices, smart devices, smart systems and sensors, personal computers, servers, laptops, game consoles, printers, etc. Many home appliances - TVs, microwave machines, washing machines, home security surveillance systems - now have Internet connectivity. There were, around 17 billion networked devices in 2016, rising to about 27 billion by 2021 (3.5 devices per capita) (CISCO, 2018). Projections indicate that network-enabled devices will be around 500 billion by 2030 (WEC, 2017). Edge devices - routers, routing switches, integrated access devices, multiplexers, metropolitan area networks (MAN), wide area networks (WAN) - all of which provide authenticated access to enterprise or provider core networks are now indispensable in industry, commerce and homes, and the world is experiencing an exponential growth in the deployment of digital technology in all other sectors of the economy. Adoptions of smart control systems and meters are leading to significant performance and efficiency gains in the power industry. Around half a billion smart meters which can be monitored and controlled remotely are helping to track and control electricity use in real time. Virtually every industrial process has advanced digital control systems which help optimize production and energy use.

There has been a surge in energy demand to power ICT systems, accounting for more than 10% of total final global energy consumption in 2016. The energy required to meet the projected growth in network-enabled devices is expected to double in the next decade. Most of the energy (up to 80%) is consumed when devices are in standby modes (ready and waiting but not performing the main function). The aggregate electronic devices in offices and homes - modems, routers, TVs, PCs, servers, set-top boxes, games consoles, printers, security equipment, etc. - constitute more than 40% of ICT electricity demand, and standby energy accounts for 1 to 2% of global electricity consumption and approximately 10% of residential electricity use. There is a significant scope for reducing this wastage and associated emissions, and equipment manufacturers are doing a lot to achieve this goal. For example, a recent European legislation limits energy consumed by an appliance to 1 watt when in standby mode compared with older equipment which consume 12-15 watts. Energy savings will become substantial as old appliances are replaced.

7.4.3 Decarbonizing energy

One of the main thrusts of global effort to mitigate energy-related environmental pollution is to decouple energy demand and greenhouse gas emissions, and the power sector offers the best opportunities. As discussed in the last section, improved efficiency has a significant impact on CO_2eq emissions: the more efficient a power plant is, the less fuel it burns for a unit power output and the lower the emissions. Much of the efficiency improvement can be achieved by adopting simple technologies that are already available (best available technologies, BAT), such as retrofitting a coal-fired power plant with more efficient burners, or enriching combustion air with oxygen. Improved efficiencies in the end-use sector and switching to lower carbon sources of primary energy (natural gas, renewables, nuclear) drive down energy intensity and the carbon footprint. However, one of the main challenges is the promotion of widespread deployment of a wide range of technologies which are already available, and new technological advances. Many new technologies are at different stages of development that could reduce significantly energy-related emissions in the future. Transformational change in the energy sector can be a very long process due largely to '*status quo*' mentality. This explains why many power plants all over the world are still operating at 30-35% efficiency when technologies that could lift efficiencies to 45% and above have been available for decades. Power plants have a life span of 30-40 years (many power plants worldwide are close) and, while many undergo retrofits over their life span to improve operating efficiencies, most of the investments in recent times have been in new power plants, featuring new advanced technologies that not only improve efficiencies, but also facilitate compliance with increasingly stringent worldwide environmental control regulations, which is in fact the main driver of change in the power industry.

The International Energy Agency (IEA, 2017d) estimates that stronger deployment of technologies that are familiar and available at commercial scale today can reduce significantly future emissions from the energy sector. Coal is the backbone of power generation in many countries and has been responsible for more than 40% of global energy-related emissions growth since 2000. Half of total CO_2eq emissions from the power sector today (around 6 giga metric tons [GT]) come from inefficient, low-technology coal power plants. Upgrading from sub-critical operation to super-critical and ultra-super-critical boiler technologies which have been available commercially for more than a decade could decarbonize fossil electricity substantially. Also, investments in renewable energy power generation technologies (in particular, wind and solar which have only 2-3% of the carbon footprint of a typical coal power plant) which are readily available commercially and becoming increasingly competitive could further mitigate pollution from the power sector. Industry, transport and buildings end-use sectors also offer substantial energy efficiency improvement opportunities which translate to lower carbon emissions, through the infusion of new technologies and fuel switching (see Figure 7.10).

7.4.4   Waste-to-energy and recycling

As discussed earlier, recycling has great potential in reducing energy consumption and therefore lowering energy intensity and emissions. The world generates about 3 million metric tons of waste daily, and the rate is projected to double by 2025 (WEC, 2016a). Nearly half is organic waste and about a third is made up of metals, plastics and paper (Figure 7.11). Around 80% is used for landfills and the balance is recycled, incinerated to generate electricity and heat, or used for composting. Waste-to-energy (WtE) or energy-from-waste (EfW) is the process of generating energy in the form of electricity or heat from the primary treatment of waste either by direct combustion, or by conversion to biofuels such as methane, methanol, hydrogen or other synthetic fuels. Modern incinerators can convert heat to electricity at 14-28% efficiencies, but if configured as part of cogeneration which combines electricity generation with space heating, efficiencies higher than 80% are possible. Also, strict emission regulations have forced manufacturers of incinerators to produce plants with low emissions of nitrogen and sulphur oxides, as well as particulates.

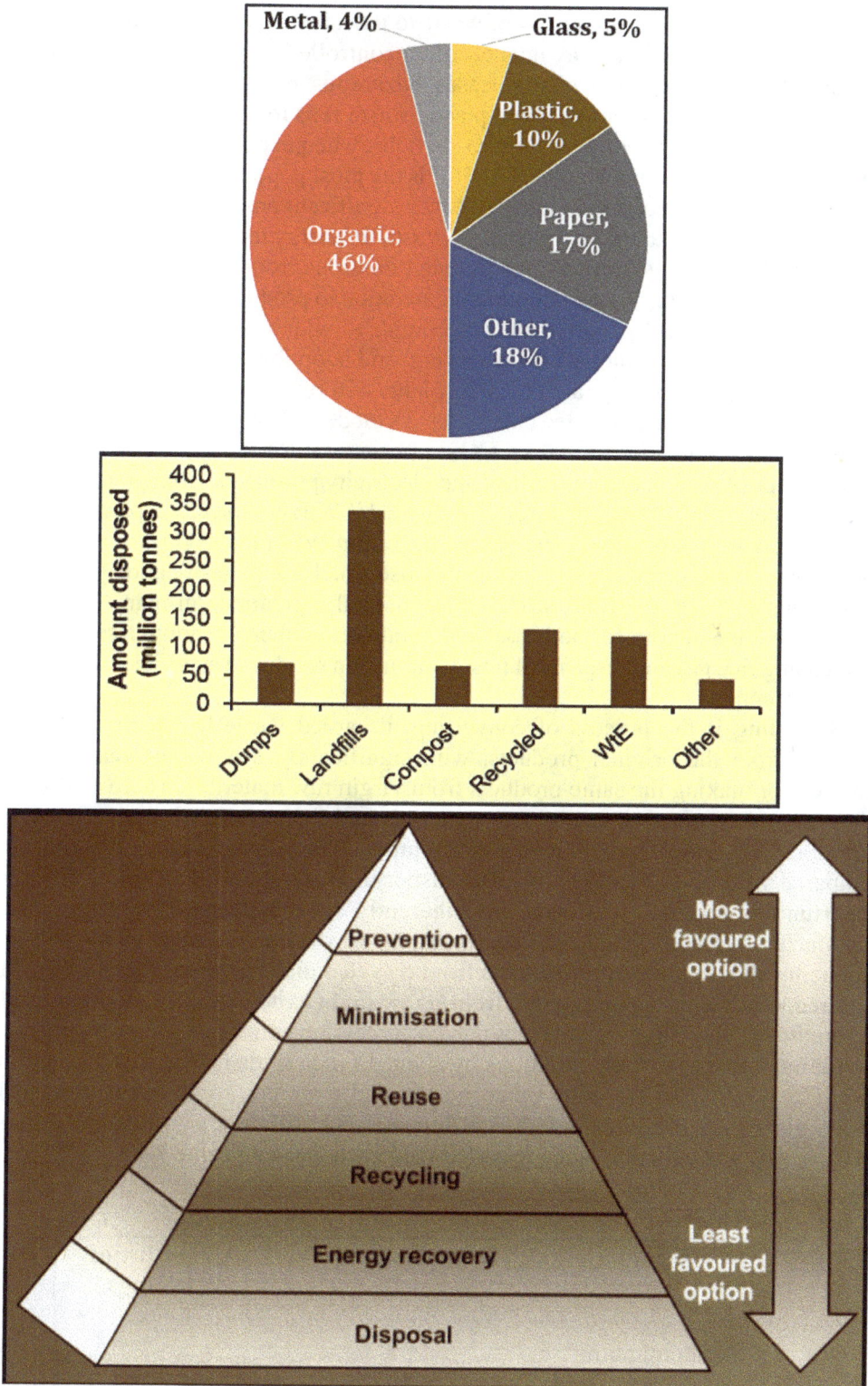

Figure 7.11 Global waste generation by type and disposal technique *(WEC, 2016a)*

Apart from the fact that conversion of waste to useful energy makes economic sense, the more important advantage is the avoidance of uncontrolled incineration or dumping of waste, particularly plastic waste which releases hazardous gaseous/particulate products into the atmosphere on incineration. Dumped plastic waste fails to degrade for decades when used for landfills, and could break up into micro-particles when exposed to ultraviolet rays of the Sun, causing new environmental issues. Landfill is the most popular disposal system worldwide, but decomposition of the organic matter releases significant amounts of carbon dioxide, methane and micro-particles into the environment. Waste-to-energy technologies can contribute to global climate mitigation by significantly reducing emissions from other methods of waste disposal. Many developed countries are adopting incineration to produce power and heat, a system which cuts CO_2eq emissions by more than 50% compared with landfill systems. New, more efficient waste gasification technologies are emerging, and adoption could cut emissions by two-thirds. Also, using the energy generated by WtE plants will reduce the demand for energy from fossil fuels and eliminate the associated emissions. A modern WtE plant can produce carbon emission savings in the range 100 to 350 kg CO_2eq per metric ton of waste processed, depending on waste composition and amount of heat and electricity produced. Even greater savings up to 800 kg CO_2eq per metric ton of waste can be achieved if WtE completely replaces landfilling (WEC, 2016a). Emissions (gas-phase and particulate) from WtE plants which feature incineration are comparable to coal-fired power plants, but those which adopt gasification produce significantly lower emissions. In any case, strict environmental regulations in many developed countries have made it mandatory for both coal and incineration plants to adopt a series of process units for cleaning flue gas and disposal of post-combustion residue in an efficient and environmentally friendly manner.

Recycling is the process of converting discarded waste to useful products, sometimes different from the original products, with significantly less energy than would have been required for making the same products from virgin raw materials. There is also the additional advantage of conserving natural raw materials and the associated production and processing emissions. The process offers higher environmental benefits and lower environmental impacts compared with other methods of waste disposal. Different products offer different recycling opportunities and gains. Metals are inherently recyclable and the same product can be manufactured from recycled material over many life cycles, with significant gains. For example, producing aluminium soft drink cans from recycled material saves around 95% of the energy required to make the same product from its virgin bauxite ore. Also, it saves four metric tons of bauxite ore and the associated mining and processing emissions. Every metric ton of recycled aluminium reduces CO2eq emissions by about 14 metric tons compared with production from virgin ore (EPA, 2011). Many other metals and materials - iron, steel, copper, lead, silver, paper, glass - are recyclable, with less dramatic but still very significant energy savings and GHG emissions reduction when recycled feedstock is used (Table 7.1)

Table 7.1 Environmental effects of recycling (*UCO, 2014; EPA, 2015*).

Material	Energy savings	Air pollution savings
Aluminium	95%	95%
Glass	5-30%	20%
Paper	50-70%	73%
Plastics (reuse)	70%	50%
Plastics (recycle)	70%	30%

Plastics offer less but still significant opportunities for recycling compared with metals and paper, primarily because of the diversity of types with very different thermo-chemical characteristics. Most plastics cannot be recycled to produce the original product, but can be reprocessed into other useful products, with significant savings in energy and reduction in anthropogenic emissions. Recycling requires significantly less energy per kilogram of material produced than primary production, and also decreases the negative impact of mining, processing and transportation of ores. Recycling slows down the need for exploiting low-grade ores, a more energy-intensive process that releases higher emissions.

All products have environmental impacts, from the extraction, upgrading and transportation of raw materials to manufacture, distribution, use and disposal. In order to conserve raw materials, reduce the life-cycle energy associated with a product, and protect the environment from indiscriminate disposal and unsustainable practices such as landfills after use, a waste management hierarchy has evolved over the years, known as R[4] – Refuse, Reduce, Reuse, Recycle – in order of priority (Figure 7.11). This model targets sensitization of people particularly on the use and dumping of plastics which has become a major global environmental problem. Literally it means: refuse to use but if you must, reduce the amount, for example by reusing, and recycle when it is no longer useful. Proper application of this waste management model can conserve raw material resources, save considerable energy, reduce life-cycle emissions associated with the product, and protect the environment from uncontrolled dumping.

Recycling one metric ton of paper saves about 17 trees, and also saves one metric ton of CO_2eq of emissions. The United States, Japan, China, and many countries in Europe have extensive WtE programs and most claim substantial reductions in anthropogenic emissions. In 2014, the United States recycled and composted about a third of the 258 million metric tons of municipal solid waste (MSW) generated, resulting in a reduction of over 181 million metric tons of carbon dioxide equivalent emissions, comparable to the annual emissions of nearly 40 million passenger cars (EPA, 2011). Over 33 million metric tons were combusted with energy recovery (which reduces fossil fuel use), while 133 million metric tons were landfilled. The United Kingdom Waste and Resources Action Program (WRAP) claims a more modest but still substantial reduction of 10-15 million metric tons a year of emissions, amounting to nearly 50 million metric tons between 2010 and 2015 (WRAP, 2010; 2018). However, many developed countries are exporting unsorted, unwashed plastic waste to emerging nations where they are sorted manually, washed and recycled or incinerated mostly by small enterprises under unhealthy conditions.

7.4.5   Carbon capture and sequestration (CCS/CCUS)

A fourth option for mitigating energy-related anthropogenic emissions is to capture and store as much of the emissions as possible at source. Technologies are already available and many are under development for the capture and safe storage of carbon dioxide in major industrial operations, in particular, power generation. Carbon capture and sequestration (CCS) is currently the only available technology that can significantly reduce GHG emissions, and a key technology option to decarbonize the power sector, especially in countries with a high share of fossil fuels in electricity production. Over twenty large-scale CCS projects are in operation or under construction, with a combined capacity to capture up to 40 million metric tons of CO_2eq per year. These projects cover a range of industries, including gas processing, power, fertilizer, steel-making, hydrogen production, and chemicals. Most of the projects are in North America and the world's first large-scale adoption of carbon capture technology in the power sector is in Canada, built in 2014. Several demonstration plants are operating in other countries and commercial projects are under construction, particularly, the United States.

Other countries - Japan, Australia, South Korea, the Middle East - are also actively considering CCS projects. Active research is also ongoing to find uses for captured emissions rather than store them, hence the modification: carbon capture, use and sequestration (CCUS). The viability of CCS/CCUS technology is not in doubt, but there are formidable obstacles to widespread deployment (WEC, 2016a). Retrofitting existing plants or including carbon capture in new plant design could drive up capital and operating costs by up to 30%, and many emerging nations that will host most of the expected new coal power plants cannot absorb the added costs. Another major constraint is what to do with the captured gas. Finding suitable sequestration sites is problematic while transportation over long distances for safe disposal is too expensive and impractical for wide adoption. Also, technologies for conversion into useful products or benign state are at early stages of development. Perhaps the best option is to focus on technologies which can convert carbon dioxide to useful or benign products and research in this area is intensive (Figure 7.12). Progress will largely determine the reductions in anthropogenic emissions that can be achieved over the next 2-3 decades through carbon capture, use and storage.

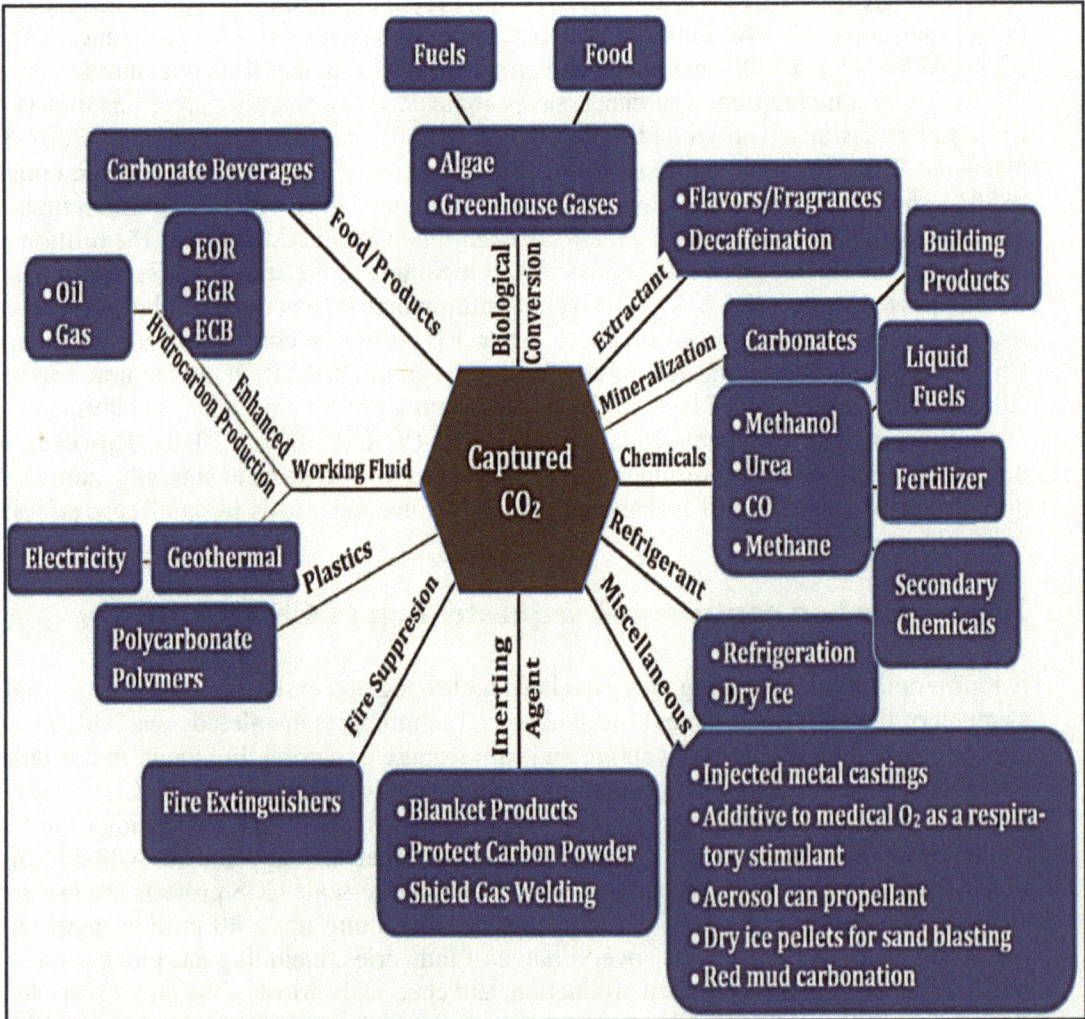

Figure 7.12 Current an potential uses of carbon dioxide. *(Damiani et al., 2011).*

7.5   OUTLOOK ON FOSSIL ENERGY USE

It is difficult to make accurate long-run energy projections because of many uncertainties. Long-range outlook on the global energy scenario especially over decades is a very complex process because of the many variables and events that could make a nonsense of predictions and projections. Market projections are based on assumptions and much uncertainty because many of the events that shape the future trajectory of energy supply and demand cannot be foreseen with certainty: global energy politics, macroeconomic growth, shifting demographics, emerging technologies, potential impacts of proposed and new energy policies and legislations, world primary energy prices, end-use energy prices, the internal political dynamics of countries, and many more. Many respected organizations publish annual model projections of global energy markets over the next two to three decades, based on assumptions and models of different scenarios of interactions of economic and demographic changes, energy supply and demand, prices, dynamics of consumer preferences, and new technologies. Modeling efforts have inherent limitations and there are significant variations in assumptions made by the different organizations, which explains why there are often significant differences in predictions. Energy outlook should therefore not be taken as a prediction of what will happen, but rather an intelligent projection of what might happen, subject to stated assumptions. Projections published recently have been reviewed in depth and, in spite of significant differences in some of the conclusions, there is general agreement on the likely trend of global economics, demographics, energy demand, and energy-relate emissions. All the outlooks reviewed show continuing domination of the primary energy scenario by fossil fuels for the next two decades, and even beyond. Also, there is a consensus on the likely trajectory of emissions and significant changes in fuel mix, with renewables (in particular, wind and solar) emerging eventually as fuels of first choice for power generation, although there will be wide variations across regions. In effect, the average global carbon intensity of energy will decline but will still be high in many emerging nations, depending on local policies on energy mix (Figures 7.13 - 7.15).

Economic growth stimulates demand for personal transport, commercial trucking, aviation, marine and rail transportation. The majority of the growth across all the sub-sectors will occur in non-OECD counties, consistent with the expected growth in gross domestic product (GDP) and population. While energy consumption is expected to grow by only 9% or so in OECD countries by 2040, the demand in non-OECD countries will rise by around 41%. The transportation sector is the main user of the liquid products of oil refineries - gasoline, diesel, jet fuel, - and around 95% of the total global consumption in 2018 was accounted for by the sector. This share is expected to reduce by 5-7% by 2040 due to a projected increase in the use of alternative fuels (biofuels, natural gas, electricity). The global fleet of vehicles will nearly double over the next 20-25 years, so will commercial trucking, air and marine transportation, and most of the increase will be in non-OECD countries which are projected to account for over 60% of the total global transportation fuel consumption by 2040. Energy demand by the transportation sector has the least growth rate of the three end-use sectors due to improvements in vehicle fuel efficiency which has doubled in the last two decades or so, notably in OECD countries. As a result, transportation energy use for the region is projected to decline in spite of substantial growth in vehicle population, but in non-OECD countries, increased demand across all modes of passenger transportation and light-commercial and heavy trucking will outpace improvements in fuel efficiency. The wide range of projections on EV future market penetration is a reflection of the numerous uncertainties that make predictions on the future prospects of EVs in decarbonizing transportation very difficult.

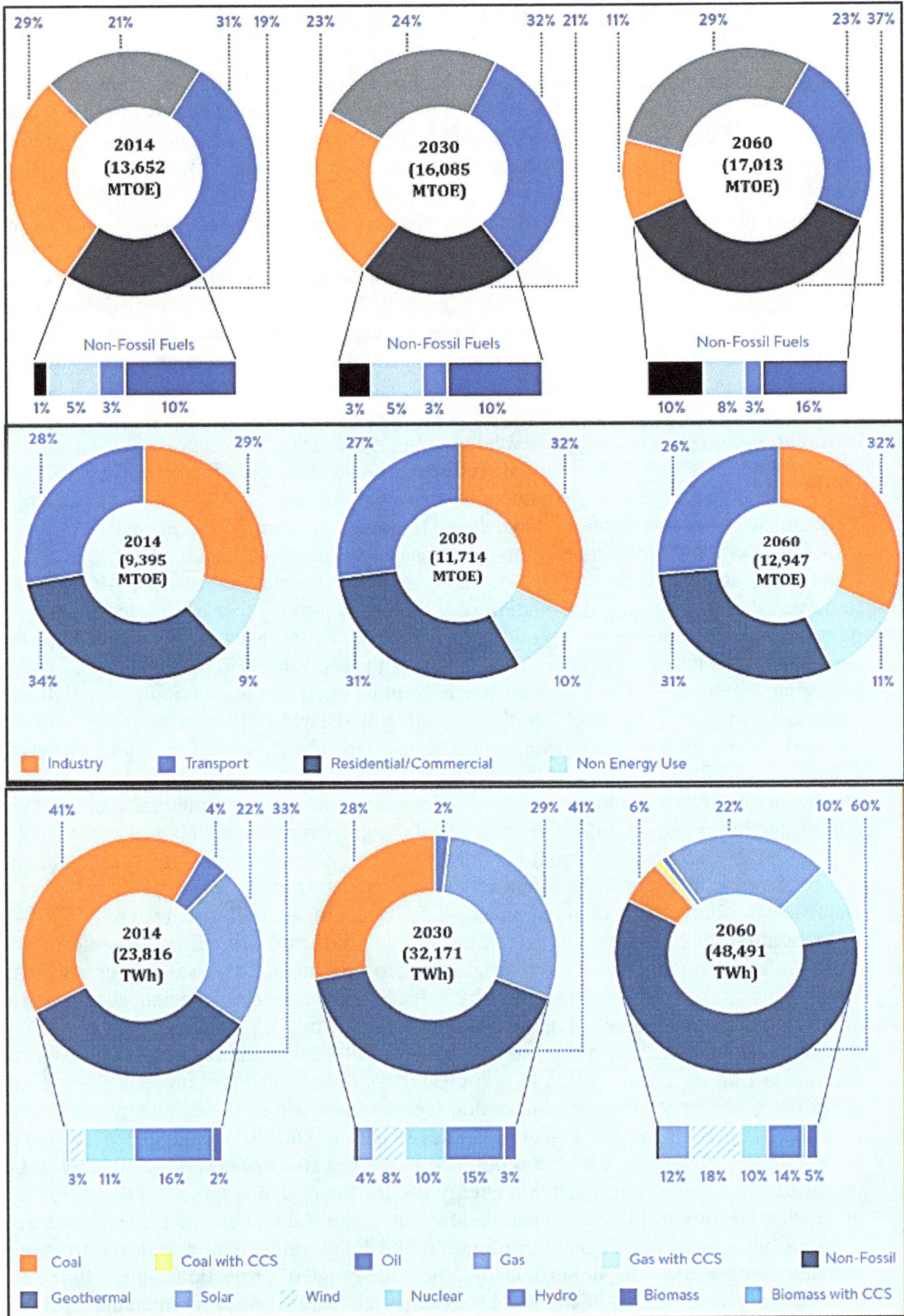

Figure 7.13 Projections of (a) Primary energy supply by fuel (b) Share of fuels by end-use (c) Electricity generation by fuel share *(WEC, 2016).*

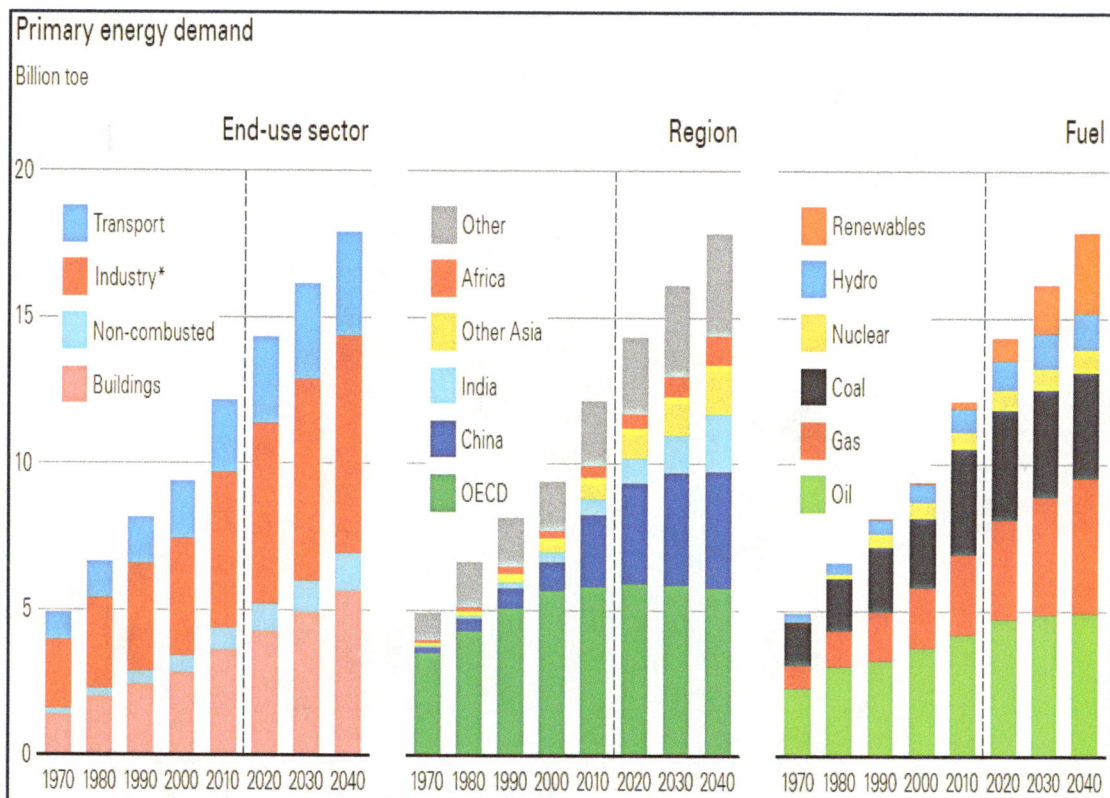

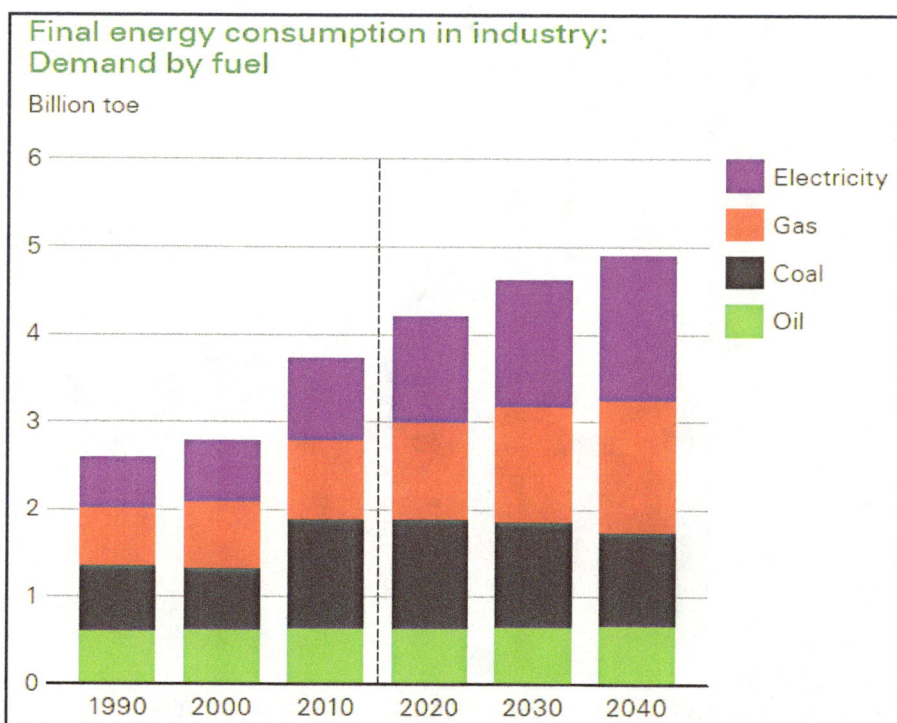

Figure 7.14a   Outlook for primary energy by end use, sector, fuel, and in industry *(BP Outlook, 2019).*

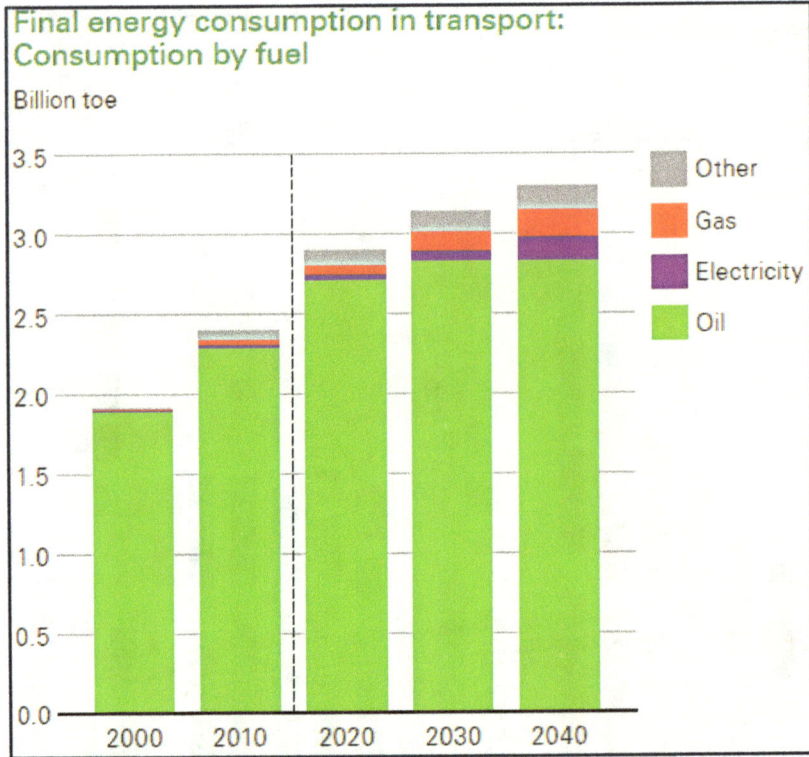

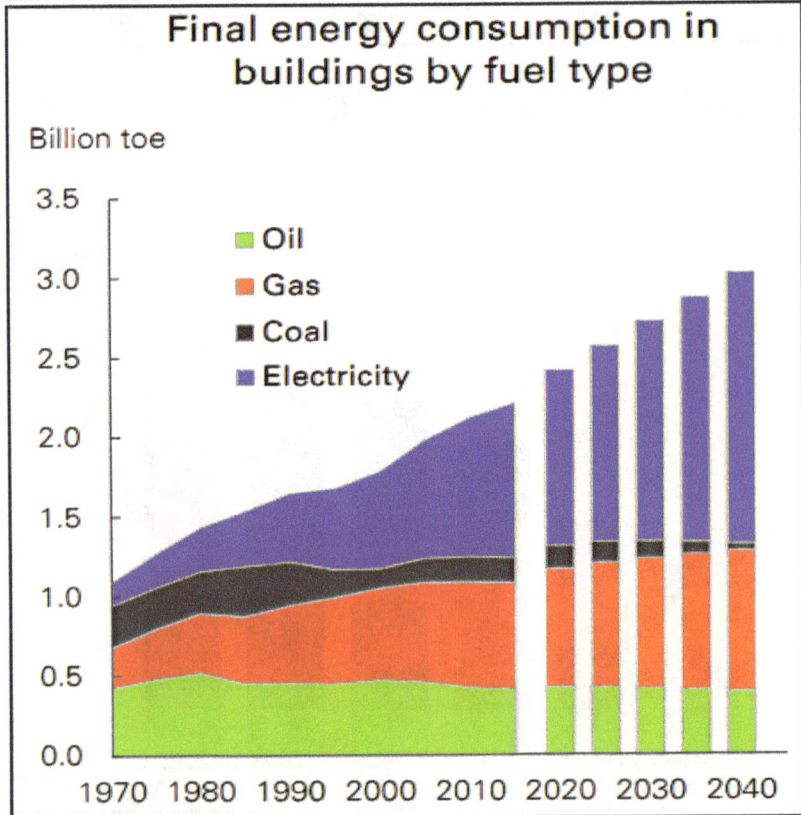

Figure 7.14b Outlook for final energy consumption in transport and building *(BP Outlook, 2019).*

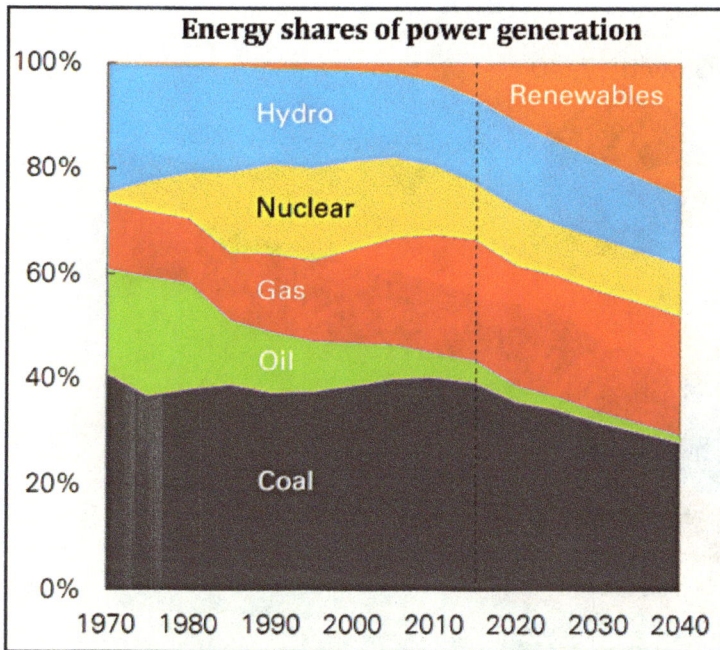

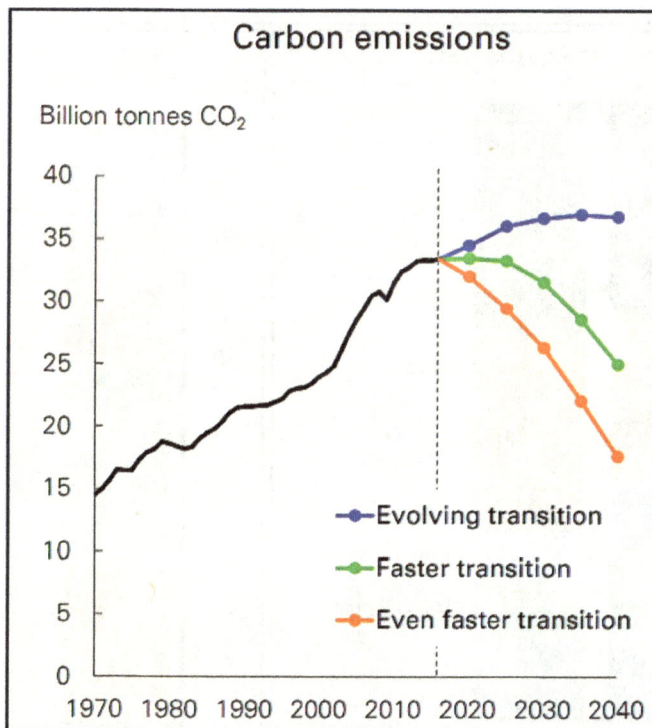

Figure 7.14c Outlook for power generation energy mix and for energy-related carbon emissions *(BP Outlook, 2019).*

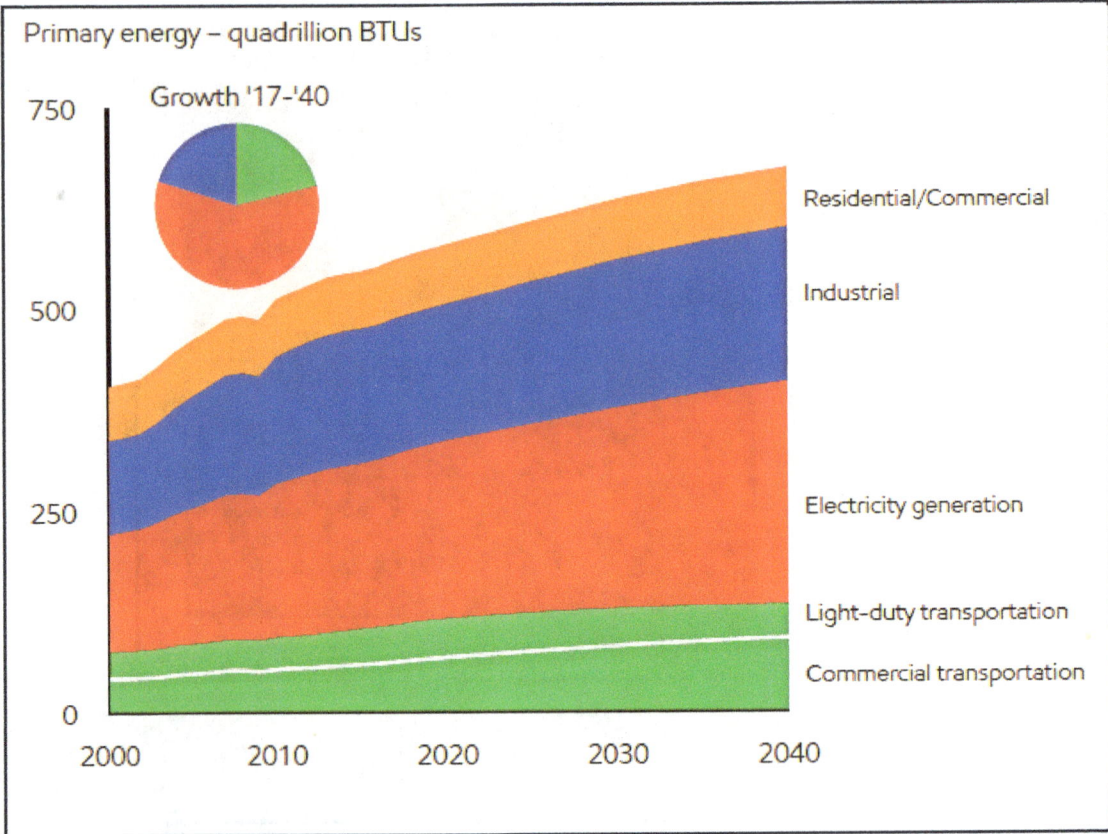

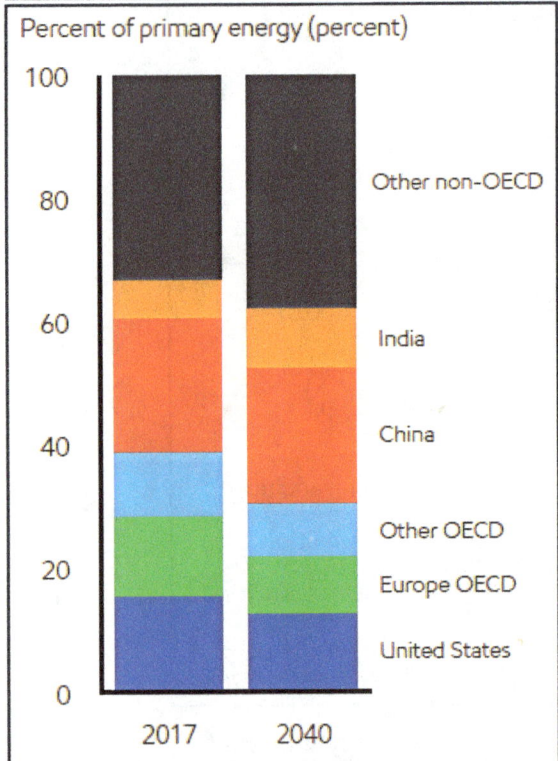

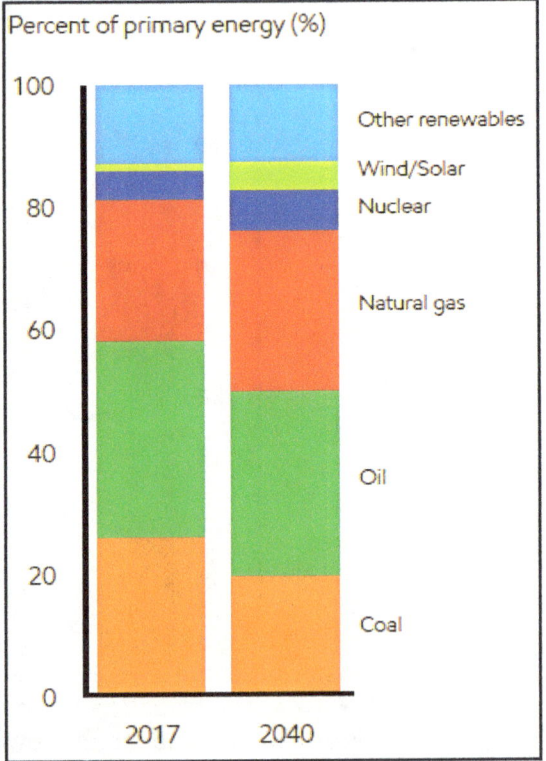

Figure 7.15a   Outlook for primary by sector, region and fuel *(ExxonMobil, 2019).*

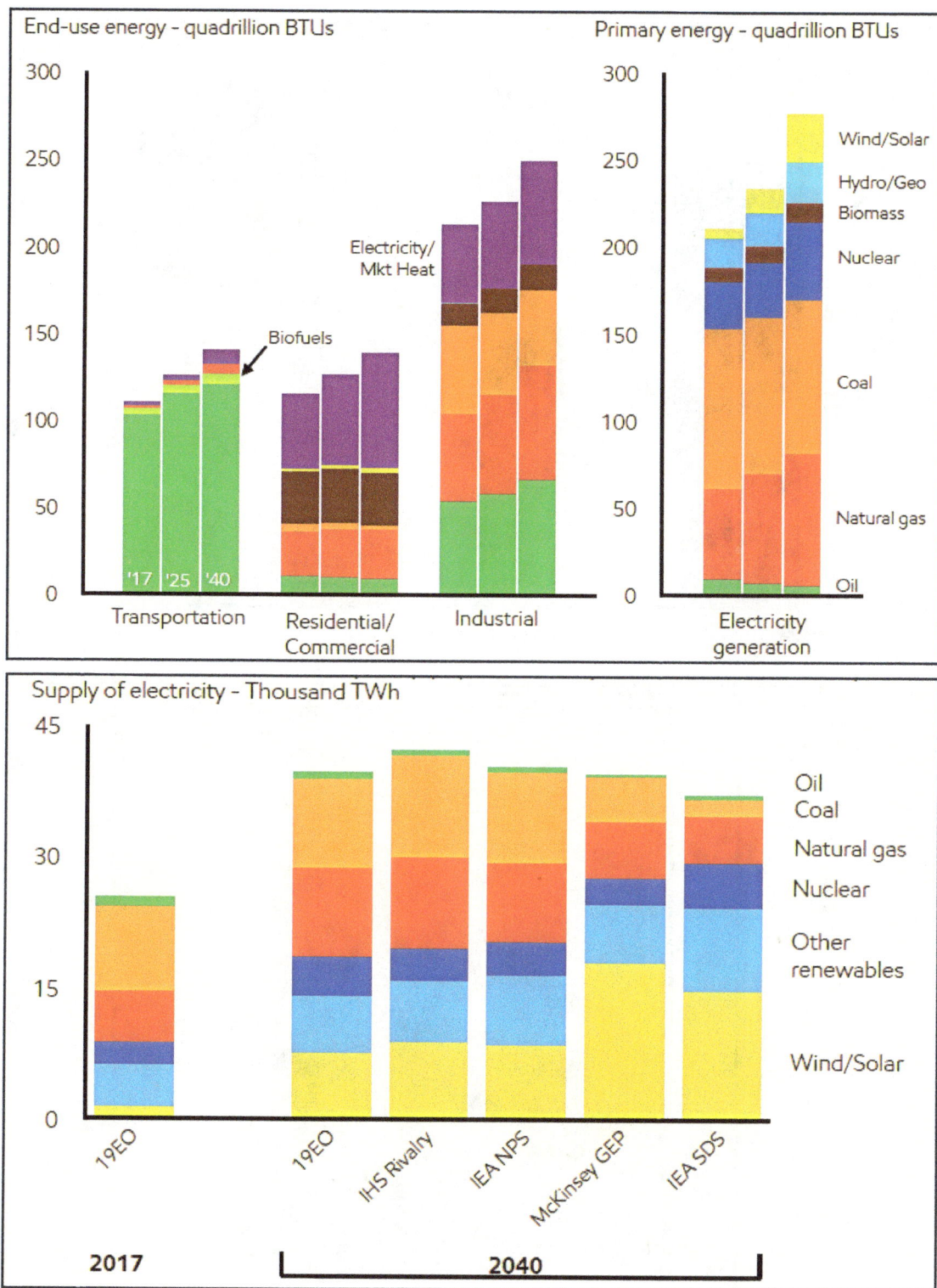

Figure 7.15b   Outlook for end-use by sector and scenarios on power generation fuel mix by different organizations *(ExxonMobil, 2019).*

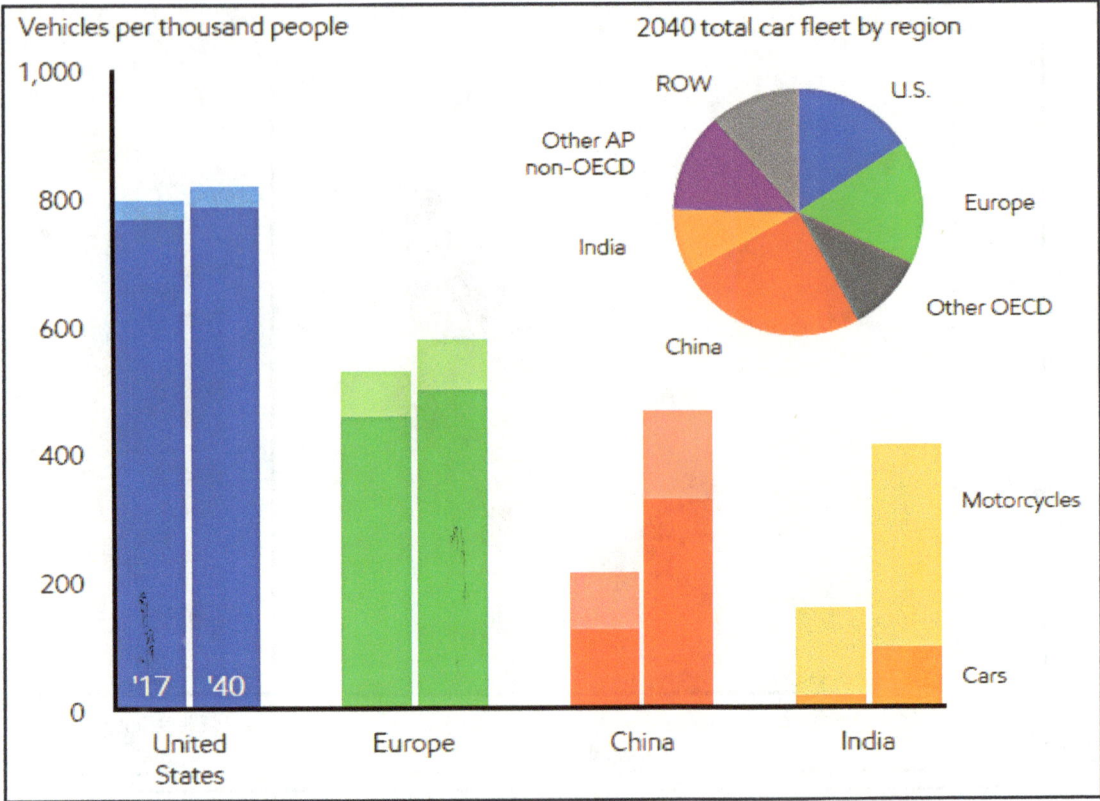

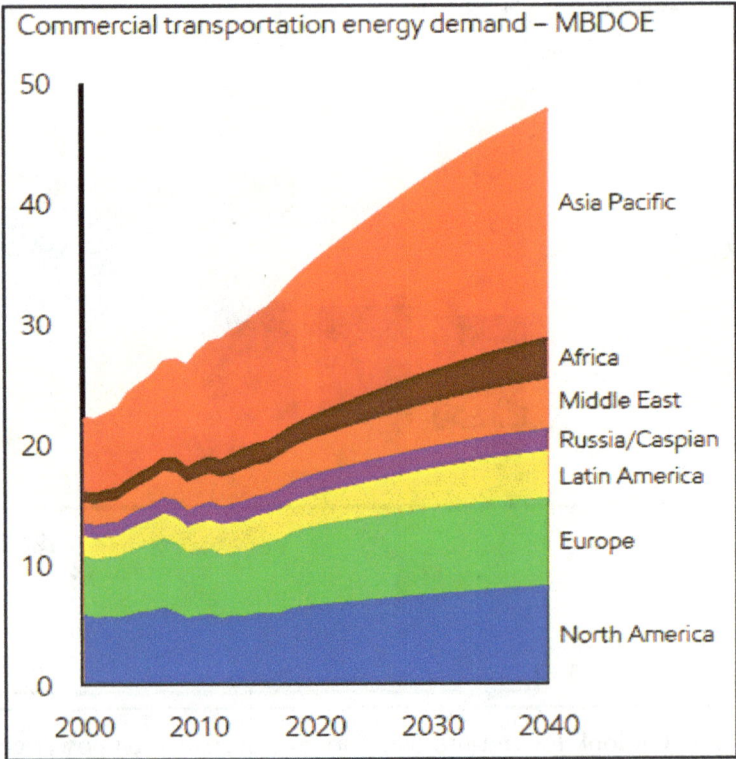

Figure 7.15c Outlook for transportation by type and region *(ExxonMobil, 2019).*

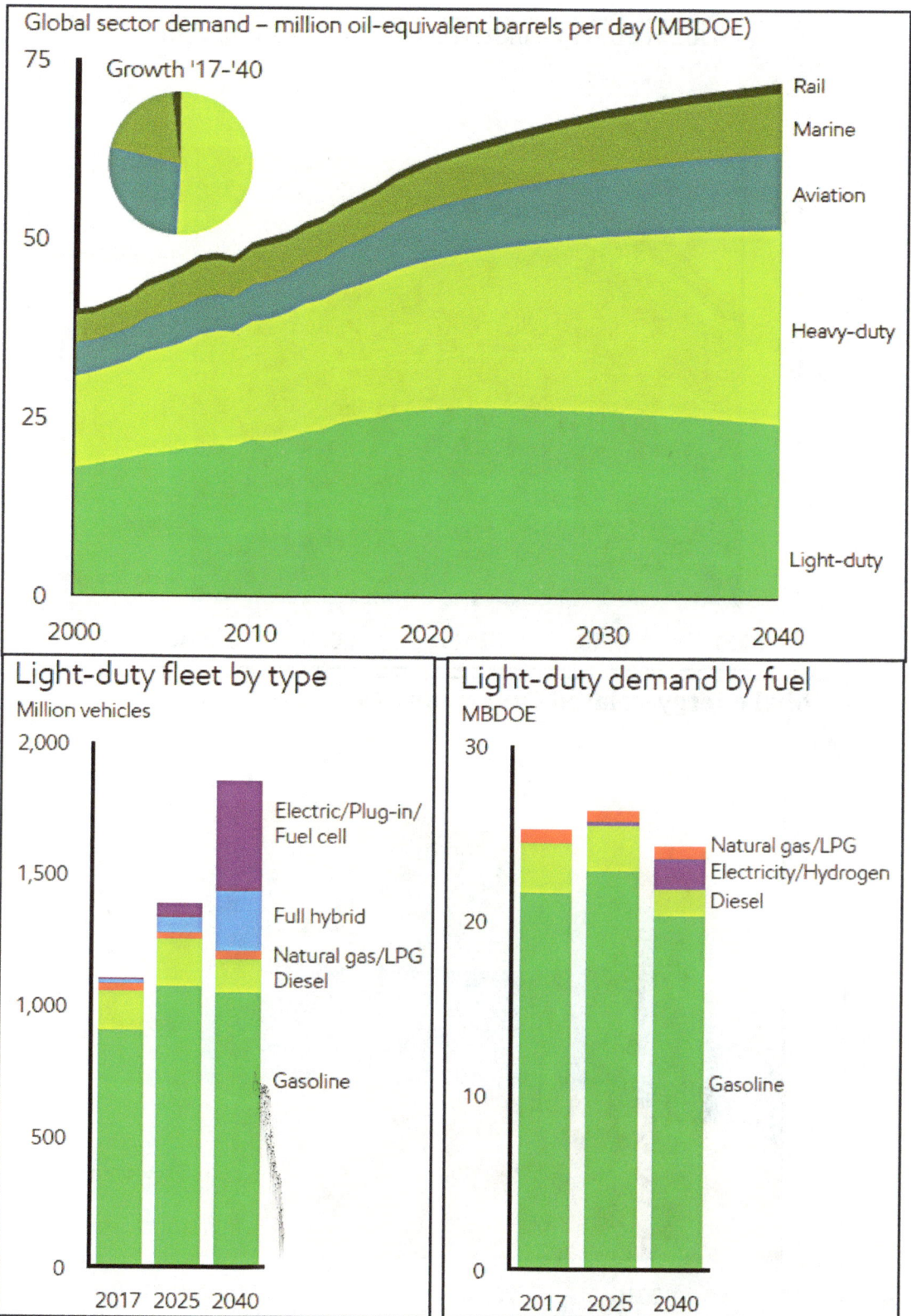

Figure 7.15d   Outlook for transportation by type and fuel *(ExxonMobil, 2019).*

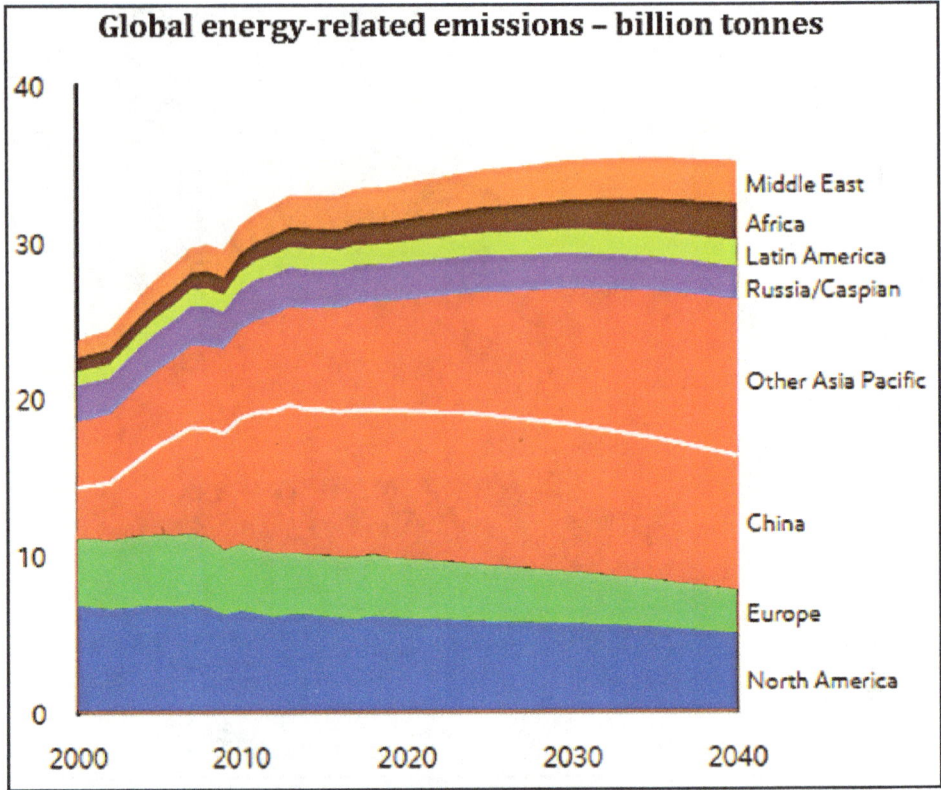

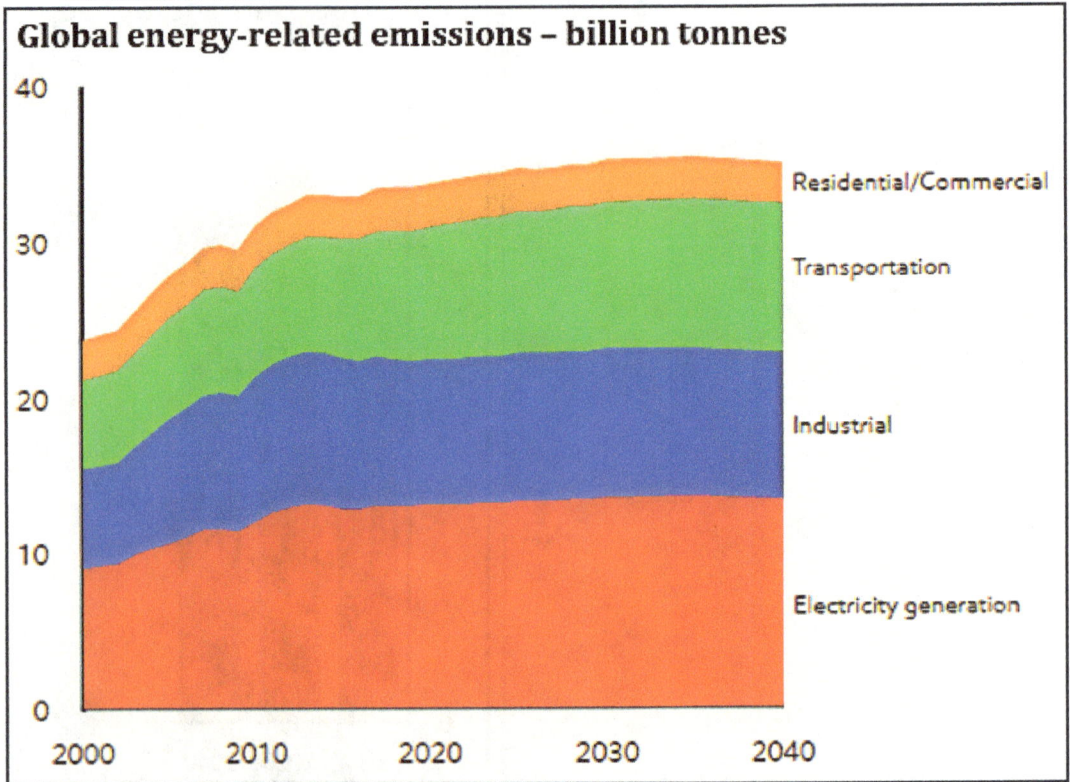

Figure 7.15e   Outlook for energy-relate emissions by region and sector *(ExxonMobil, 2019).*

Plug-in Hybrids (PHEVs) are likely to account for around a quarter of EV population because they run on a mix of electric power and oil, while the balance will be full Battery Electric Vehicles (BEVs). The pace at which EVs penetrate the global car fleet market depends on many factors, in particular, advances in battery technologies that reduce size and price substantially; policy incentives that offer attractive purchase subsidies and tax reductions, and promote proliferation of fast home, street, workplace and intra-city charging bays, prevailing global oil market prices, and above all, consumer preferences which depend on technological innovations in competing conventional ICE vehicles. State policies could provide attractive incentives to support EVs, but at the same time, the formidable research and development base of the conventional ICE vehicle industry combined with global policies targeting higher fuel efficiencies and lower emissions for conventional ICE cars will prove a challenge for the EV industry. Cars that can deliver around 25 km/liter (60 miles/US gallon) are already on the market, many of them with a very wide range of features including those that are the main selling points of EVs. Also, the latest diesel vehicles are fitted with advanced emissions control systems, and actually have similar or even lower CO_2eq emission levels per kilometer compared with petrol engines.

It is clear from the discussion above that fossil liquid fuels will remain the prime energy for the transportation sector in the foreseeable future, although the share of natural gas and electricity is projected to grow. The population of heavy trucks fueled by natural and liquefied petroleum gas is growing, so is electricity use by passenger rail, light-duty vehicles, and e-bus transportation, but the contribution to the total fuel demand by the sector will remain very small. Several European governments have recently announced policies which will eliminate the sales of new gasoline and diesel cars by 2040, and some have imposed punitive taxes on diesel vehicles. Some have also announced plans to eliminate the sale of new internal combustion engine cars by 2040. However, while the strong policy support is already helping to improve sales in Europe, the global impact on climate mitigation may not be very significant considering the global outlook. Germany is the only European auto manufacturer in the world's top six and accounted for less than 4 million of the over 90 million vehicles produced world wide in 2017, compared with China's nearly 25 million vehicles. Furthermore, the few countries that will account for around two-thirds of the future sales of electric vehicles also have carbon-intensive electricity (Figures 7.16).

The United States is the world's second largest cars/light-duty vehicles market and has projected that electric vehicles will account for only 19% of total sales of cars and light-duty vehicles in 2050 (EIA, 2018), apart from the fact that the country's electricity is still carbon-intensive. In any case, the contribution of Europe (EU-28) to global energy-related pollution has been declining for decades, accounting for only around 10% of the global total in 2015. However, even if the region only succeeds in promoting EV market penetration to just 50% of the total new car sales in member countries by 2040, the ambient pollution that causes smog over cities and impacts negatively on public health will be reduced significantly. In summary, there is no doubt that the electric car market will grow much more rapidly over the next decade or two compared with the sluggish growth in the last ten years or so, fueled by improved battery technologies and robust policy support. However, much of the climate change mitigation effect will be localized, leading to cleaner urban environment in countries that succeed in decarbonizing the electricity source. The European Union currently leads the world in low-carbon power generation but the group's global share of the EV market is likely to remain low.

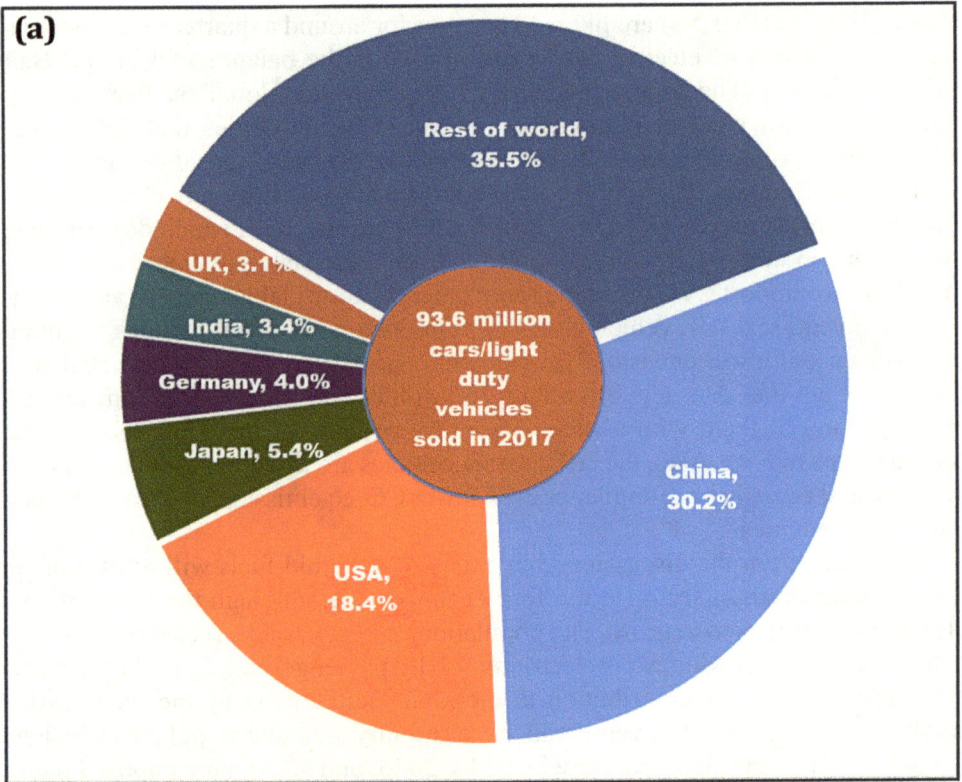

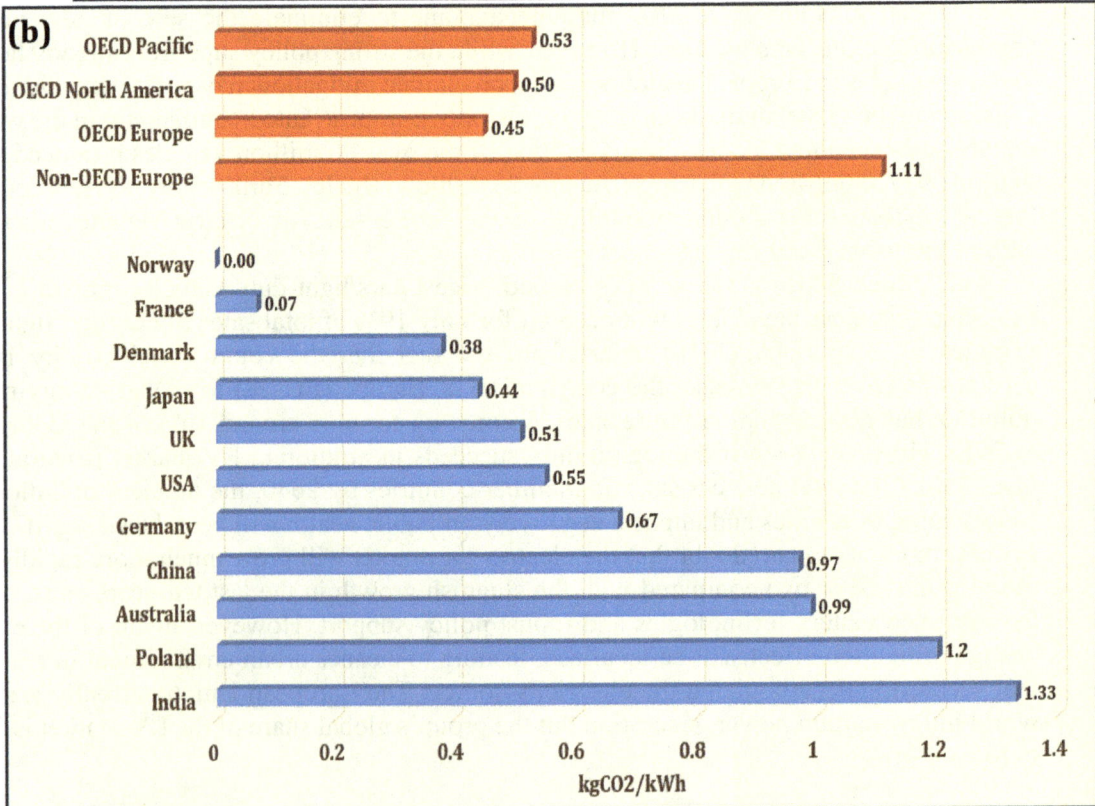

Figure 7.16   (a) Global sales of cars and light-duty vehicles in 2017.
(b) Electricity carbon intensity factors for by region and country.
(Data from statistica.com; focus2move.com; ECOMETRICA, 2011).

7.6 OUTLOOK ON CO₂ EMISSIONS

Energy-related emissions mirror the trends in energy consumption which in turn reflects global economic, population and prosperity trends. Increasing energy efficiency and decreasing energy intensity will be the key drivers of future anthropogenic emissions abatement across all regions of the world. In spite of virile economic growth and associated rise in energy demand, improved energy efficiency will continue to slow down the rate of growth of emissions, and all available projections show that emissions will fully decouple from energy demand by 2040 (Figure 7.15e). Improving efficiency and the resultant decrease in CO_2 intensity of energy use will help stem emissions growth in spite of growth in population and GDP. Even though China's GDP rose by about 1,000% from 1990 to 2015, energy efficiency gains kept the rise in CO_2 emissions to about 300%. Emissions in OECD countries have remained relatively flat for the last two decades and are projected to decline by about 20% by 2040, but emissions from the non-OECD countries will rise by around 50% over the same period despite a 40% gain in efficiency across the emerging economies. However, the growth rate less than 1% will be much lower than the 3%/year from 1990 to 2015. The prospects of lowering energy-related emissions depend largely on ability to reduce demand through improved efficiency and fuel switching (Figure 7.17), and opportunities vary across sectors of the global economy.

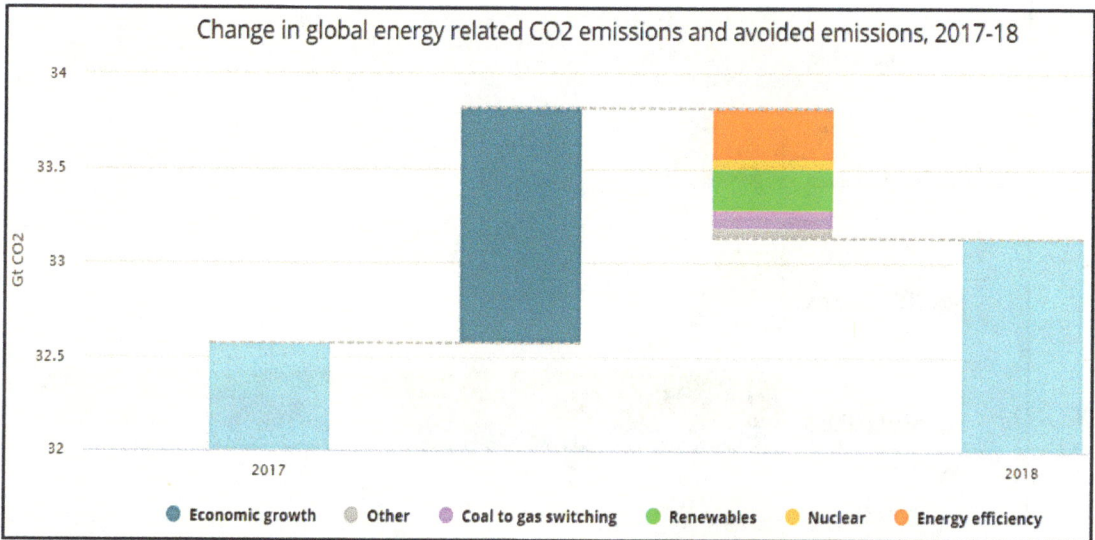

Figure 7.17 Effects of energy efficiency improvement an fuel switching on CO2 emissions in 2018 *(IEA 2019i).*

Events in two regions which contributed three-quarters of the total energy-related emissions in 2015 - Asia (53%) and North America (22%) – will largely determine the global outlook. China alone accounted for 28% while the United States contributed about 15%. China's outlook to 2050 (EIA, 2018) shows that a combination of efficiency improvements, technology implementation, fuel substitution, and emission control policies will hold emissions level flat in the next thirty years, overall, and in all sectors of the economy, with the exception of industry where an annual growth rate of around 0.6% is projected. However, Non-OECD Asia will remain the world's largest emitter of energy-related pollution in 2050 (Figure 7.18).

China's economy is slowing down and also transiting from high-intensity manufacturing to less energy-intensive production. Furthermore, growth of the renewable an nuclear energy sectors has been rising, hence a gradual decrease in energy demand rate is expected, and energy intensity will also start to decrease. Coal will account for a lower share of the primary energy demand and the share of natural gas, nuclear and renewables will rise significantly. Overall, there should be a significant fall on CO_2eq emissions of the country by 2040. Significant shifts from coal to natural gas, renewable energy and nuclear power are projected for the country, an the mitigation effect on CO_2eq emissions is expected to be substantial because of the size of the country's economy. On the contrary, energy demand by India continues to grow at more than 2.5 times that of China, representing more than a third of the global increase, and most of the growth being filled by coal and nuclear power. However, India's contribution to the global total emissions in 2015 was only a quarter of China's emissions, and projections show that China will still account for two times India's emissions in 2050.

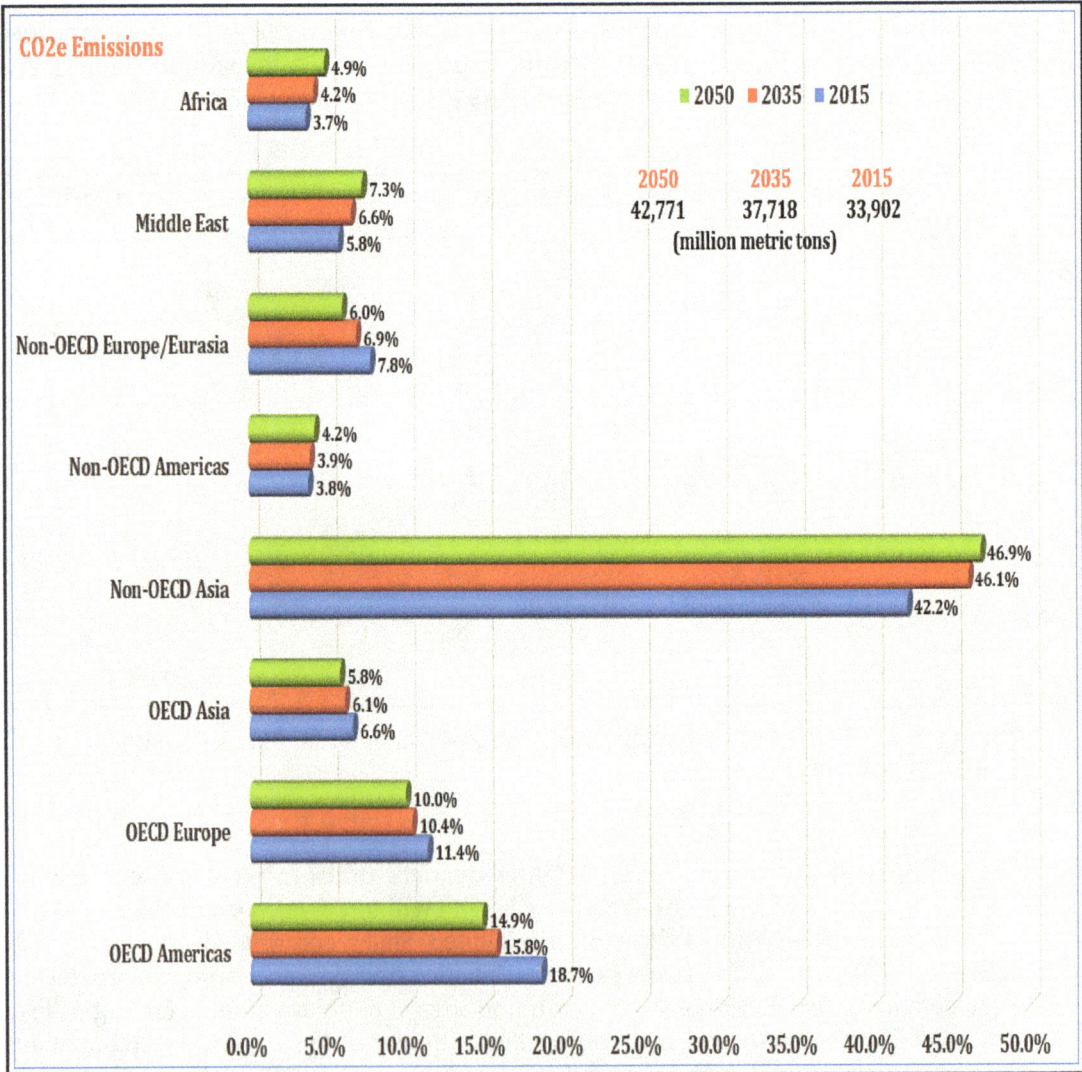

Figure 7.18 Projections of CO_2eq emissions to 2050 by region and countries. *(Data from EIA, 2017).*

7.6.1 Power sector

Decarbonized power through scaled-up deployment of available technologies and deployment of technologies currently under development is central to global strategies for clean energy transformation. Around 70% of the projected increase in global primary energy demand will be used in power generation, with power demand growing three times more quickly than any other end-use energy. However, the rapidly rising efficiency gains in the final use of electricity across the end-use sector will also translate to lower future electricity demand, particularly in OECD countries, thus helping to reduce power-related emissions. Substitution of high carbon intensity fossil fuel, in particular, coal with lower carbon sources such as natural gas and renewables has been very effective in reducing energy-related pollutions, especially in power generation. Global renewable power production increased from 2016 level by 6.3% to 25% in 2017 - mainly wind, power and hydropower additions (Figure 7.19). The trend is expected to intensify, and various projections show that renewables and nuclear energy could be producing 40-50% global electricity by 2040, coal being the main loser, accounting for just 13% of the increase in power generation over the period compared with more than 40% over the previous two decades.

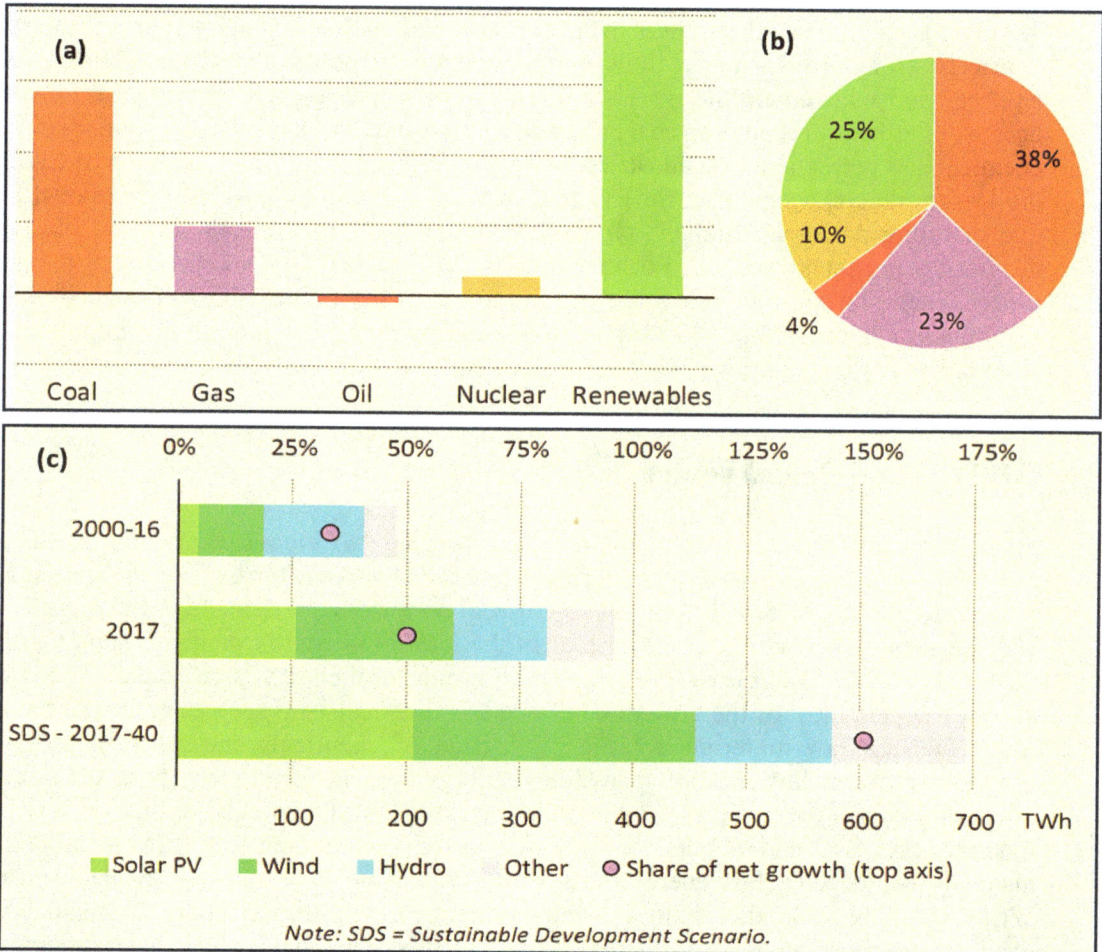

Figure 7.19 (a) Change in global electricity generation by source, 2016-2017 **(b)** Electricity generation by source in 2017 **(c)** Average annual growth in renewables-based generation by technology, historically and in the Sustainable Development Scenario *(IEA, 2018)*.

Boosted by a strong solar PV market, sharp cost reductions, and policy support, renewables energy deployment increased by 7.1% from 2017 to 2018, largely driven by solar PV deployment which grew by 50% (the growth rate in China was 11%). However, coal still remained the largest source of pollution, increasing by 3% and accounting for 30% of total global emissions, mostly coming from Asia. Despite this growth in coal use for power generation, fuel switching is making a significant impact in decarbonizing energy. Coal-to-gas switching replaces about 60 million metric tons of coal and avoided 95 million metric tons of emissions. Without this abatement, the increase in emissions would have been 15% greater. China and the United States led the power-related emissions reduction, accounting for 45 million metric tons and 40 million metric tons respectively. Increased renewable use mostly for power generation had an even greater mitigation impact on energy-related emissions, avoiding 215 million metric tons of emissions. Energy efficiency remained the largest contributor to emissions reduction in 2018 despite a continued slowdown in policy implementations, eliminating about 275 million metric tons of emissions.

China accounted for over 40% of global renewable capacity growth, and around 60% of total annual solar cell manufacturing globally in 2018. A recent projection by the International Energy Agency (IEA 2017i) shows a ten-fold rise in global solar PV capacity to 740 GW by 2022, with China alone accounting for around half of this expansion. It should be noted however that there are significant differences in individual country achievements. China leads the world, with very strong projected growth in both solar and wind power deployment, but the country has been and will remain the worlds highest source of energy-related pollution for decades because of its heavy reliance on coal power generation, and coal use by heavy primary materials, chemical and petrochemical industries. On the contrary, most of the countries of EU-28 are the least polluters, accounting for less than 10% of global total pollution emissions, largely due to strong policy instruments targeting decarbonization of energy, in particular, power generation and transportation. For example, in 2017, about 50% of the United Kingdom's power generation came from low carbon sources (25% from renewables and same proportion from nuclear power), only 7% from coal, amounting to 12% reduction in carbon emissions from electricity consumption, a saving equivalent to the almost 15% of emissions from cars currently on British roads.

7.6.2   Industrial sector

Energy-intensive industrial production is indispensable to global economic growth which requires a wide range of primary materials - iron and steel, non-ferrous metals, special alloys, polymers, chemicals, cement, etc. The industrial sector energy demand is the highest of the end-use sectors, accounting for over half of the total final energy demand. Although global GDP is expected to double by 2040, demand for industrial energy is projected to rise by only about 20%, and most of the growth will be in the chemicals sub-sector because of the use of fossil fuels as raw materials for the production of chemicals and petrochemicals. The progressive decoupling of GDP growth and energy demand is due largely to the mitigating effect of increasing efficiency. However, immense opportunities exist to further decarbonize industrial processes and reduce emissions by improving manufacturing efficiency, optimizing materials use, recycling heat energy and materials, and using more of locally available materials. Waste recycling could also result in significant reduction in primary energy demand, with the resultant decrease in energy-related lifecycle emissions that would have accompanied production from virgin ores. Many technologies that target these goals are already available and many more are under development, and faster deployment could reduce global energy-related emissions 20-40% by 2060.

7.6.3 Transport sector

Transportation accounts for about 27% of final energy consumption, and growth will be mitigated by rising transportation efficiencies. Nearly all the global transportation fuel requirements was filled by fossil oil in 2018, with renewables accounting for only 4%, and 96% of the renewable share was filled by biofuels, while electric vehicles accounted for only 2%. Projections show that growth of renewable share in transportation will remain very modest over the next decade or so. Faster growth of the electric vehicle market is a key pathway to decarbonizing transportation. Although the total global population of electric vehicles in 2018 was only 5 million, the growth rate from 2017 was phenomenal (63%), with 45% located in China, Europe 24% and United States 22%. If this momentum is sustained, the global EV population should be about 250 million in 2030, with around 90% in the light-duty vehicle category. Even this fast growth rate EVs will not be enough to get transportation on track to meet the International Energy Agency's 2 Degree Scenario, a broad set of policies summed up by IEA as "Avoid, Shift, Improve" will be required (Figure 7.20). For example, implementation of the EV30@30 Scenario (30% global market share of EVs by 2030) requires that EV sales reach around 45 million vehicles a year, raising the global population of EVs to 200-250 million by 2030. This would require very strong, globally coordinated policy instruments. However, despite strongly rising sales, the share of EVs in global electricity supply was less than 0.5% in 2018, and will likely remain below 1% over the next decade, although around 30% of the electricity consumed by EVs will come from renewables, up from the current 26%.

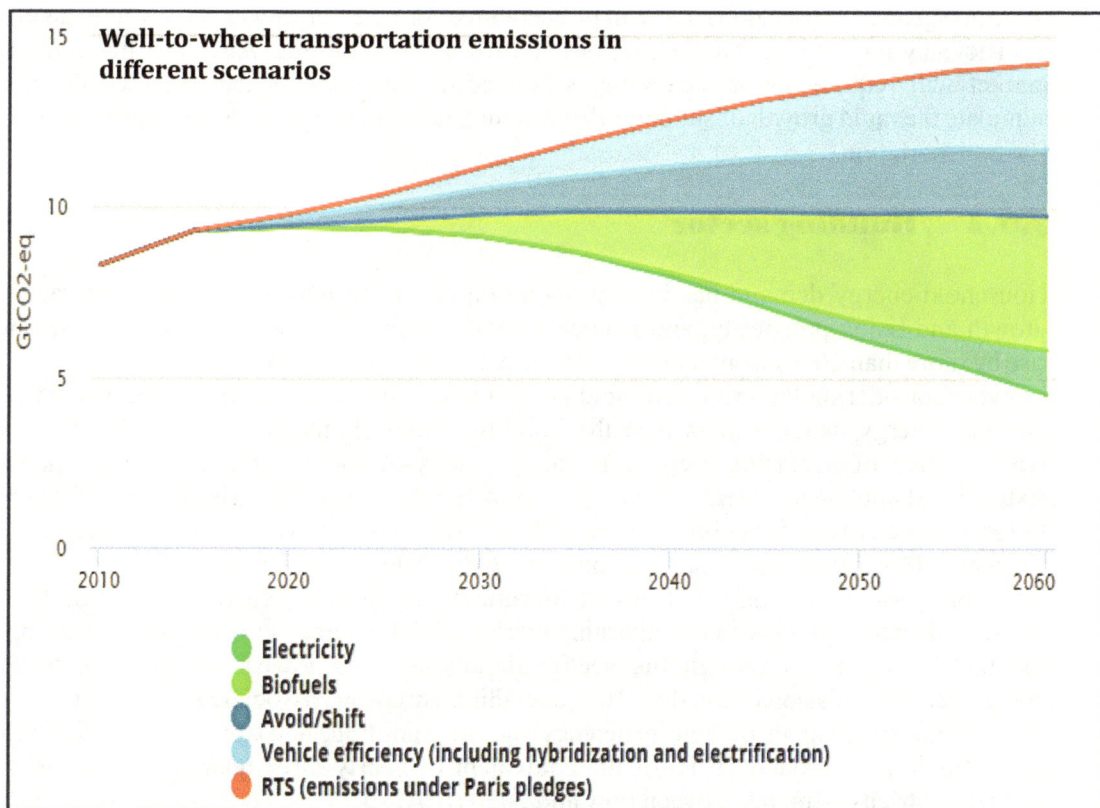

Figure 7.20a Potential impacts of different policy scenarios on transportation emissions, 2015-60 *(IEA, 2019m)*.

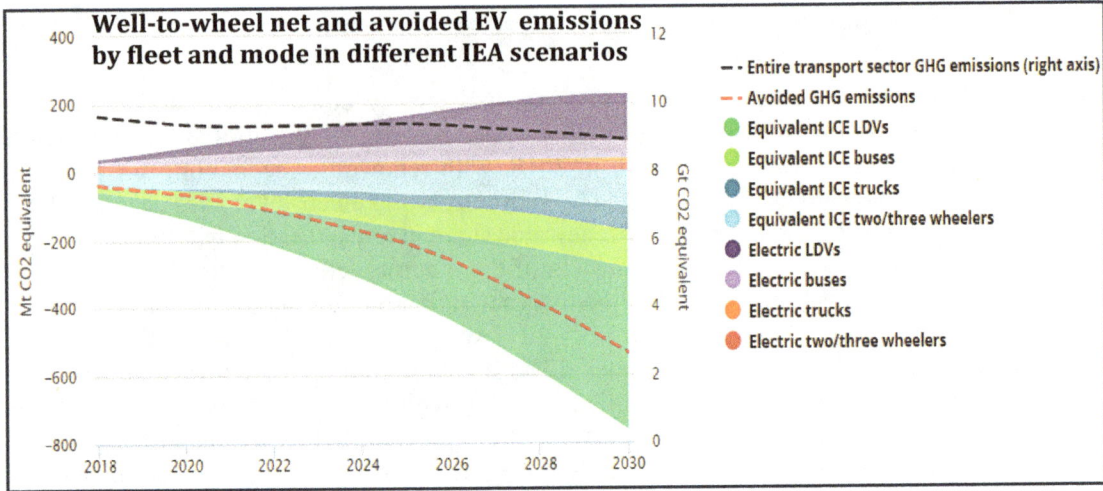

Figure 7.20b Potential impacts of different policy scenarios on EV fleet emissions by mode and total GHG emissions from the transport sector in the EV30@30 scenario *(IEA, 2019m).*

Another issue with the EVs is the fact that the market growth has been largely restricted to the European Union countries, the United States and Japan which together contribute only about a third of the total global transport-related emissions due to significant deployment of renewables in power generation. China now leads the global market in manufacture and operation of EVs which depend on carbon-intensive electricity. Furthermore, most of the projected growth in transportation will be accounted for by emerging nations where the impacts of efficiency improvement and emissions control will be minimal. Also, fast-tracking the EV market will require major technologies and policy instruments that broaden the market, stimulate the rapid growth of support infrastructure, and encourage mass (rather than individual) EV transportation.

7.6.4 Building sector

Household energy demand has been growing rapidly in response to rising global economic growth and rising prosperity, and residential and commercial energy demand is expected to rise by more than 20% through 2040, with electricity accounting for about 90% of this demand growth. Considering the projected rapid growth in the global economy and associated energy demand, energy demand growth in the building sector should be much higher but for the positive effect of increasing energy efficiency - energy-efficient building envelopes, materials, fixtures and appliances. Heating and cooling of buildings use the largest share of energy in the sector, accounting for approximately 40% of final energy consumption, and electricity will fill about 40% of the sector energy demand by 2040. Additionally, around 65% of this demand relies on fossil fuel sources. In effect, decarbonizing power generation and modernizing household energy (largely in the emerging world) hold the highest prospects for decarbonizing the building sector, although the sector already has the lowest contribution to global energy-related emissions (less than 10% excluding emissions associated with electricity use). More rapid deployment of high-efficiency lighting, cooling, and appliances in the building sector could save around 50 EJ or the equivalent of nearly three-quarters of current global annual electricity demand between now and 2030 (IEA, 2017h). The Building sector also has the greatest potential of achieving zero-net energy in the shortest possible time because stand-alone solar PV units are being deployed in residential and commercial buildings worldwide, and many units are already producing as much or even more low-carbon energy than they consume from the grid, and some are selling energy back to the grid.

7.6.5 Heat generation

Heat used for water and space buildings and for industrial processes represents almost 40% of global final energy consumption (larger than 27% for transportation). Furthermore, around 65% of this demand comes from sources fueled by coal, fuel oil, gas, solar, geothermal or traditional biomass. The share of renewable energy (mainly biomass) has been low, less than 10% in 2018 and growth over the next few years will be insignificant, therefore, decarbonizing heat production is an important mitigation option, particularly in the building sector, but remains a challenge. Much of the expected progress in decarbonizing heat energy over the next decade or so will be in China and the European Union where modern biomass-fueled heat generation plant deployment is growing. Other potential sources are renewable electricity, solar, and geothermal energy. Some combined systems which produce both heat and electricity, thereby improving operating efficiencies drastically, are also being deployed. It is estimated that energy efficiency and switching to clean final energy sources (including decarbonized electricity and district energy that produces both heat and electricity) could cut fossil fuel consumption for heating and cooling in half by 2060 compared with today.

7.7   OUTLOOK ON PARIS-2015 TARGET

The focus of the Paris 2015 Agreement was the decision of 196 countries to work together to limit global temperature rise to well below 2 degrees Celsius above pre-industrial levels, and to strive for 1.5 degrees Celsius, in order to reduce the risks and impacts of climate change. Many institutions and organizations have tried to define feasible pathways to achieving this goal, now known as 2DS (2-Degree Scenario) or 450S (450 Scenario, limiting the concentration of CO_2eq to 450 parts per million (ppm) in 2040, compared with 403 ppm in 2016 which was 40% higher than in the mid-1800s. An analysis by the International Energy Agency (IEA) concluded that current trajectory of policies and actions fall far short of what is needed to achieve this goal in a long time, much stronger and aggressive global policies are needed to reduce currently projected emissions of around 37 GT in 2040 by around 50% (see Figure 5.4). The Agency has proposed much wider deployment of innovation technologies, many of which are already in the pipeline, and stronger policy support to meet global climate ambitions. Aggressive infusion of new technologies is needed at both the supply and demand sides of the energy system: for example, the power sector would need to be heavily decarbonized by reducing the current emissions level of about 65% to no more than around 7%, and with carbon capture. Industry will need to reduce emissions by at least 20%; the electric vehicle market penetration will need to grow from the current level of around 2 million a year to 150-200 million by 2030, leading to 90% of all cars on the road being electric by 2060; and most of the electricity that will be required to power the electric cars should come from lower carbon non-fossil sources (renewables and solar). Furthermore, the building sector will account for around 40% of the total global electric power demand in 2040, hence its contribution to energy decarbonization will largely depend on the source of the electricity.

In a recent study on energy outlook by BP (2018), three different scenarios were considered: Evolving Transition (ET), based on current and expected policies; Faster Transition (FT); and Even Faster Transition, energized by much stronger technology and policy support (EFT). The last two scenarios reflect possible implications of different judgements and assumptions. None of the scenarios takes full account of all possible uncertainties, and all have equal probability of happening as projected in view of the many uncertainties surrounding energy demand forecasts (see Figure 5.5). The key conclusion of the study in respect of CO_2eq emissions is that, in the ET scenario emissions will continue to grow through to 2040, increasing about 10% above current levels. This is inconsistent with the sharp decline that is necessary to fully decouple energy use and emissions by 2040 in order to meet the Paris-2015 climate goals which in any case have been found to be inadequate. The EFT scenario projects a further reduction in emissions by around 50%, consistent with the International Energy Agency's 'Sustainable Development Scenario', also known as 'IEA-450 Scenario.' In summary current and anticipated policies will not achieve the Paris-2015 goal which requires halving current emission levels by 2040 (the IEA-450 Scenario). Much stronger and effective policies would be required to drive down emissions, and decarbonizing power generation appears to be the surest pathway.

One of the major problems with energy use and emissions outlook projections is the amount of uncertainty about major variables. Variations in global oil market are particularly disruptive. When prices are low (as they are currently), they tend to promote energy use and depress incentives for investments in efficiency upgrade or lower carbon technologies. This is evident from the latest report on global energy and CO_2 status report (IEA, 2019i). Contrary to projections, global energy consumption in 2018 increased at nearly twice the average rate of growth since 2010 driven by a strong global economy and extreme weather in some parts of the world which increased the demand for heating and cooling energy. Improvements in global energy efficiency and energy intensity have slowed down dramatically. Demand for all fuels increased, led by natural gas and higher electricity demand accounted for over half of the growth in energy needs. This unanticipated rise in global primary use increased emissions by around 2% over the 2017 level. Although deployment of renewables grew at double digit pace, fossil fuels met 70% of the growth and accounted for about 81% of the global primary energy use in 2018. Asia accounted for most of the growth in global energy demand, with India and China accounting for more than 40% of the increase. Emissions which had remained flat for the previous three years, grew significantly by 560 million tonnes, equivalent to emissions to 200 million additional cars on the roads. The trend of growing emissions is apparently as a result of the combined effect of growth in energy demand and weaker energy efficiency improvement efforts, and cuts across all regions, including most major economies, with a few exceptions such as the United States, United kingdom, Japan, and Mexico.

Several international bodies including the Intergovernmental Panel on Climate Change, and the International Energy Agency have concluded that the emerging scenario NPS) which takes into account commitments in the Paris-2015 Agreement, current and anticipated policies, probable technology innovations, will be inconsistent with sustainable environment. The IEA has proposed a faster transition/sustainable development scenario (SDS) that could put he world on the pathway to the 2 Degree goal or even lower (Figure 7.21).

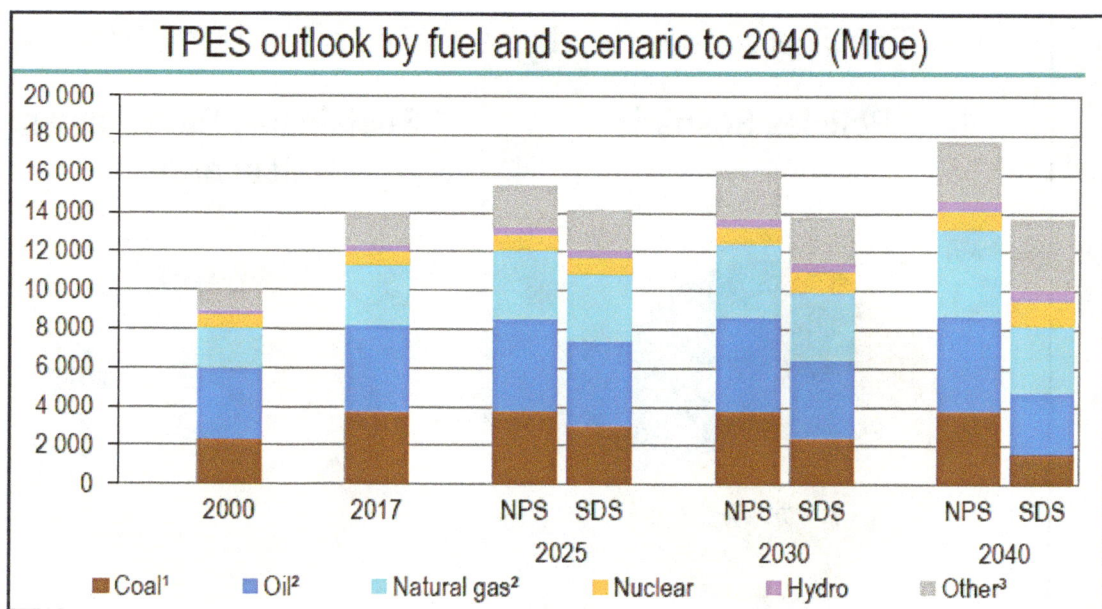

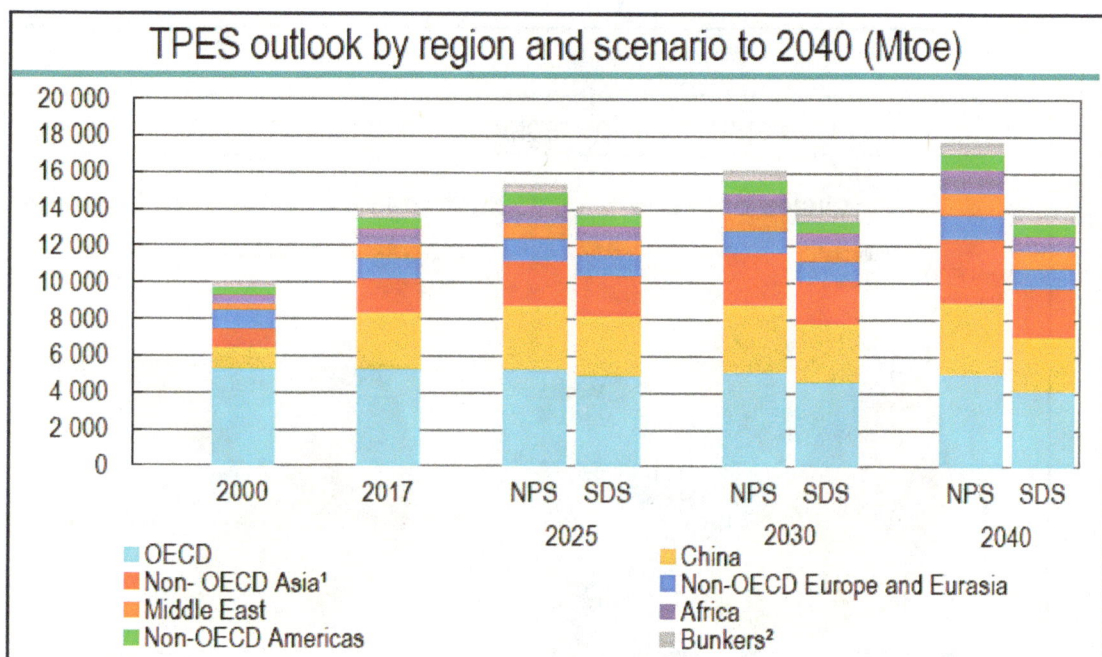

Figure 7.21a Total primary energy supply outlook by fuel, region and scenario
NPS = National Policy Scenario; SDS = Sustainable Development Scenario *(IEA, 2019n).*

Total final consumption by sector and scenario in 2040

(a)

New Policies Scenario

Non-energy use
8.8%

Industry
30.5%

Buildings & agriculture
32.0%

Transport
28.7%

12 581 Mtoe

Sustainable Development Scenario

Non-energy use
10.3%

Industry
32.1%

Buildings & agriculture
31.1%

Transport
26.5%

9 958 Mtoe

(b)

CO_2 emissions[4] by region and scenario in 2040

New Policies Scenario

Bunkers[2]
5.2%

Non-OECD Europe and Eurasia
6.9%

Non-OECD Americas
3.7%

Africa
4.8%

Non-OECD Asia[1]
22.6%

Middle East
7.0%

OECD
24.6%

China
25.2%

35 881 Mt of CO_2

Sustainable Development Scenario

Bunkers[2]
5.8%

Non-OECD Europe and Eurasia
9.1%

Non-OECD Americas
4.4%

Africa
6.4%

Non-OECD Asia[1]
23.8%

Middle East
8.4%

OECD
23.7%

China
18.4%

17 647 Mt of CO_2

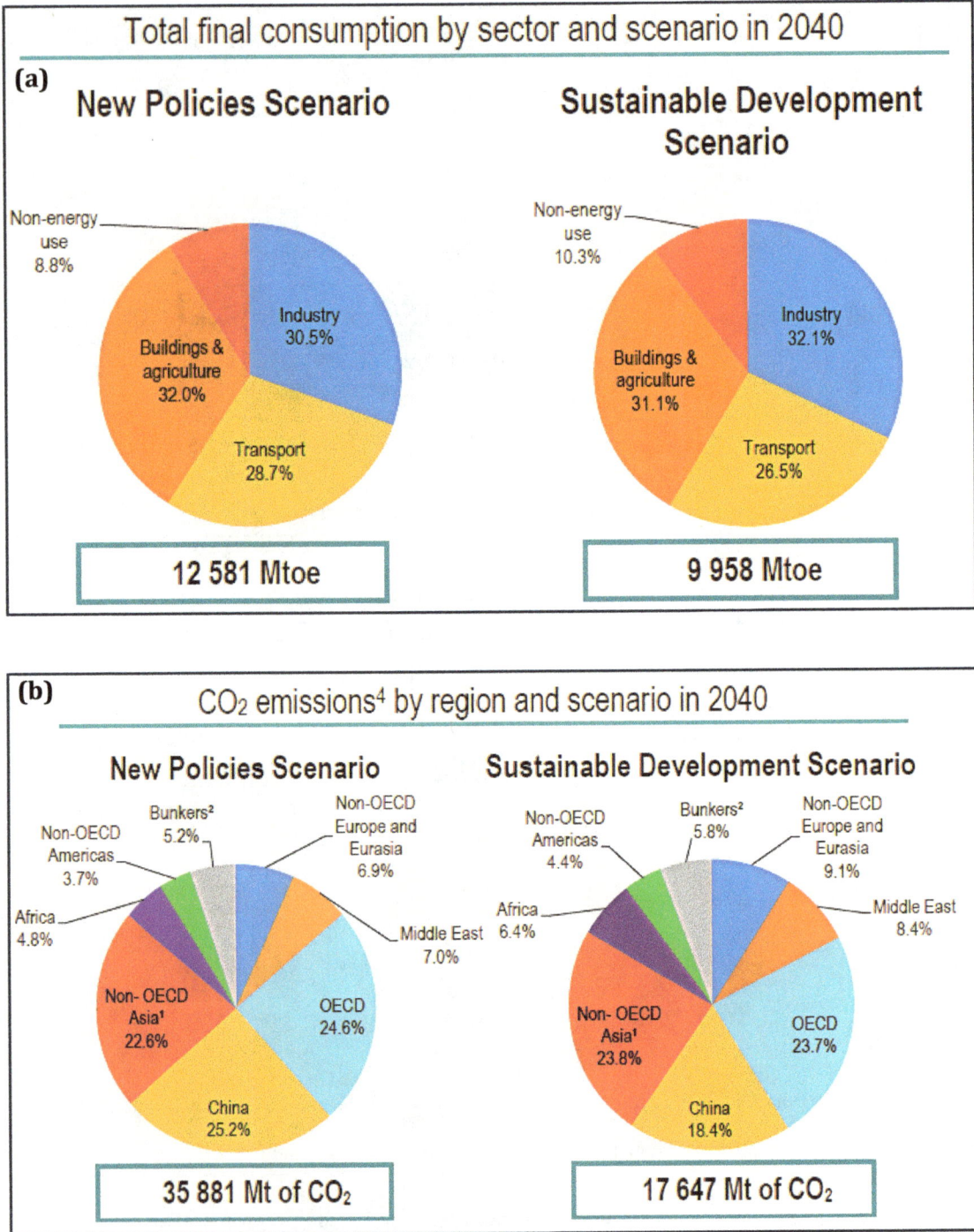

Figure 7.21b   (a) Total final energy consumption outlook by sector and scenario (b) CO_2 emissions by region and scenario.   *(IEA, 2019n).*

7.8 OUTLOOK ON PEOPLE'S MITIGATION: LIFESTYLE CHOICES

A potentially promising climate change mitigation action is people's cooperation in conserving energy thereby reducing personal carbon foot print, and there are many options such as choosing wisely when shopping for homes, vehicles and appliances, switching off idle appliances, eliminating unnecessary auto journeys, pooling vehicles, and supporting municipal recycling efforts. Effective country and local policy instruments are necessary to promote recycling, especially of reusable items such as shopping bags, metal products, plastics, paper and electronic items. It is estimated that a trillion single-use plastic bags are used worldwide annually (around 2 million every minute). More than 480 billion of plastic bottles (water, soft drinks, etc.) were used in 2016 across the world, projected to double by 2040 (Euromonitor International, 2017). Fewer than half of the bottles were collected for recycling and only 7% of those collected were turned into new bottles, the balance being discarded or used in landfill.

Most of the discarded bottles end up in rivers and oceans. It is estimated that between 5 million and 13 million tonnes of discarded plastic end up in the world's oceans, ingested by fish and other aquatic life, and some of it already showing up in the human food chain. Apart from being potentially toxic, ingested plastics cannot be usefully assimilated into the food chain: some large fish and whales found dead in recent times had large quantities of ingested plastic bags in their system which they could not digest. Another problem with plastics is the fact that they do not degrade easily. Exposure to strong ultraviolet rays from the Sun can embrittle plastic, causing it to break up and possibly end up in the atmosphere as aerosol, or in the oceans, but some can remain intact for several hundred years (see Chapter 4).

Plastics are produced from fossil fuels, and the process requires enormous energy. Strong policies that promote re-use or recycling will make a significant impact on energy-related emissions. Plastic bottles and bags are produced from polyethylene terephthalate (PET), and processing of every metric ton into bottles produces around 3-4 metric tons of CO_2eq emissions. City, state and national governments around the world, (led by Denmark as far back as 1993 and including many emerging nations) are introducing policies to control the problem of overuse or discarded plastic bags, by either limiting the use, banning some items like shopping bags and straws outright, or imposing taxes, but major drinks companies and petrochemical companies that make plastic raw materials are putting up stiff resistance, including legal action. A recent policy by the United Kingdom that introduced a charge of just 5 pence on plastic shopping bags provided by major retailers in stores and for deliveries has had a dramatic effect, cutting demand by 83% in just a six months, based on the records of seven supermarkets. Only 0.5 billion (less than 100 million a month) of plastic bags were purchased by customers in the first six months since the introduction of tax, compared with more than 7 billion (more than 600 million a month) in the preceding year when it was free (Defra, 2018). The supermarkets also reported significant cost savings. The policy is being extended to more polybag outsources, and introduction of refundable deposits on drinks bottles and cans appears imminent. Many other countries have reported similar dramatic gains, and many different plastic refuse-reduce-reuse-recycle policies are emerging all over the world, including some developing countries.

As discussed in Chapter 4, plastics are difficult to recycle because of the very wide range of types with different technical characteristics, hence intensive sorting is required. In spite of global efforts and people's cooperation, the end-of-life fate of plastics is becoming a major global environmental problem. A recent study (Geyer, *et al.* 2017) on production, use and fate of plastics ever made estimates that around 8.3 billion metric tons of virgin plastics have been produced in the world since the 1950s, 6.3 billion metric tons of plastic waste had been generated, around 9% of which had been recycled, 12% incinerated, and 79% either dumped or used as landfills. If the current production and waste management trends continue, around 12 billion metric tons of plastic waste will be in landfills or in the natural environment by 2050. Using more than three-quarters of waste plastics as landfills raises a major environmental problem because they do not degrade easily and some may remain intact for several hundred years. Those that are exposed may degrade due to exposure to the Sun's ultraviolet rays but the microparticles may end up in the atmosphere, in rivers and oceans. A small fraction of the 12% that was incinerated was used for power and heat generation but the bulk was simply burned, producing black fumes loaded with GHG gases, toxic chemical compounds and black carbon. However, the critical issue is the fact that, with the exception of a few emerging countries, most countries have no coherent policies on waste plastic recycling. The huge volumes generated in the developed world are exported to the emerging world, notably China, Hong Kong, Vietnam, Turkey, Malaysia, Senegal where the small portion that is recyclable is sorted and the bulk either incinerated or dumped. China, by far the largest destination recently closed its doors and this has caused enormous problems in the developed world, with most of plastics waste now ending up in landfills or incinerated. This development is causing municipal authorities rejecting plastics waste, thereby frustrating people who have been supporting recycling. As discussed in Chapter 4, most plastic waste is never recyclable anyway and combustion to generate electricity and heat under controlled conditions which include emissions' capture and stripping is a much more environmentally sustainable disposal method than landfill.

7.9   OUTLOOK ON VOLUNTARY COUNTRY MITIGATION (INDCs)

The main impact of stratospheric pollution is on climate change, and the effect human life and the natural ecosystem is very gradual, often taking decades to become prominent. However, in spite of the relatively short life of tropospheric (ambient) pollution, the negative impact is localized, immediate, and severe. Smog is a major issue in most cities in all regions of the world and the negative impact on human health is well documented. The sources of lower atmosphere pollution vary across countries, hence abatement solutions will differ, and this has prompted many countries to develop local action strategies to mitigate pollution. The cumulative effects of local pollutions eventually resonate in the stratosphere and enhance global warming which is the primary cause of climate change, hence local actions are also impacting positively on stratospheric pollution. The extent and diversity of voluntary mitigation actions were evident in the United Nations Paris-2015 conference - Over 150 countries submitted Action Plans for mitigating environmental pollution.

Europe leads the world in terms of regional action to balance the energy trilemma - energy equity, energy security and energy sustainability. In 2008, the European Union committed to climate and energy goals to be reached by 2020. These goals known as '20-20-20' Targets requires member countries to cut greenhouse gas (GHG) emissions by 20% compared with 1990 levels, with a 20% share of the final energy consumption coming from renewables; and a 20% increase in energy efficiency. Already, the Block has achieved an average of 25% renewable power generation and some countries within the Group have set even higher goals. For example, Denmark leads the world in renewable power generation: in 2016, wind and solar accounted for 44% (42% wind and 2% solar), and the country expects to achieve around 70% by 2022. Germany set a national emissions' reduction target of 40% and a 55% cut in emissions by 2030. Also, the share of renewable energy (mainly solar and wind) in the country's electricity consumption should rise from around 30% in 2017 to 65% by 2030, and coal-fired power generation will be phased out completely.

France derives nearly 80% of its electricity currently from nuclear power from 58 ageing nuclear reactors and is the world's largest exporter of electricity due to the very low cost of generation. In spite of the country's significantly low-carbon electricity mix arising from nuclear power generation, the country announced comprehensive plans for further decarbonization in 2015 (The Energy Transition for Green Growth Act) that should reduce the country's greenhouse gas emissions by 40% in 2030. Key targets include a reduction of nuclear capacity to 50% by 2025, and development of renewable energy. Realization of these ambitious plans will require a complete transformation of the country's energy sector and introduction of strong policy measures for renewable energy development, considering that renewables account for only 15% currently. Energy efficiency improvement across the economy would also be necessary. It should be noted also that France's energy policies have not survived political cycles and energy policy statements have been reversed several times in recent years. The current administration has recently suspended the proposal to reduce the nuclear power reactor fleet, and determined that nuclear is "the most carbon-free way to produce electricity with renewables." A progressively increasing tax regime on CO_2eq emissions was introduced to fast-track decommissioning of coal-fired power plants. This is not surprising since the nuclear industry provides employment for about 200,000 people.

Poland is one of the top ten producers of coal in the world, and the second largest user in Europe after Germany. Coal-fired power plants currently generate more than 80% of electricity in Poland and the coal industry is a major employer. For Poland, energy security remains the prime goal and balancing the energy trilemma is a major struggle. The Polish government acknowledges that coal will continue to play a prime role in the country's primary energy mix in the foreseeable future. However, the country has developed extensive plans to cut environmental pollution by increasing energy efficiency and decarbonizing the transport system. Many of the power plants are old and are being replaced by much more efficient coal-fired power plants, and nuclear power could play a significant role in the country's future energy supply. Many other countries, particularly in the industrialized world have put in place voluntary measures to reduce environmental pollution. These measures are often referred to as Intended Nationally Determined Contributors (INDCs). The United States and China lead the world as potential INDCs - they are also the two leading sources of CO_2eq pollution and, together, they accounted for over 40% of global emissions in 2018. Both countries have committed to INDCs and set targets although the United States has been reversing many of the policies.

In 2015, the United States Environmental Protection Agency (EPA) proposed a Clean Power Plan (CPP) to reduce emissions by 26-28% below 2005 levels by 2025 (EIA, 2016). Around 80% of emissions in the US are energy-related, the balance coming from other sources such as cement production, agriculture, land use and forestry. Two of the largest sources of energy-related emissions are the transportation and electric power sectors. The CPP targets power plants which are the largest sources of carbon pollution, accounting for around one-third of all greenhouse gas emissions. Emission performance rates (Best System of Emission Reduction, BSER) were established for existing fossil fuel-fired electric generating units (EGUs) - electric utility steam generating units and stationary combustion turbines. The CPP reflects the different needs of different states and each state is given the flexibility to choose how to meet the set goals. The CPP, if implemented, is projected to reduce U.S. emissions by 0.5 billion metric tons by 2040 (EIA, 2018). One major flaw in the United States CPP Action Plan which was signed into law in 2015 was its obvious focus in eliminating coal-fired power plants. It would be near impossible for existing plants to meet the stringent emission control standards, and, for the many states which depend heavily of coal for power generation, it was unacceptable. In any case, most states already have local emission mitigation policies, designed for their specific local conditions.

Many states have instituted legal actions against the federal government, and succeeded in stalling implementation, even before the recent action by the current administration to repeal the law. However, even without the CPP, various state policies, rising use of renewables, increasingly competitive natural gas pricing, and negative economic forces in the coal industry have been driving down the use of coal in power generation, with a decline of about 16% between 2010 and 2017. A further decrease of about 35% is expected by 2030, after which it levels off. Most of the decrease in coal use will be filled by renewables and natural gas. In summary, the CPP has the potential to reduce on the US power generation emissions of the United States by about 30% through 2050, but its future is in doubt, considering the fierce resistance by many states and efforts by the current administration to cancel the initiative (Figure 7.22).

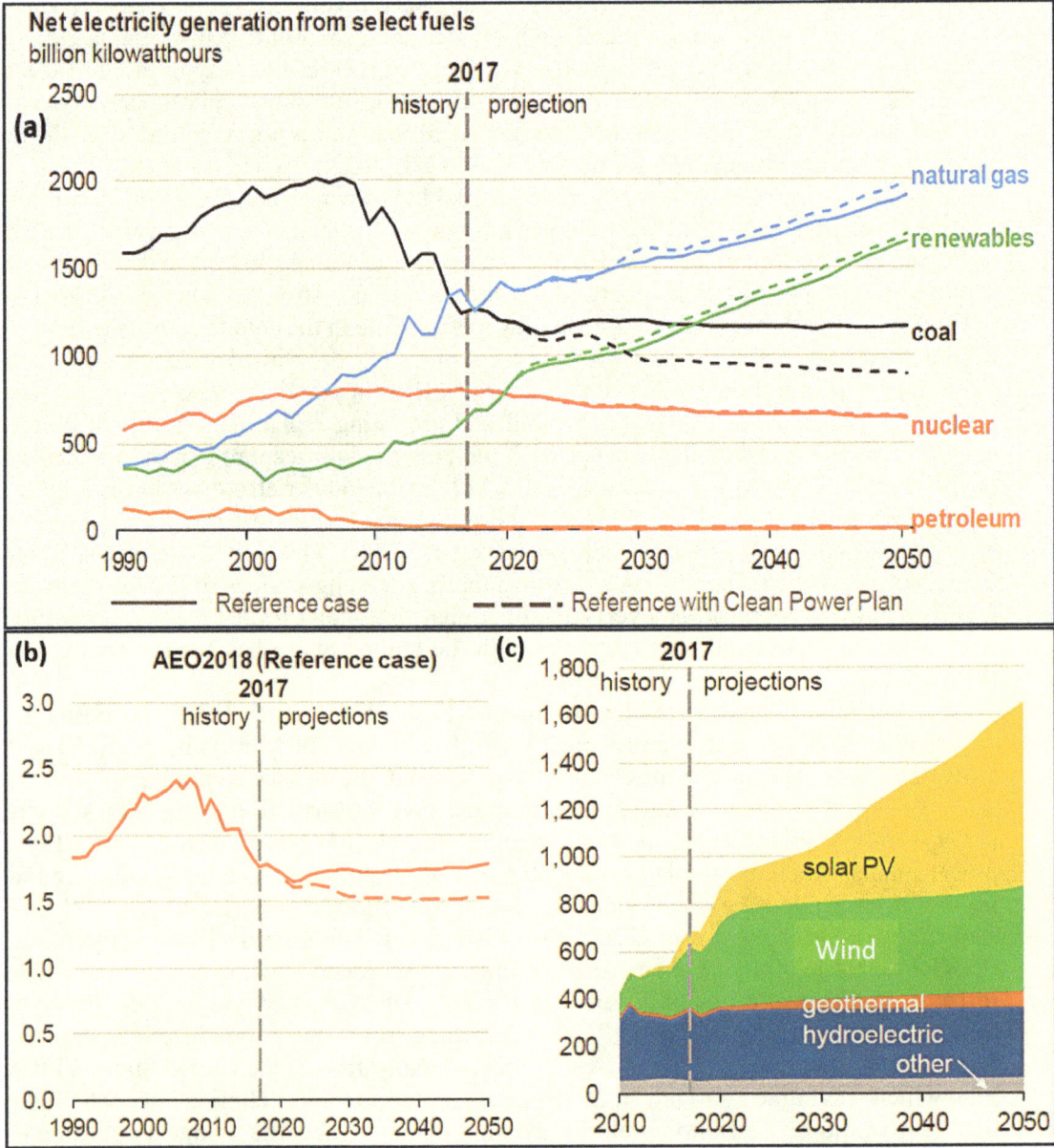

Figure 7.22 Projected net US power generation from (a) selected fuels, with and without the Clean Power Plan (b) Electricity-related carbon dioxide emissions with and without Clean Power Plan (c) Renewable electricity generation. *(EIA, 2018).*

China which surpassed the United States as the world's largest CO_2eq emitter in 2008 has also set a goal of 20% emission reduction and 20% of non-fossil energy use by 2030, and the main drivers will be solar and nuclear energy. Although the country's economy grew by around 7% in 2017, emissions increased by just 1.7%, clearly mitigated by increasing renewables deployment and faster coal-to-gas switching. Many other European OECD countries have also set ambitious INDC targets. In fact, 146 national climate change panels presented draft INDCs at the 2015 United Nations Climate Change Conference in Paris. It is unclear however whether any of the INDCs will meet or exceed the planned targets, considering the intra-country economic and political dynamics. Asia accounted for two-thirds of the growth in global emissions in 2017, due largely to the growth in the economies of the region, and all projections show an even faster growth over the next two decades or so. In spite of declared efforts to decarbonize energy in the region, it is difficult to see a clear path towards achieving this goal, since coal is the prime energy resource that fuels power and carbon-intensive primary materials production in most countries in the region.

7.10 OUTLOOK ON INTERNATIONAL ACTION ON THE ENVIRONMENT

Many international organizations are playing a key role in promoting global discourse on climate change and energy-related emissions. The United Nations, International Energy Agency (IEA), World Energy Council (WEC), Intergovernmental Panel on Climate Change (IPCC), Energy Information Administration (EIA), ExxonMobil, BP, World Health Organization (WHO), United Nations Environmental Protection Agency (UNEP), and many more have been providing extensive information and data which have become vital resources for country and international action. The World Health Organization has brought into full global focus the impact and consequences of ambient and household pollution, which have a more immediate and severe impact on the health of life on earth than climate change, accounting for (or contributing to) over seven million global deaths annually.

The 1992 United Nations' conference in Rio de Janeiro was perhaps the first major global initiative to address the problem of environmental pollution. The meeting developed an International Environmental Treaty - United Nations Framework Convention on Climate Change (UNFCCC). The thrust of the Treaty was the stabilization of greenhouse gas concentrations in the atmosphere at a level that would prevent dangerous anthropogenic interference with the climate system (UNFCCC, 1992). The parties to UNFCCC have met annually since 1995 as Conference of the Parties (COP) to assess progress in dealing with climate change. The meeting in Kyoto in 1997 developed a Kyoto Protocol which established legally binding obligations for developed countries to reduce their greenhouse gas emissions. Six greenhouse gases were identified which, if reduced could reduce global warming significantly. From 2005 the conferences have also served as the Meetings of Parties of the Kyoto Protocol (CMP).

The Kyoto Treaty was not ratified by many countries. The USA, the world's second largest emitter did not ratify the Treaty because it would cause serious harm to the US economy, and also because it did not cover developing countries (80% of the world) including China, the leading emitter in the world. Some countries which ratified the Treaty initially, for example, Canada, also withdrew in 2012 because the country was unable to achieve the 6% reduction in emission from the 1990 level. Rather, emission was 17% higher in 2012. Nevertheless, global emissions in 2012 were nearly 23% lower than 1990 levels compared with the set target of 5%, probably not because of the Treaty but because many developed countries including the United States already had policies in place to decarbonize, largely by substituting gas for coal.

Also, many developed countries adopted the 1987 Montreal Protocol to eliminate ozone-depleting gases. For example, use of fluorocarbons in refrigeration and air conditioning has been widely discontinued.

The United Nations Conference on Climate Change comprising Conference of Parties (COP) and the Meeting of Parties to the Kyoto Protocol (CMP) was held in Paris in November 2015. The meeting was attended by 197 parties and the Paris Agreement, a global agreement on the reduction anthropogenic emissions was adopted. The Agreement outlined strategies for strengthening the global response to the threat of climate change, and set a target of keeping a global temperature rise this century well below 2°C above the pre-industrial levels, while pursuing efforts to limit the increase to 1.5°C. Several other clauses in the Agreement include the enhancement of adaptive capacity on member nations by strengthening their resilience and reducing vulnerability of their people to climate change. Policy instruments were developed to help strengthen the ability of countries, especially the developing countries, to deal with the impacts of climate change. These included the establishment of financial and capacity building support. The 2015 conference also agreed on a goal of achieving zero net anthropogenic greenhouse gas emissions by the second half of the 21st century, and introduced a legal framework to compel signatories to the Agreement to adopt and domesticate the Treaty within their own legal systems. Signatories were obliged to establish National Greenhouse Gas Inventories and develop strategies for control and removal by 2020. However, many sections of the Treaty are promises, aims, goals and indefinite time frames which are not enforceable.

Furthermore, as discussed earlier, the priority of most developing countries is to provide primary energy to their growing population and most lack the wherewithal to deal effectively with pollution control. In order to help mitigate this problem, the Paris protocol established a revolving fund to help developing countries adopt non-greenhouse gas technologies. One major clause in the Paris Agreement calls for zero net anthropogenic greenhouse gas emissions to be achieved globally during the second half of the 21st century. Zero net carbon means carbon neutrality and requires balancing the carbon released into the atmosphere with an equivalent amount sequestered, offset with the use of zero carbon energy technologies, and extensive re-forestration that will suck more carbon dioxide from the atmosphere. Any shortfall could be compensated for by buying carbon credits. All parties to the Agreement are required to submit their nationally determined contributions (NDCs) and report regularly on their emissions, and on progress of their implementation efforts. The landmark accord entered into force in November, 2016, after having been signed or ratified by 197 countries. However, the agreement is non binding since there are no verification or enforcement mechanisms, and the recent withdrawal of the United States of America from the protocol and subsequent move to cancel the country's Clean Power Plan are major setbacks. However, all other major countries and many states within the U.S.A. have expressed the determination to move forward with the implementation of the agreement.

There is little doubt that the world is making some progress towards mitigation of climate change: every country that signed the agreement now has at least one policy or law in place, but it is not clear to what extent the stated goals can be achieved. What is becoming increasingly certain is the fact that, even if all the nations succeed in achieving stated NDIC goals. The effect will not be sufficient to move the world towards the Paris-2015 goal. Global emissions started increasing again by 2.1% to 33.1 gigatonnes (GT) in 2018, having stayed flat for the previous three years. Although renewable energy is projected to grow by around 2.6% per year over the period (the fastest-growing energy source), its share of global primary energy use in 2040 will still be no more than about 15% by all recent projections, but its use in power generation could be as high as 35-50%. Projections also show that fossil fuels will still meet nearly 80% of the global primary energy needs in 2040, and possibly remain dominant well into the 21st century. Emissions are projected to increase to about 36.5-37 GT in 2040, in spite of current and planned mitigation actions. However, a recent analysis by the International Energy Agency (IEA, 2017)) and several other studies have shown that emissions need to peak by around 2030 and start to reduce to about 18 GT (about 55% of current level) by 2040 in order to put the world on the pathway to the achievement of the Paris-2017 goals. A recent publication by the International Energy Agency (IEA, 2019g) examines in-depth the complexity of reconciling the divergent dimensions of the tumultuous global oil markets, geopolitical tensions, carbon emissions and climate targets, and the goal of providing electricity for the 850 million people around the world who currently lack access. The publication projects what would happen if the world continues along its present path (Current Policies Scenario), and a feasible pathway to a sustainable environment (Sustainable Development Scenario) which requires much more rapid and widespread changes across all parts of the energy system (Figures 7.23 & 7.24).

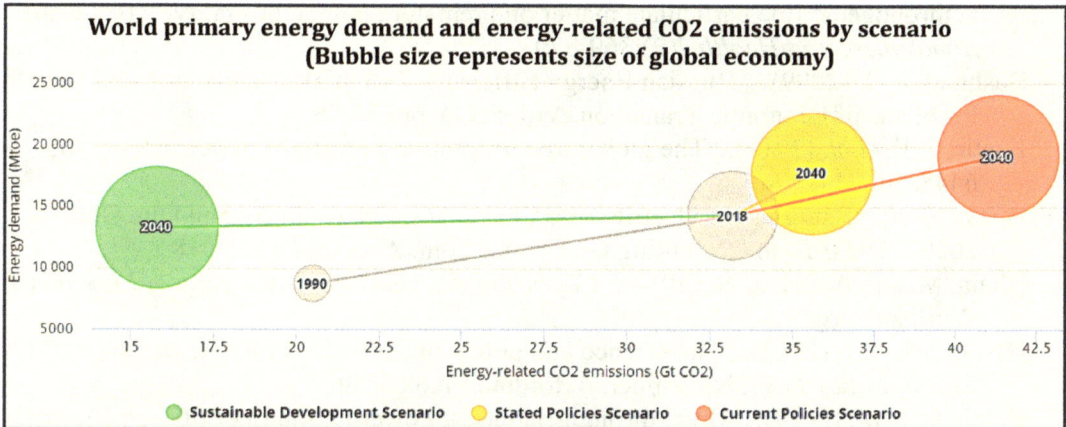

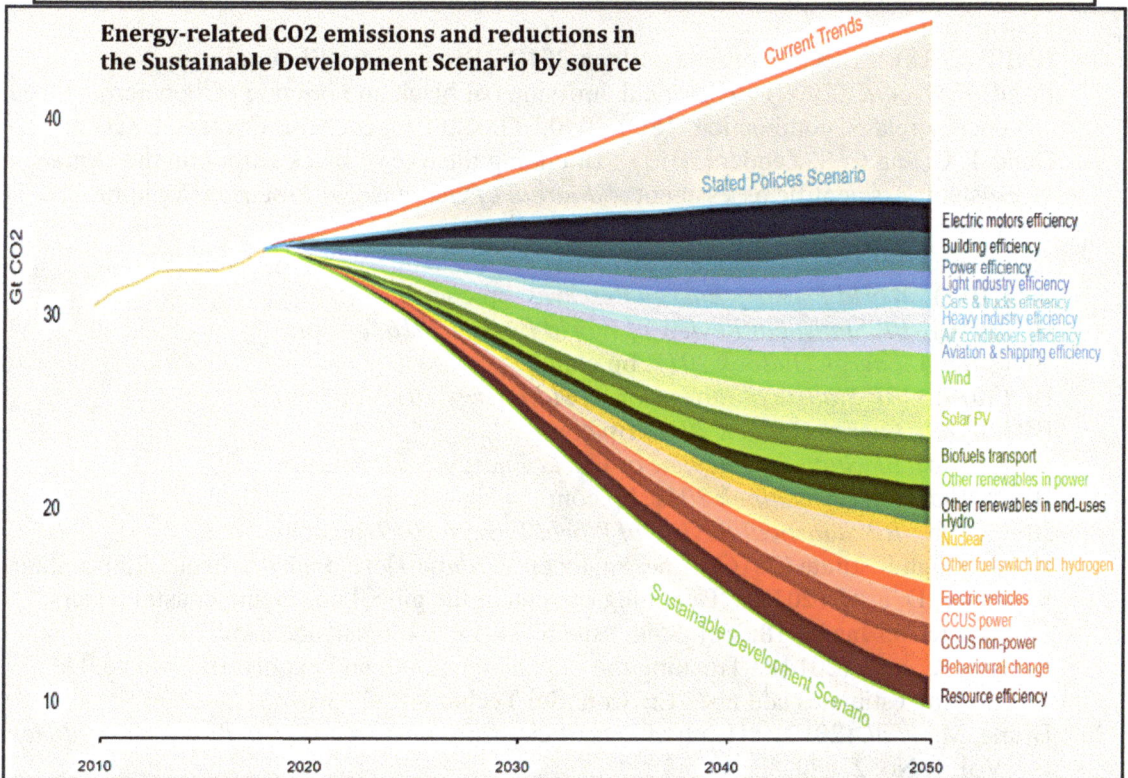

Figure 7.23 International Energy Agency Sustainable Development Scenario
(IEA, Energy Outlook, 2019).

8.   REFERENCES

Afonja, A. A., (2017). *Basic Coal Science and Technology*. SineliBooks.

Andersen, R. D. *et al.,* (2012). "Efforts to reduce flaring and venting of natural gas world-wide." Norwegian University of Science and Technology, Trondheim.

Anderson, J. O Thundiyil, J. G, and A. Stolbach, (2012). "Clearing the Air: A Review of the Effects of Particulate Matter Air Pollution on Human Health." Journal of Medical Toxicology, Vol. 8(2), pp 166-175.

Anenberg, S. C. (2012). "Global air quality and health co-benefits of mitigating near-term climate change through methane and black carbon emission controls." *Environmental Health Perspectives* 120(6);831-839.

Anenberg. S. C., Takemura, T., and F. Dentener (2014). "Impacts of long-range transport of anthropogenic fine particulate matter on premature human mortality." *Air Quality Atmosphere and Health* 7(3):369-379.

Bashmakov, I. (2009). "Russian Energy Efficiency Potential. Scale, Costs and Benefits." Problems of Economic Transition, Vol. 52(1), pp. 54-75.

Bastin, J. F. *et al.* (2019). "The global tree restoration potential." *Science*, Vol. 365, issue 6448, July 2019, p. 76.

BCG (2018) "Batteries for electric cars: Challenges, Opportunities, and the Outlook to 2020." The Boston Consulting Group. bcg.com. Accessed 412/2018.

Bhata, M. and Angelou, N. (2014). *Capturing the Multi-dimensionality of Energy Access.* Worlbank.org.

Blackstock, J. J. (2012). "The science and policy of short-lived climate pollutants." *Oxford Martin Policy Brief,* November. oxfordmartin.ox.ac.uk.

Bolch, C. *et al.* (2019). "Breakthrough Batteries: Powering the Era of Clean Electrification." rmi.org.

BNEF (2018). "Electric vehicle outlook 2018." BloombergNEF. bnef.com.

Bond, T. C. *et al.* (2007). "Historical emissions of black and organic carbon aerosol from energy-related combustion, 1850-2000." Global Biogeochem. Cycles 21, GB 2018.

Bond T. C. and C. S. Zender (2013). "Bounding the role of black carbon in the climate system: A scientific assessment ." *Journal of Geophysical Research*, Volume 118, Issue 13, pp. 5380-5552.

Boucher, O. (2015). Atmospheric Aerosols. DOI 10.1007/978-94-017-9649-1_2.

BP (2011). *World Primary Energy Reserves*. bp.com.

BP (2016). *BP Statistical Review of World Energy 2016*. bp.com

BP (2017a). *Energy Outlook 2017*. bp.com

BP (2017b). *BP Statistical Review of World Energy 2017*. bp.com

BP (2018a). *Energy Outlook 2018*. Bp.com

BP (2018b). *BP Statistical Review of World Energy 2018*. bp.com

BP (2019a). *Energy Outlook 2019*. bp.com

BP (2019b). *BP Statistical Review of World Energy 2019*. bp.com

Breakthough Institute (2016). *Energy Access Without Development*. thebreakthrough.org

Breitburg, D. *et al.* (2018). "Declining oxygen in the global ocean and coastal waters." Science 05 Jan. 2018: Vol. 359, Issue 6371. DOI:116/science.aam7240.

Breivik, K. *et.al.* (2014). "Tracking the global generation and exports of e-waste. Do existing estimates add up?" Environ. Sci Technol48:8735-8743.

Brune, M., *et al.* (2013). "Heath effects of exposure to e-waste." *The Lancet Global Health*, Vol. 1, No. 2, e70.

CEC (2012). *Oil and Gas Production: History in California.* California Energy Commission energy.ca.gov.

CEM (2017). "Electric vehicles initiative (EVI)." *Clean Energy Ministerial,* cleanenergyminesterial.org.

CISCO (2018). *Global – Device Growth Traffic Profiles.* cisco.com. Accessed 3/17/2018.

Cleland, D. (2005). "Sustainable energy use and management." *Proc. conf; People and Energy – how to use it? Christchurch Royal Society of New Zealand, Misc series,* Vol 66, pp. 96-102.

CNEV (2017). "A survey of investment and capacity planning of new energy vehicles in China." chinanev.net.

Crippa, M. *et al.* (2019). "Fossil CO2 amd GHG emissions of all world countries". Eutopean Union, 2019 Report.

DEFRA (2018). "Retailers report 83% reduction in plastic bags." Update from Department for Environment, Food and Rural Affairs, Defra, UK.

ECF (2017). "From cradle to grave: e-mobility and the French energy transition." European Climate Foundation. europeanclimate.org.

ECOMETRICA (2011). "Electricity-specific emission factors for grid electricity." emissionfactors.com.

EDGAR (2017). Emissions Database for Global Atmospheric Research. European Commission. europa.eu.

EIA (2016). *International Energy Outlook 2016.* Energy Information Administration. eia.gov.

EIA (2017). *International Energy Outlook 2017.* Energy Information Administration. eia.gov.

EIA (2018) Annual *Energy Outlook 2018.* Energy Information Administration. eia.gov.

EIA (2018b). *International Energy Outlook 2018.* Energy Information Administration. eia.gov.

Ellingsen, L. A. and A. Stromman, (2017). "Life cycle assessment of electric vehicles." *12thConcawe Symposium,* NTNU - Trondhelm.

EMF (2016). "The New Plastics Economy: Rethinking the future of plastics." Ellen Macarthur Foundation. ellenmacarthurfoundation.org.

EPA (2011). "Reducing Greenhouse Gas Emissions through Recycling an Composting." epa.gov.

EU (2016). "Advances and critical aspects in the life-cycle assessment of battery electric cars." Helmers Eckard and Weiss Martin, Publication Year: 2016. Dove Press.

Euro-Int. (2017). " Euromonitor International, 2017 global packaging trends report."

ExxonMobil (2016). *Energy Outlook 2016.* exxonmobil.com.

ExxonMobil (2017). *Energy Outlook 2017.* exxonmobil.com.

ExxonMobil (2018). *Energy Outlook 2018.* exxonmobil.com.

ExxonMobil (2019). *Energy Outlook 2019.* exxonmobil.com.

FAO (2006). "Global Forest Resource Assessment 2005. Food and Agriculture Organization of the United Nations.

Finkelman R. B. and L. Tian (2017). "The health impacts of coal use in China". *International Geology Review,* 60:5-6, RR

Flanner M. G. *et. al.* (2007). "Present-day climate forcing and response from black carbon in snow." *Journal of Geophysical Research,* Vol. 112, Issue D11.

GEA (2015). "The International Geothermal Market At a Glance – May 2015" Geothermal Energy Association. May 2015.

Geyer R. *et al.* (2017). "Production, use, and fate of all plastics ever made.) Science Advances 3(7):e1700782. July 2017.

Grant K.et al. (2013). "Health consequences of exposure to e-waste: a systematic review." http://dx.doi.org.

Grossman E. (2011). "Radioactivity in the ocean: diluted, but far from harmless." *Yale School of Forestry & Environmental Studies*. e360.yale.edu/.

HEI (2018). "State of global air/2018: A special report on global exposure to air pollution and its disease burden." *Health Effects Institute*. stateofglobalair.org.

ICCT (2018). "Effects of battery manufacturing on electric vehicle life-cycle greenhouse gas emissions." *The International Council on Clean Transportation*. theicct.org.

IAEA (2016). *Annual Report 2016*. International Atomic Energy Agency. iaea.org.

IAEA (2017). *International Status and Prospects for Nuclear Power 2017*. International Atomic Energy Agency. iaea.org.

IEA (2013a). *Resources to Reserves*. International Energy Agency. iea.org.

IEA (2013b). *Coal information*. International Energy Agency. iea.org.

IEA (2014). *Africa Energy Outlook*. International Energy Organization World Energy Outlook. iea.org.

IEA (2015a). *World Energy Outlook 2015*. International Energy Agency. iea.org.

IEA (2015b). *Electricity Information*. International Energy Agency. iea.org.

IEA (2015c). *World Energy Outlook 2015: Energy an Climate Change*. iea.org.

IEA (2016). *World Energy Outlook 2016*. International Energy Agency. iea.org.

IEA (2017e). *CO$_2$ Emissions from Fuel Combustion 2017*. International Energy Agency. iea.org.

IEA (2017a). *KeyWorldStatistics 2017*. International Energy Agency. iea.org.

IEA (2017b). *Energy Access Outlook 2017*. International Energy Agency. iea.org.

IEA (2017c). *A world in transformation: World Energy Outlook 2017*. International Energy Agency. iea.org.

IEA (2017d). *Energy Efficiency 2017*. International Energy Agency. iea.org.

IEA (2017f). *Energy Efficiency Statistics 2017*. International Energy Agency. iea.org.

IEA (2017g). *Future Scenarios for Energy Efficiency*. International Energy Agency. iea.org.

IEA (2017h). *Energy Technology Perspectives 2017*. International Energy Agency. iea.org.

IEA (2017i). *Renewables 2017*. International Energy Agency. iea.org.

IEA (2017j). *Technology Roadmap: Delivering sustainable Bioenergy*. International Energy Agency. iea.org.

IEA (2017k). *Emissions from fuel combustion highlights 2017)*. International Energy Agency. iea.org.

IEA (2017l). *Global EV Outlook 2017*. International Energy Agency. iea.org.

IEA (2018a). *Global Energy and CO2 status 2017*. International Energy Agency. iea.org.

IEA/OECD (2018b). The future of cooling. International Energy Agency. iea.org.

IEA/OECD (2018c). The future of petrochemicals. International Energy Agency. iea.org.

IEA (2019a). *Electricity Information*. International Energy Agency. iea.org.

IEA,(2019b). *Global EV Outlook 2019*. International Energy Agency. iea.org.

IEA (2019c). *Global Energy and CO2 status 2018*. International Energy Agency. iea.org.

IEA (2019d). *World Energy Investment 2019*. International Energy Agency. iea.org.

IEA (2019e). *Renewables 2018*. International Energy Agency. iea.org.

IEA (2019f). *KeyWorldStatistics 2018*. International Energy Agency. iea.org.

IEA (2019g). *World Energy Outlook 2018, & 2019*. International Energy Agency. iea.org.

IEA (2019h). *Energy Efficiency 2018: Analysis and Outlooks to 2040.* International Energy Agency. iea.org.

IEA (2019i). *Global Energy & CO2 Status Report: The latest trends in energy and emissions in 2018.* International Energy Agency. iea.org.

IEA (2019j). *Carbon capture, utilisation and sto*rage. International Energy Agency. iea.org.

IEA (2019k). *Electricity Statistics.* International Energy Agency. iea.org.

IEA (2019l). Scenarios. International Energy Agency. iea.org/weo/#scenarios

IEA (2019m). Global *Electric Vehicle Outlook 2019.* International Energy Agency. iea.org.

IEA (2019n). *KeyWorldStatistics 2019.* International Energy Agency. iea.org.

IEA (2019o). *Energy Efficiency 2018.* International Energy Agency. iea.org.

IGU (2019). World LNG Report 2019. International Gas Union. igu.org.

IHA (2017). *Briefing: 2016 Key Trends in hydropower.* International Hydropower Association. hydropower.org.

Internet Society (2017). *Global Internet Report 2016.* internetsociety.org/statistica.com, 2018.

IWS (2018). "Internet Usage and Population Statistics." internetworldstats.com

IPCC (2007). "Climate Change 2007: Energy Supply: The Physical Science Basis." Contribution of Working Group 1 the Fifth Assessment Report. ipcc.org.

IPCC (2013). "Climate Change 2013: The Physical Science Basis." Contribution of Working Group 1 to the Fifth Assessment Report. ipcc.org.

IPCC (2014a). " Climate Change 2014." ipcc.org.

IPCC (2014b). " Energy Systems and Climate Change." ipcc.org.

IPCC (2018). "Summary for Policymakers: Global Warming of 1.5°C. An IPCC Special Report." ipcc.org.

IPCC (2019). " IPCC Special Special Report on Climate Change, Desertification, Land Degradation, Sustainable Land Management, Food Security, and Greenhouse gas fluxes in Terrestrial Ecosystems: Summary for Policymakers. ipcc.org.

IRENA (2015). "Renewables and Electricity Storage: A Technology Roadmap for Remap 2030". *International Renewable Energy Agency.* irena.org.

IRENA (2016). "ReMAP: Roadmap for a renewable energy future, 2016 edition". *International Renewable Energy Agency.* irena.org.

IRENA (2017a). "Rethinking Renewable Energy." *International Renewable Energy Agency.* irena.org.

IRENA (2018a). "Renewable capacity statistics 2018." *International Renewable Energy Agency.* irena.org.

IRENA (2018b). "GET_2018 Roadmap to 2050." *International Renewable Energy Agency.* irena.org.

IRENA (2019a). "Global energy transformation 2019." *International Renewable Energy Agency.* irena.org.

IRENA (2019b). "Renewable power generation costs in 2018." *International Renewable Energy Agency.* irena.org.

ISE (2018). "Global Market Outlook For Solar Power/2017-2021." irishsolarenergy.org.

ITU (2017). "The Global E-waste Monitor 2017." International telecommunications Union. Itu.int.

Labunska I. *et al.* (2018). "Current radiological situation in areas of Ukraine contaminated by the Chernobyl accident: Part 1. Human dietary exposure to Caesium-137 and possible mitigation measures." *Environment International* Volume 117, August 2018, pp. 250-259. Elsevier.

LowCVP (2011). "Low CVP study demonstrates importance of whole life CO_2 emissions." *Low carbon vehicle partnership*. lowcvp.org.uk.

McKinsey (2010). *Energy efficiency: A compelling global resource*. McKinsey & Company. mckinsey.com.

Messagie M. (2017). "Life Cycle Analysis of the climate impact of electric vehicles." Oct 2017, VUB University, Brussels. *Transport & Environment*. transporenvironment.org.

Miotti M. *et al.* (2016). "Personal Vehicles Evaluated against Climate Change Mitigation Targets." *Environ. Sci. Technol.*, 2016, *50* (20), pp. 10795–10804.

Mok B., (2017). "Types of batteries used in electric vehicles." large.stanford.edu.

Myhre G. *et. al.* (2014). "Anthropogenic and natural radiative forcing. In Climate Change 2013: The physical Science Basis. ipcc.org.

NASA (2018). "Scientific evidence for global warming of the climate system." *National Aeronautical and Space Administration*. nasa.gov.

NAS (2014). *Climate Change: Evidence and Causes*. Washington, DC: The National Academies Press. https://doi.org/10.17226/18730.

NETL/DOE (2018). "US Department of Energy's R&D program to reduce greenhouse gas emissions through beneficial uses of carbon dioxide." netl.doe.gov. Accessed 3/2018.

NUETC (2010). "Should we be using bottled water ?". *Nottingham University Environmental Technology Centre*. Nottingham.ac.uk.

OECD/IEA (2004). *Energy and Development. World Energy Outlook 2004*. iea.org.

OECD/IEA (2012). *Measuring progress towards energy for all*. World Energy Outlook 2012. iea.org.

Pachauri *et al.* (2012). "Energy Access for Development". *Global Energy Assessment: Toward a Sustainable Future*. Eds. Team, *GEA* Writing, pp.1401-1458. Cambridge University Press.

Pozzer A. *et al.* (2017). "Impact of agricultural emission reductions on fine-particulate matter and public heath." *Atmos. Chem. Phys.*, 17. 12813-12826.

Pozzer A., *et al.* (2017). "Reducing manure and fertilizers decreases atmospheric fine particles." Max-Planck Gesellschaft, Oct. 2017. mpg.de.

Pruss-Ustun A. and M. Neira (2016). *Preventing Disease Through Healthy Environments: A global Assessment of the Burden of Disease from Environmental Risks*. World Health Organization. who.int.books.google.com.

Quinn C. H. *et. al.* (2011). "Coping with multiple stresses in rural South Africa." *Ecology and Society*, Vol. 16 No. 3, pp. 1-20.

Quinn P. K. *et al.* (2008). "Short-lived pollutants in the Arctic: their climate impact and possible mitigation strategies." *Atmos. Chem. Phys.*, 8, 1723-1735.

Raftery A. *et al.* (2017). "Less than 2°C warming by 2100 unlikely." Nature Climate Change, July 31, 2017, pp. 637-641.

REN21 (2018). "Renewables 2017: Global Status Report." *Renewable Energy Policy Network for the 21st Century*. ren21.net.

Rind D. *et al.* (2008). "Exploring the stratospheric/tropospheric response to solar forcing." *J. Geophys. Res.*, Vol. 113, D24103.

Rind D. (2009). "Do variations in the solar cycle affect our climate   system ?" nasa.gov.

Sabine C. L. *et al.* (2004). *Global Carbon Cycle: Integrating Humans, Climate and the Natural World*. C. Field, M. Raupach, Eds. Island Press, pp. 17-44.

Shindell D. T. *et. al.* (2008)."A multi-model assessment of pollution transport to the Arctic." *Atmos. Chem. Phys.*, Vol. 8, pp. 5353-5372.

Sims R. V. Gorsevski and S. Anenberg (2015). "Black Carbon Mitigation and the roe of the Global ." Environment Facility, Washington, D.C. stapgef.org.

SPE (2017). "Global Market Outlook for Solar Power, 2017-2021". Solar Power Europe. solarpowereurope.org.

T&E (2016). "Electric vehicles in Europe- 2016: Approaching adolescence." Transport and Environment. transport.environment.org.

T&E (2017). "Electric vehicles have significantly lower climate impact than diesels over their lifetime." Transport and Environment. transport.environment.org.

Tester J. W. *et al.* (2005). *"Sustainable energy: choosing among options.* The MIT Press, Cambridge, MA.

Thompson C. J. *et al.* (2009). "Plastics, the environment and human health." *Phyl. Trans. of the Royal Society B*, Vol. 364, Issue 1526.

Troy R. *et al.* (2012). "Comparative environmental life cycle assessment of conventional and electric vehicles." *Journal of Industrial Ecology.* doi.org.

United Nations (2019). "World Population Prospects 2019". un.org

UNDP (2000). *World Energy Assessment: Energy and the Challenge of Sustainability.* United Nations Development Organization. undp.org.

UNDP (2016). *Human Development Report 2016.* United Nations Development Program. undp.org.

UNEP/WMO (2011). "Integrated assessment of black carbon and tropospheric ozone." United Nations Environmental Programme and World Meteorological Organization.

UNFCC (2015a). "Paris declaration of electro-mobility and climate change and call to action." *United Nations Framework Convention of Climate Change.* unfccc.int.

UNFCC (2015b). "Adoption of the Paris Agreement." *United Nations Framework Convention of Climate Change.* unfccc.int.

USGS (2000). "Coal-bed Methane: Potential and Concerns." *United States Geological Survey.* usgs.gov.

USEPA, (2012). *Report to Congress on Black Carbon,* U/S. Environmental Protection Agency, Washington, D.C.

Velders G. J. M. *et al.* (2009). "The large contribution of projected HFH Emissions to future climate forcing." *Proc. Nat. Acad. of Sc.,* 106(27), pp. 10949-10954.

Wang, Bin *et al.* (2019). Historical change of El Nino properties sheds light on future changes of extreme El Nino." Proc. Nat. Academy of Sciences of the United States of America. Oct. 2019.

WBA (2016). *Global Bioenergy statistics, 2016.* World Bioenergy Association. worldbioenergy.org.

WEC (2004). *Energy and Development.* World Energy Council, wec.org.

- (2010). *Energy and Development.* World Energy Council, wec.org.

- (2013). *World Energy Resources.* World Energy Council, wec.org.

- (2016a). *World Energy Resources 2016.* World Energy Council, wec.org.

- (2016b). *World Energy Scenarios 2016.* World Energy Council, wec.org.

- (2016c). *Energy Efficiency; a straight path towards energy sustainability" in World.* wec.org.

- (2016d). *World Energy Trilemma 2016.* World Energy Council, wec.org.

- (2016e). *Energy Perspectives 2016.* World Energy Council, wec.org.

- (2018). "Smart connections" in *World Energy Focus Annual 2017.* World Energy Council, wec.org.

WEF/EMF (2016). "The New Plastics Economy: Rethinking the future of plastics." *Ellen Macarthur Foundation.* ellenmacarthurfoundation.org.

Wikipedia (2019). "List of countries by carbon emissions. Wikipedia,org. Accessed 9/24/2019

WHO (2006). "Preventing disease through healthy environments."
    World Health Organization. who.int.
WHO (2015). "Residential heating with wood and coal: health impacts
    and   policy options in Europe and North America." who.int.
WHO (2017b). "Ambient pollution: Pollutants." who.int.
WHO (2017c). "WHO releases country estimates on air pollution exposure and health
    impact." euro.who.int.
WHO (2017d). "Ambient (outdoor air quality and health." who.int.
WHO (2017a). "7 million premature deaths linked to air pollution." World Health
    organization. who.int. Accessed 3/28/2018.
WHO (2018). "Household air pollution and health." who.int.
WNO (2017).   "The Nuclear Fuel Cycle." World Nuclear Organization. world-nuclear.org.
WNA (2017). "World Energy Needs and Nuclear Power." World Nuclear Association.
    world-nuclear.org.
WNA (2018).   "World Nuclear Power and Uranium Requirements." World Nuclear
    Association. world-nuclear.org.
World Bank (2014a). *Access to electricity*. worldbank.org.
World Bank (2014b). *Enterprises Surveys*. World Bank: data.worldbank.org.
World Bank/IEA, (2014). Sustainable Energy for All 2013–2014: Global Tracking
    Framework. Washington, DC: World Bank. hdl.handle.net/10986/16537.
World Bank (2017). *State of Electricity Access Report 2017*. worldbank.org.
World Bank (2018). " Global Gas Flaring Reduction Partnership (GGFR)." worldbank.org.
World Bank (2019). "Zero routine flaring by 2030. Worldbank.org.
WRAP (2010). Waste & Resources Action Programme.wrap.co.uk
WRAP (2018). "WRAP's vision is a world in which resources are used sustainably."
    wrap.co.uk.
WWEA (2018). "Wind capacity reaches 539 GW, 52,6 GW added in 2017." World Wind Energy
    Association. indea.org
Zaelke D *et al*. (2013). "Short-lived climate pollutants."*Institute for Governance and
    Sustainable Development*. www.igsd.org.
Zehner O. (2012). "Green illusions: the dirty secrets of clean energy." books.google.com.
Zhang, J. Y., C. G. Zheng, and D. Y. Ren, (2004). "Distribution of potentially hazardous
    trace elements in coals from Shanxi Province, China." *Fuel,* 83:129-135.

www.ingramcontent.com/pod-product-compliance
Lightning Source LLC
Chambersburg PA
CBHW080617030426
42336CB00018B/2995

* 9 7 8 0 9 9 8 5 8 4 3 3 1 *